Methods in Molecular Biology™

Series Editor
John M. Walker
School of Life Sciences
University of Hertfordshire
Hatfield, Hertfordshire, AL10 9AB, UK

For other titles published in this series, go to
www.springer.com/series/7651

Epitope Mapping Protocols

Second Edition

Edited by

Ulrich Reineke[*] and Mike Schutkowski[†]

[*] Lead Discovery Biology Department, Jerini AG, Berlin, Germany
[†] JPT Peptide Technologies GmbH, Berlin, Germany

Humana Press

Editors
Ulrich Reineke
Lead Discovery Biology Department
Jerini AG, Berlin
Germany

Mike Schutkowski
JPT Peptide Technologies GmbH
Berlin, Germany

ISBN: 978-1-61737-924-6
e-ISBN: 978-1-59745-450-6
ISSN: 1064-3745
e-ISSN: 1940-6029
DOI: 10.1007/978-1-59745-450-6

Cover illustration: Background image from Chapter 25, Figure 3c. Other images supplied by Ulrich Reineke

Printed on acid-free paper

springer.com

Preface

More than 10 years have passed by since the first edition of *Epitope Mapping Protocols* edited by Glenn E. Morris was published as part of the Methods in Molecular Biology series. The success of the first edition clearly demonstrated the need for detailed descriptions of experimental protocols to determine epitopes, i.e., identify protein domains, sequences, or even amino acids, that are recognized by either antibodies or T-cell receptors. A multitude of disciplines require detailed knowledge about epitopes, and therefore state-of-the-art and reliable protocols. Not only immunologists who have an a priori interest depend on epitope mapping protocols, but also biologists using antibodies as research tools, structural biologists studying protein–protein interactions, clinicians investigating patients' immune responses, vaccine developers designing and testing immunogens, diagnostic labs developing and applying ELISAs, and last but not the least, biotech and pharmaceutical companies obliged to monitor the immunogenicity of novel therapeutic antibodies, proteins, and peptides, to mention only a few.

The development of new techniques as well as new applications demanded a new edition. Some of the protocols of the first edition were simply updated, while others were entirely overhauled in order to keep up with recent developments. An important extension of the scope of the book was not only to cover antibody or B-cell epitope mapping techniques but also to dedicate a total of eight protocols to T-cell epitope mapping in the second part of the volume.

However, the majority of the chapters deal with antibody epitope mapping. This part of the book starts out with two nonlaboratory protocol chapters describing general considerations and definitions of B-cell epitopes and the structural basis of antibody–antigen interactions. These chapters set the scene for the following protocols and are helpful if not necessary to interpret experimental epitope mapping results. The following chapters are arranged in four groups. The first group of eight protocols applies to whole native antigens and covers nuclear magnetic resonance (NMR), enzyme-linked immunosorbent assays (ELISAs), surface plasmon resonance (SPR), proteolytic fragmentation, chemical modification, and mass spectrometry as general methods. The second group of seven chapters addresses peptide library approaches with synthetic as well as phage-displayed peptides, antigen sequence-derived and randomly generated peptide sequences, collections of peptides derived from diverse human proteins, and peptide derivatives mimicking posttranslational modified proteins. The third group of four chapters represents a crucial, completely new part compared with the first edition. Peptides displayed on phages or on high-content microarrays are used to profile complex (auto)antibody signatures in biological fluids such as human or mice sera. Statistical analysis of results in comparison with control cohorts yields novel biomarkers for cancer, allergy, infectious, and autoimmune diseases. The last group of three protocols requires antigen expressed from recombinant DNA. The final single chapter describes B-cell epitope prediction tools.

The second part of the book focuses on techniques for T-cell epitope mapping. It starts with a chapter analyzing molecular recognition of T-cell epitopes presented by T-cell receptors. The following chapters summarize well-established techniques to identify

MHC class I and class II binding peptides such as EliSpot using peptides and peptide mixes, flow cytometry, and the tetrameric MHC-based iTOPIA epitope discovery system. Two novel powerful methods for MHC ligand identification such as the exchange of photolabile conditional MHC class I ligands by peptides and the use of peptide microarrays together with soluble MHC class II molecules complement the second part of the book. One chapter related to T-cell epitope processing was included to complete the picture of epitope presentation by MHC molecules for antigen recognition.

We hope that this book will become a standard reference for anybody interested in understanding and investigating the complexity of antigen processing, antigen presentation on cell surfaces by MHC molecules, and recognition of antigens or antigen–MHC complexes by antibodies or T-cell receptors. All chapters present well-established, state-of-the-art and cutting-edge techniques that are proven to be reliable and robust. The protocols are comprehensive and complete without cross references. Contributors to this book represent a broad spectrum of immunologists, biochemists, biologists, physicists, physicians, and mathematicians. Some have more than 30 years of experience, and most of them have published text books in their fields. Here, readers can find more complete coverage of techniques common in the diversifying field of epitope mapping compared with many immunological and molecular biological text books and manuals.

We are indebted to all the authors for their expert contributions. In addition, we thank John Walker for his editorial guidance and Humana Press for publishing this book.

Ulrich Reineke
Mike Schutkowski

Contents

Preface . . . v
Contributors . . . xi

SECTION I: B-CELL EPITOPE MAPPING

1 What Is a B-Cell Epitope? . . . 3
Marc H.V. Van Regenmortel

A: WHOLE ANTIGEN METHODS

2 Structural Basis of Antibody–Antigen Interactions . . . 23
Eric J. Sundberg

3 Epitope Mapping of Antibody–Antigen Complexes by Nuclear Magnetic Resonance Spectroscopy . . . 37
Osnat Rosen and Jacob Anglister

4 A Solid-Phase Mutual Inhibition Assay with Labeled Antigen . . . 59
Masahide Kuroki

5 Epitope Mapping by Surface Plasmon Resonance . . . 67
Pär Säfsten

6 Proteolytic Fragmentation for Epitope Mapping . . . 77
Maria R. Mazzoni, Francesca Porchia, and Heidi E. Hamm

7 Epitope Mapping by Proteolysis of Antigen–Antibody Complexes . . . 87
Suraj Dhungana, Jason G. Williams, Michael B. Fessler, and Kenneth B. Tomer

8 Identifying Residues in Antigenic Determinants by Chemical Modification . . . 103
Howard M. Reisner and Roger L. Lundblad

9 Epitope Mapping by Differential Chemical Modification of Antigens . . . 119
Suraj Dhungana, Michael B. Fessler, and Kenneth B. Tomer

B: PEPTIDE LIBRARY APPROACHES

10 Linear B-Cell Epitope Mapping Using Enzyme-Linked Immunosorbent Assay for Libraries of Overlapping Synthetic Peptides . . . 137
Michael W. Heuzenroeder, Mary D. Barton, Thiru Vanniasinkam, and Tongted Phumoonna

11 Antibody Epitope Mapping Using SPOT™ Peptide Arrays . . . 145
Ulrich Reineke and Robert Sabat

12 Peptide Microarrays for Profiling of Modification State-Specific Antibodies 169
Johannes Zerweck, Antonia Masch, and Mike Schutkowski

13 Epitope Mapping Using Phage Display Peptide Libraries 181
Volker Böttger and Angelika Böttger

14 Antibody Epitope Mapping Using *De Novo* Generated Synthetic Peptide Libraries 203
Ulrich Reineke

15 Antibody Specificity Profiling on Functional Protein Microarrays 213
Dawn R. Mattoon and Barry Schweitzer

16 Peptide Microarrays for Determination of Cross-Reactivity 225
Alexandra Thiele

C: Profiling of Antibody Signatures in Biological Fluids

17 Epitope Mapping Using Randomly Generated Peptide Libraries 237
Juliane Bongartz , Nicole Bruni, and Michal Or-Guil

18 Probing the Epitope Signatures of IgG Antibodies in Human Serum from Patients with Autoimmune Disease 247
Peter Lorenz, Michael Kreutzer, Johannes Zerweck, Mike Schutkowski, and Hans-Jürgen Thiesen

19 Microarrayed Allergen Molecules for Diagnostics of Allergy 259
Jing Lin , Ludmilla Bardina, and Wayne G. Shreffler

20 Monitoring B Cell Response to Immunoselected Phage-Displayed Peptides by Microarrays 273
Lina Cekaite, Eiving Hovig, and Mouldy Sioud

D: Antigen Expressed from Recombinant DNA

21 Epitope Mapping Using Homolog-Scanning Mutagenesis 289
Lin-Fa Wang

22 Epitope Mapping by Region-Specified PCR-Mutagenesis 305
Tsutomu Mikawa, Masayuki Ikeda, and Takehiko Shibata

23 Epitope Mapping Using Phage-Display Random Fragment Libraries 315
Lin-Fa Wang and Meng Yu

E: B-Cell Epitope Prediction

24 Prediction of Linear B-cell Epitopes 335
Ulf Reimer

Section II: T-cell Epitope Mapping

25 Molecular Recognition of Diverse Ligands by T-Cell Receptors 347
Eric J. Sundberg

26 Identification of Human MHC Class I Binding Peptides using the iTOPIA™ Epitope Discovery System 361
Markus Wulf, Petra Hoehn, and Peter Trinder

27 T-Cell Epitope Mapping in Mycobacterium tuberculosis Using PepMixes Created by Micro-Scale SPOT™ Synthesis 369
Marisa Frieder and David M. Lewinsohn

28 High-Throughput T-Cell Epitope Discovery Through MHC Peptide Exchange 383
Sine Reker Hadrup, Mireille Toebes, Boris Rodenko, Arnold H. Bakke, David A. Egan, Huib Ovaa, and Ton N.M. Schumacher

29 T-Cell Epitope Processing (The Epitope Flanking Regions Matter). 407
Alejandra Nacarino Martinez, Stefan Tenzer, and Hansjörg Schild

30 Identification of MHC Class II Binding Peptides: Microarray and Soluble MHC Class II Molecules 417
Simani Gaseitsiwe and Markus J. Maeurer

31 T-Cell Epitope Mapping. 427
Raija K.S. Ahmed and Markus J. Maeurer

32 Identification and Validation of T-Cell Epitopes Using the IFN-γ EliSpot Assay 439
Markus Wulf Petra Hoehn, and Peter Trinder

Index. 447

Contributors

RAIJA K. S. AHMED • *Microbiology and Tumor Cell Biology Center, Karolinska Institutet and the Swedish National Center for Infectious Disease Control, Stockholm, Sweden*

JACOB ANGLISTER • *Department of Structural Biology, Weizmann Institute of Science, Rehovot, Israel*

ARNOLD H. BAKKER • *Division of Immunology, The Netherlands Cancer Institute, Amsterdam, The Netherlands*

LUDMILLA BARDINA • *Mount Sinai School of Medicine, Division of Pediatric Allergy, New York, NY, USA*

MARY D. BARTON • *School of Pharmacy and Medical Sciences, University of South Australia, Adelaide, SA, Australia*

JULIANE BONGARTZ • *Systems Immunology Group, Institute for Theoretical Biology, Humboldt University Berlin, Berlin, Germany*

ANGELIKA BÖTTGER • *Department Biology II, Ludwig-Maximilians-University Munich, Planegg-Martinsried, Germany*

VOLKER BÖTTGER • *Wilex AG, Munich, Germany*

NICOLE BRUNI • *Systems Immunology Group, Institute for Theoretical Biology, Humboldt University Berlin, Berlin, Germany*

LINA CEKAITE • *Departments of Immunology and Tumor Biology, Institute for Cancer Research, University Hospital Montebello, Oslo, Norway*

SURAJ DHUNGANA • *Laboratory of Respiratory Biology, National Institute of Environmental Health Sciences, NIH, DHHS, Research Triangle Park, NC, USA*

DAVID A. EGAN • *Division of Immunology, The Netherlands Cancer Institute, Amsterdam, The Netherlands*

MICHAEL B. FESSLER • *Laboratory of Respiratory Biology, National Institute of Environmental Health Sciences, NIH, DHHS, Research Triangle Park, NC, USA*

MARISA FRIEDER • *Pulmonary & CCM, R&D 11, Portland VA Medical Center, Portland, OR, USA*

SIMANI GASEITSIWE • *Microbiology and Tumor Cell Biology Center, Karolinska Institutet and the Swedish National Center for Infectious Disease Control, Stockholm, Sweden*

SINE REKER HADRUP • *Division of Immunology, The Netherlands Cancer Institute, Amsterdam, The Netherlands*

HEIDI E. HAMM • *Department of Pharmacology, Vanderbilt University, Nashville, TN, USA*

MICHAEL W. HEUZENROEDER • *Infectious Diseases Laboratories, Institute of Medical and Veterinary Science, Adelaide, SA, Australia*

PETRA HOEHN • *Thymed GmbH, Wendelsheim, Germany*

EIVING HOVIG • *Departments of Immunology and Tumor Biology, Institute for Cancer Research, University Hospital Montebello, Oslo, Norway*

MASAYUKI IKEDA • *Nutritional Science Institute, Morinaga-Milk Industry Co. Ltd., Kanagawa, Japan*
MICHAEL KREUTZER • *Institute of Immunology, University of Rostock, Rostock, Germany*
MASAHIDE KUROKI • *Department of Biochemistry, Faculty of Medicine, Fukuoka University, Fukuoka, Japan*
DAVID M. LEWINSOHN • *Pulmonary & CCM, R&D 11, Portland VA Medical Center, Portland, OR, USA*
JING LIN • *Mount Sinai School of Medicine, Division of Pediatric Allergy, New York, NY, USA*
PETER LORENZ • *Institute of Immunology, University of Rostock, Rostock, Germany*
ROGER L. LUNDBLAD • *Department of Pathology and Laboratory Medicine, University of North Carolina, Chapel Hill, NC, USA*
ANTONIA MASCH • *JPT Peptide Technologies GmbH, Berlin, Germany*
DAWN R. MATTOON • *Protein Array Center, Invitrogen Corporation, Branford, CT, USA*
MARKUS J. MÄURER • *Microbiology and Tumor Cell Biology Center, Karolinska Institutet and the Swedish National Center for Infectious Disease Control, Stockholm, Sweden*
MARIA R. MAZZONI • *Department of Psychiatry, Neurobiology, Pharmacology and Biotechnology, University of Pisa, Pisa, Italy*
ALEJANDRA NACARINO MARTINEZ • *Institute for Immunology, University of Mainz, Mainz, Germany*
TSUTOMU MIKAWA • *Biometal Science Laboratory, RIKEN SPring-8 Center, Hyogo, Japan*
MICHAL OR-GUIL • *Systems Immunology Group, Institute for Theoretical Biology, Humboldt University Berlin, Berlin, Germany*
HUIB OVAA • *Division of Cellular Biochemistry, The Netherlands Cancer Institute, Amsterdam, The Netherlands*
TONGTED PHUMOONNA • *Department of Biochemistry and Molecular Biology, Monash University, Clayton, VIC, Australia*
FRANCESCA PORCHIA • *Department of Psychiatry, Neurobiology, Pharmacology and Biotechnology, University of Pisa, Pisa, Italy*
ULF REIMER • *Computational Chemistry Department, Jerini AG, Berlin, Germany*
ULRICH REINEKE • *Lead Discovery Biology Department, Jerini AG, Berlin, Germany*
HOWARD M. REISNER • *Department of Pathology and Laboratory Medicine, University of North Carolina, Chapel Hill, NC, USA*
BORIS RODENKO • *Division of Molecular Carcinogenesis, The Netherlands Cancer Institute, Amsterdam, The Netherlands*
OSNAT ROSEN • *Department of Structural Biology, Weizmann Institute of Science, Rehovot, Israel*
ROBERT SABAT • *Interdisciplinary Group of Molecular Immunopathology, Dermatology/Medical Immunology, University Hospital Charité, Berlin, Germany*
PÄR SÄFSTEN • *Department of Systems and Applications, Biacore AB, A GE Healthcare Company, Uppsala, Sweden*
HANSJÖRG SCHILD • *Institute for Immunology, University of Mainz, Mainz, Germany*

TON N. M. SCHUMACHER • *Division of Immunology, The Netherlands Cancer Institute, Amsterdam, The Netherlands*
MIKE SCHUTKOWSKI • *JPT Peptide Technologies GmbH, Berlin, Germany*
BARRY SCHWEITZER • *Protein Array Center, Invitrogen Corporation, Branford, CT, USA*
TAKEHIKO SHIBATA • *Cellular and Molecular Biology Laboratory, RIKEN Discovery Research Institute, Saitama, Japan*
WAYNE G. SHREFFLER • *Mount Sinai School of Medicine, Division of Pediatric Allergy, New York, NY, USA*
MOULDY SIOUD • *Departments of Immunology and Tumor Biology, Institute for Cancer Research, University Hospital Montebello, Oslo, Norway*
ERIC J. SUNDBERG • *Boston Biomedical Research Institute, Watertown, MA, USA*
STEFAN TENZER • *Institute for Immunology, University of Mainz, Mainz, Germany*
ALEXANDRA THIELE • *Max Planck Research Unit for Enzymology of Protein Folding, Halle, Germany*
HANS-JÜRGEN THIESEN • *Institute of Immunology, University of Rostock, Rostock, Germany*
MIREILLE TOEBES • *Division of Immunology, The Netherlands Cancer Institute, Amsterdam, The Netherlands*
KENNETH B. TOMER • *Laboratory of Structural Biology, National Institute of Environmental Health Sciences, NIH, DHHS, Research Triangle Park, NC, USA*
PETER TRINDER • *Thymed GmbH, Wendelsheim, Germany*
MARC H. V. VAN REGENMORTEL • *Ecole Supérieure de Biotechnologie de Strasbourg, Illkirch Cedex, France*
THIRU VANNIASINKAM • *School of Biomedical Sciences, Charles Sturt University, Wagga Wagga, NSW, Australia*
LIN-FA WANG • *CSIRO Livestock Industries, Australian Animal Health Laboratory, Geelong, VIC, Australia*
JASON G. WILLIAMS • *Laboratory of Structural Biology, National Institute of Environmental Health Sciences, NIH, DHHS, Research Triangle Park, NC, USA*
MARKUS WULF • *Thymed GmbH, Wendelsheim, Germany*
MENG YU • *CSIRO Livestock Industries, Australian Animal Health Laboratory, Geelong, VIC, Australia*
JOHANNES ZERWECK • *JPT Peptide Technologies GmbH, Berlin, Germany*

Section I

B-Cell Epitope Mapping

Chapter 1

What Is a B-Cell Epitope?

Marc H.V. Van Regenmortel

Summary

The antigenicity of proteins resides in different types of antigenic determinants known as continuous and discontinuous epitopes, cryptotopes, neotopes, and mimotopes. All epitopes have fuzzy boundaries and can be identified only by their ability to bind to certain antibodies. Antigenic cross-reactivity is a common phenomenon because antibodies are always able to recognize a considerable number of related epitopes. This places severe limits to the specificity of antibodies. Antigenicity, which is the ability of an epitope to react with an antibody, must be distinguished from its immunogenicity or ability to induce antibodies in a competent vertebrate host. Failure to make this distinction partly explains why no successful peptide-based vaccines have yet been developed. Methods for predicting the epitopes of proteins are discussed and the reasons for the low success rate of epitope prediction are analyzed.

Key words: Continuous epitope, Discontinuous epitope, Mimotope, Cryptotope, Neotope, Prediction of epitopes, Antigenic cross-reactivity, Antibody specificity, Immunogenicity, Peptide-based vaccines.

1. Introduction

Since most biologically important antigens are proteins, I will discuss only the antigenicity of proteins and will not consider carbohydrate and nucleic acid antigens. The antigenic specificity of a protein resides in restricted areas of the molecule, known as antigenic determinants or epitopes, which are recognized by the combining sites or paratopes of certain immunoglobulin molecules. Once an immunoglobulin has been shown to bind to an antigen, it becomes known as an antibody specific for that antigen. Since epitopes are able to bind antibody molecules both in their free form and as membrane-bound B-cell receptors, they are often called B-cell epitopes to distinguish them from the T-cell

Ulrich Reineke and Mike Schutkowski (eds.), *Methods in Molecular Biology, Epitope Mapping Protocols, vol. 524*

DOI: 10.1007/978-1-59745-450-6_1

epitopes, which are proteolytically cleaved peptides of the antigen that interact with the receptors of T cells. This chapter deals only with B-cell epitopes, henceforth abbreviated to epitopes.

2. Types of Epitopes

Epitopes of proteins are usually classified as either continuous or discontinuous depending on whether the amino acids included in the epitope are contiguous in the peptide chain or not. The distinction between these two types of epitopes is not clear-cut since discontinuous epitopes often contain short segments of a few contiguous residues that are able to bind to antibodies raised against the protein and which could be given the status of continuous epitopes.

2.1. Continuous Epitopes

Any linear peptide fragment of a protein that is able to bind to antiprotein antibodies is called a continuous epitope. Since the criterion for identifying such an epitope is its binding activity, continuous epitopes are actually defined in a functional manner and no evidence is presented to show that each residue in the peptide makes contact with residues of the paratope and is recognized as such by the antibody. The contribution of individual residues to the epitope can be assessed by measuring the binding capacity of peptide analogs presenting single residue replacements. When this is done, it is found that most continuous epitopes contain a number of indifferent residues that seem not to be implicated in the binding interaction and can be replaced by any other amino acid without impairing antigenic activity *(1)*. Such continuous epitopes can therefore be said to be structurally discontinuous, although it cannot be excluded that replaceable residues play a scaffolding role or are recognized through their backbone atoms. Residues that cannot be replaced in an epitope without causing a major loss in binding activity tend to be regarded as being part of a smaller entity called "functional epitope" in which each residue is assumed to contribute to the free energy of interaction *(2)*. However, measurements of perturbations are not the same as energy determinations *(3)* and substitutions of residues that do not themselves interact with the antibody may induce structural perturbations that propagate beyond the mutated region and affect the activity of a nearby epitope.

There is no reason to assume that all the residues of a continuous peptide epitope correspond to residues present in the epitope of the intact protein since only a limited degree of similarity between the two structures is sufficient to allow the peptide to bind to antiprotein antibodies. Much of our knowledge of protein antigenicity, which is derived from the study of short linear peptides, gives information on which structures are needed for a peptide to possess antigenic cross-reactivity but it does not clarify the exact structure of the actual epitopes in the intact protein. Peptide fragments of a protein are not faithful copies of antigenic regions in protein molecules, one reason being that they do not retain the conformation present in the folded protein.

Many investigators take the view that the majority of continuous epitopes of proteins described in the literature are likely to correspond to unfolded regions of denatured protein molecules and are not genuine epitopes of native proteins *(4, 5)*. They argue that it is very difficult to know whether the immunoassays used to identify continuous epitopes actually measure antibodies specific for the native state of the cognate protein or whether they measure antibodies directed to the denatured protein. Antiprotein antisera frequently contain both types of antibodies because some of the molecules used for immunization are denatured before or after being injected in the animal. It is, of course, not possible to know the exact conformation of the protein molecule when it is interacting with a B-cell receptor during the immunization process.

In the reciprocal situation where antibodies raised to peptides are allowed to react with the cognate protein, it is possible that the antibodies recognize the protein because some of the protein molecules used in the immunoassay were denatured, for instance when they were adsorbed to a solid-phase in the assay *(6)*. It is now accepted that earlier erroneous claims that immunization with peptides always elicits high levels of antibodies that cross-react with the native cognate protein *(7)* arose because it was not realized that the protein used in solid-phase immunoassays had become denatured by adsorption to plastic *(8)*.

2.2. Discontinuous Epitopes

The second type of epitope known as discontinuous epitope corresponds to the vast majority of epitopes found in proteins. They consist of atoms from surface residues of the protein that are brought together by the folding of the polypeptide chain, and their antigenic reactivity depends on the native conformation of the protein. The atomic groups that form a discontinuous epitope are not held together by internal chemical bonds and they possess a collective identity recognized by antibodies only because the entire peptide chain acts as a scaffold. If the scaffold is perturbed, the epitope ceases to exist. Although such an epitope lies in a molecule and acts like a molecule, it is not actually a molecule (*see* p. 273 in **ref.** *9)*. A discontinuous epitope, therefore, cannot be isolated as an entity independent from the rest of the molecule in which it is embedded and it cannot be shown experimentally to possess binding activity on its own, outside of the protein context.

This places severe limits on the functional characterization of discontinuous epitopes since they can only be defined in structural terms. Their structure is established by analyzing antigen–antibody complexes using X-ray crystallography (*see* Chapter "Structural Basis of Antibody–Antigen Interactions") or NMR spectroscopy (*see* Chapter "Epitope Mapping of Antibody–Antigen Complexes by Nuclear Magnetic Resonance Spectroscopy") and identifying the set of atoms of the antigen that make contact with residues of the antibody. Usually a contact between two residues is said to occur if the interatomic distance between their atoms is less than 4 Å, and this criterion leads to the conclusion that discontinuous epitopes consist of 10–22 residues. These residues originate from between two and five separate segments of the polypeptide chain that are often surface loops and are brought together by the folding of the chain. When the protein is fragmented into peptides, residues from distant parts of the sequence are scattered and the individual constituents of discontinuous epitopes are usually no longer recognized by antiprotein antibodies, although short segments of a few residues may sometimes bind to antiprotein antibodies.

Discontinuous epitopes are often called conformational epitopes because their structure depends on the intact conformation of the native protein. This terminology may lead to the erroneous conclusion that continuous epitopes, in contrast, are conformation-independent. This is, of course, not the case since linear peptides constituting continuous epitopes necessarily also have one or, more likely, a number of different conformations. Peptide fragments of a protein very rarely retain the conformation initially present in the corresponding residues of the correctly folded protein. On the other hand, it is equally unlikely that if a discontinuous epitope could somehow be excised from a protein, it would retain its original conformation.

Studies with monoclonal antibodies (Mabs) have shown that only about 10% of the Mabs that react with a native protein are able to bind short peptide fragments of the protein. Since the range of specificities observed with a panel of Mabs is very similar to that found in a polyclonal antiserum raised against the same antigen *(10)*, it is reasonable to assume that also about 10% of the antibodies present in an antiprotein antiserum are able to recognize peptide fragments of the protein *(11, 12)*. These peptides are the ones that will be considered continuous epitopes of the antigen. Although much less common than discontinuous epitopes, continuous epitopes have been studied extensively because they have many applications, for instance as diagnostic reagents to replace infectious agents in immunoassays *(13)*, as immunogens to obtain antibody reagents useful for isolating gene products *(14, 15)*, or as potential synthetic vaccines *(16–18)*.

2.3. Additional Epitope Types

The quaternary structure of polymerized proteins introduces an additional level of antigenic complexity in macromolecular assemblies such as viruses. This has led to the recognition of two other types of epitopes, cryptotopes and neotopes, which are nowadays easily identifiable by using Mabs *(19)*. Cryptotopes are epitopes hidden in polymerized proteins or in virus particles because they are present on the surface of the protein subunits that become buried when the subunits aggregate. Cryptotopes of viruses are antigenically active only after dissociation of virus particles. Cryptotopes are fairly conserved in a group of related viruses because the intersubunit surfaces which control virus assembly tend to vary less than the outer surface of virions. As a result, antibodies to cryptotopes will emphasize antigenic similarities between the members of a virus family and they are therefore useful diagnostic reagents for detecting a wide range of related viruses instead of a single member of a virus genus or family *(20, 21)*.

Neotopes are epitopes that are specific for the quaternary structure of virus particles and are absent in dissociated viral subunits. Neotopes may arise from the juxtaposition of residues from neighboring subunits that are recognized by the antibody as a single epitope. For instance, one epitope of the serotype 1 of poliovirus consists of residues 221–226 of protein VP1, together with residues 164–172 and 270 of protein VP2 *(22)*. Neotopes can also arise through the conformational changes in protein subunits that result from intersubunit interactions. Because the quaternary structure of virus particles is not static but can undergo major rearrangements following small changes in pH and temperature *(23)*, neotopes are often transitional epitopes *(24)* that can assume different conformations. This is one of the reasons why it is difficult to mimic them by chemical synthesis, for instance when attempts are made to develop synthetic vaccines *(17)*.

3. Mimotopes

The term mimotope coined by Geysen *(25)* was originally defined as a peptide able to bind to a particular antibody, although unrelated in sequence to the protein antigen used to induce the antibody, usually because the antibody is directed to a discontinuous epitope. Currently, the term mimotope is applied to any epitope mimic irrespective of whether the protein epitope being mimicked is continuous or discontinuous. Mimotopes are usually identified by testing combinatorial peptide libraries obtained by chemical synthesis (*see* Chapters "Antibody Epitope Mapping Using De Novo Generated Synthetic Peptide Libraries" and "Epitope Mapping Using Randomly Generated Peptide Libraries") or phage display and selecting peptides able to bind antiprotein Mabs (*see* Chapter "Epitope Mapping Using Phage Display Peptide Libraries"). When mimotopes are selected for use as reagents in the diagnosis of virus infections, it is possible to screen phage libraries with sera collected from inviduals who recovered from a viral infection and had mounted an immune response against the infectious agent, in the absence of any knowledge of which antigens are involved *(26)*.

Mimotopes often show a limited amount of sequence similarity with the protein immunogen that gave rise to the antibodies used in the mimotope screening. It may also happen that the mimotope shows no sequence similarity whatsoever with the protein immunogen, although it is able to induce antibodies that cross-react with the protein *(27)*. Such findings underline the fact that the recognition between epitope and paratope does not take place at the level of whole residues but at the level of individual atoms. It is always a minority of the atoms of a given residue that participate in the interaction, a situation that is obscured by the division of epitopes into continuous and discontinuous classes which may give the impression that the elementary units of recognition are amino acid residues.

The presence of hydropathic complementarity between short segments of residues in an epitope and in the CDRs of a paratope may be sufficient to give rise to an antigen–antibody interaction. It has been demonstrated, for instance, that peptides corresponding to short sequences present in one CDR loop of an antilysozyme antibody were able to bind the antigen with similar specificity as the whole antibody, albeit with lower affinity *(28)*. In analogy with the binding observed by so-called continuous epitopes that are actually part of more complex discontinuous epitopes, these short segments in paratopes have been called continuous paratopes *(29)*. Hydropathic complementarity arises from an inverted hydropathic pattern in two short peptide sequences and is caused by the attraction between hydrophilic and hydrophobic groups *(30, 31)*. Peptide analogs that retain the original hydropathic profile present in a continuous epitope but possess no longer any sequence similarity with it may still bind the same antibody, a phenomenon that could be responsible for the binding activity of certain mimotopes *(32, 33)*.

To qualify as a mimotope, a peptide should not only be able to bind to a particular antibody but it should also be capable of eliciting antibodies that recognize the epitope being mimicked. This requirement stems from the fact that a single immunoglobulin molecule always harbors a number of partly overlapping or nonoverlapping paratopes, each one capable of binding to related or unrelated epitopes. The potential binding pocket of an immunoglobulin comprises as many as 50–70 hypervariable residues distributed over the six CDRs, although each individual paratope consists of only 10–20 CDR residues. This means that about two thirds of the CDR residues could bind to other epitopes that bear little or no resemblance to the first epitope, a situation that explains the considerable multispecificity of antibodies. Furthermore, different paratopes may partly overlap, in which case binding to one epitope may prevent a second unrelated epitope from being accommodated at a nearby location. Therefore, when a peptide is labeled a mimotope of epitope A because of its capacity either to bind to an anti-A antibody or to inhibit the binding of epitope A to this antibody, it cannot be excluded that the so-called mimotope actually binds to a different paratope from the one that interacts with epitope A. This is why it is necessary to show that a peptide is also able to induce antibodies that cross-react with epitope A, to demonstrate that it really is a mimotope of epitope A *(34)*.

Although the study of mimotopes has been of little value for investigating the structure of protein epitopes, it has led to the identification of many antigenically active peptides that are useful as immunochemical reagents and could lead to synthetic vaccines *(35–37)*.

4. Epitopes Are Relational Entities and Not Intrinstic Features of Proteins

It is important to realize that the epitope nature of a set of amino acids can only be established if an immunoglobulin able to bind to it has been found. In the same way, the antibody nature of an immunoglobulin defined by its paratope becomes apparent only when a complementary epitope has been identified. The immunoglobulin is then called an antibody specific for the antigen that harbors the epitope. This sometimes leads to confusion since most antigens possess many different epitopes and an antibody cannot be specific for the multiepitopic antigen as a whole but only for one of its epitopes *(38)*.

Epitopes and paratopes are relational entities defined by their mutual complementarity and they depend on each other to acquire a recognizable identity. This means that an epitope is not an intrinsic structural feature of a protein that could be identified in the absence of a particular interaction with a paratope.

Since epitopes acquire an identity by virtue of a relational nexus with complementary paratopes, the number of epitopes in a protein can be equated with the number of different Mabs that can be raised against it. In the case of the small insulin molecule, this number was estimated to be around 100 *(39)*. It is now accepted that the entire accessible surface of a protein harbors many overlapping epitopes, which can be recognized only if a sufficiently large panel of Mabs is available *(40, 41)*. Because of this relational dependence, analyzing the antigenic diversity of a protein corresponds to analyzing the size of the immunological repertoire of a host immunized with that protein.

The same residues at the surface of a protein can be part of different overlapping epitopes recognized by different paratopes and it is therefore not possible to draw clear boundaries between individual epitopes. There is no clear-cut minimum difference in atomic positions at epitope–paratope interfaces or in the binding affinity of interactions, that could be used as an absolute yardstick for deciding that two epitopes are the same or not. Epitopes have been called "fuzzy" binding sites *(38)*, and they share this fuzziness with all protein binding sites. As pointed by Moodie et al. *(42)*, the shape and electrostatic complementarity between two binding sites can be achieved by not just a single arrangement of amino acids but by a large number of alternative arrangements. Continuous epitopes have fuzzy boundaries because there are many ways to interpret the effect that removing or adding residues has on the antigenic activity of peptides. For instance, a longer peptide may be more active because the added residues are part of the epitope or because they induce a more active conformation in a nearby epitope. However, longer peptides are not necessarily more active than shorter peptides *(43)*. Sometimes the shortest peptide that retains significant binding activity in an immunoassay is called the epitope, but this is also unsatisfactory since different immunoassays have very different sensitivity thresholds and can induce different conformations in the peptide *(43)*.

5. Antigenic Cross-Reactivity and Antibody Specificity

Antigenic cross-reactivity is a common phenomenon caused by the ability of an antibody to recognize not only the epitope against which it was elicited but also a variety of related epitopes that possess some structural similarity with it. This type of cross-reactivity has been called "true cross-reactivity" *(44)*, to distinguish it from the less common "shared cross-reactivity," which occurs when an antibody recognizes the same epitope in two different multi-epitopic antigens. Usually, a paratope reacts with higher affinity with the homologous epitope used for raising the antibody than with cross-reacting epitopes, although a paratope can also bind more strongly to heterologous epitopes, a phenomenon known as heterospecificity or heteroclitic binding *(45)*. Heterospecificity is potentially widespread but is only observed when it is looked for, for instance if an antibody is tested against a series of analogs related to the epitope used for immunization *(46–48)*. Heterospecificity can be of practical use when attempts are made to obtain from a single hybridoma fusion experiment a number of Mabs specific for different members of a family of related proteins *(49)*. Heterospecificity is due to the fact that the clonal selection of a B cell, which eventually leads to antibody secretion, can be triggered by an immunogen endowed with only moderate affinity for the B-cell receptor. Because a high degree of fit between epitope and paratope is not required for initiating B-cell differentiation and because of antibody multispecificity, there is nothing strange in the finding that low-affinity antibodies may react better with related epitopes endowed with a superior degree of complementarity with the paratope. It can even happen that the antibody has such a low affinity for the immunogen that it does not react with it at all and only binds to a related antigen. This was commonly observed with antibodies raised against tobacco mosaic virus (TMV), which reacted with a mutant harboring a single proline residue substitution in the viral coat protein but not with TMV itself (*see* p. 198 in **refs.** *50; 51, 52)*.

As discussed in **Subheading 2**, the description of protein antigenicity in terms of continuous epitopes is based on the ability of short peptides to cross-react with antiprotein antibodies. These cross-reactions occur even when only a few of the peptide residues correspond to interacting residues in an epitope of the protein immunogen and in spite of major differences in conformation between the peptide and the corresponding region in the intact protein. The structural basis of antigenic cross-reactivity has been studied extensively *(53)*, and it is known that the flexibility of CDR loops greatly facilitates the ability of antibodies to adapt to a variety of epitopes *(54)*.

Biological specificity has been defined as the exact complementary relationship between an agent and something acted on (*see* p. 199 in **ref.** *55)*, and such a definition is valid for the specificity of enzyme–substrate, receptor–ligand, and antigen–antibody interactions. The term specificity is derived from the word species and describes what is characteristic of a species. Biologists believed for centuries that biological species were separated by clear-cut discontinuities, and bacteriologists later turned to serology in the hope that it would allow them to distinguish between different species of bacteria by using specific antisera raised against different bacteria. A belief in the absolute separation between bacterial species led Paul Ehrlich to believe in the absolute immunological specificity of antibacterial antibodies *(38, 56)*. This view was questioned by Landsteiner *(57)*, who demonstrated experimentally that serological cross-reactions between different cell types were caused by antibodies that reacted to different degrees with a wide range of cells. He showed that there was no one-to-one relationship between an antigen and its antibody and that antigens were able to elicit a wide spectrum of antibodies capable of cross-reacting with many related antigens.

The widespread occurrence of cross-reactions between epitopes and paratopes may seem to contradict the accepted view that antigen–antibody interactions are very specific. However, a perfect fit between an epitope and its paratope is not a meaningful concept, since it would imply that heterospecific binding or additional affinity maturation of the antibody during prolonged immunization could not occur. Antibody specificity is often believed to be correlated with high affinity, since it is expected that highly specific antibodies will possess a better stereochemical complementarity with their antigens than will antibodies of lower affinity. However, there is no necessary link between affinity and specificity, and antibodies of low affinity may in fact discriminate better between two antigens than do antibodies of high affinity. The reason is that low-affinity antibodies may detect fewer cross-reactions than do antibodies of high affinity since weaker cross-reactions will tend more quickly to be below the level of detection in the case of low-affinity antibodies *(38)*.

It is generally more meaningful to speak of the discrimination potential of antibodies rather than of their specificity. Since proteins harbor many different epitopes, different degrees of cross-reactivity will be found depending on the epitope that is singled out by a particular Mab. It is, in fact, the wish of the investigator to differentiate between two proteins that provide the criterion for deciding whether a particular antibody is specific or not, usually because it recognizes an epitope present in only one of the two proteins. If the antibody recognizes an epitope present in both proteins, it would be called nonspecific. Antibody reagents are thus considered specific if they achieve the level of discrimination that is required in any particular case and the same antibody will be considered specific or nonspecific depending on what the investigator is trying to achieve *(38)*. Antibodies, of course, are only specific for individual epitopes and not for antigens.

Although epitope–paratope recognition phenomena possess a fair degree of specificity, this is not due to the existence of specific “immunological” bonds different from the bonds observed in other protein interactions. Extensive shape complementarity at the surface of the two partners in an antigen–antibody complex leads to the formation of physicochemical bonds consisting of electrostatic forces and polar forces, such as van der Waals and hydrogen bonds that are present in all protein interactions.

The equilibrium affinity constants of antibodies lie in the range of 10^6 to 10^{10} L/mol *(58)*. The interpretation of antigen–antibody binding energies is complicated by the role played by entropy and hysteresis in the interaction. It was initially assumed that all water molecules are extruded from antigen–antibody interfaces in a complex, but it was later found that many interstitial water molecules remained at the interface because of imperfect steric complementarity. The reaction is in fact often driven by enthalpy and not by entropy *(59)*.

Hysteresis, the phenomenon whereby more energy is needed to dissociate most antigen–antibody bonds than is required to prevent their formation, is caused by the formation of additional secondary bonds subsequent to the initial primary bonds. For an extensive discussion of antigen–antibody bonds, see Oss *(60)*.

It is important to appreciate that the discrimination potential of antibodies is optimal only in a limited range of experimental conditions. When they are used at high concentrations, antibodies tend to react nonspecifically with many molecules, especially when the antigen is also present at a high local concentration, for instance on a solid-phase or in an immunoblot assay *(61, 62)*. One method for ascertaining the specificity of an interaction is to measure its stoichiometry for instance by biosensor assays, since nonspecific interactions do not possess a unique stoichiometric binding ratio when tested at different concentrations of the two reactants. However, it is always good practice to include various controls in any immunoassay to establish that the observed reaction is specific for the molecule of interest.

6. Antigens vs. Immunogens

The antigenicity of proteins is a chemical property describing interactions between epitopes and paratopes in terms of chemical and structural complementarity. Such a description takes the existence of antibodies for granted and does not consider the biological origin and synthesis of antibodies by the immune system. In contrast, immunogenicity, which is the ability of a protein to give rise to an immune response in a competent, vertebrate host, is a biological property definable only in the biological context of an immune system *(34)*. Immunogenicity always depends on extrinsic factors such as the host immunoglobulin repertoire and self-tolerance, the production of chemokines and cytokines as well as numerous cellular and regulatory mechanisms of the immune system.

The difference between antigens and immunogens is a crucial one in immunology but is often not sufficiently appreciated. For instance, when a peptide fragment of a protein is found to cross-react antigenically with antibodies raised against the protein, this does not in any way guarantee that it will be able to elicit antibodies that cross-react with the protein *(5)*.

Most peptides are immunogenic in the sense that they readily elicit antibodies that react with the peptide immunogen. However, this type of immunogenicity is mostly irrelevant since the purpose of peptide immunization is usually to obtain antibodies that cross-react with the cognate, native protein. What is needed, therefore, is so-called cross-reactive immunogenicity, i.e., the ability to induce antibodies that cross-react with the cognate protein. In addition, if the peptide is to have vaccine potential, it must elicit antibodies that neutralize the infectivity of the pathogen harboring the cognate protein; i.e., it must also possess so-called cross-protective immunogenicity *(5, 34)*. This type of immunogenic capacity is not necessarily present when the peptide is antigenically active and able to bind to a neutralizing antibody. An antiprotein antibody used in an immunoassay may be able to select one of the conformations present in the peptide or may induce such a cross-reactive conformation by an induced fit mechanism, whereas a B-cell receptor that possesses no prior specificity for the cognate protein will not be able during the immunization process to carry out such a selection or induction. The peptide will bind to various B-cell receptors that recognize some of the peptide conformations, but it will not preferentially bind to those rare receptors which, in addition to recognizing the peptide, also cross-react with the epitope present in the native protein. There is thus no reason why most of the elicited antipeptide antibodies should also react with the cognate protein.

Since B-cell epitopes are usually defined as regions of the antigen that bind both free and membrane-bound antibodies, little attention is given to the fact that the structural context in which an epitope recognizes a free antibody molecule or a B-cell receptor embedded in a membrane is not the same. This difference is one of the reasons why knowledge of the structure of a viral epitope bound to an antibody does not necessarily provide relevant information on its immunogenic potential since this depends on a reaction occurring in the membrane environment of a B-cell receptor. It has been pointed out by Zwick *(63)*, for instance, that the type of lipid that surrounds the gp41 membrane-proximal external region (MPER) antigen of HIV, when it is presented to membrane-bound B-cell receptors, is likely to affect its immunogenicity in ways that are not predictable from the structure of MPER antigen–antibody complexes studied by X-ray crystallography outside such a lipid environment.

Furthermore, the structure of epitopes and paratopes seen in a complex may be different from the structure of the respective binding sites in the free antigen and antibody molecules, before they have been altered by the mutual adaptation that occurs during the binding interaction *(64, 65)*. As a result the structure of an epitope after complexation with a neutralizing Mab may be an unreliable guide for identifying the exact epitope structure that was recognized by B-cell receptors during the immunization process and which should be present in a vaccine immunogen intended to elicit the same type of neutralizing antibodies.

7. Epitope Prediction

The main purpose of predicting epitopes (*see* Chapter "Prediction of Linear B-cell Epitopes") is to replace the epitope regions of intact antigen molecules by linear synthetic peptides that could be used as reagents for detecting antiprotein antibodies in an immunoassay *(13)* or as immunogens for raising antipeptide antibodies able to cross-react with the protein *(15)*. A further purpose is to develop synthetic peptide vaccines in which case the predicted epitopes must also be able to elicit antibodies that neutralize the infectivity of the pathogen harboring the protein antigen *(5)*. Unfortunately, neutralization epitopes that elicit antibodies that protect against infections do not have special physicochemical properties that allow them to be recognized and predicted separately from other epitopes.

Most attempts at predicting epitopes have been restricted to continuous epitopes since any predicted structure could then easily be synthesized chemically or inserted into a recombinant protein. Since the epitopes of native proteins are located on the surface of the molecules, initial prediction attempts analyzed protein sequences using amino acid propensity scales which identified segments of the protein that protruded at the surface, were hydrophilic, and possessed a high mobility *(66–69)*.

Many different scale-based prediction methods were developed and compared *(70, 71)*, but none of them, even when used in combination, gave high rates of successful prediction *(72–74)*. The apparent success rate of a prediction depends very much on the method used to measure its effectiveness *(75)*, and at present, a method measuring the values of the area under the receiver operating characteristic curve (A_{roc}) is the one most commonly used *(74, 76)*.

As an increasing number of 3D structures of proteins is becoming available, new prediction methods are being developed which incorporate information from docking algorithms and 3D structures *(71)*. Unfortunately, many investigators claim that they are able to predict discontinuous epitopes when in fact they only predict that certain surface residues are likely to be part of a discontinuous epitope. They do not predict which minimum set of combined residues must be assembled in a defined configuration to achieve a structure with antigenic or immunogenic activity. Such an alleged prediction of discontinuous epitopes is actually a misnomer since it does not entail predicting that a particular collection of atoms or residues possesses the characteristic immunological activity of an epitope.

Since the usual purpose of discontinuous epitope prediction is to be able to replace the protein epitope by a synthetic construct that possesses the same immunological activity, many investigators have attempted to design linear peptides that mimic the binding activity of a surface patch of the protein. One such approach is the Mapitope strategy *(77)*, which consists in assembling clusters of connected amino acid pairs that lie within the footprint of an epitope. Residue pairs originating from distant regions of the protein sequence are affinity-selected from a random peptide library and retained if they are present at the protein surface. Other approaches have also been used to reconstitute discontinuous epitopes by aligning putative epitope residues along a synthetic peptide *(78)*. However, it may be insufficient to include only the solvent-exposed surface of proteins when mapping epitopes, since certain buried residues can be involved in the paratope interaction following conformational rearrangements *(79)*.

Since it is well-known that synthetic peptides are less rapidly degraded and more immunogenic if they are constrained into stable secondary structures by cyclization or other chemical procedures, there have been many attempts to constrain peptides into helix, turn, or cyclic conformations *(18, 80, 81)*. However, constraining peptides will not necessarily make them adopt a conformation that closely mimics the immunogenic structure in the cognate protein. In some cases unconstrained peptides possessing intrinsic disorder can be superior immunogens, possessing because they present some of the conformations found in the cognate protein *(82)*. For the same reason, terminal segments of proteins are often correctly predicted to be continuous epitopes because these regions tend to be surface-oriented *(83)* and are more hydrophilic and mobile than internal regions *(84)*.

As discussed elsewhere, there are many reasons for the low success rate of epitope predictions *(71, 74, 75)*. Unsuccessful epitope prediction may partly explain why, after several decades of intensive research efforts, no synthetic peptide vaccine has yet been developed. More than a thousand synthetic peptides have been examined as potential vaccines in numerous preclinical studies. About 125 peptides have progressed to phase I clinical trials and about 30 to phase II trials but not a single peptide vaccine passed phase III trials and is presently marketed for human use *(18)*. This striking lack of success leads to the inescapable conclusion that some of our underlying assumptions regarding what constitutes an effective "protective" B-cell epitope must have been incorrect *(17, 85)*. One can only hope that the new bioinformatics tools being developed at present *(71)* will improve our understanding of the nature of epitopes and allow us in future to develop better immunological intervention strategies.

References

1. Getzoff, E. D., Tainer, J. A., Lerner, R. A., and Geyson, H. M. (1988) The chemistry and mechanism of antibody binding to protein antigens. *Adv. Immunol.* 43, 1–98.
2. Cunningham, B. C., and Wells, J. A. (1993) Comparison of a structural and a functional epitope. *J. Mol. Biol.* 234, 554–563.
3. DeLano, W. L. (2002) Unraveling hot spots in binding interfaces: progress and challenges. *Curr. Opin. Struct. Biol.* 12, 14–20.
4. Laver, W. G., Air, G. M., Webster, R. G., and Smith-Gill, S. J. (1990) Epitopes on protein antigens: misconceptions and realities. *Cell* 61, 553–556.
5. Van Regenmortel, M. H. V. (2006) Immunoinformatics may lead to a reappraisal of the nature of B cell epitopes and of the feasibility of synthetic peptide vaccines. *J. Mol. Recognit.* 19, 183–187.
6. Van Regenmortel, M. H. V. (1999) Molecular dissection of protein antigens and the prediction of epitopes, in *Synthetic Peptides as Antigens* Van Regenmortel, M. H. V., and Muller, S. eds., Elsevier, Amsterdam, The Netherlands, pp. 1–78.
7. Lerner, R. A. (1984) Antibodies of predetermined specificity in biology and medicine. *Adv. Immunol.* 36, 1–44.
8. Darst, S. A., Roberston, C. R., and Berzofsky, J. A. (1988) Adsorption of the protein antigen myoglobin affects the binding of conformation-specific monoclonal antibodies. *Biophys. J.* 53, 533–539.
9. Rosen, R. (1991) *Life Itself.* Columbia Univ Press, New-York.
10. Quesniaux, V. F. J., Schmitter, D., Schreier, M., and Van Regenmortel, M. H. V. (1990) Monoclonal antibodies to Cyclosporine are representative of the major antibody populations present in antisera of immunized mice. *Mol. Immunol.* 27, 227–236.
11. Barlow, D. J., Edwards, M. S., and Thornton, J. M. (1986) Continuous and discontinuous protein antigenic determinants. *Nature* 322, 747–748.
12. Van Regenmortel, M. H. V. (1996) Mapping epitope structure and activity: from one-dimensional prediction to four-dimensional description of antigenic specificity. *Methods* 9, 465–472.
13. Leinikki, P., Lehtinen, M., Hyöty, H., Parkkonen, P., Kantanen, M. L, and Hakulinen, J. (1993) Synthetic peptides as diagnostic tools in virology. *Adv. Virus Res.* 42, 149–186.
14. Walter, G. (1986) Production and use of antibodies against synthetic peptides. *J. Immunol. Methods* 88, 149–161.
15. Muller, S. (1999) Use of antipeptide antibodies in molecular and cellular biology, in *Synthetic Peptides as Antigens* (Van Regenmortel, M. H. V., and Muller, S. eds., Elsevier, Amsterdam, The Netherlands, pp. 215–235.
17. Van Regenmortel, M. H. V. (1999) Molecular design versus empirical discovery in peptide-based vaccines. Coming to terms with fuzzy recognition sites and ill-defined structure–function relationships in immunology. *Vaccine* 18, 216–221.
18. Hans, D., Young, P. R., and Fairlie, D. P. (2006) Current status of short synthetic peptides as vaccines. *Med. Chem.* 2, 627–646.
19. Van Regenmortel, M. H. V. (1992) The conformational specificity of viral epitopes. *FEMS Microbiol. Lett.* 100, 483–488.
20. Shepard, J. F., Secor, G. A., and Purcifull, D. E. (1974) Immunochemical cross-reactivity between the dissociated capsid proteins of PVY group plant viruses. *Virology* 58, 464–475.
21. Dougherty, W., Willis, L., and Johnston, R. E. (1985) Topographic analysis of tobacco etch virus capsid protein epitopes. *Virology* 144, 66–72.
22. Boeye, A., and Rombaut, B. (1992) The proteins of poliovirus. *Prog. Med. Virol.* 62, 139–166.
23. Bothner, B., Dong, X. F., Bibbs, L., Johnson, J. E., and Suizdak, G. (1998) Evidence of viral capsid dynamics using limited proteolysis and mass spectrometry. *J. Biol. Chem.* 273, 673–676.
24. Zolla-Pazner, S. (2004) Identifying epitopes of HIV-1 that induce protective antibodies. *Nat. Rev. Immunol.* 4, 199–210.
25. Geysen, H. M., Rodda, S. J., and Mason, T. J. (1986) A priori delineation of a peptide which mimics a discontinuous antigenic determinant. *Mol. Immunol.* 23, 709–715.
26. Folgori, A., Tafi, R., Meola, A., Felici, F., Galfre, G., Cortese, R., Monaci, P., and Nicosia, A. (1994) A general strategy to identify mimotopes of pathological antigens using only random peptide libraries and human sera. *EMBO J.* 13, 2236–2243.
27. Delmastro, P., Meola, A., Monaci, P., Cortese, R., and Galfrè, G. (1997) Immunogenicity of filamentous phage displaying peptide mimotopes after oral administration. *Vaccine* 15, 1276–1285.
28. Laune, D., Molina, F., Ferrieres, G., Mani, J. C., Cohen, P., and Simon, D. (1997) Systematic exploration of the antigen binding activity of synthetic peptides isolated from the variable regions of immunoglobulins. *J. Biol. Chem.* 272, 30937–30944.

29. Van Regenmortel, M. H. V. (1998) Mimotopes, continuous paratopes and hydropathic complementarity: novel approximations in the description of immunological specificity. *J. Disper. Sci. Technol.* 19, 1199–1219.
30. Blalok, J. (1990) Complementarity of peptides specified by "sense" and "antisense" strands of DNA. *Trends Biotechnol.* 8, 140–144.
31. Tropsha, A., Kizler, J. S., and Chaiken, I. M. (1992) Making sense of antisense: a review of experimental data and developing ideas on sense–antisense recognition. *J. Mol. Recognit.* 5, 43–54.
32. Boquet, D., Déry, O., Forbert, Y., Grassi, J., and Couraud, J. Y. (1995) Is hydropathic complementarity involved in antigen–antibody binding? *Mol. Immunol.* 32, 303–308.
33. Hanin, V., Déry, O., Boquet, D., Sagot, M. A., Crémion, C., Courand, J. Y., and Grassi, J. (1997) Importance of hydropathic complementarity for the binding of the neuropeptide substance P to a monoclonal antibody: equilibrium and kinetic studies. *Mol. Immunol.* 34, 829–838.
34. Van Regenmortel, M. H. V. (2001) Antigenicity and immunogenicity of synthetic peptides. *Biologicals* 29, 209–213.
35. Meloen, R. H., Puijk, W. C., and Slootstra, J. W. (2000) Mimotopes: realization of an unlikely concept. *J. Mol. Recognit.* 13, 352–359.
36. Mullen, L. M., Nair, S. P., Ward, J. M., Rycroft, A. N., and Henderson, B. (2006) Phage display in the study of infectious diseases. *Trends Microbiol.* 14, 141–147.
37. Larralde, O. G., Martinez, R., Camacho, F., Amin, N., Aguilar, A., Talavera, A., Stott, D. I., and Perez, E. M. (2007) Identification of hepatitis A virus mimotopes by phage display: antigenicity and immunogenicity. *J. Virol. Methods* 140, 49–58.
38. Van Regenmortel, M. H. V. (1998) From absolute to exquisite specificity. Reflections on the fuzzy nature of species, specificity and antigenic sites. *J. Immunol. Methods* 218, 37–48.
39. Schroer, J. A., Bender, T., Feldmann, T., and Kim, K. J. (1983) Mapping epitopes on the insulin molecule using monoclonal antibodies. *Eur. J. Immunol.* 13, 693–700.
40. Benjamin, D. C., Berzofsky, J. A., East, I. J., Gurd, F. R. N., Hannum, C., Leach, S. J., Margoliash, E., Michael, J. G., Miller, A., Prager, E. M., Reichlin, M., Sercaz, E. E., Smith-Gill, S. J., Todd, P. E., and Wilson, A. C. (1984) The antigenic structure of proteins: a reappraisal. *Ann. Rev. Immunol.* 2, 67–101.
41. Berzofsky, J. A. (1985) Intrinsic and extrinsic factors in protein antigenic structure. *Science* 229, 932–940.
42. Moodie, S. L., Mitchell, J. B. O., and Thornton, J. M. (1996) Protein recognition of adenylate: an example of a fuzzy recognition template. *J. Mol. Biol.* 263, 486–500.
43. Muller, S., Plaué, S., Couppez, M., and Van Regenmortel, M. H. V. (1986) Comparison of different methods for localizing antigenic regions in histone H2A. *Mol. Immunol.* 23, 593–561.
44. Berzofsky, J. A., Schechter, A. N. (1981) The concepts of crossreactivity and specificity in immunology. *Mol. Immunol.* 18, 751–763.
45. Mäkelä, O. (1965) Single lymph node cells producing heteroclitic bacteriophage antibody. *J. Immunol.* 95, 378–386.
46. Al Moudallal, Z., Briand, J. P., and Van Regenmortel, M. H. V. (1982) Monoclonal antibodies as probes of the antigenic structure of tobacco mosaic virus. *EMBO J.* 1, 1005–1010.
47. Underwood, P. A. (1985) Theoretical considerations of the ability of monoclonal antibodies to detect antigenic differences between closely related variants, with particular reference to heterospecific reactions. *J. Immunol. Methods* 85, 295–307.
48. Harper, M., Lema, F., Boulot, G., and Poljak, R. J. (1987) Antigen specificity and cross-reactivity of monoclonal anti-lysozyme antibodies. *Mol. Immunol.* 24, 97–108.
49. Frison, E. A., and Stace-Smith, R. (1992) Cross-reacting and heterospecific monoclonal antibodies produced against arabis mosaic nepovirus. *J. Gen. Virol.* 73, 2525–2530.
50. Van Regenmortel MHV (1982). *Serology and Immunochemistry of Plant Viruses.* Academic Press, New-York.
51. Loor, F. (1971) On the existence of heterospecific antibodies in sera from rabbits immunized against tobacco mosaic virus determinants. *Immunology* 21, 557–564.
52. Sengbusch, P., and Wittmann, H. G. (1965) Serological and physicochemical properties of the wild strain and two mutants of tobacco mosaic virus with the same amino acid exchange in different positions of the protein chain. *Biochem. Biophys. Res. Commun.* 18, 780–787.
53. Roberts, V. A., Getzoff, E. D., and Tainer, J. A. (1993) Structural basis of antigenic cross-reactivity, in *Structure of Antigens, Vol. 2* (Van Regenmortel, M. H. V., ed.), CRC, Boca Raton, FL, pp. 31–53.
54. James, L. C., Roversi, P., and Tawfik, D. S. (2003) Antibody multispecificity mediated by conformational diversity. *Science* 299, 1362–1367.
55. Medawar, P. B., and Medawar, J. S. (1978) *The Life Science.* Granada Publishing, London.
56. Mazumder, P. H. (1995). *Species and Specificity.* Cambridge University Press, Cambridge.

57. Landsteiner, K. (1947). *The Specificity of Serological Reactions.* Harvard University Press, Cambridge, Mass.
58. Foote, J., and Eisen, H. N. (1995) Kinetic and affinity limits on antibodies produced during immune responses. *Proc. Natl. Acad. Sci. USA* 92, 1254–1256.
59. Braden, B. C., and Poljak, R. J. (1995) Structural features of the reactions between antibodies and protein antigens. *FASEB J.* 9, 9–16.
60. Van Oss, C. J. (1995) Hydrophobic, hydrophilic and other interactions in epitope-paratope binding. *Mol. Immunol.* 32, 199–211.
61. Ghosh, G., and Cambell, A. M. (1986) Multispecific monoclonal antibodies. *Immunol. Today* 7, 217–222.
62. Zimmermann, D., and Van Regenmortel, M. H. V. (1989) Spurious cross-reactions between plant viruses and monoclonal antibodies can be overcome by saturating ELISA plates with milk proteins. *Arch. Virol.* 106, 15–22.
63. Zwick, M. B. (2005) The membrane-proximal external region of HIV-1 gp41: a vaccine target worth exploring. *AIDS* 19, 1725–1737.
64. Wilson, I. A., and Stanfield, R. L. (1994) Antigen–antibody interactions: new structures and new conformational changes. *Curr. Opin. Struct. Biol.* 4, 857–867.
65. Halperin, I., Ma, B., Wolfson, H., and Nussinov, R. (2002) Principles of docking: an overview of search algorithms and a guide to scoring functions. *Proteins* 47, 409–443.
66. Hopp, T. P., and Woods, K. R. (1981) Prediction of protein antigenic determinants from amino acid sequences. *Proc. Natl. Acad. Sci. USA* 78, 3824–3828.
67. Westhof, E., Altschuh, D., Moras, D., Bloomer, A. C., Mondragon, A., Klug, A., and Van Regenmortel, M. H. (1984) Correlation between segmental mobility and the location of antigenic determinants in proteins. *Nature* 311, 123–126.
68. Thornton, J. M., Edwards, M. S., Taylor, W. R., and Barlow, D. J. (1986) Location of 'continuous' antigenic determinants in the protruding regions of proteins. *EMBO J.* 5, 409–413.
69. Novotny, J., Bruccoleri, R. E., Carlson, W. D., Handschumacher, M., and Haber, E. (1987) Antigenicity of myohemerythrin. *Science* 238, 1584–1586.
70. Pellequer, J. L., Westhof, E., and Van Regenmortel, M. H. (1991) Predicting the location of continuous epitopes in proteins from their primary structures. *Methods Enzymol.* 203, 176–201.
71. Ponomarenko, J. V., Van Regenmortel, M. H. V., (2009). B cell epitope prediction. In: *Structural Bioinformatics*, 2nd edition (Bourne, P. E., and Gu J., eds). John Wiley, Hoboken, NJ. pp 849–879.
72. Odorico, M., and Pellequer, J. L. (2003) BEPITOPE: predicting the location of continuous epitopes and patterns in proteins. *J. Mol. Recognit.* 16, 20–22.
73. Blythe, M. J., and Flower, D. R. (2005) Benchmarking B cell epitope prediction: underperformance of existing methods. *Protein Sci.* 14, 246–248.
74. Greenbaum, J. A., Andersen, P. H., Blythe, M., Bui, H. H., Cachau, R. E., Crowe, J., Davies, M., Kolaskar, A. S., Lund, O., Morrison, S., Mumey, B., Ofran, Y., Pellequer, J. L., Pinilla, C., Ponomarenko, J. V., Raghava, G. P., van Regenmortel, M. H., Roggen, E. L., Sette, A., Schlessinger, A., Sollner, J., Zand, M., and Peters, B. (2007) Towards a consensus on datasets and evaluation metrics for developing B-cell epitope prediction tools. *J. Mol. Recognit.* 20, 75–82.
75. Van Regenmortel, M. H. V., and Pellequer, J. L. (1994) Predicting antigenic determinants in proteins: looking for unidimensional solutions to a three-dimensional problem? *Pept. Res.* 7, 224–228.
76. Haste Andersen, P., Nielsen, M., and Lund, O. (2006) Prediction of residues in discontinuous B-cell epitopes using protein 3D structures. *Protein Sci.* 15, 2558–2567.
77. Bublil, E. M., Freund, N. T., Mayrose, I., Penn, O., Roitburd-Berman, A., Rubinstein, N. D., Pupko, T., and Gershoni, J. M. (2007) Stepwise prediction of conformational discontinuous B-cell epitopes using the Mapitope algorithm. *Proteins* 68, 294–304.
78. Timmerman, P., Beld, J., Puijk, W. C., and Meloen, R. H. (2005) Rapid and quantitative cyclization of multiple peptide loops onto synthetic scaffolds for structural mimicry of protein surfaces. *ChemBioChem* 6, 821–824.
79. Alexander, H., Alexander, S., Getzoff, E. D., Tainer, J. A., Geysen, H. M., and Lerner, R. A. (1992) Altering the antigenicity of proteins. *Proc. Natl. Acad. Sci. USA* 89, 3352–3356.
80. Shepherd, N. E., Hoang, H. N., Abbenante, G., and Fairlie, D. P. (2004) Single turn peptide alpha helices with exceptional stability in water. *J. Am. Chem. Soc.* 127, 2974–2983.
81. Sundaram, R., Lynch, M. P., Rawale, S. V., Sun, Y., Kazanji, M., and Kaumaya, P. T. (2004) De novo design of peptide immunogens that mimic the coiled coil region of human T-cell leukemia virus type-1 glycoprotein 21 transmembrane subunit for induction

of native protein reactive neutralizing antibodies. *J. Biol. Chem.* 279, 24141–24151.

82. Uversky, V. N., Oldfield, C. J., and Dunker, A. K. (2005) Showing your ID: intrinsic disorder as an ID for recognition, regulation and cell signaling. *J. Mol. Recognit.* 18, 343–384.
83. Thornton, J. M., and Sibanda, B. L. (1983) Amino and carboxy-terminal regions in globular proteins. *J. Mol. Biol.* 167, 443–460.
84. Pellequer, J. L., Westhof, E., and Van Regenmortel, M. H. V. (1994) Epitope predictions from the primary structure of proteins, in *Peptide Antigens: A Practical Approach* (Wisdom, G. B., ed.), JRL, Oxford, UK, pp. 7–25.
85. Van Regenmortel, M. H. V. (2007) The rational design of biological complexity: a deceptive metaphor. *Proteomics* 7, 965–975.

Subsection A

Whole Antigen Methods

Chapter 2

Structural Basis of Antibody–Antigen Interactions

Eric J. Sundberg

Summary

Antibody molecules can be regarded as products of a protein engineering system for the generation of a virtually unlimited repertoire of complementary molecular surfaces. This extreme structural heterogeneity is required for recognition of the nearly infinite array of antigenic determinants. This chapter discusses the structures of antibodies and their specific recognition of antigens, the binding energetics of these interactions, the cross-reactivity and specificity of antibody–antigen interactions, the role of conformational flexibility in antigen recognition, and the structural basis of the antibody affinity maturation process.

Key words: Antibody, Antigen, X-ray crystallography, Binding energetics, Affinity maturation.

1. Structural Overview of Antibodies

The basic building blocks of antibodies are small protein domains, each composed of two antiparallel β-sheets and belonging to the immunoglobulin (Ig) fold superfamily *(1)*. **Fig. 1** provides an overview of the structural characteristics of Ig domains, how they are assembled to form functional antibodies, and how they generally recognize antigenic molecules. Antibody molecules are composed of two identical polypeptide chains of ~500 amino acids (the heavy or H chains) covalently linked through disulfide bridges to two identical polypeptide chains of roughly 250 residues (the light or L chains) (**Fig. 1a**). The H and L chains may be divided into N-terminal variable (V) and C-terminal constant (C) portions. Each H chain contains four or five Ig domains (V_H, C_H1, C_H2, $C_H3 \pm C_H4$, depending on the antibody isotype), while each L chain consists of two such domains (V_L, C_L). The V_L and C_L domains are disulfide-linked with the V_H and C_H1

Ulrich Reineke and Mike Schutkowski (eds.), *Methods in Molecular Biology, Epitope Mapping Protocols, vol. 524*

DOI: 10.1007/978-1-59745-450-6_2

domains, respectively, to form the Fab region (large oval, **Fig. 1a**) of the antibody, which is linked through a hinge region to the Fc domain, formed by noncovalent association of the C_H2–3/4 domains from both chains.

The variable domains of antibodies (V_H and V_L), which together form what is referred to as the Fv (small oval, **Fig. 1a**), each contain three segments, which connect the β-strands and are highly variable in length and sequence *(4)*. These so-called complementarity-determining regions (CDRs) lie in close spatial proximity on the surface of the V domains and determine the conformation of the combining site (**Fig. 1b, c**). In this way, the CDRs confer specific binding activity to apical regions of the Ig domain. The central

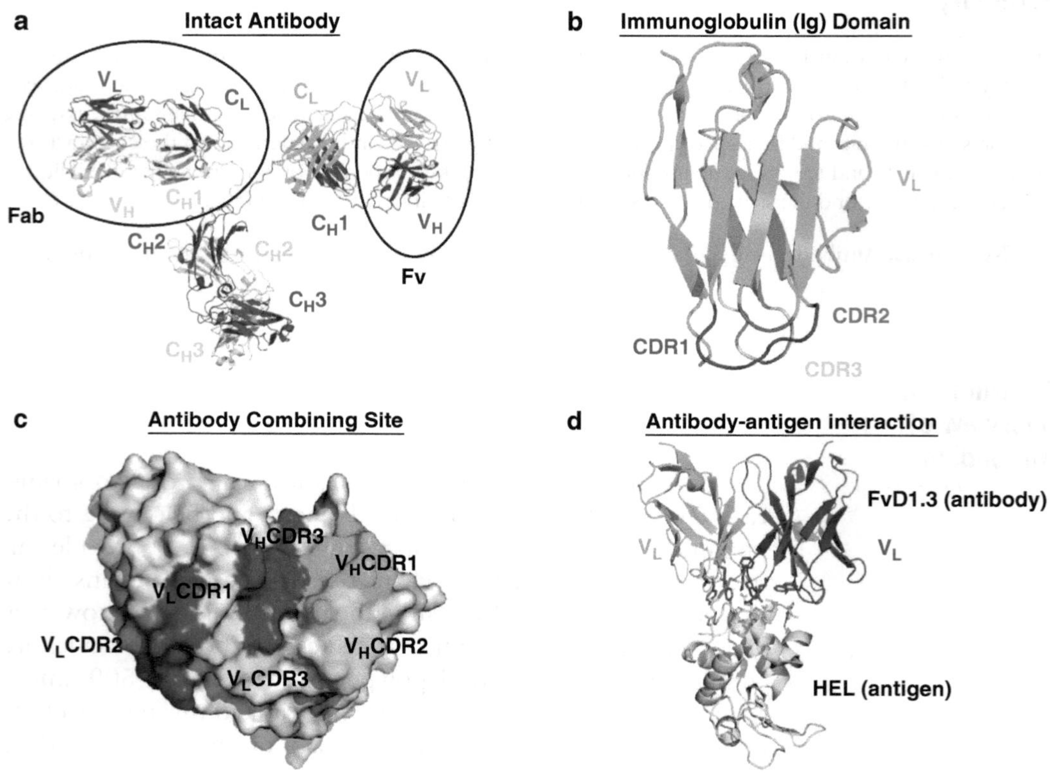

Fig. 1. Structural overview of antibodies. (**a**) Structure of the intact murine IgG2a monoclonal antibody, Mab231 *(2)*, including two light chains, each composed of a variable (V_L) and a constant (C_L) immunoglobulin (Ig) domain (*red* and *green*), and two heavy chains, each composed of a variable (V_H) and three constant (C_H1, C_H2, and C_H3) domains (*blue* and *yellow*) *(3)*. The common fragments, Fab (*large oval*) and Fv (*small oval*), are indicated. (**b**) Ribbon diagram of a single Ig domain, V_L, of Mab231 highlighting its antiparallel β-sheet secondary structure. The complementarity-determining region loops are marked, CDR1 (*blue*), CDR2 (*magenta*), and CDR3 (*yellow*). (**c**) Molecular surface of the antibody-combining site of Mab231 formed by the intersection of the apical regions of V_L and V_H. The CDR loops provide a contiguous surface for antigen recognition. Colors are as follows: V_LCDR1 (*blue*), V_LCDR2 (*magenta*), V_LCDR3 (*yellow*), V_HCDR1 (*green*), V_HCDR2 (*cyan*), V_HCDR3 (*red*). (**d**) Ribbon diagram of the FvD1.3–hen egg lysozyme (HEL) antibody–antigen complex. Colors are as follows: HEL (*yellow*), D1.3 V_L domain (*green*), and D1.3 V_H domain (*blue*). Residues of HEL and D1.3 involved in interactions in the antigen–antibody interface are *cyan* and *red*, respectively (*see Color Plates*).

paradigm of antigen recognition is that the three-dimensional structure formed by the six CDRs recognizes and binds a complementary surface (epitope) on the antigen (**Fig.** 1d).

Although CDR loops are hypervariable, they adopt a limited number of canonical structures in antibodies *(5)*. Usage of the six CDR loops that confer antigen-binding specificity varies, especially for antibodies. Antibodies to smaller antigens, such as haptens and peptides, commonly do not utilize all six CDRs *(6, 7)*, while antiprotein antibodies generally do. Camelid antibodies that have no light chains *(8)* can nonetheless bind protein antigens with nanomolar affinities using as few as two CDR loops *(9)*. Cartilaginous fish, such as sharks, are the oldest living organisms that express components of the vertebrate adaptive immune system. These animals can recognize antigens using a single Ig domain that is similar to camelid heavy chain V domains *(10)*. Indeed, some of the contacts to various mammalian antibody CDR loops by protein antigens, while confirmed as structurally belonging to the molecular interface, are energetically meaningless. Additionally, both polyclonal and monoclonal antibodies (mAbs) raised against small (8- to 15-mer) peptides often bind to both the peptide and to the whole correlate protein, sometimes with higher affinity than antibodies raised directly against the latter *(11–13)*. Framework regions are commonly invoked in antigen recognition to varying degrees, and can comprise up to 15% of the buried surface area of an antibody–antigen complex *(14)*. The V_HCDRs, and V_HCDR3 in particular, generally make more extensive contacts than V_LCDRs, and the geometrical center of the antibody–antigen interface tends to lie near V_HCDR3. There exists a strong correlation between residues that do not form contacts with antigen and those residues that are important in defining the canonical backbone structures of the CDR loops *(15)*. These residues tend to pack internally and are therefore less exposed on the antibody-combining site surface.

Antibody–antigen complexes exhibit a high degree of both shape and chemical complementarity at their interacting surfaces *(16)*. The combined solvent-accessible surfaces buried in antiprotein antibody–antigen complexes range from ~1,400 to 2,300 Å^2, with roughly equal contributions from antigen and antibody, while smaller antigens, such as haptens and peptides, generally bury less overall surface area when bound to the antibody. The surface topography of the antigen-contacting surface, as well as other general structural features, of antibodies can vary significantly according to antigen size *(17)*. While the percentage of the antigen surface buried in the interface with the antibody is always high and their surfaces are complementary, the antibody contact surface becomes more concave as the antigen becomes smaller. Thus, although the combining sites of antibodies that recognize large protein antigens are generally planar, and are often more planar than a number of other types of protein–protein interfaces *(18)*, antibodies that recognize medium-sized antigens, such as peptides, DNA, and carbohydrates, often have a grooved antigen-contacting surface, while even smaller antigens (haptens) are recognized by antibodies with distinct cavities *(19)*. A common feature of antipeptide antibody–antigen interactions is a β-turn motif of the peptide buried deeply into the combining site *(20–22)*. The amount of surface area on the antibody molecule buried by the antigen decreases with antigen size, as less of the antibody surface is utilized to envelop the smaller antigens. Large antigens often contact antibody residues at the edge of the combining site and interact with the more apical portions of the CDR loops, while the interactions of smaller antigens are more restricted to the central portion of the antibody-combining site *(17)*.

2. Binding Energetics of Antigen Recognition

There exists a functional affinity window for antigen recognition. Antibodies undergo affinity maturation upon encountering their specific antigens (addressed later in this chapter). Below, the binding properties of fully matured antibodies are discussed.

Most mature antibodies have affinities for their specific antigens in the range of $10^7 – 10^8$ M^{-1}, although many functional antibodies that recognize carbohydrates and bacterial polysaccharides fail to reach affinity levels of 10^6 M^{-1}. It has been proposed *(23)* that, owing to diffusion rates and the residence time required for antibody internalization controlling on- and off-rates, there exists an affinity ceiling for antibody–antigen interactions of ~10^{10} M^{-1}. Antibodies with antigen affinities above this threshold, presumably, would possess no further advantage over their lower affinity counterparts in the antibody selection process in vivo. The existence of this affinity ceiling has been demonstrated for antigen-specific B-cell transfectants, and more important, an affinity window for effective B-cell response has been revealed for which a minimum affinity of 10^6 M^{-1} and half-life of 1 s were required for detectable B-cell triggering that reached a plateau for affinities beyond 10^{10} M^{-1} *(24)*. Not surprisingly, when primary response antibodies exhibit affinities for their specific antigens approaching this affinity ceiling, they neither require nor undergo further affinity maturation *(25)*. This effective affinity window, however, appears to shift to a range of lower affinities, with an affinity ceiling of ~10^6 M^{-1}, when the antigen is in particulate form, presumably because of avidity effects. Conversely, the range of the affinity window for extraction of antigen from a noninternalizable surface remains quite broad with an affinity ceiling similar to that of soluble antigens *(26)*. Antigens in these nonsoluble forms are thought to more closely mimic the properties of antigens in vivo.

As the overall affinity of antibody–antigen interactions can vary by several orders of magnitude, so too can the kinetics of these interactions. In a number of kinetic analyses of antiprotein antibodies *(27–30)*, both association and dissociation rates vary by greater than 2 log-fold. Thermodynamically, the formation of many antibody–antigen complexes reflects an enthalpically driven process with some compensating negative entropy component, alluding to an important role for the release of bound water molecules. In fact, a strong correlation between decreases in water activity and association constants in an antibody–protein-antigen complex has been observed by calorimetric binding analyses done in the presence of cosolutes with polarities lower than that of water *(31)*. Although other antibody–protein-antigen *(32)* and antibody–carbohydrate-antigen *(33)* interactions also appear enthalpically driven, this may not be the general rule for antibody–antigen associations because of the limited number of such systems whose thermodynamics have been rigorously determined. In accordance with the significance of water activity on antigen recognition, antibodies binding to both protein and hapten antigens have exhibited a thermodynamic dependence on the solvent pH and ionic strength *(27, 34–36)*.

3. Antigen Cross-Reactivity and Specificity

Although specific recognition of foreign vs. self material is tantamount to proper immune function, antibodies and TCRs are frequently involved in spurious interaction events. While antibodies are commonly highly specific for a single antigen, it is not at all uncommon for them to cross-react with many, structurally similar, yet distinct, antigenic molecules. In some cases, cross-reactivity has been shown to be involved in autoimmune and allergic reactions *(37, 38)*.

Certain antibodies can bind better to antigens not used in challenging the immune system than to the original immunogen, a phenomenon known as heteroclitic binding. For example, the mAb D11.15, raised against hen egg lysozyme (HEL), interacts with higher affinity with several other avian lysozymes, and the molecular basis for this cross-reactivity has been elucidated *(6)*. FvD11.15 binds eight different avian lysozymes, and all of these exhibit high affinities for the antibody. Two of these, pheasant egg-white lysozyme and guinea fowl egg-white lysozyme, exceed the affinity of the interaction with HEL, and another, Japanese quail egg-white lysozyme, exhibits a slightly lower affinity than that of HEL. Crystal structures of these antibody–antigen complexes reveal that distinct structural mechanisms, such as displacement of a loop region or an increase in hydrophobic surface on the antigen, are the cause of these heteroclitic binding events. Another anti-HEL antibody, D1.3, binds only its immunogen and one other avian lysozyme, bobwhite quail egg-white lysozyme, with high affinity. Much of the sequence variability between these eight avian lysozymes occurs at HEL residue Gln-121. For the highly cross-reactive D11.15, lysozyme residue 121 is located at the periphery of the antigenic epitope. Conversely, for the highly specific D1.3, this residue is located centrally to the binding interface and acts as a hot spot in binding for the D1.3–HEL complex *(39)*.

Anti-idiotopic antibodies *(40, 41)* recognize an antigenic determinant that is unique to an antibody or group of antibodies, or idiotope. An idiotope is defined functionally by the interaction of an anti-idiotopic antibody (Ab2) with an antibody (Ab1) bearing the idiotope. Conventional Ab2 antibodies recognize idiotopes outside of the antibody-combining site paratope, while internal image Ab2 antibodies are able to mimic the molecular surface encountered by Ab1, thereby mimicking stereochemically the antigen specific for Ab1. Numerous efforts have been made to use these molecular mimics as therapeutics, similar to vaccines. The D1.3 antibody binds to two structurally distinct ligands – its cognate antigen, HEL, and the anti-idiotypic antibody E5.2 – and these interactions exhibit molecular mimicry. The crystal structures of the complexes formed by FvD1.3 with both HEL *(42)* and FvE5.2 *(43, 44)* have been determined under high resolution. FvD1.3 contacts HEL and FvE5.2 through essentially the same set of combining site residues and most of the same atoms. Of the 18 FvD1.3 residues that contact FvE5.2 and the 17 that contact HEL, 14 are in contact with both FvE5.2 and HEL. These 14 FvD1.3 residues make up 75% of the total contact area with FvE5.2 and 87% of that with HEL. Furthermore, the positions of the atoms of FvE5.2 that contact FvD1.3 are close to those of HEL that contact FvD1.3, and 6 of the 12 hydrogen bonds in the FvD1.3–FvE5.2 interface are structurally equivalent to hydrogen bonds in the FvD1.3–HEL interface.

Perhaps the most striking example of antigen cross-reactivity of an antibody is that of SPE7, a mouse monoclonal IgE antibody raised against the hapten 2,4-dinitrophenol (DNP), to which it binds with relatively high affinity (K_D = 20 nM) *(45)*. SPE7 also binds to other small molecules with widely ranging affinities *(46)*, as well as to a structurally unrelated protein antigen, Trx-Shear3, selected using a directed evolution strategy. The Fv portion of this antibody crystallized in two different conformations in its unbound form, one of which resembled the structure of the antibody when bound by DNP any of a number of small molecules, while the other resembled the antibody structure in its protein-antigen-bound form *(47)*. Analysis of the presteady-state kinetics of complex formation between SPE7 and DNP revealed that the antibody exists in two distinct isomers, only one of which is capable of binding to the small molecule antigens. Thus, there appears to exist an equilibrium between preexisting SPE7 isomers that have the ability to bind different antigens. This diversity in the conformational ensemble space in the unbound form may serve to increase the repertoire of functional antibodies.

4. Conformational Flexibility in Antibody–Antigen Interactions

The kinetics of antibody–antigen interactions is commonly temperature-dependent. In some cases this may be indicative of the structural plasticity involved in antigen binding. Indeed, the binding kinetics of several anti-HEL antibodies have been shown to conform to a two-state model describing induced fit, with distinct association steps for molecular encounter and docking *(48, 49)*. Although numerous hypotheses concerning the correlation between antibody flexibility and signaling have been proposed over the years, the establishment of molecular flexibility as a component of signaling, beyond the antigen recognition event, remains elusive.

For smaller antigens, notably peptides and DNA, antibody plasticity is generally more pronounced than for protein antigens, although associations with the latter commonly involve a nominal degree of molecular flexibility and cannot necessarily be classified as “lock-and-key” interactions. Two types of backbone movements within the antibody-combining site have commonly been observed upon antibody–antigen complex formation, including concerted movements of multiple residue segments of CDR loops and more heterogeneous rearrangements of CDR residues. For example, upon binding antigen, heavy chain CDR loops in the antipeptide Fab8F5 undergo essentially rigid-body movements in which the unliganded loop conformations are conserved, while changes in the main chain conformation of the light chain are insignificant *(50)*. The culmination of concerted heavy chain CDR movements towards the light chain reduces the volume of the antigen-binding site by some 3% relative to the unbound Fab8F5. Other examples of segments of CDR loops moving en masse towards antigen have been observed *(21)*. In Fab17/9, a significant rearrangement of the V_HCDR3 loop is induced by binding of its peptide antigen, for which the largest backbone changes are 5 Å *(20)*. Restructuring of CDR loop regions from both the heavy and light chains of the anti-DNA antibody FabBV04–01 has also been observed *(51)*. Induced CDR loop movements upon antigen binding seem to be less extreme for antiprotein antibodies. Generally, these are small, concerted displacements of less than 3 Å *(42, 52–56)*.

Molecular flexibility is not limited to a single side of the interface, as a number of structural studies have shown varying degrees of protein plasticity for antigens upon recognition by antibodies. HEL can be crystallized in several space groups *(57–59)*. Comparison of the structures reveals significant flexibility of several loops at the molecular surface, including a number of Cα atom displacements greater than 3 Å between HEL molecules from different space groups. Between crystal structures of HEL bound to different antibodies, some main chain movements become more pronounced *(6, 42, 54, 60)*. Increased antigen flexibility, however, is not always beneficial to epitope recognition by antibodies. For instance, in order to produce mimics of the N-terminal sequence of a transforming growth factor alpha epitope recognized by the mAb tAb2, peptides required cyclicization to constrain their conformations to ones that are suitable for binding *(61)*.

5. Antibody Affinity Maturation

The function of the immune system is dependent on the recognition of essentially any antigenic material, yet the structural diversity of antigens greatly outweighs the genetic diversity encoded by immune system genes. Thus, molecular recognition of diverse antigens is accomplished by producing antibodies with specificity for almost any antigen via recombination and imprecise joining of antibody gene segments. This focuses molecular diversity at the contiguous molecular surface formed by the CDR loops, the combining site for antigen recognition. This results in germline antibodies of relatively low affinity and specificity *(62)*. This junctional diversity in the primary repertoire can produce CDR loops of different lengths and varying structures *(63, 64)*.

The affinity requirements for functional antibodies (approximately K_Ds in the nanomolar range) necessitate a secondary process for improving affinity and specificity once diversity has been established. The somatic hypermutation of antibody V regions spreads structural diversity generated by gene segment recombination to regions at the periphery of the binding site *(65)*. Selective expansion of antibody clones on the basis of antigen affinity produces mature antibodies that are high in both affinity and specificity *(66)*. Somatic hypermutation is primarily a point mutation process in gene regions that are highly conserved in the primary repertoire that can result, at times, in codon insertions or deletions *(65)*. It has been shown that the presence or absence of certain V_HCDR3 junctional amino acids can determine the affinity maturation pathway of an antibody by biasing subsequent amino acid replacements by somatic hypermutation *(67)* and that these effects are correlated to the structure and flexibility of the V_HCDR3 loop in the germline antibodies *(68)*.

Structural and energetic studies comparing germline and mature antibodies bound to the same antigen have advanced our understanding of the effects of somatic hypermutation on antibody affinity maturation. The mature Fab48G7 and its germline counterpart, Fab48G7g, both bind a nitrophenyl phosphonate transition-state analog, but with a 30,000-fold difference in affinity, primarily due to a decrease in the dissociation rate *(69)*. The sequence differences between the Fabs are limited to nine somatic hypermutations, six in V_H and three in V_L, located up to 15 Å from the bound hapten. Crystal structures of the unliganded germline Fab48G7g and its complex with hapten *(69)* reveal large conformational changes induced upon antigen binding, while crystal structures of the mature Fab48G7 *(70, 71)* in its free and hapten-bound forms exhibit very few conformational changes upon complex formation. The conformational changes induced upon antigen binding by Fab48G7g are later observed in the mature Fab structure even in the absence of antigen, and thus it appears, at least in the case of the Fab48G7 system, that the affinity maturation process is driven in large part by a mechanism of preorganizing the antibody-combining site into a conformation that is favorable for binding its hapten antigen.

Through the introduction of forward and back site-directed mutations in the germline and mature Fabs and measurements of binding affinities, the effects of the nine somatic hypermutations on the affinity maturation pathway of Fab48G7 have been dissected *(72)*. In this system, the effect on binding of the individual mutations was either positive or neutral, yet their additive changes in affinity were not equal to the overall change in affinity between the germline and mature Fabs. Double mutations revealed a high degree of cooperativity between mutations, not only between individually neutral mutations but also between even the two most positive individual mutations.

Cooperativity between somatic hypermutations, however, does not appear to be a required mechanism for affinity maturation. For Fab39-A11, which catalyzes a Diels–Alder reaction, only two somatic mutations exist between the germline and mature counterparts, of which only one contributes the majority of binding affinity to mature Fab *(73)*. Another catalytic antibody, AZ-28, which catalyzes an oxy-Cope rearrangement, has six somatic mutations, five of which contribute to differences in affinity between germline and mature antibodies in a strictly additive way *(74)*. In the affinity maturation of an antiprotein antibody,

FvD1.3, the five somatic hypermutations have also been shown to be energetically additive *(75)*. In this system, changes in antigen affinity are dominated by the only mutated amino acid that is in direct contact with the antigen, HEL.

The quantity and cooperativity of somatic hypermutations may be dependent on the affinity differences between the germline and mature antibodies. The affinity discrepancy between Fab48G7 and Fab48G7g is 30,000-fold *(69)*, while FabAZ-28, with only five significant somatic mutations has an antigen affinity only 40-fold greater than its germline counterpart *(74)*. Furthermore, Fab39-A11 and Fab39-A11g, with only one significant amino acid difference, both bind nine haptens, for most of which the difference in affinity is within an order of magnitude *(73)*. Germline and mature FvD1.3 also differ by only five amino acids and by 60-fold in affinity *(75)*. If one considers that mature antibodies must break a minimum affinity threshold for antigen binding through a limited number of somatic mutations to be functional in vivo, then it follows that the number of somatic mutations will increase as the difference in affinities between germline and mature antibodies gets larger and cooperativity between the somatic mutations will be utilized in cases where the affinity maturation process must overcome extreme germline–mature affinity discrepancies. Precise affinity ranges for the lack or presence of cooperativity associated with somatic hypermutation may or may not actually exist.

Recently, the crystal structures of four closely related anti-HEL antibodies (HyHEL8, HyHEL10, HyHEL26, and HyHEL63), representing different stages of affinity maturation, were determined bound to the same site on HEL *(76)*, revealing that enhanced binding is achieved by the burial of increasing amounts of apolar surface, at the expense of polar surface, accompanied by improved shape complementarity. The increase in hydrophobic interactions, which can fully account for the 30-fold affinity improvement in these anti-HEL antibodies according to an experimental estimate of the hydrophobic effect in protein–protein interactions *(77)*, is the consequence of subtle, yet highly correlated, structural rearrangements in antibody residues at the periphery of the interface with the antigen, adjacent to the central energetic hot spot, whose structure remains unaltered. While increasing hydrophobic interactions and improving the fit at peripheral sites that have not been optimized for binding, and whose plasticity and ability to accommodate mutations render them permissive to such optimization, constitute effective strategies for maturing antiprotein antibodies, other, as yet unobserved, mechanisms may be utilized by various antibodies for affinity maturation.

Some of the energetic factors involved in the preorganization of mature antibodies through somatic hypermutation of germline antibodies have been elucidated recently using surface plasmon resonance techniques in which different binding characteristics at various temperatures of the same complex provide information relative to the enthalpic and entropic contributions to the interaction. The affinities of panels of early primary and secondary response mAbs for a model synthetic 40-mer peptide were determined at two temperatures *(78)*. The effects of temperature on the dissociation step of the interaction were similar for mAbs in both panels, while opposite temperature effects on association were observed for each panel of mAbs. For primary mAbs, complex association was enthalpically highly favorable but entropically unfavorable, while dissociation was enthalpically unfavorable and entropically favorable. The equilibrium binding for primary mAbs was enthalpically driven with a large entropic cost of complex formation, resulting in relatively low affinity. Conversely, in secondary mAbs, association was enthalpically unfavorable but the entropic costs had been reduced markedly. Because the dissociation step of the reaction was similar to that for primary mAbs, equilibrium binding in the secondary mAbs was essentially independent of enthalpy effects, and instead, was driven by entropic changes. Thus, the relatively high affinity of the secondary mAbs is derived exclusively from the nearly complete abolishment of any entropic costs of complex association in comparison to the primary mAbs. While these experiments seem to confirm the idea of antibody affinity maturation through paratope preorganization, at least for an antipeptide antibody, it is intriguing to note that the increased affinities in the antihapten Fab48G7 and the antiprotein FvD1.3 systems derive nearly entirely from decreases in the dissociation phases of the reactions *(69, 75)*. Although similar experiments examining enthalpy and entropy effects on antigen binding to germline and mature Fab48G7 and FvD1.3 have not been done, it is likely that these types of experiments would reveal that these complexes are stabilized because of large entropic barriers to dissociation in the mature vs. germline antibodies.

References

1. Amzel, L. M., and Poljak, R. J. (1979) Three-dimensional structure of immunoglobulins. *Annu. Rev. Biochem.* 48, 961–997.
2. Harris, L. J., Larson, S. B., Hasel, K. W., and McPherson, A. (1997) Refined structure of an intact IgG2a monoclonal antibody. *Biochemistry* 36, 1581–1597.
3. Harris, L. J., Skaletsky, E., and McPherson, A. (1998) Crystallographic structure of an intact IgG1 monoclonal antibody. *J. Mol. Biol.* 275, 861–872.
4. Wu, T. T., and Kabat, E. A. (1970) An analysis of the sequences of the variable regions of Bence Jones proteins and myeloma light chains and their implications for antibody complementarity. *J. Exp. Med.* 132, 211–250.
5. Al-Lazikani, B., Lesk, A. M., and Chothia, C. (1997) Standard conformations for the canonical structures of immunoglobulins. *J. Mol. Biol.* 273, 927–948.
6. Chitarra, V., Alzari, P. M., Bentley, G. A., Bhat, T. N., Eiselé, J. L., Houdusse, A., Lescar,

J., Souchon, H., and Poljak, R. J. (1993) Three-dimensional structure of a heteroclitic antigen–antibody cross-reaction complex. *Proc. Natl. Acad. Sci. USA* 90, 7711–7715.

7. Wilson, I. A., and Stanfield, R. L. (1993) Antibody–antigen interactions. *Curr. Opin. Struct. Biol.* 3, 113–118.
8. Hamers-Casterman, C., Atarhouch, T., Muyldermans, S., Robinson, G., Hamers, C., Songa, E. B., Bendahman, N., and Hamers, R. (1993) Naturally occurring antibodies devoid of light chains. *Nature* 363, 446–448.
9. Decanniere, K., Desmyter, A., Lauwereys, M., Ghahroudi, M. A., Muyldermans, S., and Wyns, L. (1999) A single-domain antibody fragment in complex with RNase A: non-canonical loop structures and nanomolar affinity using two CDR loops. *Struct. Fold. Des.* 7, 361–370.
10. Stanfield, R. L., Dooley, H., Flajnik, M. F., and Wilson, I. A. (2004) Crystal structure of a shark single-domain antibody V region in complex with lysozyme. *Science* 305, 1770–1773.
11. Chersi, A., Galati, R., Ogino, T., Butler, R. H., and Tanigaki, N. (2002) Anti-peptide antibodies that recognize conformational differences of HLA class I intracytoplasmic domains. *Hum. Immunol.* 63, 731–741.
12. Hewer, R., and Meyer, D. (2003) Peptide immunogens based on the envelope region of HIV-1 are recognized by HIV/AIDS patient polyclonal antibodies and induce strong humoral immune responses in mice and rabbits. *Mol. Immunol.* 40, 327–335.
13. Metaxas, A., Tzartos, S., and Liakopoulou-Kyriakide, M. (2002) The production of anti-hexapeptide antibodies which recognize the S7, L6 and L13 ribosomal proteins of *Escherichia coli. J. Pept. Sci.* 8, 118–124.
14. Wilson, I. A., and Stanfield, R. L. (1994) Antibody–antigen interactions: new structures and new conformational changes. *Curr. Opin. Struct. Biol.* 4, 857–867.
15. Chothia, C., Lesk, A. M., Tramontano, A., Levitt, M., Smith-Gill, S. J., Air, G., Sheriff, S., Padlan, E. A., Davies, D., and Tulip, W. R. (1989) Conformations of immunoglobulin hypervariable regions. *Nature* 342, 877–883.
16. Conte, L. L., Chothia, C., and Janin, J. (1999) The atomic structure of protein–protein recognition sites. *J. Mol. Biol.* 285, 2177–2198.
17. MacCallum, R. M., Martin, A. C., and Thornton, J. M. (1996) Antibody–antigen interactions: contact analysis and binding site topography. *J. Mol. Biol.* 262, 732–745.
18. Jones, S., and Thornton, J. M. (1996) Principles of protein–protein interactions. *Proc. Natl. Acad. Sci. USA* 93, 13–20.
19. Webster, D. M., Henry, A. H., and Rees, A. R. (1994) Antibody–antigen interactions. *Curr. Opin. Struct. Biol.* 4, 123–129.
20. Rini, J. M., Schulze-Gahmen, U., and Wilson, I. A. (1992) Structural evidence for induced fit as a mechanism for antibody–antigen recognition. *Science* 255, 959–965.
21. Stanfield, R. L., Fieser, T. M., Lerner, R. A., and Wilson, I. A. (1990) Crystal structures of an antibody to a peptide and its complex with peptide antigen at 2.8 Å. *Science* 248, 712–719.
22. Garcia, K. C., Ronco, P. M., Verroust, P. J., Brünger, A. T., and Amzel, L. M. (1992) Three-dimensional structure of an angiotensin II-Fab complex at 3 Å: hormone recognition by an anti-idiotypic antibody. *Science* 257, 502–507.
23. Foote, J., and Eisen, H. N. (1995) Kinetic and affinity limits on antibodies produced during immune responses. *Proc. Natl. Acad. Sci. USA* 92, 1254–1256.
24. Batista, F. D., and Neuberger, M. S. (1998) Affinity dependence of the B cell response to antigen: a threshold, a ceiling, and the importance of off-rate. *Immunity* 8, 751–759.
25. Roost, H. P., Bachmann, M. F., Haag, A., Kalinke, U., Pliska, V., Hengartner, H., and Zinkernagel, R. M. (1995) Early high-affinity neutralizing anti-viral IgG responses without further overall improvements of affinity. *Proc. Natl. Acad. Sci. USA* 92, 1257–1261.
26. Batista, F. D., and Neuberger, M. S. (2000) B cells extract and present immobilized antigen: implications for affinity discrimination. *EMBO J.* 19, 513–520.
27. Xavier, K. A., McDonald, S. M., McCammon, J. A., and Willson, R. C. (1999) Association and dissociation kinetics of bobwhite quail lysozyme with monoclonal antibody HyHEL-5. *Protein Eng.* 12, 79–83.
28. Gerstner, R. B., Carter, P., and Lowman, H. B. (2002) Sequence plasticity in the antigen-binding site of a therapeutic anti-HER2 antibody. *J. Mol. Biol.* 321, 851–862.
29. Rajpal, A. and Kirsch, J. F. (2000) Role of the minor energetic determinants of chicken egg white lysozyme (HEWL) to the stability of the HEWL antibody scFv-10 complex. *Proteins* 40, 49–57.
30. England, P., Bregegere, F., and Bedouelle, H. (1997) Energetic and kinetic contributions of contact residues of antibody D1.3 in the interaction with lysozyme. *Biochemistry* 36, 164–172.
31. Goldbaum, F. A., Schwarz, F. P., Eisenstein, E., Cauerhff, A., Mariuzza, R. A., and Poljak, R. J. (1996) The effect of water activity on the association constant and the enthalpy of reaction between lysozyme and the specific

antibodies D1.3 and D44.1. *J. Mol. Recognit.* 9, 6–12.

32. Kelley, R. F., O'Connell, M. P., Carter, P., Presta, L., Eigenbrot, C., Covarrubias, M., Snedecor, B., Bourell, J. H., and Vetterlein, D. (1992) Antigen binding thermodynamics and antiproliferative effects of chimeric and humanized anti-p185HER2 antibody Fab fragments. *Biochemistry* 31, 5434–5441.
33. Sigurskjold, B. W., Altman, E., and Bundle, D. R. (1991) Sensitive titration microcalorimetric study of the binding of Salmonella O-antigenic oligosaccharides by a monoclonal antibody. *Eur. J. Biochem.* 197, 239–246.
34. Gibas, C. J., Subramaniam, S., McCammon, J. A., Braden, B. C., and Poljak, R. J. (1997) pH dependence of antibody/lysozyme complexation. *Biochemistry* 36, 15599–15614.
35. Omelyanenko, V. G., Jiskoot, W., and Herron, J. N. (1993) Role of electrostatic interactions in the binding of fluorescein by anti-fluorescein antibody 4-4-20. *Biochemistry* 32, 10423–10429.
36. De Genst, E., Areskoug, D., Decanniere, K., Muyldermans, S., and Andersson, K. (2002) Kinetic and affinity predictions of a protein–protein interaction using multivariate experimental design. *J. Biol. Chem.* 277, 29897–29907.
37. Kaplan, M. H., and Meyeserian, M. (1962) An immunological cross-reaction between group-A streptococcal cells and human heart tissue. *Lancet* 1, 706–710.
38. Oldstone, M. B. (1998) Molecular mimicry and immune-mediated diseases. *FASEB J.* 12, 1255–1265.
39. Dall'Acqua, W., Goldman, E. R., Eisenstein, E., and Mariuzza, R. A. (1996) A mutational analysis of the binding of two different proteins to the same antibody. *Biochemistry* 35, 9667–9676.
40. Poljak, R. J. (1994) An idiotope–anti-idiotope complex and the structural basis of molecular mimicking. *Proc. Natl. Acad. Sci. USA* 91, 1599–1600.
41. Pan, Y., Yuhasz, S. C., and Amzel, L. M. (1995) Anti-idiotypic antibodies: biological function and structural studies. *FASEB J.* 9, 43–49.
42. Bhat, T. N., Bentley, G. A., Boulot, G., Greene, M. I., Tello, D., Dall'Acqua, W., Souchon, H., Schwarz, F. P., Mariuzza, R. A., and Poljak, R. J. (1994) Bound water molecules and conformational stabilization help mediate an antigen–antibody association. *Proc. Natl. Acad. Sci. USA* 91, 1089–1093.
43. Braden, B. C., Fields, B. A., Ysern, X., Dall'Acqua, W., Goldbaum, F. A., Poljak, R. J., and Mariuzza, R. A. (1996) Crystal structure of an Fv-Fv idiotope-anti-idiotope complex at 1.9 Å resolution. *J. Mol. Biol.* 264, 137–151.
44. Fields, B. A., Goldbaum, F. A., Ysern, X., Poljak, R. J., and Mariuzza, R. A. (1995) Molecular basis of antigen mimicry by an anti-idiotope. *Nature* 374, 739–742.
45. Eshhar, Z., Ofarim, M., and Waks, T. (1980) Generation of hybridomas secreting murine reaginic antibodies of anti-DNP specificity. *J. Immunol.* 124, 775–780.
46. Varga, J. M., Kalchschmid, G., Klein, G. F., and Fritsch, P. (1991) Mechanism of allergic cross-reactions – I. Multispecific binding of ligands to a mouse monoclonal anti-DNP IgE antibody. *Mol. Immunol.* 28, 641–654.
47. James, L. C., Roversi, P., and Tawfik, D. S. (2003) Antibody multispecificity mediated by conformational diversity. *Science* 299, 1362–1367.
48. Li, Y., Lipschultz, C. A., Mohan, S., and Smith-Gill, S. J. (2001) Mutations of an epitope hot-spot residue alter rate limiting steps of antigen–antibody protein–protein associations. *Biochemistry* 40, 2011–2022.
49. Lipschultz, C. A., Li, Y., and Smith-Gill, S. J. (2000) Experimental design for analysis of complex kinetics using surface plasmon resonance. *Methods* 20, 310–318.
50. Tormo, J., Blaas, D., Parry, N. R., Rowlands, D., Stuart, D., and Fita, I. (1994) Crystal structure of a human rhinovirus neutralizing antibody complexed with a peptide derived from viral capsid protein VP2. *EMBO J.* 13, 2247–2256.
51. Herron, J. N., He, X. M., Ballard, D. W., Blier, P. R., Pace, P. E., Bothwell, A. L., Voss, E. W., Jr., and Edmundson, A. B. (1991) An autoantibody to single-stranded DNA: comparison of the three-dimensional structures of the unliganded Fab and a deoxynucleotide–Fab complex. *Proteins* 11, 159–175.
52. Prasad, G. S., Earhart, C. A., Murray, D. L., Novick, R. P., Schlievert, P. M., and Ohlendorf, D. H. (1993) Structure of toxic shock syndrome toxin 1. *Biochemistry* 32, 13761–13766.
53. Mylvaganam, S. E., Paterson, Y., and Getzoff, E. D. (1998) Structural basis for the binding of an anti-cytochrome c antibody to its antigen: crystal structures of FabE8-cytochrome c complex to 1.8 Å resolution and FabE8 to 2.26 Å resolution. *J. Mol. Biol.* 281, 301–322.
54. Li, Y., Li, H., Smith-Gill, S. J., and Mariuzza, R. A. (2000) Three-dimensional structures of the free and antigen-bound Fab from monoclonal antilysozyme antibody HyHEL-63. *Biochemistry* 39, 6296–6309.
55. Braden, B. C., Souchon, H., Eiselé, J. L., Bentley, G. A., Bhat, T. N., Navaza, J., and Poljak, R. J. (1994) Three-dimensional structures of the free and the antigen-complexed

Fab from monoclonal anti-lysozyme antibody D44.1. *J. Mol. Biol.* 243, 767–781.

56. Faelber, K., Kirchhofer, D., Presta, L., Kelley, R. F., and Muller, Y. A. (2001) The 1.85 Å resolution crystal structures of tissue factor in complex with humanized Fab D3h44 and of free humanized Fab D3h44: revisiting the solvation of antigen combining sites. *J. Mol. Biol.* 313, 83–97.
57. Kurinov, I. V., and Harrison, R. W. (1995) The influence of temperature on lysozyme crystals. Structure and dynamics of protein and water. *Acta Crystallogr. D* 51, 98–109.
58. Harata, K. (1994) X-ray structure of a monoclinic form of hen egg-white lysozyme crystallized at 313 K. Comparison of two independent molecules. *Acta Crystallogr. D* 50, 250–257.
59. Ramanadham, M., Sieker, L. C., and Jensen, L. H. (1990) Refinement of triclinic lysozyme: II. The method of stereochemically restrained least squares. *Acta Crystallogr. B* 46, 63–69.
60. Padlan, E. A., Silverton, E. W., Sheriff, S., Cohen, G. H., Smith-Gill, S. J., and Davies, D. R. (1989) Structure of an antibody–antigen complex: crystal structure of the HyHEL-10 Fab-lysozyme complex. *Proc. Natl. Acad. Sci. USA* 86, 5938–5942.
61. Hahn, M., Winkler, D., Welfle, K., Misselwitz, R., Welfle, H., Wessner, H., Zahn, G., Scholz, C., Seifert, M., Harkins, R., Schneider-Mergener, J., and Höhne, W. (2001) Cross-reactive binding of cyclic peptides to an anti-TGFalpha antibody Fab fragment: an X-ray structural and thermodynamic analysis. *J. Mol. Biol.* 314, 293–309.
62. Tonegawa, S. (1983) Somatic generation of antibody diversity. *Nature* 302, 575–581.
63. Chothia, C., Lesk, A. M., Gherardi, E., Tomlinson, I. M., Walter, G., Marks, J. D., Llewelyn, M. B., and Winter, G. (1992) Structural repertoire of the human VH segments. *J. Mol. Biol.* 227, 799–817.
64. Tomlinson, I. M., Cox, J. P., Gherardi, E., Lesk, A. M., and Chothia, C. (1995) The structural repertoire of the human V kappa domain. *EMBO J.* 14, 4628–4638.
65. Tomlinson, I. M., Walter, G., Jones, P. T., Dear, P. H., Sonnhammer, E. L., and Winter, G. (1996) The imprint of somatic hypermutation on the repertoire of human germline V genes. *J. Mol. Biol.* 256, 813–817.
66. Rajewsky, K. (1996) Clonal selection and learning in the antibody system. *Nature* 381, 751–758.
67. Furukawa, K., Akasako-Furukawa, A., Shirai, H., Nakamura, H., and Azuma, T. (1999) Junctional amino acids determine the maturation pathway of an antibody. *Immunity* 11, 329–338.
68. Furukawa, K., Shirai, H., Azuma, T., and Nakamura, H. (2001) A role of the third complementarity-determining region in the affinity maturation of an antibody. *J. Biol. Chem.* 276, 27622–27628.
69. Wedemayer, G. J., Patten, P. A., Wang, L. H., Schultz, P. G., and Stevens, R. C. (1997) Structural insights into the evolution of an antibody combining site. *Science* 276, 1665–1669.
70. Wedemayer, G. J., Wang, L. H., Patten, P. A., Schultz, P. G., and Stevens, R. C. (1997) Crystal structures of the free and liganded form of an esterolytic catalytic antibody. *J. Mol. Biol.* 268, 390–400.
71. Patten, P. A., Gray, N. S., Yang, P. L., Marks, C. B., Wedemayer, G. J., Boniface, J. J., Stevens, R. C., and Schultz, P. G. (1996) The immunological evolution of catalysis. *Science* 271, 1086–1091.
72. Yang, P. L., and Schultz, P. G. (1999) Mutational analysis of the affinity maturation of antibody 48G7. *J. Mol. Biol.* 294, 1191–1201.
73. Romesberg, F. E., Spiller, B., Schultz, P. G., and Stevens, R. C. (1998) Immunological origins of binding and catalysis in a Diels-Alderase antibody. *Science* 279, 1929–1933.
74. Ulrich, H. D., Mundorff, E., Santarsiero, B. D., Driggers, E. M., Stevens, R. C., and Schultz, P. G. (1997) The interplay between binding energy and catalysis in the evolution of a catalytic antibody. *Nature* 389, 271–275.
75. England, P., Nageotte, R., Renard, M., Page, A. L., and Bedouelle, H. (1999) Functional characterization of the somatic hypermutation process leading to antibody D1.3, a high affinity antibody directed against lysozyme. *J. Immunol.* 162, 2129–2136.
76. Li, Y., Li, H., Smith-Gill, S. J., and Mariuzza, R. A. (2003) X-ray snapshots of the maturation of an antibody response to a protein antigen. *Nat. Struct. Biol.* 10, 482–488.
77. Sundberg, E. J., Urrutia, M., Braden, B. C., Isern, J., Tsuchiya, D., Fields, B. A., Malchiodi, E. L., Tormo, J., Schwarz, F. P., and Mariuzza, R. A. (2000) Estimation of the hydrophobic effect in an antigen–antibody protein–protein interface. *Biochemistry* 39, 15375–15387.
78. Manivel, V., Sahoo, N. C., Salunke, D. M., and Rao, K. V. (2000) Maturation of an antibody response is governed by modulations in flexibility of the antigen-combining site. *Immunity* 13, 611–620.

Chapter 3

Epitope Mapping of Antibody–Antigen Complexes by Nuclear Magnetic Resonance Spectroscopy

Osnat Rosen and Jacob Anglister

Summary

Nuclear magnetic resonance (NMR) is a very powerful tool for determining the boundaries of peptide epitopes recognized by antibodies. NMR can be used to study antibodies in complexes that exhibit a wide range of binding affinities from very weak and transient to very tight. Choice of the specific method depends upon the dissociation constant, especially the ligand off-rate.

Epitope mapping by NMR is based on the difference in mobility between the amino acid residues of a peptide antigen that interact tightly with the antibody and residues outside the epitope that do not interact with the antibody. The interacting peptide residues become considerably immobilized upon binding. Their mobility will resemble that of the antibody's residues. Several NMR methods were developed based on these characteristics. In this chapter we discuss some of these methods, including dynamic filtering, comparison of ^{1}H-^{15}N HSQC peaks' intensities, transverse relaxation time, measurements of ^{1}H-^{15}N nuclear Overhauser effect (NOE) values, and measurements of $T_{1\rho}$ relaxation time.

Key words: Epitope mapping, NMR, Antibody, V3, gp120, Acetylcholine receptor, α-bungarotoxin, Dynamic filtering, Relaxation times, Peptide antigen.

1. Introduction

Nuclear magnetic resonance (NMR) spectroscopy has become a powerful tool in the study of protein–protein interactions and the dynamics of protein–ligand complexes. NMR can be used to study protein complexes exhibiting a wide range of binding affinities from very weak and transient binding to very tight. The particular method that is selected depends upon the dissociation constant and especially on the ligand off-rate. NMR can also be used to determine the structure of antibody–antigen complexes

Ulrich Reineke and Mike Schutkowski (eds.), *Methods in Molecular Biology, Epitope Mapping Protocols, vol. 524*

DOI: 10.1007/978-1-59745-450-6_3

and to study antibody–antigen interactions. However, for complete structure determination, the antibody Fv fragment is used, and the analysis is limited by the size of the Fv–antigen complex. The development of *E. coli* systems for expression of Fv fragments, combined with uniform isotope labeling with ^{15}N and ^{13}C, has enabled structure determination of a lyzozyme–antilyzozyme Fv complex *(1)* and a complex of the Fv of a HIV (human immunodeficiency virus)-1-neutralizing antibody with a third variable loop (V3) peptide derived from the HIV-1 envelope protein gp120 *(2)*. It also allowed structure determination of different V3 peptides bound to the Fv fragment of a HIV-1-neutralizing antibody *(3–6)*.

Even in the absence of an Fv fragment, NMR can still be used to study the interactions of the larger Fab antibody fragment with peptide antigens or haptens. However, structure determination of such Fab–antigens complexes is beyond current capabilities of NMR spectroscopy. Transferred nuclear Overhauser effect (NOE) was used to study the structure of a cholera-toxin peptide bound to three different antibodies *(7, 8)*. The structure of a V3 peptide bound to the Fab fragment of 0.5β, an HIV-1-neutralizing antibody, was studied using specific deuteration and NOE spectrometry (NOESY) difference spectroscopy *(9–11)*.

NMR is a very powerful tool for determining the boundaries of epitopes recognized by antibodies, and, for this purpose, a Fab fragment or perhaps even the entire antibody molecule can be used. Epitope mapping by NMR is based on the difference in mobility between those peptide antigen residues that interact tightly with the antibody and the residues that are outside the epitope recognized by the antibody. The peptide residues that interact with the antibody become considerably immobilized upon binding, their mobility being comparable to that of the antibody residues. As a result, their transverse relaxation time, T_2, will be shortened considerably in comparison with the free peptide. Peptide residues outside the epitope retain considerable mobility and their T_2 relaxation times are noticeably longer than those of the protons of the peptide interacting with the antibody. Different NMR techniques can differentiate between mobile and rigid segments of a protein or peptide on the basis of differences in T_2 and $T_{1\rho}$ relaxation times. The section below discusses the application of these techniques for epitope mapping.

1.1. Dynamic Filtering

The dynamic filtering approach uses the homonuclear Hartmann Hahn (HOHAHA or TOCSY) and rotating-frame Overhauser enhancement spectroscopy (ROESY) experiments to differentiate between mobile and immobile residues of the antigen. Both are two-dimensional (2D) homonuclear spectra that do not require isotopic labeling of either the antibody or the antigen. The signal-to-noise ratio of the observed cross-peaks in these spectra depends on the $T_{1\rho}$ relaxation time of the protons that is practically equal to T_2 in most cases. The mixing period in the NMR experiments is tuned to discriminate between ligand protons that interact with the antibody and are thus immobilized and ligand protons that do not interact with the antibody and are flexible. When a long enough mixing period is used in HOHAHA and ROESY experiments, most of the cross-peaks of the protein protons as well as those of the immobilized residues of the epitope are canceled out, so that only cross-peaks of the mobile residues are observed. Thus, the dynamic filtering technique enables us to map accurately the segment of the peptide antigen interacting with the antibody or another protein. Measurement of HOHAHA and ROESY spectra in H_2O takes advantage of the dispersion in the amide protons' chemical shift and enables the sequential assignment of the mobile segments of the peptide antigen. In addition, peptide residues interacting with the antibody undergo significant changes in chemical shift upon binding to the antibody or another protein. Residues that do not participate in binding have identical or nearly identical chemical shifts in the free and antibody-bound form of the peptide antigen. Hence, peptide residues inside the epitope will have broader peaks and exhibit different chemical shift whereas those residues outside the interacting region will give narrow resonance lines and will have chemical shifts identical to those of the free ligand.

The dynamic filtering approach was applied to map the epitopes of several V3 peptides in complex with anti-gp120 HIV-neutralizing antibodies. To map the antigenic determinant recognized by 0.5β antibody, a complex of this antibody with a 24-residue $V3_{IIIB}$ (V3 of the HIV-1 IIIB strain) peptide corresponding to residues N301-G324 of $gp120_{IIIB}$ was studied using NMR *(10)*. A combination of HOHAHA and ROESY experiments of the free peptide as well as of the complex with 0.5β Fab were measured in H_2O. Comparison of the HOHAHA spectrum of the free peptide (**Fig. 1a**) with that of the complex (**Fig. 1b**) allowed us to assign the residues that are outside the epitope recognized by the 0.5β antibody. Superposition of the HOHAHA and ROESY spectra of the complex enabled sequential assignment of the segments that are outside the epitope and retained considerable mobility (**Fig. 2**). In this manner, a 14-residue segment corresponding to residues S306-T319 of $gp120_{IIIB}$ was shown to be immobilized beyond detection in the HOHAHA and ROESY spectra and therefore was determined as being part of the antigenic determinant recognized by the antibody 0.5β (the numbering is according to the HXB2 HIV-1 strain). K305 and I320 were found to retain considerable mobility in the bound peptide while their amide protons underwent significant change in chemical shift upon binding. This observation suggested that these two residues were at the boundaries of the determinant recognized by the antibody

(10, 12). The NMR structure determination of the $V3_{IIIB}$–peptide complex with 0.5β Fv indeed verified that K305 and I320 were at the boundaries of the epitope and had a few interactions with the antibody *(2)*.

In another study, the same procedure was applied for epitope mapping of a different V3 peptide, $V3_{MN}$ (V3 of the HIV-1 MN strain), in complex with 447–52D anti-gp120 HIV-neutralizing antibody *(5)*. In this case, seven residues of the C-terminal region of the V3 peptide corresponding to the segment T319-G325 of $gp120_{MN}$ and two of the N-terminal segment, N302 and R304,

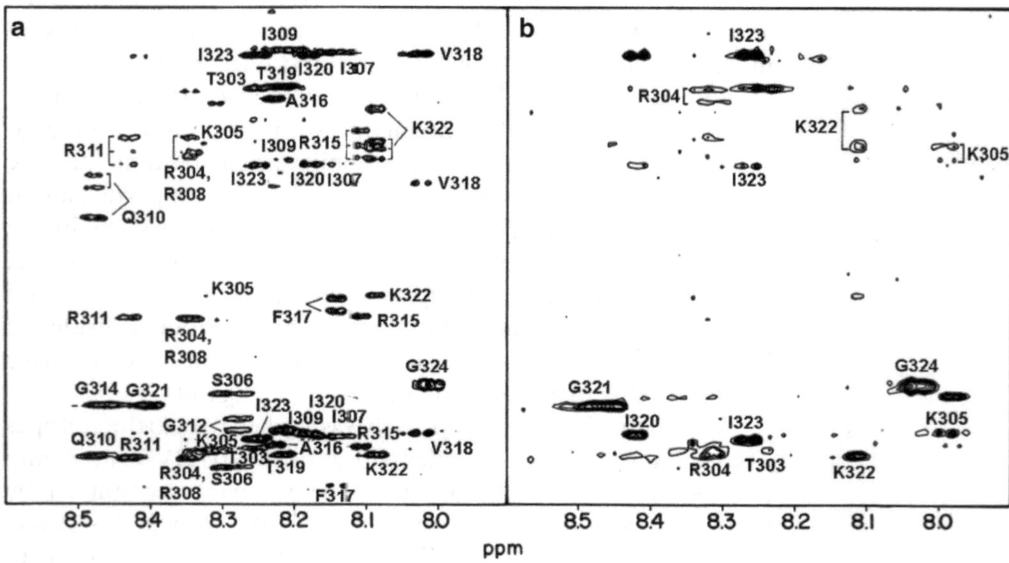

Fig. 1. HOHAHA spectra of free and antibody-bound $V3_{IIIB}$ peptide showing amide proton connectivities with amino acid side-chains. (**a**) Free $V3_{IIIB}$ peptide. (**b**) Complex of $V3_{IIIB}$ peptide with 0.5β Fab. (Reproduced from *(10)*).

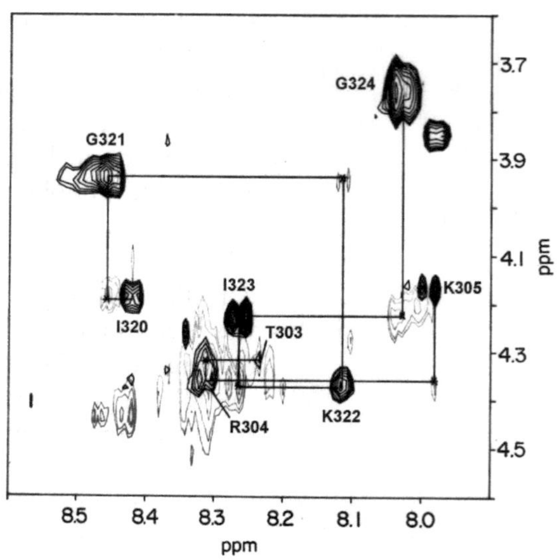

Fig. 2. Superposition of the NH-CαH region of HOHAHA (*black*) and ROESY (*gray*) spectra of $V3_{IIIB}$ peptide in complex with 0.5β Fab showing sequential connectivities. (Reproduced from *(10)*).

were observed. The proton chemical shifts of these residues were identical to those observed for the free peptide, confirming that they do not interact or at most have only very minor interactions with the antibody. The HOHAHA cross-peaks of the residues in the peptide segment corresponding to K305-R315 of gp120$_{MN}$ were undetectable in the spectra, implying strong interactions with the 447–52D antibody. The cross-peaks of residues A316-Y318 of gp120$_{MN}$ were weak, indicating that these three residues are part of the epitope. By this method, the epitope recognized by the 447–52D Fv was mapped to K305-Y318 and was later confirmed by other methods.

The dynamic filtering approach can be used to map segments of large proteins, such as membrane proteins, that are recognized by other proteins. The power of the approach is illustrated by mapping the determinant of the nicotinic-acetylcholine-receptor α-subunit (α-AChR) that is recognized by the snake neurotoxin α-bungarotoxin (α-BTX) *(13)*. In this study we used two overlapping synthetic peptides corresponding to segments $^{\alpha\text{-AChR}}$K-79-D200 and $^{\alpha\text{-AChR}}$R182-T202 that were complexed with α-BTX. Two glutamic acid residues were added on each side of the latter peptide to increase its solubility. To locate the N-terminus of the α-AChR epitope recognized by α-BTX, a set of HOHAHA spectra with different mixing times was acquired for the $^{\alpha\text{-AChR}}$K179-D200/α-BTX complex. The cross-peaks of the mobile part of the $^{\alpha\text{-AChR}}$K179-D200 retained good signal-to-noise ratio in the HOHAHA and ROESY spectra measured with a mixing time of 400 ms, while the contribution of the α-BTX toxin cross-peaks to the spectra was minimal (**Fig. 3**). Using these spectra, five residues, $^{\alpha\text{-AChR}}$E180-W184, as well as the cross-peaks arising from $^{\alpha\text{-AChR}}$K185-H$^{N\varepsilon}$, could easily be assigned. Proton chemical shifts of residues $^{\alpha\text{-AChR}}$E180-G183 were practically identical to those of the free peptide, indicating that these residues were flexible and did not participate in binding. The chemical shifts of $^{\alpha\text{-AChR}}$W184-H$^{\alpha}$ and $^{\alpha\text{-AChR}}$K185-H$^{N\varepsilon}$ differed from those of the free peptide, and their HOHAHA cross-peaks were very weak, indicating that these residues were within the AChR determinant recognized by α-BTX. The cross-peaks of $^{\alpha\text{-AChR}}$H186-Y198 were undetectable in the spectra. We therefore concluded that N-terminal residues $^{\alpha\text{-AChR}}$K179-G183 lie outside the determinant recognized by α-BTX. To locate the C-terminal residues of the α-AChR epitope recognized by α-BTX, a set of HOHAHA spectra with different mixing times was acquired for the $^{\alpha\text{-AChR}}$R182-T202/α-BTX complex. A 250-ms mixing time yielded a spectrum showing a number of peptide cross-peaks with high signal-to-noise ratio while only a limited number of α-BTX cross-peaks were observed. The peptide cross-peaks corresponding to residues $^{\alpha\text{-AChR}}$E180-W184 and $^{\alpha\text{-AChR}}$I201-T202 (and the EE tag) could be assigned. We therefore concluded that the determinant recognized by α-BTX comprised residues $^{\alpha\text{-AChR}}$W184-D200.

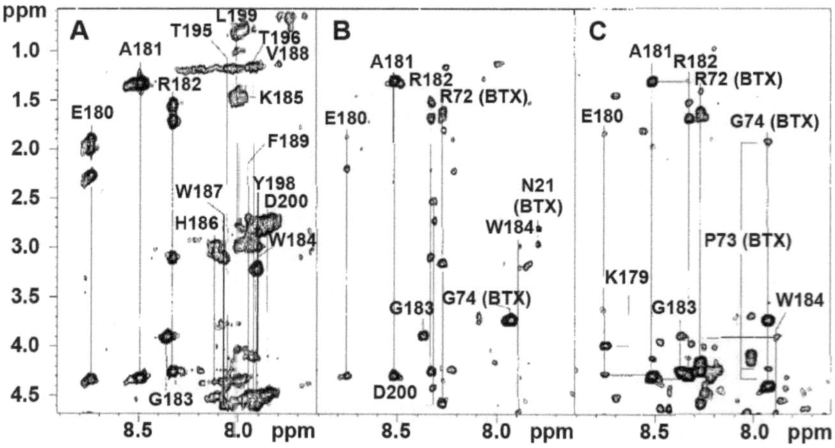

Fig. 3. Dynamic filtering spectra. (**a**) HOHAHA spectrum of free $^{\alpha\text{-AChR}}$K179-D200 peptide acquired with a mixing time of 150 ms. (**b**) HOHAHA spectrum of the α-BTX/$^{\alpha\text{-AChR}}$K179-D200 complex acquired with a mixing time of 400 ms. (**c**) ROESY spectrum of the α-BTX/$^{\alpha\text{-AChR}}$K179-D200 complex acquired with a mixing time of 400 ms. The sequential assignment for the mobile segment $^{\alpha\text{-AChR}}$K179-W184 is presented. (Reproduced from *(13)*).

1.2. Epitope Mapping by Comparison of ^{1}H-^{15}N HSQC Peaks' Intensities

Expression of peptide antigens in *E. coli* enables uniform labeling with ^{15}N and ^{13}C. This labeling allowed us to use several heteronuclear NMR experiments to map the segment of peptide antigens recognized by antibodies and eventually enabled structure determination of peptide antigen bound to the antibody Fv. The simplest experiment is the edited ^{1}H-^{15}N HSQC of uniformly ^{15}N-labeled peptide in its free form and in complex with the antibody Fv or the Fab, followed by comparison of the cross-peaks' intensities. The edited experiments show only the cross-peaks of the labeled heteronuclei and the hydrogen atoms bonded to them. Since the peptide is labeled and the Fv is not, the edited spectra show only the cross-peaks originating from the peptide while cross-peaks of the Fv are canceled out. A prerequisite for this approach is that all ^{1}H-^{15}N cross-peaks are assigned to the corresponding antigen residues. The sequential backbone assignment can be accomplished using conventional isotope-edited 3D experiments such as HNCO, CBCACONH, HNCA, and HNCACB *(14)*. The sequential assignment must be carried out for both the free and Fv-bound peptide. In these experiments the Fv is unlabeled.

The signal-to-noise ratio and the line-width of the cross-peaks in the ^{1}H-^{15}N HSQC spectrum of the ^{15}N-labeled peptide in complex with the Fv depend on the T_2 relaxation times of the peptide's amide protons. Examination of the ^{1}H-^{15}N HSQC spectra of the Fv–peptide complex revealed that residues that did not interact with the Fv exhibit intense and narrow cross-peaks that did not change their chemical shifts in comparison with the free peptide. However, peptide residues that interact with the antibody Fv exhibit considerably broader and weaker ^{1}H-^{15}N HSQC cross-peaks and also exhibit chemical shift changes in comparison with the cross-peaks of the free peptide.

In our studies on the interactions between $V3_{IIIB}$ peptide and 447–52D antibody we used the above method to determine the epitope recognized by the antibody as shown in **Fig. 4**. This measurement indicates that the core epitope S306-F317 is immobilized and K322 is completely outside the epitope. The C-terminal segment V318-G321, although part of the epitope, is more mobile than the N-terminal half of the $V3_{IIIB}$ peptide, which is part of the core epitope. This is due to much less interactions between the C-terminal half of the $V3_{IIIB}$ peptide and the antibody Fv in comparison with the large number of interactions observed between the N-terminal half of the $V3_{IIIB}$ peptide and the Fv. K305 exhibits reduced intensity because of faster solvent exchange of its amide proton. K305 is the third residue of the peptide. The amide protons of the first two residues T303-R304 are not observed at all due to of fast exchange with the solvent.

1.3. Mapping of the Epitope by Transverse Relaxation Time, T_2 Measurements

Measurements of NMR relaxation times can provide detailed information about the dynamics of proteins and the changes in dynamics upon ligand binding. In recent years, ^{15}N relaxation has been used extensively to describe the dynamics of the protein backbone. NMR relaxation times are influenced by the global tumbling of the molecules as well as by local motions. Transverse relaxation time, T_2, also known as spin–spin relaxation, also depends upon conformational and chemical exchange. Thus,

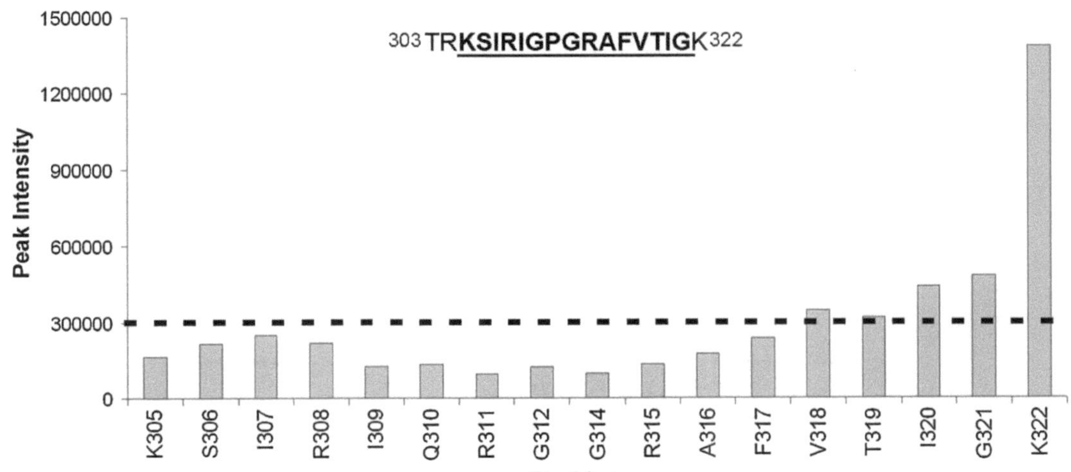

Fig. 4. The variations in the $^{1}H/^{15}N$ cross-peak intensities in the ^{1}H-^{15}N HSQC spectrum recorded with uniformly ^{15}N-labeled $V3_{IIIB}$ peptide bound to unlabeled 447–52D Fv. The intensity is given in arbitrary units. The $V3_{IIIB}$ sequence is shown and the epitope is in *bold* and *underlined*. The first two residues are not observed in the spectrum. The arbitrary 300,000 threshold, used to differentiate between residues interacting with the Fv and those that do not, is drawn as a *horizontal dashed line*. According to this analysis I320 and G321 are at the border of the epitope and K322 is outside.

measurements of T_2 can provide more quantitative insight into the dynamics of proteins in solution and the changes that occur upon ligand binding. We used this technique and determined the T_2 values of all V3$_{IIIB}$ residues. As shown in **Fig. 5**, when applied to the peptide corresponding to residues T303-K322 of gp120$_{IIIB}$ in complex with 447–52D, short ^{15}N T_2 relaxation times were measured for residues S306-T319, indicating backbone immobilization for these residues upon binding the antibody. K305, I320, and G321 exhibit slightly increased T_2 relaxation times, indicating that they are at the "borders" of the epitope. The much longer relaxation time of K322 suggests that this residue is flexible and completely outside the antibody epitope.

1.4. Epitope Mapping by ^1H-^{15}N NOE Measurements

The heteronuclear 2D ^{15}N-{^1H} nuclear Overhauser effect (hetNOE) is the most commonly used NMR experiment to study protein dynamics on fast time scales (picoseconds to nanoseconds). hetNOE allows quantification of thermal fluctuations in a protein on a per residue basis. The ^1H-^{15}N NOE values can range from −3.5 for very flexible segments to 1 for very rigid segments *(15)*. This wide range of the NOE effect allows sensitive discrimination between different residues in a peptide antigen complexed with an Fv or Fab molecule according to their mobility. The NOE values are determined by taking the ratio of the ^{15}N signal intensities recorded in the presence and absence of proton saturation prior to excitation of ^{15}N magnetization *(15)*. At least three critical factors require special attention for reliable NOE measurements: First, water–amide proton exchange can affect NOE values if the water-flip pulse is not applied. Second, ^1H saturation is also important for quantitatively reliable NOE values. Third, complete magnetization recovery during the pulse repetition delay is critical for accurate measurements.

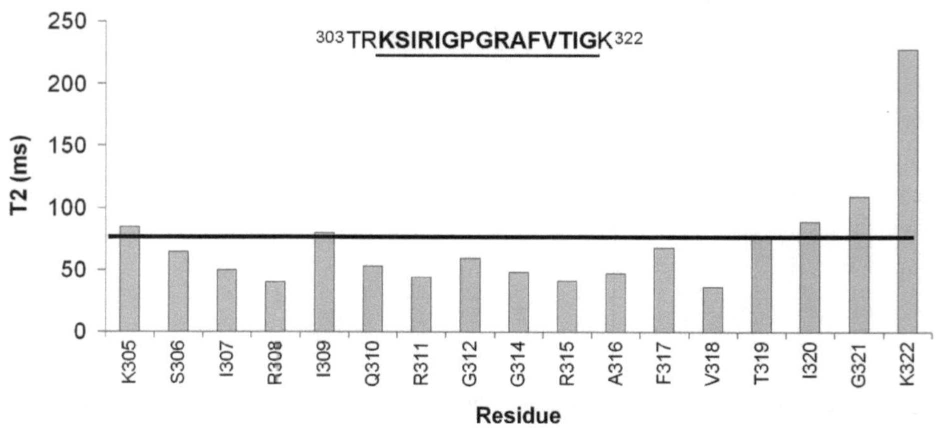

Fig. 5. The variations of the ^{15}N T_2 relaxation times of the V3$_{IIIB}$ peptide bound to 447–52D Fv along the peptide sequence. The 75-ms threshold for T_2, used to differentiate between residues interacting with the Fv and those that do not, is drawn as a *horizontal line*. According to this analysis K305, I320, and G321 are at the border of the epitope and K322 is outside.

The measurements of ^{1}H-^{15}N NOE of a $V3_{IIIB}$ peptide bound to the 447–52D, as shown in **Fig. 6**, revealed that only the core epitope encompassing residues I307-V318 is strongly immobilized. K305-S306 and T319-I320 revealed some mobility as a result of their location at the edge of the epitope. K322 is completely outside the epitope.

1.5. Quantitative Measurements of $T_{1\rho}$ for Epitope Mapping Using Homonuclear Spectra

Dynamic studies usually focus on measurements of the relaxation parameters of ^{15}N and ^{13}C nuclei bonded to ^{1}H using labeled proteins, as described above. Proton relaxation times in unlabeled proteins have not been investigated as thoroughly. This can be attributed to several factors, including the difficulty in measurements and data analysis due to spectral overlap. It is also difficult to interpret the T_2 data and analyze the type and time scale of motions contributing to relaxation because each proton may relax by several different mechanisms such as scalar couplings and dipole–dipole interactions with multiple nearby protons *(16)*. In our laboratory, we developed a simple homonuclear 2D method for measuring proton $T_{1\rho}$ relaxation times based on the HOHAHA experiment. This method can be used for epitope mapping without requiring ^{15}N or ^{13}C labeling. This technique was applied to α-BTX in complex with α-AChR peptide *(17)*. To calculate the $T_{1\rho}$ values, the decay in the intensity of the H_N-H_α cross-peaks as a function of the duration of the spin-lock pulse was fitted to a monoexponential curve with minor deviations. Long relaxation times, 155 ms on average, were measured for the free peptide. After binding, peptide residues outside the binding determinant, namely, $^{\alpha\text{-AChR}}$E181-R182, $^{\alpha\text{-AChR}}$W184, and $^{\alpha\text{-AChR}}$I201-E204 (and the EE tag), exhibited $T_{1\rho}$ values above an

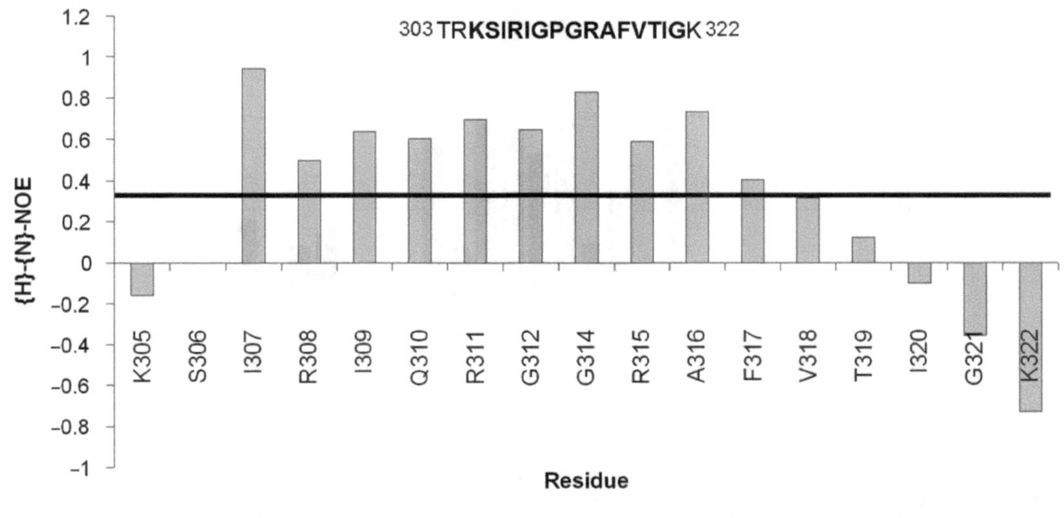

Fig. 6. The variations of the ^{1}H-^{15}N NOE ratios of the $V3_{IIIB}$ peptide bound to 447–52D Fv along the peptide sequence. The 0.35 threshold used to differentiate between the core epitope and the rest of the peptide is drawn as a *horizontal line*.

arbitrary threshold value of 45 ms. Residues within the binding determinant, $^{\alpha\text{-AChR}}$K185-D200, with the exception of $^{\alpha\text{-AChR}}$D195, displayed $T_{1\rho}$ values of <45 ms. In general, the change in the exposed surface upon binding of the AChR peptide is correlated with the change in $T_{1\rho}$ times, as presented in **Fig. 7.**

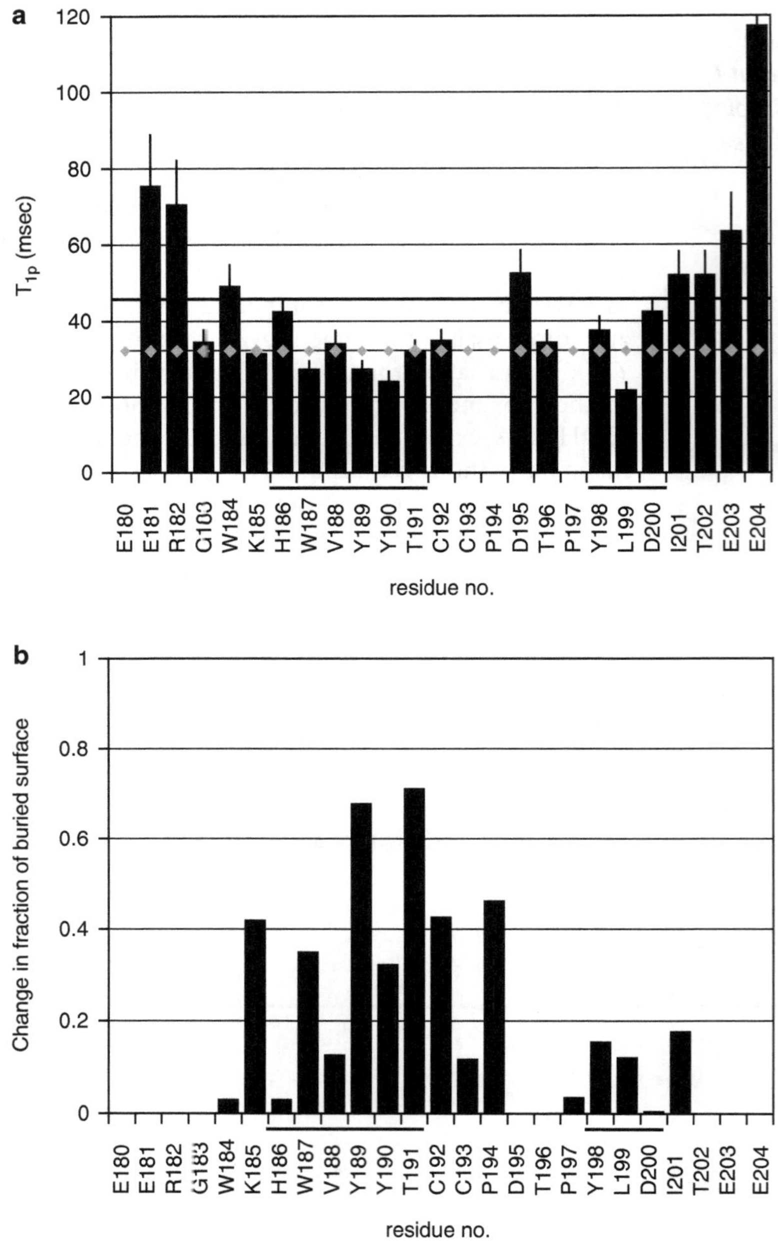

Fig. 7. $T_{1\rho}$ values of Hα protons in the bound AChR peptide and changes in the exposed surface upon binding. (**a**) $T_{1\rho}$ values for each residue in the AChR peptide in complex with the toxin. The 45-ms threshold used to differentiate between the peptide residues interacting with the toxin and those that do not is drawn as a *horizontal line*. (**b**) Fractional decrease in the exposed surface for residues in the AChR peptide. (Reproduced from *(17)*).

2. Materials

2.1. Antibody's Fv Expression and Purification

1. BL21-Gold(DE3)pLysS strain (Stratagene, La Jolla, CA)
2. pET-27b vector (Novagen, Madison, WI)
3. SBP medium
4. Isopropyl-β-D-thiogalactoside (IPTG): 1 M stock solution
5. Ethylene diamine tetraacetic acid (EDTA): 0.5 M stock solution
6. Isotope-enriched Celtone-rich medium (Martek Biosciences, Columbia, MD)
7. RPMI 1640 vitamin cocktail (Gibco-Invitrogen, Paisley, SC)
8. *N*-ethylmaleimide, iodoacetamide, benzamidine, and *N*-tosyl-L-phenylalanine chloromethyl ketone (Sigma-Aldrich, Israel) as protease inhibitors
9. 25 mM Tris-HCl buffer, pH 8.0
10. 50 mM Tris-HCl buffer (pH 8.0), 0.05% NaN_3, 5 mM EDTA
11. 50 mM Tris-HCl buffer (pH 8.0), 0.05% NaN_3, 0.5 M NaCl
12. 25 mM Tris-HCl buffer (pH 7.5), 125 mM NaCl, 0.02% NaN_3
13. 25 mM Tris-HCl
14. Acetic acid
15. Triton X-100
16. Buffer A: 25 mM acetate buffer, pH 4
17. Washing buffer I: 50 mM Tris (pH 7.9), 0.5 mM EDTA, 50 mM NaCl, and 5% glycerol
18. Guanidine hydrochloride (GuHCl): 6 M solution
19. Chromatography columns (Q-Sepharose Fast Flow, Sepharose 4 Fast Flow, SP-Sepharose, HiLoad Superdex 75 26/60), beads for the columns, and ÄKTA chromatography system (Amersham Biosciences, Uppsala, Sweden)
20. Vivaspin concentrator with a 10-kDa cutoff membrane (Vivascience, Westfold, MA)

2.2. Peptide Expression and Purification

1. BL21-Gold(DE3)pLysS strain (Stratagene)
2. Isotope-enriched Celtone-rich medium and RPMI 1640 vitamin cocktail (*see* **Subheading 2.1, item 7**)
3. Ampicillin and chloramphenicol
4. Isopropyl-β-D-thiogalactoside (IPTG): 1 M stock solution

5. Protease inhibitor stock solutions: aprotinin (10 mg/mL), leupeptin (10 mg/mL), and PMSF (17.4 mg/mL)
6. Triton X-100
7. 8 M urea, 50 mM glycine, and 450 mM NaCl in water
8. Washing buffer II: 50 mM Tris-HCl (pH 8.0), 0.5 mM EDTA, 0.5 M NaCl, and 5% glycerol
9. NH_4HCO_3: 50 mM solution in water
10. Cyanogen bromide (CNBr)
11. Formic acid: at least 70%
12. Trifluoro acetic acid: at least 70%
13. Dimethyl sulfoxide
14. Vydacpolymer(polystyrene–divinylbenzene)reversed-phased semipreparative (25 × 250 mm^2), Vydac (10 × 250 mm^2), and analytical Vydac (4 × 250 mm^2) columns (Amersham Biosciences)
15. pET11d vector (Novagen)
16. Restriction enzymes and *E. coli* DNA polymerase I (New England Biolabs, Beverly, MA)
17. T4 DNA ligase (Gibco Life Technologies, Gaithersburg, MD)
18. Vivaspin concentrator (*see* **Subheading 2.1, item 14**)

2.3. Fv–Peptide Complex and NMR Sample Preparation

1. Vivaspin concentrator (*see* **Subheading 2.1, item 15**)
2. Shigemi NMR tube (Shigemi, Allison Park, PA)

2.4. Binding Interface Mapping Protocol

1. DMX-500 Bruker Avance NMR spectrometer equipped with triple resonance probe with Z-gradient
2. DRX-800 Bruker Avance NMR spectrometer equipped with triple resonance inverse cryoprobe with Z-gradient
3. Software to process and analyze spectra: NMRView, NMRDraw, and NMRPipe

3. Methods

3.1. Antibody's Fv Expression and Purification

3.1.1. Expression and Purification of 447–52D V_L

1. Transform BL21-Gold(DE3)pLysS cells with the 447Lb27b plasmid.
2. Inoculate a 1-L flask containing 50 mL of SBP medium with starter culture that was diluted 1:100 and incubated with constant shaking at 37°C.
3. Add IPTG to a final concentration of 1 mM (1 mL of 1 M stock to 1-L culture) when the culture reaches an optical

density of 0.7–0.9 at 595 nm and incubate the culture overnight at 20°C with shaking at 225 rpm.

4. Pellet the bacteria by centrifugation at 3,200×*g* for 30 min at 4°C and collect the culture medium containing the V_L protein. Inhibit endogenous protease activity by adding 2 mL of each: 0.5 M EDTA, 0.1 M *N*-ethylmaleimide, 0.1 M iodoacetamide, 0.1 M benzamidine, and 4 mL of 3 mg/mL of *N*-tosyl-l-phenylalanine chloromethyl ketone to every 50 mL culture medium.
5. Adjust the culture medium solution to pH 8 and centrifuge at 18,500×*g* for 30 min at 4°C; then dilute the supernatant fourfold with HPLC-grade water.
6. Purify the protein on a Q-Sepharose Fast Flow anion exchange column equilibrated with 25 mM Tris-HCl, pH 8. Collect the flow-through, adjust to pH 4 with acetic acid, and centrifuge at 18,500×*g* for 30 min at 4°C.
7. Collect the supernatant and load onto a Sepharose 4 Fast Flow column connected in series with a SP-Sepharose column. Equilibrate both columns with buffer A. Wash the SP-Sepharose column with 5 bed volumes of buffer A containing 100 mM NaCl.
8. Elute the protein with 15 column volumes of a 0.1–0.7 M NaCl linear gradient in buffer A. Adjust the pH of the eluted light-chain protein to pH 8 by adding Tris-HCl solution at pH 9 to a final concentration of 73 mM of the buffer to prevent adverse effects of the acidic pH on the protein.

3.1.2. Expression and Purification of 447–52D V_H

1. Transform BL21-Gold(DE3)pLysS cells with the 447Hb27b plasmid.
2. Inoculate a 1-L flask containing 50 mL of SBP medium with starter culture that was diluted 1:100 and incubate with constant shaking at 37°C.
3. Add IPTG to a final concentration of 1 mM when the culture reaches an OD_{595} of 0.7–0.9 and incubate the culture for 4 h at 25°C with shaking at 225 rpm.
4. Pellet the bacteria by centrifugation at 3,200×*g* for 30 min and collect the V_H protein, which sequesters in inclusion bodies, by suspending the cell harvest in 5 mL washing buffer I (50 mM Tris (pH 7.9), 0.5 mM EDTA, 50 mM NaCl, and 5% glycerol).
5. Sonicate for 5 min on ice and remove membrane components by adding Triton X-100 to a final concentration of 1%. Then incubate the slurry on ice for 10 min.
6. Centrifuge at 18,500×*g* for 10 min at 4°C. Wash the pellet containing the inclusion bodies twice with 5 mL washing

buffer I containing 1% Triton X-100 and a third time with washing buffer I only.

7. Suspend the inclusion bodies from 50 mL of bacterial growth in 1.5 mL of 6 M guanidine-hydrochloride (GuHCl) and centrifuge the extracted protein.
8. Purify the V_H by size fractionation of the denaturated protein using a HiLoad Superdex 75 26/60 column, with 4 M GuHCl as a running buffer.

3.1.3. Recombination of the 447–52D V_L and V_H Chains

1. Combine light- and heavy-chain proteins by mixing the V_H protein (dissolved in 4 M GuHCl) and the V_L protein (in the SP-Sepharose elution buffer) in a molar ratio of 1:1 (equal to 1:1.25 mg ratio).
2. Dialyze the Fv overnight against 50 mM Tris-HCl (pH 8), 0.05% NaN_3, and 5 mM EDTA at 4°C.
3. Concentrate to 0.6 mg/mL using a Vivaspin concentrator and purify the Fv further by size-exclusion chromatography on a HiLoad Superdex 75 26/60 column equilibrated with 125 mM NaCl, 25 mM Tris-HCl (pH 7.5), and 0.02% NaN_3.
4. Load the eluted Fv onto a benzamidine column equilibrated with 0.5 M NaCl, 50 mM Tris-HCl (pH 8), and 0.05% NaN_3 to remove serine proteases. The Fv fraction is eluted in the flow-through.

3.2. Peptide Expression and Purification

3.2.1. Expression of $V3_{MN}$ Fusion Protein in E. coli

1. Transform *E. coli* competent cells of the BL21-Gold(DE3) pLysS strain with the pM4-$V3_{MN}$ plasmid containing the sequence for the $V3_{MN}$ peptide.
2. Inoculate growth medium at 1:100 dilution and grow the cells at 37°C in Celtone containing 100 μg/mL ampicillin, 25 μg/mL chloramphenicol, RPMI 1640 vitamin cocktail, 1 mM $MgSO_4$, and 50 mM K_2HPO_4 to a cell density of OD_{595} 1.0–1.2.
3. Induce expression by adding 1 mM IPTG, followed by incubation for 3 h at 37°C and shaking at 225 rpm. Then collect the bacteria by centrifugation (8,000×*g* for 30 min). The over-expressed protein is obtained in large quantities in the form of inclusion bodies.
4. Suspend the cells in 20 mL of washing buffer II and add a mixture of protease inhibitors (4 μL of 10 mg/mL aprotinin, 4 μL of 10 mg/mL leupeptin, and 200 μL of 17.4 mg/mL PMSF) to the cooled buffer, and sonicate the mixture for 90 s on ice.
5. Add 1% Triton X-100 in order to remove membrane components and incubate the cell lysate on ice for 10 min.
6. Spin the cell lysate at 20,200×*g* for 10 min at 4°C and wash the pellet containing the inclusion bodies twice with 20 mL

washing buffer II and 1% Triton X-100 to dispose of cell debris. Remove traces of Triton X-100 by a third wash with the washing buffer II.

7. Gradually solubilize the inclusion bodies in 8 M urea, 50 mM glycine, and 450 mM NaCl in washing buffer II (*see* **Note 1**). Collect the protein from the supernatant by centrifugation at 20,200×g for 10 min and subject the supernatant to an additional centrifugation at 30,900×*g* for 30 min to pellet the remaining DNA.
8. Dialyze the supernatant against 50 mM NH_4HCO_3 (four changes) and lyophilize. This yields a 90% pure preparation of the fusion protein.

3.2.2. $V3_{MN}$ Peptide Cleavage and Purification

1. Release the peptide from the fusion protein using CNBr (*see* **Note 2**), which cleaves peptide bonds at the C-terminus of a methionine residue *(18)*. Cleavage requires 24 h in 70% formic acid at room temperature in a darkened flask (*see* **Note 3**).
2. Add 10 volumes of HPLC-grade water upon completion of the reaction and lyophilize to complete dryness. Then dissolve in 100% dimethyl sulfoxide.
3. Dilute the cleaved protein solution to 30% dimethyl sulfoxide and purify by HPLC on a Vydac polymer reversed-phased semipreparative column (25 × 250 mm^2) and then on a Vydac (10 × 250 mm^2) column.
4. Lyophilize and detect the purity on an analytical Vydac column (4 × 250 mm^2), using an acetonitrile water gradient. All solvents contain 0.1% trifluoroacetic acid.

3.2.3. Construction of the $V3_{IIIB}$ Expression Vector

1. Use the unique NdeI and BamHI restriction sites, found at the 5′ and 3′ ends of the $V3_{MN}$ construct in pM4-$V3_{MN}$, for the removal of the $V3_{MN}$ gene and the introduction of the $V3_{IIIB}$.
2. Prepare the nucleotide sequence of the $V3_{IIIB}$ gene, then digest it with *Nde*I and *Bam*HI and ligate with the pM4-$V3_{MN}$ vector, previously cut with the same enzymes.
3. Transform *E. coli* TG1 strain with this new construct to obtain the recombinant clones.
4. Transform *E. coli* BL21(DE3)pLysS strain with the plasmid and express in a similar manner as the $V3_{MN}$ peptide.
5. For $V3_{IIIB}$, which has threonine as the first amino acid, carry out the CNBr cleavage in 70% trifluoro acetic acid instead of 70% formic acid. The efficiency of the cleavage in formic acid when a threonine residue follows methionine is very low *(19)*.

3.3. Fv–Peptide Complex and NMR Sample Preparation

1. Prepare the complex of the 447–52D Fv fragment and a $^{13}C/^{15}N$ V3 peptide (~28 kDa) by shaking the Fv and the labeled peptide in a 1:1.2 molar ratio respectively for 3 h at 37°C.
2. Remove excess peptide and concentrate the sample by subjecting to centrifugal filtration in Vivaspin tubes (Vivascience) with a 10-kDa membrane cutoff.
3. Insert the recovered solution into a Shigemi NMR tube (total volume of 350 μL). All samples contained 10 mM deuterated acetate buffer (pH 5.0), 95% H_2O/5% D_2O, 1 mM EDTA, and 0.02% NaN_3.

3.4. Binding Interface Mapping Protocol

3.4.1. NMR Dynamic Filtering

1. The experimental and processing parameters are given in **Note 4** (for α-BTX) and **Note 5** (for V3 peptides).
2. Perform sequential assignment of $^{\alpha\text{-AChR}}$K179-D200 and $^{\alpha\text{-AChR}}$R182-T202 peptides according to the well-established method of Wüthrich *(20)*. This is achieved by measuring HOHAHA, ROESY, and NOESY spectra of the free peptides.
3. Measure HOHAHA and ROESY spectra of the α-BTX/$^{\alpha\text{-AChR}}$K179-D200 and α-BTX/$^{\alpha\text{-AChR}}$R182-T202 complexes with mixing times of 400 and 250/200 ms respectively.
4. Compare the HOHAHA spectrum of the complex with that of the free peptide to identify cross-peaks corresponding to the peptide residues that do not interact with the receptor, thus retaining their mobility and exhibiting chemical shift values similar to those found for the free peptide. As a result of the long mixing period used in the HOHAHA and ROESY spectra, the cross-peaks of peptide protons interacting with the receptor as well as of most receptor protons vanish.
5. For the V3: Measure HOHAHA, ROESY, and NOESY spectra for the free V3 peptide in H_2O. Completely assign all cross-peaks observed for the free peptide according to the well-established method of Wüthrich *(20)*.
6. Measure ROESY and HOHAHA spectra with long mixing times (90 ms) for the peptide–Fv (or peptide–Fab) complex in H_2O. The mixing time is adjusted to discriminate between cross-peaks of peptide protons that are immobilized in the complex because of interactions with the antibody and therefore have a short $T_{1\rho}$ relaxation time and those protons that do not interact with the Fv and, therefore, retain considerable mobility and have a long $T_{1\rho}$. The combination of the HOHAHA and ROESY spectra is used for sequential assignment of the mobile segments of the peptide in the Fv–peptide complex.
7. Compare the HOHAHA spectrum of the complex **(step 6)** with that of the free peptide **(step 5)** to identify cross-peaks corresponding to the peptide residues that do not interact with the Fv, thus retaining their mobility and exhibit chemical

shift values similar to those found for the free peptide. Cross-peaks of peptide protons interacting with the Fv as well as of most Fv protons vanish as a result of the long mixing period used in the HOHAHA and ROESY spectra.

3.4.2. ^{1}H-^{15}N HSQC Intensities

1. Measure the ^{15}N TROSY HSQC spectrum of the uniformly ^{15}N-labeled peptide in complex with unlabeled Fv.
2. Measure cross-peak intensities and draw intensities vs. peptide residues. Examine the variation in peak intensity. Those residues interacting with the antibody have low intensity while those outside the epitope have higher values.

3.4.3. ^{15}N T_2 Relaxation Times

1. The experimental and processing parameters are given in **Note 6**.
2. Measure a series of HSQC spectra modified in a way that the magnetization will remain on the nitrogen for a different amount of time in each measurement. The delay time, during which the ^{15}N relaxes, is increased gradually in the series of experiments.
3. Process the spectra in a manner similar to that of HSQC spectra and then extract the T_2 relaxation time values by fitting the decline in signal strength over time to a decreasing exponential.
4. The major factor influencing T_2 is the tumbling rate of the protein. Rapid tumbling results in a long T_2, whereas slow tumbling results in a short T_2. Therefore, V3 residues interacting with the Fv will tumble slower and have short T_2.

3.4.4. Proton $T_{1\rho}$ Relaxation Times

1. The experimental and processing parameters are given in **Note 7**.
2. Measure HOHAHA, ROESY, and NOESY spectra for the toxin and its peptide complex in H_2O. Sequentially assign all cross-peaks observed according to the well-established method of Wüthrich *(20)*.
3. Measure a series of HOHAHA spectra for the unlabeled complex. The HOHAHA experiment is based on the HOHAHA pulse sequence *(21)* and incorporates a spin lock (SL) pulse after the initial 90° pulse. Measure six experiments with SL durations varying from 0 to 25 ms in 5-ms increments.
4. Determine proton $T_{1\rho}$ values by fitting the decay in cross-peak intensity as a function of the duration of the SL pulse to a single-exponential decay [$I(t) = I_0 \exp(-t/T_{1\rho})$] using the modelXY package *(22)*. Obtain error values to be included in the calculation from signal-to-noise ratios.
5. A decrease in the relaxation times and the mobility of residues involved in binding is exhibited, while residues not implicated in binding retain considerable mobility. The quantitative $T_{1\rho}$ measurements enable to corroborate the mapping of boundaries of binding determinant by other methods.

3.4.5. ^1H-^{15}N NOE

1. Record pairs of interleaved ^{1}H-^{15}N NOE spectra with and without proton saturation during the recycle delay. Since the longitudinal magnetization of the heteroatom is to be measured for calculating the hetNOE, magnetization must originate from the ^{15}N spin. Magnetization is subsequently transferred via a refocused INEPT sequence to the directly coupled NH proton for observation.
2. hetNOEs are evaluated by calculating the ratio of intensities in spectra recorded with and without proton saturation.

4. Notes

1. The yield of the purified V3 peptide varied from 6–13 mg/L of Celtone medium. To obtain higher yields, the inclusion bodies (**Subheading 3.2.1**, **step 7**) can be repeatedly solubilized.
2. Since CNBr is volatile and toxic (**Subheading 3.2.2**, **step 1**), the V3 peptide cleavage reaction from the fusion protein should be carried out in a hood and all tips and tubes used must be neutralized with 10 M NaOH.
3. CNBr cleavage should be done in a darkened flask using a 400-fold molar excess of CNBr over methionine residues (**Subheading 3.2.2**, **step 1**).
4. All NMR spectra were acquired on Bruker DMX 500-MHz and DRX 800-MHz spectrometers (**Subheading 3.4.1**, **steps 2**, **3**). HOHAHA, ROESY, and NOESY spectra of $^{\alpha\text{-AChR}}$K179-D200 were acquired at 20°C using mixing times of 150, 150, and 300 ms, respectively. For $^{\alpha\text{-AChR}}$R182-T202 peptide, HOHAHA spectra were acquired at 30 and 47°C. The HOHAHA and ROESY spectra of the α-BTX/$^{\alpha\text{-AChR}}$K179-D200 peptide and α-BTX/$^{\alpha\text{-AChR}}$R182-T202 complexes were acquired with varying mixing times of 100–400 ms, 2–4K points in the F2 and 256–600 increments in the F1 dimensions at 20 and 30°C. For the complete sequential assignment of the α-BTX/$^{\alpha\text{-AChR}}$R182-T202 complex, HOHAHA and NOESY spectra with 8K data points in F2 and 800 increments in F1 were acquired with mixing times of 70 and 150 ms, respectively, at 30 and 37°C. For assignment of the aliphatic region in 99.99% D_2O, HOHAHA and NOESY spectra with water presaturation were acquired at 37°C with mixing times of 70 and 150 ms, respectively.
5. The long mixing time (90 ms) in the ROESY and HOHAHA spectra (**Subheading 3.4.1**, **step 6**) was adjusted to discriminate

between cross-peaks of peptide protons that are immobilized in the complex because of interactions with the antibody and have a short $T_{1\rho}$ relaxation time and those of protons that do not interact with the Fv and, therefore, retain considerable mobility and have a long $T_{1\rho}$. Hence, the combination of the HOHAHA and ROESY spectra was used for sequential assignment of the mobile segments of the peptide in the Fv–peptide complex. NMR spectra were acquired at 35°C on Bruker DMX 500 and DRX 800 spectrometers. Two-dimensional ROESY and HOHAHA spectra of the unlabeled complex were measured at 30°C, 20°C, and 10°C and at pH 7, 5, and 4.25.

6. $T_2{}^{15}N$ relaxation time measurements (**Subheading 3.4.3, step 2**) were carried out with a total of 182 transients. Six time points were collected with parametric delays of 8, 16, 24, 32, 48, and 72 ms on a Bruker DRX-800 MHz spectrometer, with a 2-s delay between scans.

7. NMR spectra of α-BTX and its complex were acquired at 30°C. The $T_{1\rho}$-filtered HOHAHA spectra were measured using the following pulse sequence: $90_x°$-SL_y-evolution (t_1)-(WALTZ)-acquisition (t_2) (**Subheading 3.4.4, step 3**). Isotropic mixing was realized using a WALTZ *(23)* pulse sequence with a short duration of 30 ms. The spectra were acquired using sensitivity enhancement and TPPI, and the water signal was suppressed by the WATERGATE pulse sequence (water suppression by gradient-tailored excitation) *(24)*. The numbers of complex points acquired on the 500-MHz spectrometer were 2,048 and 256 in the F_2 and F_1 dimensions, respectively, with spectral width of 6,000 Hz. The numbers of complex points acquired on the 800-MHz spectrometer were 8,192 and 800 in the F_2 and F_1 dimensions, respectively, with spectral widths of 11,160 Hz. All spectra were processed and analyzed using NMRDraw and NMRPipe *(22)*. Unresolved peaks at frequencies close to that of water required additional baseline correction using polynomial water subtraction in all experiments. Linear prediction in the F_1 dimension was also done to increase resolution and improve the automated peak picking. Automated peak picking in these spectra was achieved using NMRPipe and in-house Perl scripts.

Acknowledgments

We thank Drs. Avraham Samson, Anat Zvi, Irina Kustanovic, Michal Sharon, and Naama Kessler, who did some of the studies described in this chapter. We gratefully acknowledge help from

Dr. Tali Scherf in maintaining the NMR spectrometers and setting up some of the experiments. We thank Dr. Sandy Livnat for editorial assistance. This study was supported by the National Institute of Health Grant GM 53329 to Jacob Anglister who is the Dr. Joseph and Ruth Owades Professor of Chemistry.

References

1. Williams, D. C., Jr., Rule, G. S., Poljak, R. J., and Benjamin, D. C. (1997) Reduction in the amide hydrogen exchange rates of an antilysozyme Fv fragment due to formation of the Fv–lysozyme complex. *J. Mol. Biol.* 270, 751–762.
2. Tugarinov, V., Zvi, A., Levy, A., Hayek, Y., Matsushita, S., and Anglister, J. (2000) NMR structure of an anti-gp120 antibody complex with a V3 peptide reveals a surface important for co-receptor binding. *Struct. Fold. Des.* 8, 385–395.
3. Rosen, O., Chill, J., Sharon, M., Kessler, N., Mester, B., Zolla-Pazner, S., and Anglister, J. (2005) Induced fit in HIV-neutralizing antibody complexes: evidence for alternative conformations of the gp120 V3 loop and the molecular basis for broad neutralization. *Biochemistry* 44, 7250–7258.
4. Rosen, O., Sharon, M., Quadt-Akabayov, S. R., and Anglister, J. (2006) Molecular switch for alternative conformations of the HIV-1 V3 region: implications for phenotype conversion. *Proc. Natl. Acad. Sci. USA* 103, 13950–13955.
5. Sharon, M., Kessler, N., Levy, R., Zolla-Pazner, S., Gorlach, M. and Anglister, J. (2003) Alternative conformations of HIV-1 V3 loops mimic beta hairpins in chemokines, suggesting a mechanism for coreceptor selectivity. *Structure* 11, 225–236.
6. Sharon, M., Rosen, O., and Anglister, J. (2005) NMR studies of V3 peptide complexes with antibodies suggest a mechanism for HIV-1 co-receptor selectivity. *Curr. Opin. Drug Discov. Dev.* 8, 601–612.
7. Scherf, T., Hiller, R., Naider, F., Levitt, M., and Anglister, J. (1992) Induced peptide conformations in different antibody complexes: molecular modeling of the three-dimensional structure of peptide–antibody complexes using NMR-derived distance restraints. *Biochemistry* 31, 6884–6897.
8. Zilber, B., Scherf, T., Levitt, M., and Anglister, J. (1990) NMR-derived model for a peptide–antibody complex. *Biochemistry* 29, 10032–10041.
9. Zvi, A., Feigelson, D. J., Hayek, Y., and Anglister, J. (1997) Conformation of the principal neutralizing determinant of human immunodeficiency virus type 1 in complex with an anti-gp120 virus neutralizing antibody studied by two-dimensional nuclear magnetic resonance difference spectroscopy. *Biochemistry* 36, 8619–8627.
10. Zvi, A., Kustanovich, I., Feigelson, D., Levy, R., Eisenstein, M., Matsushita, S., Richalet Secordel, P., Regenmortel, M. H., and Anglister, J. (1995) NMR mapping of the antigenic determinant recognized by an anti-gp120, human immunodeficiency virus neutralizing antibody. *Eur. J. Biochem.* 229, 178–187.
11. Zvi, A., Tugarinov, V., Faiman, G. A., Horovitz, A., and Anglister, J. (2000) A model of a gp120 V3 peptide in complex with an HIV-neutralizing antibody based on NMR and mutant cycle-derived constraints. *Eur. J. Biochem.* 267, 767–779.
12. Scherf, T., and Anglister, J. (1993) A T1 rho-filtered two-dimensional transferred NOE spectrum for studying antibody interactions with peptide antigens. *Biophys. J.* 64, 754–761.
13. Samson, A. O., Chill, J. H., Rodriguez, E., Scherf, T., and Anglister, J. (2001) NMR mapping and secondary structure determination of the major acetylcholine receptor alpha-subunit determinant interacting with alpha-bungarotoxin. *Biochemistry* 40, 5464–5473.
14. Sattler, M., Schleucher, J., and Griesinger, C. (1999) Heteronuclear multidimensional NMR experiments for the structure determination of proteins in solution employing pulsed field gradients. *Prog. Nucl. Magn. Reson. Spectrosc.* 34, 93–158.
15. Kay, L. E., Torchia, D. A., and Bax, A. (1989) Backbone dynamics of proteins as studied by 15N inverse detected heteronuclear NMR spectroscopy: application to staphylococcal nuclease. *Biochemistry* 28, 8972–8979.
16. Ishima, R., Wingfield, P. T., Stahl, S. J., Kaufman, J. D., and Torchia, D. A. (1998) Using amide 1H and 15N transverse relaxa-

tion to detect millisecond time-scale motions in perdeuterated proteins: application to HIV-1 protease. *J. Am. Chem. Soc.* 120, 10534–10542.

17. Samson, A. O., Chill, J. H., and Anglister, J. (2005) Two-dimensional measurement of proton T1rho relaxation in unlabeled proteins: mobility changes in alpha-bungarotoxin upon binding of an acetylcholine receptor peptide. *Biochemistry* 44, 10926–10934.
18. Gross, E. (1967) The cyanogen bromide reaction, in Methods in Enzymology (Hirs, C. H. W.), Academic, New York, NY, pp. 238–255.
19. Kaiser, R., and Metzka, L. (1999) Enhancement of cyanogen bromide cleavage yields for methionyl-serine and methionyl-threonine peptide bonds. *Anal. Biochem.* 266, 1–8.
20. Wuthrich, K. (ed.) (1986) NMR of proteins and nucleic acids. Wiley, New York, NY, pp. 130–161.
21. Braunschweiler, L., and Ernst, R. R. (1983) Coherence transfer of isotropic mixing: application to proton correlation spectroscopy. *J. Magn. Reson.* 53, 521–528.
22. Delaglio, F., Grzesiek, S., Vuister, G. W., Zhu, G., Pfeifer, J., and Bax, A. (1995) NMRPipe: a multidimensional spectral processing system based on UNIX pipes. *J. Biomol. NMR* 6, 277–293.
23. Shaka, A. J., Keeler, J., and Freeman, R. (1983) Evaluation of a new broadband decoupling sequence: WALTZ-16. *J. Magn. Reson.* 53, 313–340.
24. Piotto, M., Saudek, V., and Sklenar, V. (1992) Gradient-tailored excitation for single-quantum NMR spectroscopy of aqueous solutions. *J. Biomol. NMR* 2, 661–665.

Chapter 4

A Solid-Phase Mutual Inhibition Assay with Labeled Antigen

Masahide Kuroki

Summary

A widely applicable method for the determination of the epitope specificities of a large number of monoclonal antibodies (MAbs) is presented. The method is based on the solid-phase mutual inhibition assay using 96-well plates coated with the respective MAbs, competitor MAbs, biotinylated antigen, and avidin–peroxidase conjugate. Using carcinoembryonic antigen (CEA) as a model antigen, the method was applied to determine epitope specificities of anti-CEA MAbs. A constant amount of biotinylated CEA was incubated with a given MAb immobilized on wells of 96-well plates in the presence of increasing amounts of soluble competitor MAbs. The biotinylated CEA bound to the immobilized antibody was then reacted with avidin–peroxidase conjugate and the activity of the bound peroxidase was determined by using *o*-phenylenediamine and hydrogen peroxide. The method used alleviates the laborious procedures of labeling all antibodies to be tested and the confusion caused by differential labeling among different MAbs. It is a convenient method for mapping analysis of many MAbs if the corresponding purified antigen is available.

Key words: Monoclonal antibody, Epitope mapping, Epitope specificity, Solid-phase, Mutual inhibition assay, Labeled antigen

1. Introduction

Epitope specificities of monoclonal antibodies (MAbs) have usually been determined using the competitive solid-phase assay in which the antigen is immobilized and a radiolabeled antibody as well as competing unlabeled antibodies are mixed in solution (**Fig. 1a**) *(1, 2)*. Although this method facilitates separation of free from bound antibody, there is the problem of labeling all antibodies to be tested. Since the number of MAbs to be screened is usually large, this method is time-consuming and tedious, and the instability of radiolabels represents a significant drawback.

Ulrich Reineke and Mike Schutkowski (eds.), *Methods in Molecular Biology, Epitope Mapping Protocols, vol. 524*

DOI: 10.1007/978-1-59745-450-6_4

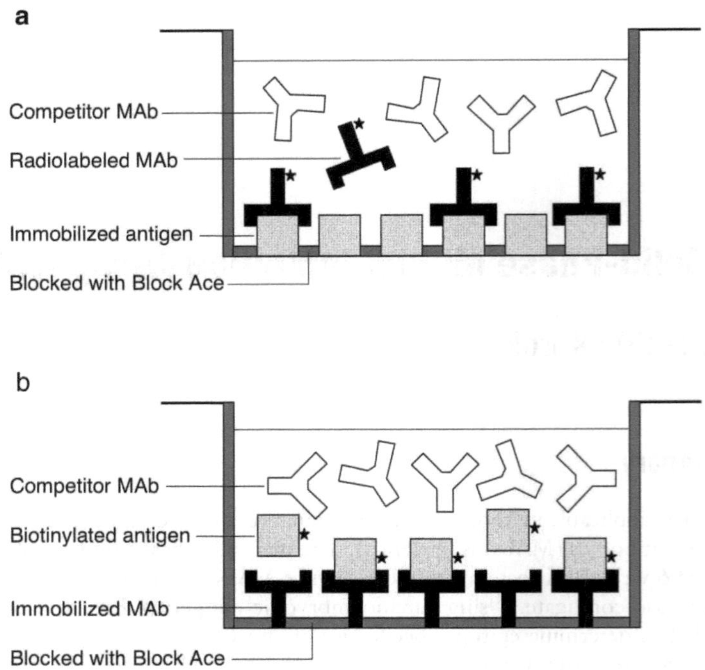

Fig. 1. Diagrammatic representation of a conventional competition immunoassay with labeled antibody (**a**) and a competition immunoassay with labeled antigen (**b**) used in epitope mapping. In the competition assay with labeled antibody, the antigen is immobilized and a radiolabeled antibody as well as competing unlabeled antibodies are mixed in solution. Finally, the radiolabeled antibody bound to the immobilized antigen is detected in a gamma counter. Currently, biotinylation is often used instead of radiolabeling. On the other hand, in the competition assay with labeled antigen, a constant amount of biotinylated antigen is incubated with a given MAb immobilized on wells of 96-well plates in the presence of increasing amounts of soluble competitor MAbs. The biotinylated antigen bound to the immobilized antibody is then reacted with avidin–peroxidase conjugate and the activity of the bound peroxidase is determined.

In addition, radioactive hazards have to be taken into account. Currently, nonisotopic tracers, such as biotin *(3, 4)* and fluorescein isothiocyanate *(4)*, are introduced for determination of epitope specificities of MAbs, but these methods still have the problem of labeling all antibodies to be tested.

This chapter describes a solid-phase mutual competition assay for determination of epitope specificities of MAbs by using 96-well plates coated with MAbs, competitor MAbs, biotinylated antigen, and avidin–peroxidase conjugate (**Fig. 1b**) *(5)*. A constant amount of biotinylated antigen is incubated with a given MAb immobilized on wells of 96-well plates in the presence of increasing amounts of soluble competitor MAbs. The biotinylated antigen binds to the immobilized antibody, is then reacted with avidin–peroxidase conjugate, and the activity of the bound peroxidase is determined by using *o*-phenylenediamine (OPD) and hydrogen peroxide (H_2O_2).

The exchange of competing and immobilized antibodies in this assay system does not demand additional labeling procedures *(6)*. Thus, the competition method described here alleviates the laborious procedures of labeling all antibodies to be tested and the confusion caused by differential labeling among different MAbs. It is a convenient method for mapping analysis of many MAbs if the corresponding purified antigen is available.

The typical competition curves of one group D anti-CEA MAb (F33–13) are shown in **Fig. 2**. Of the seven MAbs used as competitors, four MAbs, including MAb F33–13 itself, showed more than 80% inhibition at the highest input levels. Three MAbs yielded only marginal inhibition and could not exhibit more than 50% inhibition in this assay. To quantify the inhibitory effect of each MAb, the amounts of competitor MAbs required to inhibit the biotinylated antigen binding by 50% to each MAb dried on wells are determined from the respective inhibition curves. **Table 1** summarizes the results of mutual competition assays among 7 group D anti-CEA MAbs. This presentation allows for the comparison of the ability of each MAb to inhibit the binding of other MAbs to CEA (carcinoembryonic antigen) with the reciprocal competition of each MAb binding to CEA by the other MAbs *(7)*. Nonreciprocal cross competitions could result from (a) recognition by an antibody of several structurally related sites, only some of which may be recognized by other antibodies, (b) steric hindrance of an epitope by a second antibody bound to a different site, or (c) conformational change in the antigen molecule by binding of one antibody, which may affect binding of the second antibody.

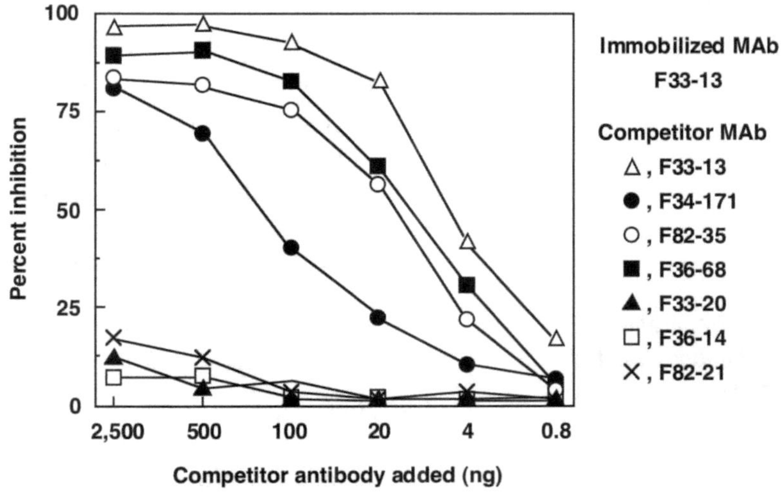

Fig. 2. Mutual competition assays among Group D anti-CEA MAbs by using biotinylated CEA and purified MAb preparations. One MAb of Group D, F33–13, was dried onto wells of 96-well plates. Purified IgG's from the Group D MAbs were used as competitors at the indicated quantities.

Table 1
Mutual competition assays among group D MAbs

Competitor MAb	Immobilized MAb						
	F33–13	F34–171	F82–35	F36–68	F33–20	F36–14	F82–21
F33–13	3+	3+	3+	3+	–	–	–
F34–171	2+	3+	3+	3+	–	–	–
F82–35	3+	3+	1+	3+	–	–	–
F36–68	3+	3+	3+	3+	–	–	–
F33–20	–	–	–	–	3+	–	–
F36–14	–	–	–	–	–	3+	–
F82–21	–	–	–	–	–	–	3+
Epitope	D-a	D-a	D-a	D-a	D-b	D-c	D-d

The group D anti-CEA MAbs recognized the epitopes on the domain A1-B1 of the CEA molecule *(7)*

The amount of competitor MAb required to give half-maximal inhibition of binding of biotinylated CEA was determined from the respective inhibition cures

3+, half-maximal inhibition at <100 ng; 2+, 100–500 ng; +, 500–2,500 ng; –, no 50% inhibition even at the highest amount (2,500 ng) of competitor antibody

2. Materials

2.1. Biotinylation of Antigen

1. 1-mL reaction vials with internal cone (Reacti-Vial™, Pierce Chemical, Rockford, IL).
2. PBS: 0.1 M sodium phosphate buffer (pH 7.0), 0.9% NaCl.
3. Antigen solution: 1 mg/mL in PBS.
4. *N*-hydroxysuccinimidobiotin (NHSB, mol. wt. = 341.4) solution: 2.0 mg/mL (5.86 mM) in dimethylformamide. Add 4.0 mg of NHSB, with stirring, to 0.5 mL dimethylformamide in a Reacti-Vial, and dilute the solution up to 2 mL with deionized distilled water. Prepare freshly just before use (*see* **Note 1**).

2.2. Preparation of Antibody-Coated Plates

1. 96-well polystyrene plates
2. BBS: 0.01 M borate-buffer (pH 8.0), 0.9% NaCl
3. Blocking solution: Block Ace (Dainihon Chemical, Osaka, Japan) which includes casein and some other proteins from bovine milk (*see* **Note 2**)
4. Washing buffer: 0.05% Nonidet P-40 (NP-40) in BBS

2.3. Mutual Competition Assay

1. Assay buffer: 1% bovine serum albumin, 0.1% methyl *p*-hydroxybenzoate, 0.01% propylhydroxybenzoate/in BBS (*see* **Note 3**).

2. Competitor antibody solutions: Make serial fivefold dilutions of the MAb to be tested in the sample buffer. The starting concentration of each MAb is 100 μg/mL.
3. Biotinylated antigen solution: 200 ng/mL in the assay buffer (*see* **Subheading 3.1**).
4. Horseradish peroxidase (HRP)–avidin D solution: 0.25 μg/mL in the assay buffer.
5. Citrate/phosphate buffer (CPB): 0.05 M citrate, 0.1 M phosphate (pH 5.0).
6. Substrate stock: 4% OPD in methanol. Store in aliquots at −70°C.
7. 30% H_2O_2.
8. Substrate solution: 0.04% OPD, 0.006% H_2O_2 in CPB. Dilute 150 μL of 4% OPD and 3 μL of 30% H_2O_2 up to 15 mL with CPB for one plate. This should be prepared freshly.
9. Stopping solution: 8 N H_2SO_4.

3. Method

3.1. Biotinylation of Antigen

1. Add 25 μL (50 μg; 146 nmol) of freshly prepared NHSB solution to 0.1 mL (100 μg) of antigen solution in a 1-mL Reacti-Vial with rapid stirring (*see* **Note 4**).
2. After incubation for 2 h at room temperature, dialyze exhaustively at 4°C against BBS.
3. Determine the protein concentration of the biotin-labeled antigen by reading the OD at 280 nm (if the extinction coefficient at 280 nm of the antigen is available) or by another method, such as the bicinchoninic acid method *(8)*. Also, *see* Vol. 32 of this series.
4. Store in aliquots at −20°C (*see* **Note 5**).

3.2. Preparation of Antibody-Coated Plates

1. Dilute the MAbs to be tested in BBS at concentrations of 0.5–5 μg/mL (*see* **Note 6**).
2. Add 50 μL of each MAb solution into each well of 96-well plates and dry down at 37°C overnight (*see* **Note 7**).
3. Block nonspecific protein absorption by adding 200 μL of the blocking solution into each well and incubating for 1 h at 37°C.
4. Remove the blocking solution, and wash the plates three times with the washing buffer.

3.3. Mutual Competition Assay

1. To each well of the 96-well plates previously coated with a given MAb, add increasing amounts of competitor MAbs in 25 μL of the sample buffer and 5 ng of biotinylated antigen in 25 μL of the same buffer (*see* **Notes 8** and **9**), and shake the plate for 20 s on a plate shaker.
2. Incubate for 1 h at 37°C, remove the mixed solutions, and wash three times with the washing buffer.
3. Add 100 μL of the HRP–avidin solution and incubate for 1 h at room temperature.
4. Remove the HRP–avidin solution and wash three times with the washing buffer.
5. Add 150 μL of the substrate solution and incubate for 20–30 min at room temperature.
6. Terminate the reaction by adding 20 μL of the stopping solution and read the OD of each well at 492 nm in a plate reader.
7. Determine the amount of competitor MAb required to give half-maximal inhibition of binding of biotinylated antigen from the respective inhibition curves (*see* **Fig. 2**).

4. Notes

1. NHSB is frequently replaced with long-chain homologs, such as sulfosuccinimidyl-6-(biotinamido) hexanoate *(9)*. The addition of a spacer in biotinylating reagents facilitates subsequent interaction with avidin probes. Water-soluble analogs of NHSB and its derivatives, i.e., the sulfosuccinimide reagents, are also available from Pierce Chemical Co. In certain cases, these may be favorable, especially when working with proteins that are sensitive to organic solvents, such as *N,N*-dimethylformamide.
2. 5% Bovine serum albumin in BBS can be also used for blocking, but Block Ace™ is more effective for preventing nonspecific protein binding.
3. Instead of sodium azide, methyl *p*-hydroxybenzoate and propylhydroxybenzoate are used as preservatives that do not affect the color reaction of OPD. The assay buffer containing these preservatives can be stored for 2–3 months at 4°C.
4. Biotin can be readily conjugated to a variety of molecules, such as antibodies, enzymes, nucleic acids, and so forth. The small size (mol. wt. = 341.4) of the biotin molecule prevents the biotinylation procedure from modifying the chemical,

physical, or immunological properties of the molecules to which biotin is conjugated. Moreover, multiple biotinylation of the same molecule can be done without any adverse effect *(5, 10)*. It is sometimes difficult, however, to determine the exact number of biotin molecules per antigen (protein) molecule. We usually use an NHSB/antigen molar ratio of 1,000 for preparing biotinylated antigen, because when CEA was biotinylated, this ratio did not affect the immunoreactivity of CEA and gave the maximum avidin-binding activity of CEA *(5)*.

5. Most biotinylated antigens are very stable and can be stored at 4°C for at least 2 years.
6. The concentrations of MAbs used for coating the plates should be those at which the biotinylated antigen used gave an absorbance ranging from 1.0 to 1.2 (<1.5) in the absence of competitor antibody, resulting in good inhibition.
7. When the plates are coated with antibody and dried overnight, ensure that they are completely dry. Otherwise, maximum antibody binding will not be obtained, and some binding protein will be lost from the plates during subsequent procedures.
8. The competitor MAb solutions should be added into the antibody-coated plates before adding the biotinylated antigen solutions, also resulting in good inhibition results.
9. Usually, only 1–10 ng of biotinylated antigen is enough for this competition assay.

References

1. Kaufman, B. M., and Goldsby, R. A. (1982) Epitope ratio analysis (ERA): a simple radioimmunological method using monoclonal antibodies for the simultaneous analysis of several antigens. *J. Immunol. Methods* 54, 1–7.
2. Wagener, C., Yang, Y. H. J., Crawford, F. G., and Shively, J. E. (1983) Monoclonal antibodies for carcinoembryonic antigen and related antigens as a model system: a systematic approach for the determination of epitope specificities of monoclonal antibodies. *J. Immunol.* 130, 2308–2315.
3. Bayer, E. A., and Wilchek, M. (1990) Protein biotinylation. *Methods Enzymol.* 184, 138–160.
4. Harlow, E., and Lane, D. (1989) Antibodies: A Laboratory Manual, 2nd ed., Cold Spring Harbor Lab., Cold Spring Harbor, NY.
5. Kuroki, M., Wakisaka, M., Murakami, M., Haruno, M., Arakawa, F., Higuchi, H., and Matsuoka, Y. (1992) Determination of epitope specificities of a large number of monoclonal antibodies by solid-phase mutual inhibition assays using biotinylated antigen. *Immunol. Invest.* 21, 523–538.
6. Kuroki, M., Greiner, J. W., Simpson, J. W., Primus, F. J., Guadagni, F., and Schlom, J. (1989) Serologic mapping and biochemical characterization of the carcinoembryonic antigen epitopes using fourteen monoclonal antibodies. *Int. J. Cancer* 44, 208–218.
7. Kuroki, M., Arakawa, F., Haruno, M., Murakami, M., Wakisaka, M., Higuchi, H., Oikawa, S., Nakazato, H., and Matsuoka, Y. (1992) Biochemical characterization of 25 distinct carcinoembryonic antigen (CEA) epitopes recognized by 57 monoclonal antibodies and categorized into 7 groups in terms of domain structure of the CEA molecule. *Hybridoma* 11, 391–407.
8. Smith, P. K., Krohn, R. I., Hermanson, G. T., Mallia, A. K., Gartner, F. H., Provenzano,

M. D., Fujimoto, E. K., Goeke, N. M., Olson, B. J., and Klenk, D. C. (1985) Measurement of protein using bicinchoninic acid. *Anal. Biochem.* 150, 76–85.

9. Savage, M. D., Mattson, G., Desai, S., Nielander, G. W., Morgensen, S., and Conklin, E. J. (1994) Avidin–Biotin Chemistry: A Handbook, 2nd ed., Pierce Chemical Co., Rockford, IL.

10. Wilchek, M., and Bayer, E. A. (1990) Introduction to avidin–biotin technology. *Methods Enzymol.* 184, 5–13.

Chapter 5

Epitope Mapping by Surface Plasmon Resonance

Pär Säfsten

Summary

Biacore™ systems (Biacore AB) enable label-free detection of molecular interactions in real time using surface plasmon resonance detection. Epitope mapping of antibodies is usually performed in a pairwise fashion where one antibody is used to capture the antigen from solution and the binding of a secondary antibody is monitored. In contrast to alternative approaches, the method allows for mapping of large matrices of antibody pairs without the need for cumbersome labeling steps, and the real-time analysis enables a better ranking of complex stability when compared with end-point assays.

Key words: Biacore, Surface plasmon resonance, Epitope mapping, Antibodies.

1. Introduction

1.1. Biacore Systems

After the first Biacore™ instrument (Biacore AB) was introduced in 1990, it has been followed by a range of analytical systems to study molecular interactions, all built around the same principle. Each system consists of three main parts, an optical detection unit, a flow system, and a sensor chip *(1)*. The system is controlled using a computer, and the software has continuously been developed to be more and more user-friendly. The detection is based on surface plasmon resonance, an optical phenomenon that can measure mass concentration-dependent changes in refractive index close to a sensor surface. This allows label-free detection of molecular interactions in real time. The basis of the detection and the underlying technical principle are described in **Fig. 1**.

The sensor chip consists of a glass slide covered on one side with a thin (50 nm) layer of gold. A linker layer is used to cover the gold and for attachment of a flexible dextran matrix to the

Ulrich Reineke and Mike Schutkowski (eds.), *Methods in Molecular Biology, Epitope Mapping Protocols, vol. 524*

DOI: 10.1007/978-1-59745-450-6_5

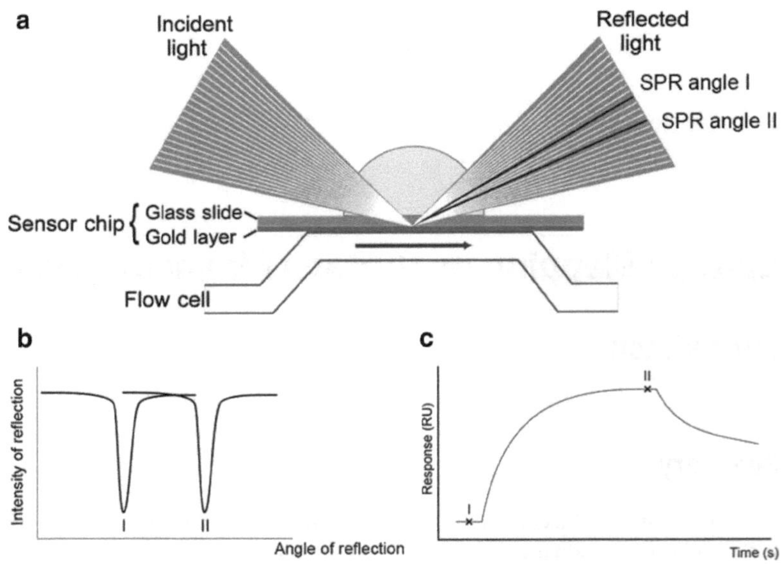

Fig. 1. (**a**) Principle of surface plasmon resonance detection. Monochromatic light is focused through a prism into a wedge-shaped beam onto the gold surface of the sensor chip providing a continuous interval of incident light angles. (**b**) The reflected light is detected, and at a certain angle, the resonance angle, a dip in intensity of the reflected light is seen. The resonance angle depends on the mass concentration close to the surface, and an increase in mass will change the angle. (**c**) The resonance angle is plotted against time to generate a sensorgram. The sensorgram in this figure shows a baseline in buffer (I), after which a solution with an analyte that binds to the surface is injected. The binding can be followed in real time until (II) where the interaction is at steady state and the analyte solution is replaced with buffer again. The final part of the sensorgram shows the dissociation of the analyte from the surface.

sensor surface. The dextran is carboxymethylated, which allows immobilization of molecules to the sensor chip surface using a variety of chemistries *(2)*.

The flexible hydrogel and the broad spectrum of immobilization chemistries make it possible to immobilize proteins while keeping a high activity and to study subsequent interactions with other molecules in a bufferlike environment. The dextran matrix also shows low nonspecific binding of components present in hybridoma supernatants and other complex media.

1.2. Antibody Screening and Characterization

Since the advent of monoclonal antibody technology *(3)* there has been a continuously growing need for the molecules it can produce. The methods to screen for antibodies with desired properties are constantly improving, and new parameters are taken into account increasingly early in the screening process. The most common technology for primary screening is still based on Enzyme-Linked ImmunoSorbent Assay (ELISA) (*see* Chapter “A Solid-Phase Mutual Inhibition Assay with Labeled Antigen”). The result is evaluated using a single response that is correlated to binding of the antibody to the desired antigen. The hits from the

ELISA are usually first confirmed by another method that preferably has the antigen in solution to minimize the risk for masked or changed epitopes due to immobilization of the antigen to a surface. The hits are then further characterized with respect to kinetic properties and the binding epitope. The kinetics of binding are usually expressed in terms of the rate constants k_a (1/M s) and k_d (1/s) of the interaction or as their ratio the affinity, K_D (M). These can easily be determined using Biacore systems and this has been extensively described in many good articles *(4–6)*. The rest of this chapter focuses on how to find pairs of antibodies with different binding epitopes using Biacore instruments (*see* **Note 1**). Although the methods already described work effectively and are well established, there is a growing need for higher content information earlier in the screening process. The main driver is to enable better-informed decisions about the selection of antibodies from the screening step. This can be achieved by looking at not only affinity but also kinetic properties, to account for the dynamic nature of the interaction as well as trying to find antibodies with unique epitope specificities. We have described a method for high throughput kinetic ranking using Biacore systems that enables screening of 384 antibodies in 12 h *(7)*. It is likely that the method can be improved and extended to allow for ten 96-well plates in 24 h.

1.3. Analysis of Epitope Mapping

The most common way to perform epitope mapping in Biacore instruments *(8)* is via pairwise mapping, using a sandwich assay where the primary antibody is captured to the surface. Antigen is injected and its binding to the primary antibody is monitored. A secondary antibody is then injected across the primary antibody–antigen complex which can bind only to a different epitope compared to the primary. Failure to bind is either due to it not recognizing the antigen at all, or because the binding epitope on the antigen is masked by the primary antibody (*see* **Note 2**). The analysis is repeated for all possible pairs among the antibodies of interest and the results are collated into a matrix, in which the positive and negative pairs are marked. The matrix is then used to map which antibodies recognize which epitopes based on how they can form pairs with other antibodies (*see* **Note 3**). Pairwise epitope mapping with labeling technologies such as ELISA and radio immuno assays requires time-consuming labeling methods that can be difficult to optimize for all the reactants involved. This is particularly true if the number of antibodies to be mapped is large. The labeling step commonly requires considerable amounts of purified antibodies for the protocols to work, which could be a limiting factor. However, protocols to circumvent these limitations are described (*see* Chapter "A Solid-Phase Mutual Inhibition Assay with Labeled Antigen"). Using Biacore systems for analysis removes the need for labeling and there is no need for purification of the interactants. (The analysis can be performed directly in hybridoma supernatants.) The fact that Biacore systems are automated makes it possible to perform large mapping matrices overnight if needed.

As an example of how to perform a pairwise epitope mapping, an analysis of eight antibodies was recently carried out using Biacore A100 (*see* **Note 4**). This instrument has four flow cells, each with five measurement spots. The flow cells are addressed in parallel enabling up to 20 different interactions to be studied per analysis cycle. The goal of the study was to find antibody pairs that bind the antigen (creatine kinase-MB) simultaneously and to rank the pairs based on binding complex stability. The analysis was done using a specific software module for epitope mapping that was installed on the system. This module allows for the binding complex to be studied with a minimum of artifacts by using the hydrodynamic addressing capabilities of the system (i.e., directed flow over specific regions of the flow cell) in a dedicated analysis sequence (**Fig. 2**).

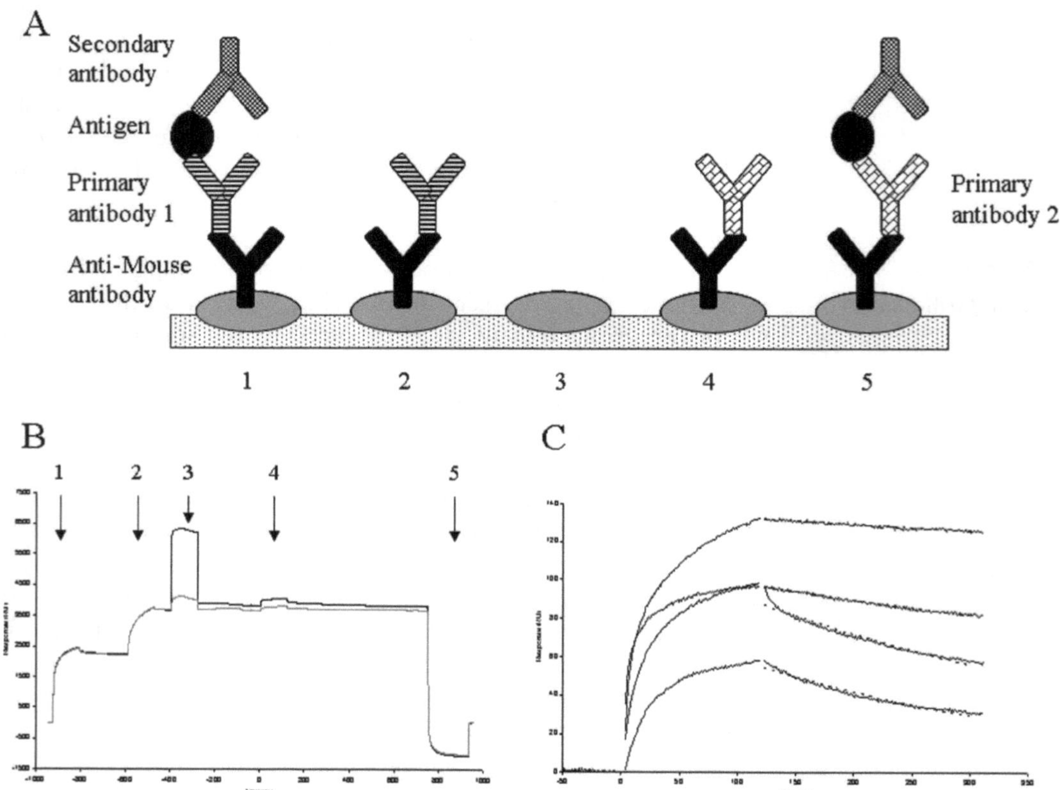

Fig. 2. (**a**) The assay format utilizes a capturing antibody covalently attached to the sensor surface on spot 1, 2, 4, and 5. The antibody is used to capture two different primary antibodies (spots 1–2 and 4–5). Antigen is then injected on spots 1 and 5. All spots are then blocked by an injection of a mixture of nonspecific antibodies. Finally, two secondary antibodies are injected on spots 1–2 and 4–5. This is done in parallel for 4 flow cells testing 8 antibody pairs per cycle. Spot 2 is used as a reference for spot 1 and spot 4 for spot 5. (**b**) Sensorgrams from spots 1 (*black*) and 2 (*grey*) from a typical analysis cycle: (1) capture of primary antibody, (2) blocking, (3) antigen, (4) secondary antibody, (5) regeneration. (**c**) The reference-subtracted sensorgram from the secondary antibody injection when 1277 was used as primary antibody. Note the differences in apparent off-rates for the different secondary antibodies, from *top* to *bottom* 1285, 1257, 1275, and 1274.

The assay format provides perfect referencing in terms of nonspecific binding and drift since the active and the reference spot are subject to exactly the same conditions except for the antigen injection. The results from the assay were evaluated in a four-step evaluation procedure using the dedicated epitope mapping module available in Biacore A100. The first step of the evaluation checks whether primary antibody has been captured to the surface and the second establishes whether it binds the antigen. Positive binding in both steps is required if the binding of the secondary antibody is to be evaluated. In the third step positive pairs are identified, and finally in the last step, the apparent off-rate for the complex is determined for each positive pair by fitting a first-order exponential decay function to the dissociation phase of the secondary antibody. This off-rate is the sum of the dissociation of the antigen–secondary antibody complex from the primary antibody and the dissociation of the secondary antibody from the antigen. The result of the analysis is presented in **Table 1**.

The results from epitope mapping experiments such as these can be difficult to interpret. From **Table 1** it is easy to state, for example, that antibodies 1275, 1277, and 1285 recognize three different epitopes since they form pairs regardless of which of them is used as primary or secondary antibody. On the other hand, 1257 and 1275 seem to recognize the same epitope since they form pairs with the same antibodies (1000, 1208, 1277,

Table 1
Results of the epitope mapping presented as a matrix, with primary antibodies in rows and secondary in columns

	1000	1208	1246	1257	1274	1275	1277	1285
1000				**$7*10^{-4}$**	$2*10^{-3}$	**$2*10^{-3}$**	$5*10^{-3}$	$4*10^{-4}$
1208				**$8*10^{-4}$**	$2*10^{-3}$	**$2*10^{-3}$**	$2*10^{-3}$	$2*10^{-4}$
1246	–	–	–	–	–	–	–	–
1257	**$2*10^{-3}$**	**$2*10^{-3}$**	$5*10^{-3}$	–	$8*10^{-3}$	$6*10^{-3}$	**$2*10^{-3}$**	**$7*10^{-4}$**
1274	–	–	–	–	–	–	–	–
1275	**$1*10^{-3}$**	**$1*10^{-3}$**					**$2*10^{-3}$**	**$8*10^{-4}$**
1277				**$9*10^{-4}$**	$3*10^{-3}$	**$2*10^{-3}$**		**$3*10^{-4}$**
1285				**$1*10^{-3}$**		**$2*10^{-3}$**	**$1*10^{-3}$**	

The diagonal is marked to show the negative control. The apparent off-rate measured during the dissociation phase after the injection of the secondary antibody is included for the positive pairs. Pairs that are positive regardless of which antibody is primary or secondary are given in bold. Negative pairs are left blank, and analyses that failed because of the first antibody not binding the antigen are marked with "–"

and 1285) but not with each other. Note that when 1257 is used as primary antibody a weak binding of 1275 is indicated, but this is a typical example where the results would need to be confirmed with a second experiment or another assay. Another type of behavior that is common in epitope mapping assays is represented by 1274, which does not recognize the antigen when used as primary antibody, but does bind when used as secondary antibody. This behavior could be due to a conformational change of the antigen that is induced by the primary antibody binding. Ranking of the stability of the antibody–antigen–antibody complexes is achieved in this case by looking at the apparent off-rate during dissociation. **Fig. 2c** shows the corresponding sensorgrams when antibody 1277 was used as primary antibody. There are two secondary antibodies (1274 and 1275) that give less stable complexes, which is clearly visible by the faster off-rate during the dissociation phase of the sensorgram. In contrast, 1257 and 1285 produce much slower dissociation phases. This type of ranking gives an indication of how the antibody pair would work in different applications. For sandwich assays with high demands on sensitivity, it is common to look for antibody pairs that give as stable complexes as possible.

2. Materials

2.1. Biacore Systems

1. Biacore A100 system (Biacore AB, Uppsala, Sweden) equipped with the Antibody Extension Package.
2. Series S Sensor Chip CM5, covered with a carboxymethylated dextran matrix suitable for immobilization of proteins (Biacore AB, Uppsala, Sweden).

2.2. Immobilization

1. Mouse antibody capture kit containing anti-mouse IgG antibodies, 10 mM sodium acetate buffer (pH 5.0) for immobilization, and 10 mM glycine (pH 1.7) for regeneration (Biacore AB).
2. Amine Coupling Kit type 2, containing the reagents *N*-ethyl-*N′*-(3-dimethylaminopropyl)carbodiimide hydrochloride, *N*-hydro-xysuccinimide, and ethanolamine-HCl suitable for coupling of proteins to the sensor chip surface via amine groups (Biacore AB).

2.3. Epitope Mapping by Pairwise Binding

1. 10× Stock of 10 mM HEPES, 150 mM NaCl, 3.4 mM EDTA, 0.05% polysorbate 20, pH 7.4 (HBS-EP+), used as running and sample dilution buffer (Biacore AB).
2. Normalization solution, 70% glycerol (Biacore AB).

3. Antibodies from an internal library directed towards creatine kinase-MB. The antibodies were used at a concentration of 50 μg/mL.
4. Creatine kinase-MB antigen diluted to 16 μg/mL (BiosPacific, Emeryville, CA).
5. Blocking solution: a mix of murine IgG1, IgG2a, IgG2b, and IgG3, at 100 μg/mL. (*see* **Note 5**)

3. Methods

The Biacore A100 contains predefined methods for the most common assay types, and in this case the method for epitope mapping is used. The manuals and help functions of the system are extensive and provide guidance on how to modify the methods to fit the specific needs of a particular experiment.

3.1. Preparation of the Instrument

1. Switch on the instrument and computer, start the control software, and follow the instructions to get the system up and running.
2. Prepare buffer according to the instructions for use. If you prepare your own buffer make sure to filter it properly.
3. Take out the sensor chip from the refrigerator and allow it to equilibrate to room temperature before inserting it into the instrument.
4. In Biacore A100, perform normalization and hydrodynamic addressing procedures according to the instructions in the software.

3.2. Immobilization

The antimouse antibody in the mouse antibody capture kit is immobilized to the sensor chip using amine coupling. A predefined method for amine coupling with a 10-min activation time is included in the system used. (Other methods are available for coupling; see the manual delivered with the instrument.). The execution of the method is fully automated and consists of the following steps:

1. A 10-min activation of spots 1 and 2 in each flow cell using a mixture of 200 mM *N*-ethyl-*N'*-(3-dimethylaminopropyl) carbodiimide hydrochloride/50 mM *N*-hydroxysuccinimide.
2. A 7-min injection of antimouse antibody diluted to 30 μg/mL in 10 mM acetate (pH 5.0) over spots 1 and 2 in each flow cell.
3. A 7-min deactivation of spots 1 and 2 in each flow cell using 1 M ethanolamine-HCl, pH 8.5.
4. Steps 1–3 are repeated for spots 4 and 5 in each flow cell.

The result is a functionalized sensor chip with antimouse antibody immobilized to ~10,000 RU in spots 1, 2, 4, and 5 in each of the four flow cells of the Biacore A100 instrument (*see* **Note 6**).

3.3. Epitope Mapping by Pairwise Binding

3.3.1. Execution

The previously prepared surface with antimouse antibody is utilized for a sandwich assay in which pairs of antibodies are tested against each other according to a mapping matrix (*see* **Table 1**). The execution of the method is fully automated and each analysis cycle consists of the following steps:

1. Antibody 1 is injected across spots 1 and 2.
2. Antibody 2 is injected across spots 4 and 5.
3. Blocking antibody is injected across all spots.
4. Antigen is injected across spot 1.
5. Antigen is injected across spot 5.
6. Antibody 3 is injected across spots 1 and 2.
7. Antibody 4 is injected across spots 4 and 5 and the dissociation of complex from the surface is followed during buffer flow.
8. Regeneration solution is injected across all spots (*see* **Note 7**).

This is done in parallel in all four flow cells until all possible combinations of pairs of antibodies have been tested.

3.3.2. Results

The results are evaluated using the epitope mapping module in the evaluation software. In short the steps are as follows:

1. All cycles where the capture level of the primary antibody is too low (<300 RU) for an analysis to be done are removed.
2. All cycles where the antigen binding level (<20 RU) is too low for an analysis to be done are removed.
3. The binding responses from the secondary antibodies are compared to the negative controls (same antibody used as primary and secondary). The level for positive pairs should be set at the level of the negative controls.
4. An exponential decaying function is fitted to the dissociation phase of the sensorgrams for the positive pairs.
5. The results are presented in a matrix (*see* **Table 1** for an example).

4. Notes

1. Epitope mapping on Biacore instruments can also be performed against immobilized peptides; with Biacore A100 up to 16

peptides can be immobilized per chip and the binding of antibodies (or other proteins) can be measured. This could easily be adapted for classical alanine scanning experiments or to find a sequential epitope in a protein by utilization of overlapping amino acid sequences.

2. Negative results could be the effect of a fast dissociation of antigen from the primary antibody and are not necessarily an effect of the two antibodies binding the same epitope. This can easily be assessed by looking at the sensorgrams.

3. Stoichiometry calculations can be a good tool to confirm the activity of the components in the assay. The response for proteins correlates to the molecular weight. If 1,000 RU of a monovalent protein is immobilized it can bind 1,000 RU of another protein with the same molecular weight, provided that it is 100% active. If the second protein has a molecular weight that is 10% of the immobilized protein the theoretical max response would instead be 100 RU.

4. The majority of Biacore instruments have a serial flow system with four flow cells. These can also be used for epitope mapping but require a slightly different assay design. The reference spot is usually omitted, creating an increased risk for misinterpretation of the results due to drift and nonspecific binding. One useful alternative is to run each pair of antibodies in two consecutive cycles one with antigen and one without, and subtract the cycle without antigen from the one with antigen in order to obtain a reference that accounts for the drift and nonspecific binding. Regardless of the precise method used, epitope mapping on Biacore instruments works well, as witnessed by references in the literature.

5. The immobilized capturing antibody recognizes both the primary and the secondary antibody in a pair. Since capture of secondary antibody gives the same increase in signal as binding to the antigen, this can lead to misinterpretation of the results. The blocking solution is therefore an important component of the assay and it should be confirmed that it blocks the surface completely.

6. The immobilization level can be adjusted by changing antibody concentration and contact time. Approximately 10,000 RU is a suitable immobilization level for capturing antibodies used in epitope mapping experiments.

7. Since the surface is used multiple times it is important that the regeneration of bound antibodies is complete, without destroying the surface. Confirm that there is no significant increase in baseline level during the assay and that sufficient levels of antibodies are captured throughout the assay.

References

1. Karlsson, R., Michaelsson, A., and Mattsson, L. (1991) Kinetic analysis of monoclonal antibody–antigen interaction with a new biosensor-based analytical system. *J. Immunol. Methods* 145, 229–240.
2. Johnsson, B., Löfås, S., Lindquist, G., Edström, A., Müller Hillgren, R. M., and Hansson, A. (1995) Comparison of methods for immobilization to carboxymethyl dextran sensor surfaces by analysis of the specific activity of monoclonal antibodies. *J. Mol. Recognit.* 8, 125–131.
3.. Kohler, G., and Milstein, C. (1975) Continuous cultures of fused cells secreting antibody of predefined specificity. *Nature* 256, 495–497.
4. Papalia, G. A., Baer, M., Luehrsen, K., Nordin, H., Flynn, P., and Myszka, D. G. (2006) High-resolution characterization of antibody fragment/antigen interactions using Biacore T100. *Anal. Biochem.* 359, 112–119.
5. Karlsson, R., and Larsson, A. (2004) Affinity measurement using surface plasmon resonance. *Methods Mol. Biol.* 248, 389–415.
6. Karlsson, R., and Falt, A. (1997) Experimental design for kinetic analysis of protein–protein interactions with surface plasmon resonance biosensors. *J. Immunol. Methods* 200, 121–133.
7. Säfsten, P., Klakamp, S. L., Drake, A. W., Karlsson, R., and Myszka, D. G. (2006) Screening antibody–antigen interactions in parallel using Biacore A100. *Anal. Biochem.* 353, 181–190.
8. Fägerstam, L. G., Frostell, A., Karlsson, R., Kullman, M., Larsson, A., Malmqvist, M., and Butt, H. (1990) Detection of antigen–antibody interactions by surface plasmon resonance. Application to epitope mapping. *J. Mol. Recognit.* 3, 208–214.

Chapter 6

Proteolytic Fragmentation for Epitope Mapping

Maria R. Mazzoni, Francesca Porchia, and Heidi E. Hamm

Summary

The use of antigen fragments generated by specific proteolytic cleavage is a relatively simple "library" approach for epitope mapping in which possible overlapping fragments are screened with the antibody on Western blots. Proteolytic fragmentation with numerous proteases having different cleavage specificites can be carrried out on native and denaturated proteins, generating a small and large number of fragments, respectively. To determine the antigenic site of a monoclonal antibody, we have examined the limited proteolytic digestion of the transducin α-subunit with four different proteases and detected antibody binding to fragments by Western blot. Using this approach, the epitope for this antibody was localized within the amino-terminal 17 residues of transducin α-subunit.

Key words: Epitope, Monoclonal antibody, Protease, Proteolytic fragmentation, Western blot.

1. Introduction

Epitope mapping defines the identification process of the molecular determinants for antibody–antigen recognition. Identification of the epitope is a key step in the characterization of monoclonal antibodies (MAb), especially those used in therapeutic strategies *(1)*. Several methods have been developed for mapping protein epitopes of MAbs, which involve competition assay (*see* Chapter "A Solid-Phase Mutual Inhibition Assay with Labeled Antigen"), partial proteolysis (*see also* Chapter "Epitope Mapping by Proteolysis of Antigen–Antibody Complexes"), expressed fragments (*see* Chapter "Epitope Mapping Using Phage-Display Random Fragment Libraries"), peptide libraries (*see* Chapters "Antibody Epitope Mapping Using De Novo Generated Synthetic Peptide Libraries" and "Epitope Mapping Using Randomly

Ulrich Reineke and Mike Schutkowski (eds.), *Methods in Molecular Biology, Epitope Mapping Protocols, vol. 524*

DOI: 10.1007/978-1-59745-450-6_6

Generated Peptide Libraries"), mass spectrometry (MS) (*see* Chapters "Epitope Mapping by Proteolysis of Antigen–Antibody Complexes" and "Epitope Mapping by Differential Chemical Modification of Antigens") and structure resolution by nuclear magnetic resonance (*see* Chapter "Epitope Mapping of Antibody–Antigen Complexes by Nuclear Magnetic Resonance Spectroscopy") and X-ray crystal diffraction (*see* Chapter "Structural Basis of Antibody–Antigen Interactions"). The gold standard for epitope definition is surely the X-ray crystal diffraction analysis of antigen–antibody complexes, but it is not readily applicable to many antigens and antibodies, and requires a very high degree of sophistication and expertise.

Proteolytic fragmentation has been used by us *(2)* and others *(3)* for mapping protein epitopes of MAbs. A protein can be cleaved chemically or enzymatically to generate various internal peptides. The number of peptides that are produced depends on whether the protein is cleaved completely at many sites or at a limited number of sites. Small numbers of large fragments are produced by limited proteolysis of native proteins. Low ratios of protease achieve efficient limited digestion of native globular proteins since cleavages tend to occur between compact structural domains. The large fragments are easily separated and purified by either SDS-polyacrylamide gel electrophoresis (SDS-PAGE) or high-performance liquid chromatography (HPLC). Proteolytic fragments separated by SDS-PAGE can be electroblotted to an immobilizing membrane and probed with the MAb. The antibody will recognize those fragments containing the epitope. Proteolytic enzymes with different substrate specificity can be used to perform limited digestion of a protein antigen, and antibody binding to the fragments can be analyzed by Western blot. However, the sequential appearance and origin of proteolytic fragments should be known. When the primary sequence of the protein is known, the origin of the proteolytic fragments is easily determined after HPLC separation by electrospray ionization time of flight mass spectrometry or N-terminal sequencing of peptides using repeated cycles of the Edman degradation reaction. The use of different proteolytic enzymes is important for an accurate identification of an antigenic determinant. A limitation to this procedure is that the denaturated protein and its proteolytic fragments bound to the immobilizing membrane may no longer contain the same conformational and structural antigenic determinants present in the native protein. This technique is not reliable to identify complex conformational epitopes but it is still useful to detect epitopes consisting of a specific amino acid sequence. In the case of conformational epitopes, the combination of immunoprecipitation and limited proteolytic digestion of the protein antigen, followed by matrix-assisted laser desorpion ionization time of flight mass spectrometry (MALDI-TOF MS) analysis, can be useful for localization of the antibody-binding region.

Another approach to epitope mapping is complete proteolytic digestion of the target protein. The choice of complete vs. limited proteolysis to localize functional epitopes depends on several factors. If the unknown region is relatively big, then a limited digest with subsequent generation of large fragments is preferable. Large protein fragments are also more likely to maintain a more native-like conformation, which may be important for epitope identification. However, many proteins have no clear patterns following limited proteolytic fragmentation, and instead generate multiple products of partial cleavage. The complex patterns produced by limited proteolysis can be difficult to interpret, making complete proteolytic cleavage advantageous. Complete proteolytic fragmentation is also instrumental when the goal is to localize a specific epitope to the smallest number of amino acid residues possible. Nowadays, the origin of the proteolytic fragments is easily identified using either ESI-TOF or MALDI-TOF MS.

Here, we report the methodology used by us *(2)* to identify the epitope of the monoclonal antibody 4A (MAb4A) on the α-subunit ($G\alpha_t$) of the heterotrimeric G protein transducin (G_t). Reaction times and reagents presented in this chapter may require some modification when a different protein antigen is under investigation.

2. Materials

2.1. Buffer and Enzymes

1. Buffer A: 10 mM 3-[*N*-morpholino]propanesulfonic acid (MOPS), pH 7.5, 200 mM NaCl, 2 mM $MgCl_2$, 1 mM dithiothreitol (*see* **Note 1**).
2. L-1-Tosylamido-2-phenylethyl chloromethyl ketone (TPCK)-treated trypsin (Worthington Biochem., Lakewood, NJ) stock solution: 2 mg/mL in 0.1 mM HCl stored at 4°C. Diluted solution: 0.08 mg/mL of TPCK-treated trypsin in buffer A containing 25% glycerol freshly made from stock solution (*see* **Note 2**).
3. Endoproteinase Arg-C and Lys-C (Roche Diagnostics, Indianapolis, IN) stock solution: aliquots of 3 mg/mL solution of either endoproteinase Arg-C or Lys-C prepared in distilled water and stored at –20°C. Diluted solutions: 1 mg/mL of endoproteinase Arg-C and 0.01 mg/mL of endoproteinase Lys-C in buffer A freshly made from stock solutions.
4. Endoproteinase Glu-C (V8 protease) (Roche Diagnostics) stock solution: 2 mg/mL freshly made in distilled water.

Diluted solution: 0.012 mg/mL of endoproteinase Glu-C in buffer A freshly made from stock solution.

5. 1-Chloro-3-tosylamido-7-amino-2-heptanone hydrochloride (TLCK)-treated chymotrypsin (Worthington Biochem.) stock solution: 2 mg/mL freshly made in distilled water. Diluted solution: 0.08 mg/mL in buffer A freshly made from stock solution.

2.2. Protease Inhibitors

1. Soybean trypsin inhibitor (Worthington Biochem.): aliquots of 8 mg/mL solution prepared in distilled water and stored at –20°C.
2. TLCK (Roche Diagnostics): stock solution (18 mg/mL) prepared freshly in distilled water.
3. TPCK (Roche Diagnostics): aliquots of 50 mM stock solution prepared in ethanol and stored at –20°C in the dark.
4. Phenylmethanesulfonyl fluoride (Sigma-Aldrich, St. Louis, MO): aliquots of 100 mM stock solution prepared in ethanol and stored at –20°C. Diluted solution (5 mM in ethanol) freshly made from stock solution.

2.3. Electroblotting

1. SDS-polyacrylamide gels of varying percentages should be prepared according to Laemmli *(4)*, using reagents of the highest quality available, and Nalgene (Rochester, NY) filters (0.2 μm) should be used to filter all electrophoresis solutions (*see* **Note 3**).
2. Electrophoresis buffer: 3 g Tris-base (Tris ultra pure, MP Biomedicals, Solon, OH), 14.4 g glycine (glycine electrophoresis grade, MP Biomedicals), and 2 g SDS (sodium dodecyl sulfate ultrapure, MP Biomedicals) in 1 L of distilled water.
3. Stock solution of sodium thioglycolate (Sigma-Aldrich): 0.1 M in distilled water. A diluted solution (0.1 mM) of sodium thioglycolate is freshly made in electrophoresis buffer from the stock solution.
4. The electroblot transfer should be made to nitrocellulose (0.1 μm) or PVDF-type membranes.
5. Electroblotting buffer: 25 mM Tris-base, 192 mM glycine, 0.2% SDS, 20% (v/v) methanol, pH 6.5.
6. TBS: 50 mM Tris-HCl (pH 8.5), 150 mM NaCl.
7. DTBS: TBS containing 0.1% Nonidet™ P 40 substitute or Igepal® CA-630 (Sigma-Aldrich).
8. OTBS: 50 mM Tris-HCl (pH 8.5), 150 mM NaCl, 3% ovalbumin centrifuged at 14,000×*g* for 20 min at 4°C. Stock solution (15%) of ovalbumin made by diluting 150 g of ovalbumin (Sigma-Aldrich) and 1 g of NaN_3 in 600 mL of distilled water. Stir overnight at room temperature and bring the final volume up to 1 L with distilled water. Centrifuge the stock solution at 14,000×*g* for 40 min at 4°C, collect the supernatant, and store aliquots at –20°C (*see* **Note 4**).

9. ^{125}I-protein A affinity purified for blotting applications (Amersham Radiochemicals, GE Healthcare UK Ltd, Little Chalfont, Buckinghamshire, UK): stock solution as provided by manufacturer, 100 μCi/mL (specific activity, 30 μCi/μg).
10. Kodak XAR-2 film (Eastman Kodak, Rochester, NY).
11. Electroblotting stock buffer CAPS: 22.13 g 3-cyclohexylamino-1-propanesulfonic acid (CAPS) (Sigma-Aldrich) in 1 L of distilled water titrated with NaOH to pH 11.0 and stored at 4°C.
12. Electroblotting buffer (10 mM CAPS in 10% MeOH): mix 200 mL of CAPS stock solution, 200 mL of methanol, and 1,600 mL of distilled water.
13. Ponceau S staining solution: 0.2% Ponceau S in 1% acetic acid. Dissolve 0.4 g of Ponceau S in 198 mL of distilled water and stir for 30 min. Add 2 mL of acetic acid to the mixture.

3. Methods

To determine the MAb4A epitope, $G\alpha_t$ both in the GDP- or GTP-bound conformation was digested with different proteases and limited proteolytic patterns were examined for antibody binding by Western blots. After purification of the proteolytic fragments, the identity was determined by amino-terminal sequencing of peptides using repeated cycles of the Edman degradation reaction. A schematic representation of limited proteolytic digestion of $G\alpha_t$ by endoproteinase Glu-C is shown in **Fig. 1**. MAb4A did not bind to the $G\alpha_{t38}$ fragment which lacks the amino-terminal 21 residues of $G\alpha_t$ (**Fig. 1**, right panel; **Fig. 2**). As indicated in **Fig. 2**, proteolytic cleavage sites of $G\alpha_t$ are located in three regions: near the amino-terminus at Leu^{15}-Lys^{25}, in the switch region II at Arg^{204}-Trp^{207}, and near the carboxyl-terminus at Arg^{310}. Limited proteolytic cleavage of $G\alpha_t$ with trypsin, chymotrypsin, and endoproteinase Arg-C gave more complex proteolytic patterns but the requirement of the amino-terminal residues for antibody binding was confirmed. More specifically, the MAb4A epitope is localized within $G\alpha_t$ amino-terminal region, from Gly^2 to Lys^{17}. Further studies using synthetic peptides in a competition ELISA confirmed the epitope localization and indicated that the myristoyl moiety bound Gly^2 is part of the antigenic determinant (unpublished observations).

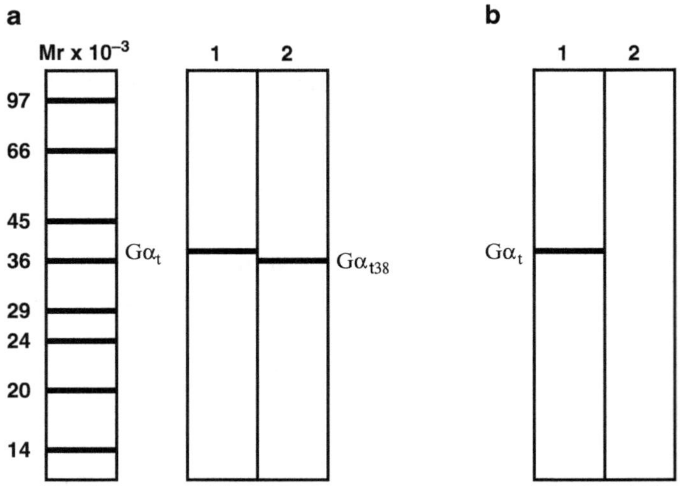

Fig. 1. Schematic representation of the limited proteolytic digestion pattern of $G\alpha_t$ by endoproteinase Glu-C. (**a**) Coomassie-blue-stained gel; (**b**) immunoblot with MAb4A. *Lane 1:* uncleaved $G\alpha_t$. *Lane 2:* $G\alpha_t$ cleaved with endoproteinase Glu-C.

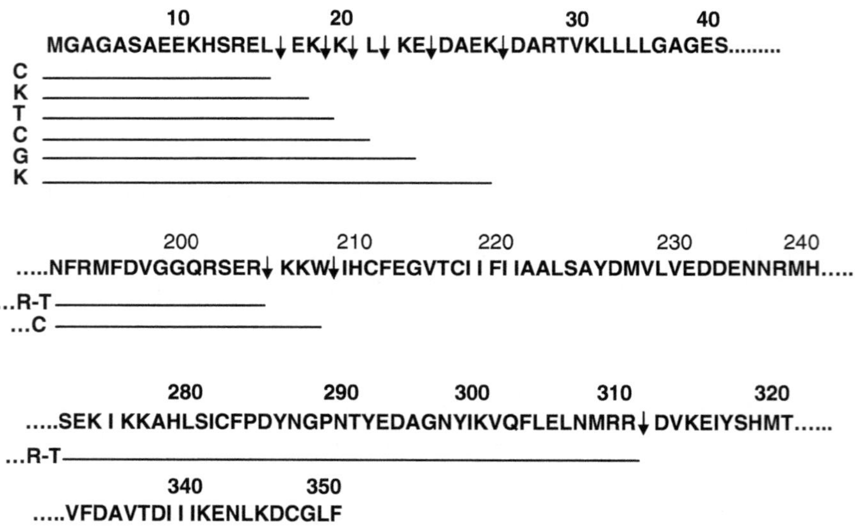

Fig. 2. Proteolytic cleavage sites on native $G\alpha_t$. *C* chymotrypsin, *G* endoproteinase Glu-C, *K* endoproteinase Lys-C, *R* endoproteinase Arg-C, *T* trypsin.

3.1. Limited Proteolytic Digestion

3.1.1. Trypsin

1. Incubate the protein in buffer A containing 25% glycerol with an equal volume of TPCK-treated trypsin (protease-to-protein molar ratio, 1:7) at 0°C.
2. Stop the reaction at specific time-points by incubating an aliquot with trypsin inhibitor at a trypsin-to-trypsin inhibitor ratio of 1:10 (w/w) (*see* **Note 5**).
3. Incubate the aliquots at 0°C for 5 min (*see* **Note 6**), add electrophoresis sample buffer *(4)*, and boil the samples for 5 min before analysis by SDS-PAGE.

3.1.2. Endoproteinase Arg-C and Lys-C

1. Perform the proteolytic digestion with endoproteinase Arg-C at a protease-to-protein molar ratio of 1:1.8 and with endoproteinase Lys-C at a protease-to-protein molar ratio of 1:40 in buffer A at room temperature.
2. At each time-point, remove an aliquot and stop the reaction by the addition of TLCK to a final concentration of 54 μg/mL.
3. After 5 min at 0°C (*see* **Note 6**), add electrophoresis sample buffer *(4)* and boil the samples for 5 min before separation by SDS-PAGE (*see* **Note 7**).

3.1.3. Endoproteinase Glu-C

1. Perform the proteolytic digestion in buffer A at a protease-to-protein molar ratio of 1:32 for 2 h at room temperature.
2. Stop the reaction by the addition of TLCK and phenylmethanesulfonyl fluoride (final concentration 2.4 and 2.3 mg/mL, respectively).
3. After 5 min at 0°C (*see* **Note 6**), add the electrophoresis sample buffer *(4)* and boil the samples for 5 min before separation by SDS-PAGE.

3.1.4. Chymotrypsin

1. Incubate the protein in buffer A with an equal volume of TLCK-treated chymotrypsin (protease-to-protein molar ratio, 1:7.5) at 37°C.
2. At each time point, remove an aliquot of the mixture and stop the reaction by addition of phenylmethanesulfonyl fluoride to a final concentration of 0.7 mM.
3. After 5 min at 0°C (*see* **Note 6**), add electrophoresis sample buffer *(4)* and boil the samples for 5 min.

3.2. Immunoblotting

1. Separate proteolytic fragments by SDS-PAGE as described earlier.
2. Remove the gel from the electrophoresis cell and soak it in 100 mL of electroblotting buffer, as well as the nitrocellulose to be used.
3. Carry out the electroblot transfer overnight at 4°C and at a constant voltage (30 V) in a transblot tank (*see* **Note 8**).
4. After transfer, incubate the immunoblot in OTBS for 3 h at room temperature to block nonspecific binding, followed by OTBS containing the appropriate primary antibody (50 μg/mL) (*see* **Note 9**) for 3 h to overnight, with constant orbital shaking.
5. After two washes in TBS, one wash in DTBS, and one wash in TBS, all with vigorous shaking, incubate the immunoblot in OTBS containing ^{125}I-protein A (0.5 μCi/mL) for 4 h at room temperature (*see* **Note 10**).
6. Rinse the immunoblot as described above, dry between Whatman filter paper, and then expose overnight to an autoradiography film (Kodak XAR-2 film) with an intensifying screen at –70°C.

3.3. Separation of Proteolytic Fragments by SDS-PAGE and Electroblotting to PVDF-Type Membrane

1. Carry out the limited proteolytic cleavage of the protein to completion as described above.
2. Prepare SDS-polyacrylamide gels, age them at 0°C for at least 24 h, pre-run with 0.1 mM sodium thioglycolate (a scavenger of free radicals) for 4 h, and then load the samples (*see* **Note 11**).
3. Following electrophoresis, soak the gel in 100 mL of electroblotting buffer (CAPS buffer) for 5 min. Meanwhile, soak the PVDF membrane in 100% methanol for a few seconds, followed by electroblotting buffer (CAPS buffer).
4. Carry out the electroblot transfer in a cell tank using chilled CAPS buffer at a constant voltage (50 V) at room temperature for 30 min (*see* **Note 12**).
5. After transfer is complete, remove the membrane from the transblotting sandwich, and rinse with distilled water before staining.
6. Detect proteins on PDVF-type membrane with a conventional staining technique, such as Coomassie brilliant blue, Ponceau S, or Amido black (*see* **Note 13**).
7. After destaining, excise protein bands using a clean, sharp razor (*see* **Note 14**), and place them in the sequencing machine.
8. Analyze amino acid sequence of electroblotted peptides using a pulsed liquid protein sequencer (*see* **Notes15** and **16**).

3.4. HPLC Purification of Proteolytic Fragments

1. Purify the proteolytic fragments by reverse-phase HPLC using a linear acetonitrile gradient in 0.1% trifluoroacetic acid.
2. After purification, analyze amino acid sequencing using a pulsed liquid-phase protein sequencer (*see* **Notes 15** and **16**).

4. Notes

1. MOPS buffer can be substituted with HEPES buffer if more convenient for the protein under investigation. Since $G\alpha_t$ contains a Cys residue in the carboxyl-terminal region which must be in the reduced form for the functional activity of the protein, buffer A contains 1 mM dithiothreitol as reducing agent. Therefore, the addition of dithiothreitol to the buffer is not required for other proteins under investigation.
2. We have found that the addition of glycerol slows down the tryptic digestion, leading to the appearance of transient fragments of $G\alpha_t$.

3. Precasted polyacrylamide gels (Invitrogen, Carlsbad, CA) can be also used choosing the most appropriate for the investigation and following the manufacturer's instructions. However, precasted minigels are not the best choice to obtain an efficient separation of numerous proteolytic fragments with similar molecular mass.
4. In our experience a phosphate-buffered saline (PBS: 10 mM NaH_2PO_4, pH 7.5, 0.9% NaCl) containing 3% nonfat dry milk (w/v) and 0.2% Tween-20 (v/v) (PBS/milk buffer) is also suitable as blocking solution for immunoblots. Alternatively, nonfat dry milk and ovalbumin can be substituted with 3% bovine serum albumin.
5. Alternatively, the reaction may be stopped by the addition of TLCK to a final concentration of 40–50 μg/mL.
6. Incubation of samples with protease inhibitors is an extremely important step. To prevent extra cleavages of the protein, it is necessary to block the enzyme activity completely before the addition of electrophoresis sample buffer.
7. If the proteolytic fragments are separated/purified by reverse-phase HPLC, stop the reaction by the addition of 1% trifluoroacetic acid/6 M guanidine HCl (final concentration) before loading onto the HPLC column.
8. Alternatively, the transfer may be carried out for a shorter time using a semidry electroblotting procedure. An appropriate transfer buffer should be used and transfer is usually complete within 30 min. After transfer, the gel should be stained to check the presence of residual proteins and monitor the efficiency of transfer. In our experience, the use of a tank apparatus is preferable, since the proteolytic fragments can be produced in low amounts and quantitation is important.
9. The suitable concentration of the primary antibody must be determined for each type of antibody used.
10. As an alternative to ^{125}I-labeled protein A, antigen can be visualized directly on the transfer membrane using an enzyme-conjugated secondary antibody, directed against the IgG of the species from which the primary antibody was obtained. If a peroxidase-labeled secondary antibody is used, the immunoblot signal can be visualized by the enhanced chemiluminescence detection method (ECL, GE Healthcare, UK). Using ^{125}I-labeled protein A, we obtain clean and strong signals by autoradiography.
11. Enough protein should be loaded in a well so that at least 10 pmol of sample are in a single band on the blot. We routinely separate proteins and peptides using 0.75-mm-thick full-size gels. However, minigel systems are also suitable. For maximum

separation of peptide bands, the bromophenol blue dye should be allowed to run within 1 cm of the end of the gel.

12. The use of semidry electroblotters is also satisfactory. However, the transfer time should be determined for each protein since the yield of the protein from electroblotted samples is affected by the efficiency of transfer. The electroblotting time varies with the gel thickness, molecular weight of the protein, and amperage of transfer. The transfer time can be judged empirically by staining the gel after electroblotting.
13. We use a 0.2% Ponceau S solution in 1% acetic acid. Membranes are stained in the Ponceau S staining solution with constant orbital shaking. Protein bands usually appear within 1 min and membranes are destained by rinsing with distilled water.
14. Excess of membrane should be carefully trimmed away from the stained band to give $2 \times 4\text{-mm}^2$ segments. Stained bands can be stored dry in Eppendorf tubes at –20°C.
15. Further information about the use of N-terminal sequence analysis of proteins and peptides can be obtained from Matsudaira *(5)*.
16. Proteolytic fragments can be readily identified by accurate mass measurement by ESI- or MALDI-TOF MS. MS allows de novo characterization of peptide fragments and their sequence confirmation.

References

1. Nelson, P. N., Reynolds, G. M., Waldron, E. E., War, E., Giannopoulos, K., and Murray, P. G. (2000) Monoclonal antibodies. *Mol. Pathol.* 53, 111–117.
2. Mazzoni, M. R., Malinski, J. A., and Hamm, H. E. (1991) Structural analysis of rod GTP-binding protein, Gt. *J. Biol. Chem.* 266, 14072–14081.
3. Pacholczyk, T., and Sweadner, K. J. (1997) Epitope and mimotope for an antibody to the Na, K-ATPase. *Protein Sci.* 6, 1537–1548.
4. Laemmli, U. K. (1970) Cleavage of structural proteins during the assembly of the head of bacteriophage T4. *Nature* 227, 680–685.
5. Matsudaira, P. (1990) Limited N-terminal sequence analysis. *Methods Enzymol.* 182, 602–603.

Chapter 7

Epitope Mapping by Proteolysis of Antigen–Antibody Complexes

Suraj Dhungana, Jason G. Williams, Michael B. Fessler, and Kenneth B. Tomer

Summary

The ability to accurately characterize an epitope on an antigen is essential to understand the pathogenesis of an infectious material, and for the design and development of drugs and vaccines. Emergence of a new contagious microbial or viral variant necessitates the need for robust identification and characterization of the antigenic determinant. Recent advances have made mass spectrometry (MS) a robust and sensitive analytical tool with high mass accuracy. The use of MS to characterize peptides and proteins has gained popularity in the research arena involving protein–protein interactions. Combining the modern mass spectrometric principles of protein–protein interaction studies with the classical use of limited proteolysis, a linear epitope on a peptide or a protein antigen can be accurately mapped in a short time, compared with other traditional techniques available for epitope mapping. Additionally, complete MS analyses can be achieved with very little sample consumption. Here we discuss the overall approach to characterize the detailed interaction between a linear antigen (either a peptide or a protein antigen) and its corresponding monoclonal antibody by using MS. The steps involved in epitope excision, epitope extraction, and indirect immunosorption are outlined thoroughly. Conditions required for MS analysis using either matrix assisted laser desorption ionization (MALDI) or electrospray ionization (ESI) sources are summarized, with special emphasis on the experimental protocols.

Key words: Antibody, Antigen, Epitope mapping, Linear epitope, Epitope excision and extraction, Limited proteolysis, Epitope foot-printing, Mass spectrometry, MALDI-MS, ESI-MS.

1. Introduction

Antibodies are produced by the immune system as a response to an infection *(1)*. The areas between the antigen and antibody involved in the recognition are designated as the epitope on the antigen and the paratope on the antibody (*see* Chapter

Ulrich Reineke and Mike Schutkowski (eds.), *Methods in Molecular Biology, Epitope Mapping Protocols, vol. 524*

DOI: 10.1007/978-1-59745-450-6_7

"What Is a B-Cell Epitope?"). The antigenic epitopes vary significantly as several different classes of molecules, including carbohydrates, nucleic acids, peptides, and proteins, can act as antigens *(2)*. In an antibody–antigen system consisting of a peptide or a protein antigen, the epitope is the smallest number of the amino acids needed for the interaction with the antibody *(3)*. However, the antigenic determinant is often much larger than the minimum amino acid residues that directly bind to the antibody. A protein antigen with a tertiary structure has the potential to interact with its antibody in a wide variety of ways. These interactions can involve a simple short amino acid sequence (linear epitope) or a series of amino acid sequences from different parts of the antigen that are constrained together by the protein's conformation (conformational or discontinuous epitopes) *(4, 5)*.

Epitopes on protein antigens can be classified into three groups on the basis of their interaction with the antibody. The first class of epitope, the linear epitope, has a short linear sequence of amino acids that elicits the antibody response *(3)*. Epitopes that are composed of a longer linear sequence of amino acids can be conformationally constrained by disulfide bridges on the N- and C-terminal ends of the epitope or be restrained, e.g., in a helix with a hydrophobic face. This gives rise to the second group of epitopes, the conformational epitopes. The conformational constraints are not always positioned within the epitope, but often are adjacent to the epitope and critical to the antigenic determinant. The largest group of epitopes is classified as discontinuous epitopes. Discontinuous epitopes are formed when several short amino acid sequences from different parts of the antigen spati-ally organize to generate an epitope with a characteristic three-dimensional conformation *(4, 5)*. Regardless of the nature of the epitope, linear, conformational, or discontinuous, the association constant for an antigen–antibody complex is typically greater than 10^9 M^{-1} *(6)*. This high affinity constant is routinely utilized in conjunction with various analytical and biochemical techniques to decipher the epitope on an antigen and characterize antigen–antibody interactions.

The experimental approach used to map the epitope is determined by the type of the epitope present on an antigen. Given the importance and interest in understanding and identifying antigenic epitopes, a number of methods have been developed for epitope mapping. X-ray diffraction analysis of the antigen–antibody complex is by far the most precise method for characterizing the antigenic determinant *(7–11)* (*see* Chapter "Structural Basis of Antibody–Antigen Interactions"). However, growing suitable crystals of antibody-bound antigen for X-ray analysis is extremely difficult. The absence of a single method to characterize the epitope on a natively folded protein at atomic resolution has necessitated the development of different approaches that utilize a wide variety of analytical techniques, including mass spectrometry (MS) *(2, 12–19)*, nuclear magnetic resonance *(20, 21)* (*see* Chapter

"Epitope Mapping of Antibody–Antigen Complexes by Nuclear Magnetic Resonance Spectroscopy"), enzyme-linked immunosorbent assay *(22)* (*see* Chapter "A Solid-Phase Mutual Inhibition Assay with Labeled Antigen"), surface plasmon resonance *(23)* (*see* Chapter "Epitope Mapping by Surface Plasmon Resonance"), and phage display *(24)* (*see* Chapters "Epitope Mapping Using Phage Display Peptide Libraries," "Humoral Response to Immunoselected Phage-Displayed Peptides by Microarray-Based Method," and "Epitope Mapping Using Phage-Display Random Fragment Libraries"). Epitope mapping by MS-based approaches has proven to be one of the most promising of these techniques as it allows for mapping of all forms of unknown epitopes on proteins *(12–16, 25)*. Here we outline the principles of linear epitope mapping by limited proteolysis of protein antigen–antibody complexes and discuss the use and advantages of MS as an analytical technique for characterizing protein epitopes.

Proteolytic cleavage of antigen–antibody complexes provides a simple and straightforward approach to identify a linear epitope on a protein antigen *(26)*. The basic principle of limited proteolysis epitope mapping is the proteolysis of antigen when it is noncovalently bound to the antibody. Two factors contribute to the success of this principle: (1) high association constant of the antigen–antibody complex *(6)*, and *(2)* the relatively high resistance of antibodies, especially the antigen-binding domains, to proteolysis *(27)*. The high affinity between antigen and antibody allows for highly selective binding of the epitope to the antibody and allows for minimal nonspecific interactions. Moreover, proteolytic digestion of an antigen tightly bound to the antibody usually differs from that of a free antigen because of the presence of the antibody. Areas of the antigen that are shielded by the antibody are digested more slowly by proteases than areas that are exposed. This results in protection of the antigen in the immediate vicinity of the epitope. Consequently, the peptides that comprise the epitope are generated more slowly during the digestion of the antigen–antibody complex than during the digestion of the free antigen. The kinetic difference in the antigen digestion, the basis of limited proteolysis, is used to profile the peptides corresponding to the epitope and map the complete epitope on an antigen *(26)*.

Technical advances in the field of MS have made this analytical technique the method of choice for protein and peptide analysis *(28–32)*. The high sensitivity and high mass-resolving power of modern mass spectrometers have made them attractive platforms for high throughput analysis. Mass spectrometers with either matrix assisted laser desorption ionization (MALDI) or electrospray ionization (ESI) sources can be used for analyzing proteolytic peptides of antigen–antibody complexes *(2)*. The kinetic profile of antigen digestion, as a free antigen or antibody-bound complex, can easily be constructed by monitoring the molecular weight of antigen at different time points during proteolysis. Furthermore, a mass spectrometer capable of tandem mass spectrometric analysis (MS/MS) can analyze proteolytic peptides and provide amino acid sequence information. By using different concentrations of protease, varying the incubation time, and/or using different proteases, one can generate the kinetic profiles of antigen fragments consisting of peptides with varying lengths and different degradation patterns, but all containing the same epitope *(33)*. This wealth of information can then be used to accurately map the linear epitope on an antigen.

Linear protein epitope mapping by MS can be achieved by two experimental approaches, with each involving limited proteolysis of the antigen *(2, 13–16, 25)*. The first of the two approaches is epitope excision. During epitope excision, the antigen is digested by a protease while it is noncovalently bound to an immobilized antibody. The proteolytically robust paratope on the antibody protects the antigen from protease and slows its digestion. Epitope excision relies on the differential digestion of antibody-bound antigen compared to the unbound antigen. **Fig. 1** summarizes the experimental approach for the epitope excision process designed for MALDI-MS analysis. The second approach, epitope extraction, on the other hand, relies on the high affinity between the antibody and the antigenic determinant. When the antigen is digested with a different protease and passed over a column of immobilized antibody, peptides containing the antigenic determinant are bound tightest to the antibody and are retained on the column with the antibody. **Fig. 2** summarizes the primary steps involved in the epitope extraction process. For both

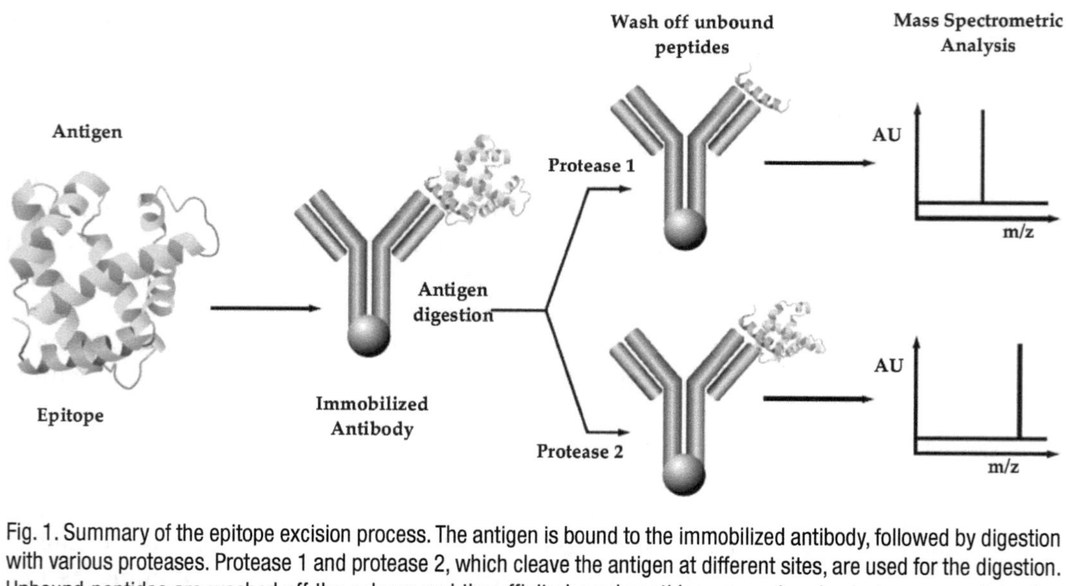

Fig. 1. Summary of the epitope excision process. The antigen is bound to the immobilized antibody, followed by digestion with various proteases. Protease 1 and protease 2, which cleave the antigen at different sites, are used for the digestion. Unbound peptides are washed off the column and the affinity-bound peptides are analyzed using mass spectrometry. For MALDI-MS analysis, the immobilized antibody–epitope complex can be directly analyzed, while the ESI-MS requires elution of epitope-containing peptides from the immune complex prior to analysis (*see Color Plates*).

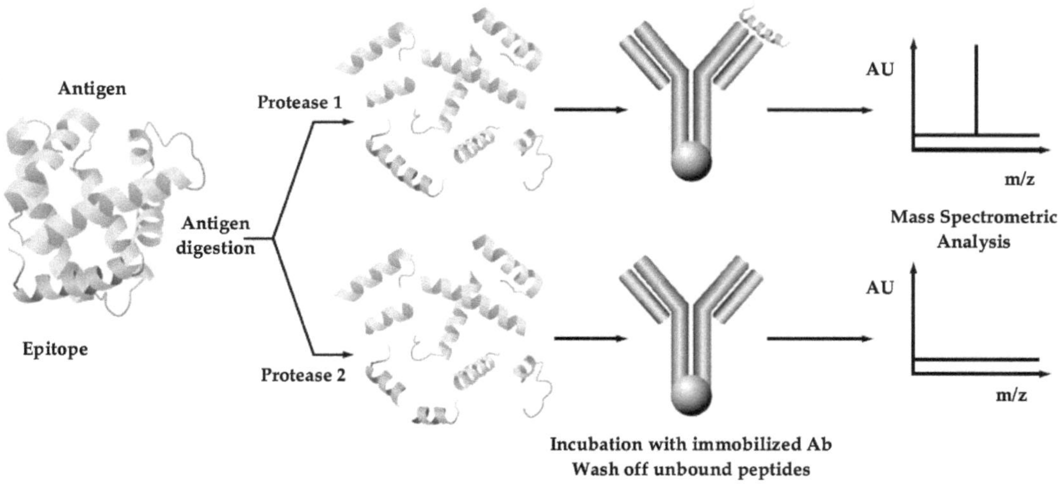

Fig. 2. Summary of the epitope extraction process. Protease 1 and protease 2, which cleave the antigen at different sites, are used to digest the antigen. The antigen digest is passed through separate columns containing immobilized antibody. The antigenic determinants of the antigen remain bound to the antibody, while the unbound peptides are washed off from the column. The affinity-bound peptides are analyzed using mass spectrometry. For MALDI-MS analysis, the immobilized antibody–epitope complex can be directly analyzed, while the ESI-MS requires elution of epitope-containing peptides from the immune complex prior to analysis (*see Color Plates*).

epitope excision and extraction, the robustness of MALDI-MS analyses allows the direct spotting of the immobilized antibody on the Sepharose beads onto the MALDI target for MS analysis *(13–16, 18, 25)*. Thus, MALDI-MS provides a rapid, clean, and straightforward way to analyze the samples. Mass spectrometers equipped with electrospray ionization mass spectrometry (ESI-MS) are also routinely used for epitope mapping experiments *(33–35)*. However, the physical requirements associated with ESI-MS prevent the use of this ionization technique to directly analyze the Sepharose-bound antibody–antigen complex. Therefore, the immune complex, consisting of immobilized antibody and the antigenic determinants, must be dissociated from the affinity matrix. The released epitope peptides can then be collected and analyzed using ESI-MS. During the analysis of eluted peptides by ESI-MS or MALDI-MS, it is extremely important to analyze both the epitope peptides and the nonepitope peptides (wash). Characterization of the peptides present in the wash is used to determine which part of the antigen is not interacting with the antibody. This information complements that acquired from the direct analysis of the epitope-containing peptides and together this information is useful for mapping the interaction between the antigen and the antibody as it allows for a confident assignment of the amino acid residues that are the antigenic determinant.

The power of mass spectrometric analysis lies in its ability to detect ions at very low sample concentrations (e.g., femtomole to picomole range). During MS analyses, however, ions corresponding to trace amounts of impurities can also be detected along with the sample ions, and the interpretation of such a data set can become difficult. Antibodies of interest are not always available as ultrapure samples. Crude solutions of antibodies, such as ascites or cell supernatants, contain impurities that will interfere not only with mass spectrometric analysis, but also with the immobilization of the antibody to the activated Sepharose beads. An improperly coupled antibody can also have the stereochemistry at its paratope altered, resulting in poor binding of the antigen. To address these challenges, Peter and Tomer have adapted indirect immunosorption in which the Fc region of the primary antibody is captured by a secondary antibody *(16)*. This approach selectively enriches for the primary antibody in the proper orientation. As with all immobilized antibody columns, once immobilized, the contaminants can be washed off the antibody column. With the secondary antibody recognizing the Fc region of the primary antibody, the latter is optimally positioned with its paratope exposed to the solution for interaction with the antigen. The antibody column for indirect immunosorption is completed by cross-linking the primary and the secondary antibodies using a lysine-specific cross-linker. This provides stability to the immunocomplex and furthermore reduces the complexity of mass spectrometric analysis by reducing further potential proteolysis of the antibodies when using the most common proteolytic enzyme, trypsin, which cannot cleave at the derivatized lysines. Major steps involved in the indirect immunosorption, followed by epitope mapping by limited proteolysis and MS, are summarized in **Fig. 3**.

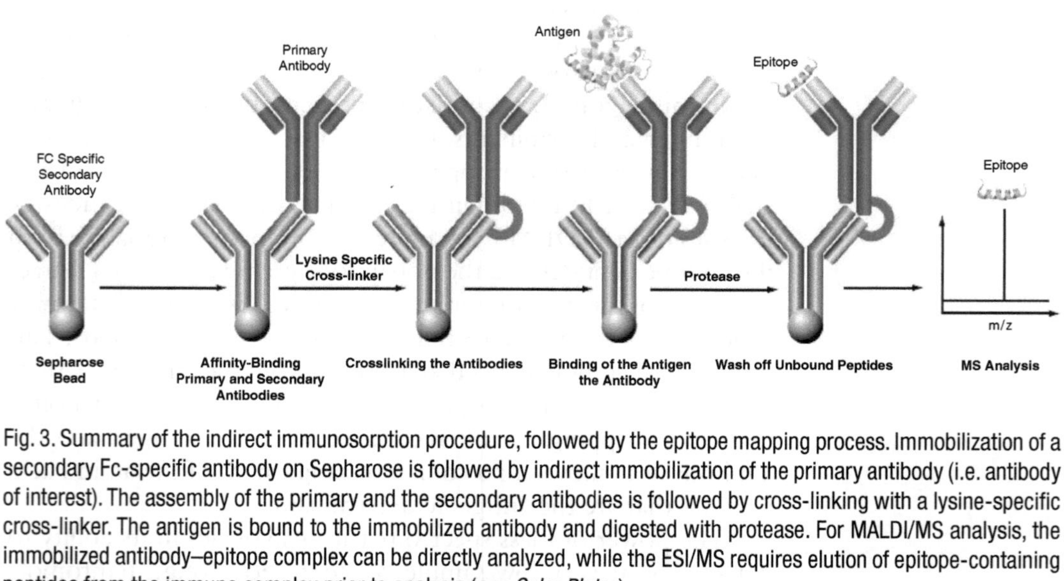

Fig. 3. Summary of the indirect immunosorption procedure, followed by the epitope mapping process. Immobilization of a secondary Fc-specific antibody on Sepharose is followed by indirect immobilization of the primary antibody (i.e. antibody of interest). The assembly of the primary and the secondary antibodies is followed by cross-linking with a lysine-specific cross-linker. The antigen is bound to the immobilized antibody and digested with protease. For MALDI/MS analysis, the immobilized antibody–epitope complex can be directly analyzed, while the ESI/MS requires elution of epitope-containing peptides from the immune complex prior to analysis (*see Color Plates*).

2. Materials

1. Purified antigen and antigen-specific monoclonal antibody.
2. Anti-Fc-specific IgG antibody as secondary antibody.
3. High-purity proteases, e.g., modified trypsin (Promega, Madison, WI) and Glu-C (Roche, Indianapolis, IN).
4. Cyanogen bromide (CNBr)-activated Sepharose 4B beads (Amersham Pharmacia, Uppsala, Sweden).
5. Compact reaction columns (CRCs) with 35-μm column filters (USB, Cleveland, OH).
6. Antibody coupling buffer: 0.1 M $NaHCO_3$, 150 mM NaCl, pH 8.2.
7. Quenching buffer: 0.1 M Tris-HCl, pH 8.0.
8. Unreacted antibody washing buffer: 0.1 M sodium acetate, 0.5 M NaCl, pH 4.0.
9. Phosphate-buffered saline (PBS): 100 mM sodium phosphate buffer, 150 mM NaCl, pH 7.2.
10. 10 mM bis(sulfosuccinimidyl)suberate (Pierce, Rockford, IL) in PBS, pH 7.2.
11. MALDI-MS matrix solution: saturated solution of recrystallized α-cyano-4-hydroxycinnamic acid in ethanol/water/concentrated formic acid (45/45/10, v/v/v).
12. 0.1% Formic acid or trifluroacetic acid.
13. Acetonitrile for LC separation.

3. Methods

3.1. Preparation of Immobilized Antibody Column

The cross-linking protocol described below is applicable to the initial immobilization of both the primary and the secondary antibody onto Sepharose beads.

1. Place ~0.2 g of dry CNBr-activated Sepharose 4B beads into a Falcon– tube and allow it to swell in 10 mL of deionized water (*see* **Note 1**).
2. For each sample, pipet 20 μL of beads slurry into each of two compact reaction columns (CRC).
3. Wash CNBr-activated Sepharose columns six times with 0.8 mL 1 mM HCl and six times with 0.4 mL 0.1 M $NaHCO_3$, pH 8.2. At the end of the washes, drain all liquid from the beads.
4. To the first CRC, add 80 μL of antibody-coupling buffer (0.1 M $NaHCO_3$/150 mM NaCl, pH 8.2), and 20 μL of

antibody (50 μg/mL) of choice (a primary antibody for the direct immunosorption or a secondary antibody for the indirect immunosorption). To the second CRC, which will serve as the control, add 100 μL of PBS. Incubate at room temperature with slow rotation for 1.5 h.

5. Drain the liquid from the CRCs and block any unreacted sites with quenching buffer (0.1 M Tris-HCl, pH 8.0). Then, rinse the beads once with 0.4 mL of quenching buffer, followed by the incubation of the beads in a second aliquot of 0.4 mL of quenching buffer for 1 h at room temperature with slow rotation.
6. Use a series of washing steps to remove unbound antibody. First, wash the column with 0.4 mL of washing buffer and then with 0.4 mL of quenching buffer. Repeat the process three times.
7. Follow up by further washing the column three times with 0.4 mL PBS.
8. At this point, MALDI-MS spectra of the beads from either CRC can be obtained. No ions should be detected if the antibody is fully cross-linked and any free antibody is thoroughly washed off the column (*see* **Notes 2 and 3**).

3.2. Epitope Excision

3.2.1. Preparation of Immobilized Antibody–Antigen Complex

1. Take 50 μg of antigen solution in 200 μL and incubate with the "immobilized antibody" column and the "control" column. Rotate slowly at room temperature for 2 h.
2. Drain the liquid and wash the beads three times with 0.4 mL PBS.
3. Obtain MALDI-MS spectra of 0.5-μL aliquots of the antibody-containing beads and of the control beads. The antibody-containing beads will bind the antigen; thus the MALDI-MS spectrum will show ions corresponding to the antigen. The control beads should not bind any antigen and should not show any peak on the MALDI-MS spectrum. If antigen peaks are observed in the MALDI-MS spectrum of the control beads, it suggests insufficient washing and **step 2** above should be repeated.

3.2.2. On-Column Proteolysis

1. The on-column digestion of affinity-bound antigens is carried out by adding a protease of choice to the CRC columns and rotating the sample at the temperature desired for maximal protease activity. The concentration of protease used in the digestion, the length of digestion time (*see* **Note 4**), and the buffer used to prepare the protease stock solution are largely dependent on the protease chosen for the experiment (*see* **Note 5**).
2. Drain the liquid and wash the beads three times with 0.4 mL PBS.

3. Obtain MALDI-MS spectra of a 0.5-μL aliquot of the beads containing the immobilized antibody and the antigenic determinant and that of the control beads. As an additional control, obtain MALDI-MS spectra of a 0.5-μL aliquot of the wash. If using LC/ESI-MS for the analysis, the antigenic peptides must be eluted from the CRCs prior to analysis (*see* **Subheading 3.5.2**).

3.3. Epitope Extraction

3.3.1. Proteolysis of Antigen

1. Take two aliquots, ca. 50 μg of antigen in solution (200 μL), in two different vials and digest using two different proteases.
2. Digest the antigen by adding the protease of choice to the CRC columns and rotating the sample at the temperature desired for maximal protease activity. The concentration of protease used in the digestion, the length of digestion time (*see* **Note 4**), and the buffer used to prepare the protease stock solution are largely dependent on the protease chosen for the experiment (*see* **Note 5**).
3. To determine the autodigest products of the proteases, prepare the protease solutions in a manner identical as for the antigen and incubate under identical conditions. The autodigest products can be analyzed and characterized by either MALDI-MS/MS or LC/ESI-MS/MS (*see* **Subheading 3.5** for the details). Independent identification of autodigest products assists in the interpretation of data from mass spectrometric analyses.

3.3.2. Preparation of Immobilized Antibody–Antigen Complex

1. Take each entire volume of antigen digest obtained above (**Subheading 3.3.1**) and divide in two aliquots.
2. Incubate the first aliquot with immobilized antibody CRC. Following the preparation of the antibody-immobilized CRC column (**Subheading 3.1**), the immobilized antibody can be divided into subaliquots to make additional CRCs, as needed. Each new CRC can be made by adding 10 μL of initial bead slurry.
3. As a control, also incubate the second aliquot of the digested antigen in a "Control CRC," containing just the Sepharose 4B beads and not the antibody.
4. Carry out the incubation for 2 h at room temperature while slowly rotating the sample.
5. Drain the liquid and wash the beads three times with 0.4 mL PBS.
6. Obtain a MALDI-MS spectrum of a 0.5-μL aliquot of the antigen-containing beads and of the control beads and the "wash/drain." If using LC/ESI-MS for analysis and characterization, the antigenic peptides must be eluted from the CRCs prior to analysis (*see* **Subheading 3.5.2**) (*see* **Note 6**).

3.4. Indirect Immunosorption

1. Immobilize the secondary antibody, anti-Fc-specific, to the CNBr-activated Sepharose 4B following the steps outlined in **Subheading 3.1**. Prepare a sufficient amount such that antibody-immobilized beads can be divided into two or more CRCs with immobilized secondary antibody (for multiple experiment and control columns).
2. Add 50 μL of primary antibody solution (equivalent to 50 μg of primary antibody) to the first CRC column containing the immobilized anti-Fc-specific secondary antibody. To the second CRC (the control column) add 50 μL PBS. Incubate at room temperature with slow rotation for 1 h.
3. Drain the liquid from the beads and save the solution (*see* **Note 7**).
4. Wash the beads three times with 0.4 mL PBS.
5. Obtain MALDI-MS spectra of a 0.5-μL aliquot of the beads to determine whether or not the primary antibody was successfully affinity-bound to the secondary antibody. At this point, ions corresponding to the primary antibody should be observed.
6. Cross-link the primary and the secondary antibodies by adding 10 μL of freshly prepared 10 mM solution of bis(sulfosuccinimidyl)suberate in PBS to each CRC and incubate 45 min at room temperature with slow rotation. Treat both the experimental and the control vial with the cross-linker.
7. To quench the cross-linking reaction, drain the beads and wash twice with the quenching buffer.
8. Wash the beads three times with 0.4 mL PBS.
9. Obtain spectra of a 0.5-μL aliquot of the beads by MALDI-MS. No ions from the antibodies should be observed in the spectrum as the primary and the secondary antibodies are fully cross-linked and free antibody is thoroughly washed off the column.
10. From the first CRC vial containing the cross-linked primary and the secondary antibodies, now cross-linked, transfer about one fourth of the beads into a new CRC and add 50 μL of PBS. This vial will serve as the control for the proteolysis experiments.
11. Incubate the remaining beads in the first CRC with 200 μL of antigen solution containing 50 μg of antigen. Rotate slowly for 2 h at room temperature. Concurrently, incubate the control beads without antigen.
12. Drain the beads and rinse with PBS.
13. Perform the proteolysis experiments by adding the protease of choice to both CRCs (CRC containing the antigen with

the cross-linked antibodies as well as the control CRCs without the antigen). Proceed with the steps involved in the proteolysis, the subsequent wash, and the MS analysis, as outlined in **Subheading 3.2.2**.

Representative mass spectra obtained during indirect immunosorption experiment are summarized in **Fig. 4**.

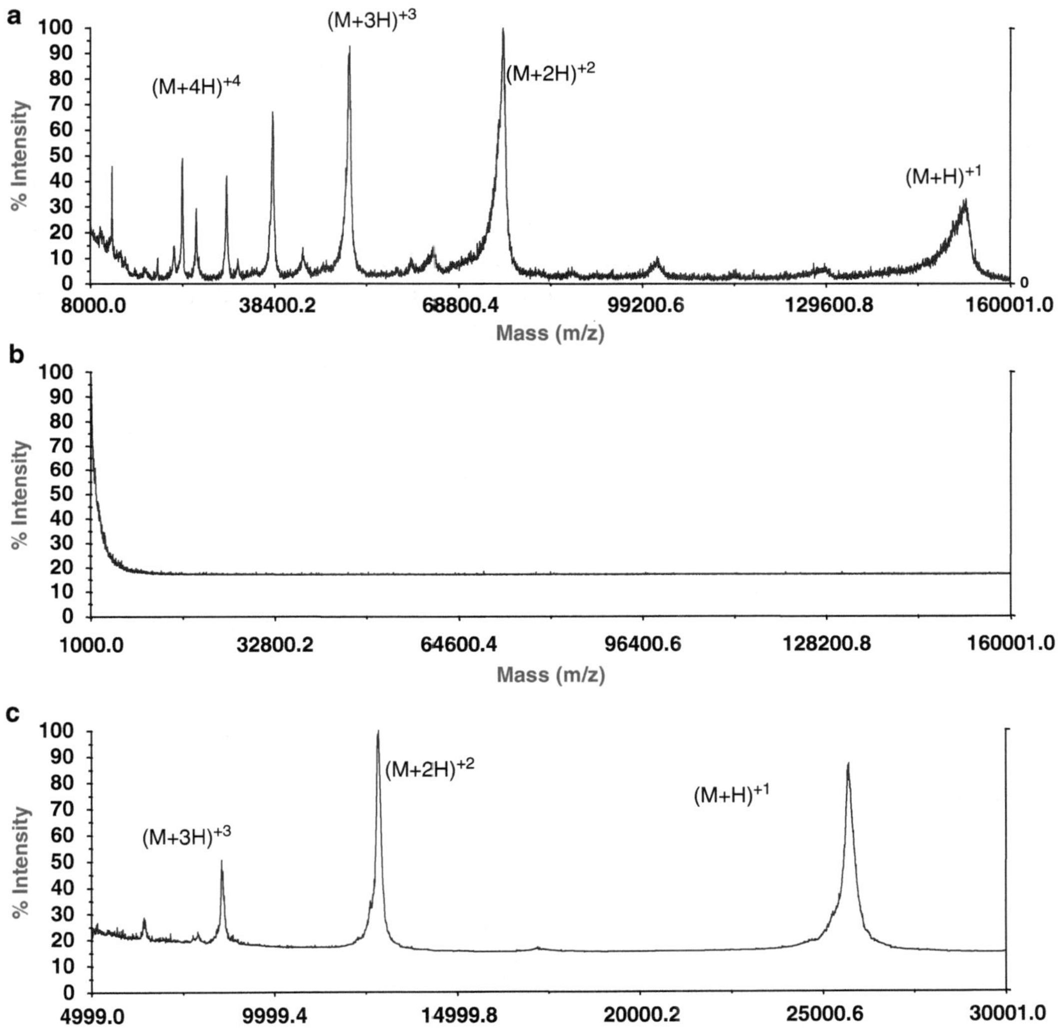

Fig. 4. Representative mass spectra obtained during indirect immunosorption. Rabbit anti-human Fc polyclonal antibodies were immobilized onto cyanogen-bromide-activated Sepharose 4B. The antibody of interest, in this example a monoclonal human antibody, 241D, against HIV p24 protein, is captured by the previously immobilized anti-human antibodies. (**a**) A MALDI-TOF spectrum of the direct analysis of the captured 241D antibody. Following the capture of the antibody of interest, the two antibodies are covalently linked with bis(sulfosuccinimidyl)suberate so that they do not interfere with subsequent analyses. (**b**) A MALDI-TOF spectrum of the prepared antibody beads after bis(sulfosuccinimidyl)suberate cross-linking. The prepared beads are now competent for capturing the antigen of interest. (**c**) A MALDI-TOF spectrum obtained by direct analysis of the prepared beads after capturing p24.

3.5. Mass Spectro-metric Analysis

3.5.1. MALDI-MS Analysis

1. All MALDI-MS analyses can be performed on a MALDI mass spectrometer that can tolerate the direct ionization off Sepharose beads. For our experiments, a Voyager DE STR mass spectrometer from Applied Biosystems was used for the direct analysis from Sepharose beads.
2. Spot a 0.5-μL aliquot of the bead slurry or the drained liquid on a stainless steel target and mix with an equal volume of the MALDI-MS matrix solution.
3. Allow the sample on the target to air dry at room temperature.
4. For highest mass accuracy, use an external calibrant to calibrate the instrument over the mass range of interest.

3.5.2. LC/ESI-MS Analysis

1. Sepharose-bound antibodies cannot be analyzed using LC/ESI-MS analysis. Therefore, all antigenic determinants that are captured by the immobilized antibody need to be eluted prior to analysis (*see* **Note 8**).
2. (Following **Subheading 3.2.2**, **step 2**; **Subheading 3.3.2**, **step 5**; and **Subheading 3.3**, **step 12**) To dissociate the antigenic peptide from the immune complex, add 500 μL of 0.1% formic acid or trifluroacetic acid.
3. Shake the column gently for 15 min and collect the released epitope peptides in a micro reaction tube.
4. The samples can then be lyophilized and stored at –80°C until mass spectrometric analysis.
5. A typical LC gradient (water/acetonitrile gradient starting from 3% acetonitrile to 50% acetonitrile over 1 h) can be used for the LC separation of peptides (*see* **Note 9**).

4. Notes

1. Carefully refer to manufacturer's recommendations for handling and activation of CNBr-activated Sepharose 4B beads. There will be slight variations among protocols across different manufacturers.
2. During MALDI-MS analyses, any unbound antibody will be ionized and the ions will be detected. Absence of detectable ions corresponding to the antibody is a good indication that the antibody is immobilized on the beads and that there are no un-cross-linked antibodies nonspecifically interacting with the beads. If ions corresponding to the antibody are detected, then it is necessary to repeat the wash and follow-up with a second MALDI-MS analysis. If any ions are still visible after multiple washes, the CNBr-activated Sepharose 4B beads

cross-linking protocol should be reevaluated according the manufacture's guidelines.

3. This CNBr-activated Sepharose 4B beads and antibody cross-linking protocol can be used to immobilize all antibodies. The same immobilization protocol should be used in both epitope excision and epitope extraction methods discussed in this chapter.
4. The concentration of the protease used in the digestion and the time of the digestion can be varied to obtain a good kinetic profile of the proteolytic peptides. A high concentration of a protease or a longer digestion time will result in a complete digestion of the antigen. By controlling the concentration and the time, it is possible to control the digestion kinetics and consequently obtain peptides of different masses over time. This will allow for the acquisition of the most amount of information on the immune complex being studied. Additionally, different protease sources and/or manufactured batches have different catalytic activities. Therefore, the concentration of the protease used for the digestion should be adjusted accordingly.
5. It is important to note that different proteases function optimally at different conditions (pH, buffer, etc.). Thus, it is recommended that optimal proteolysis conditions are thoroughly determined prior to the epitope excision or epitope extraction. Since proteolysis is the heart of epitope excision and extraction, the protease used in these experiments must be ultra pure. Impurities present in the protease, antigen, or antibody will appear as ions in the mass spectrum and will confound the data analysis.
6. It is important to note that ions that are observed during the MS analysis of the "control" should be considered as background ions and should not be interpreted as ions corresponding to the epitope.
7. It is important to save the drained solution. The yield of primary antibody capture can be low and a longer incubation time may be required to get optimal results.
8. If LC/ESI-MS is used for the analysis of the antigenic and the nonantigenic peptides, it is important to save the last wash (**Subheading 3.2.2**, **step 2**; **Subheading 3.3.2**, **step 5**; and **Subheading 3.3**, **step 12**) that contains the peptides that are not interacting with the antibody paratope.
9. The immobilized antibody column can be regenerated if the epitope is eluted using formic acid or trifluroacetic acid. For the regeneration of the column, wash the column with 10 mL of 0.1% formic acid or trifluroacetic acid followed by 20 mL of PBS. Affinity microcolumns, when regenerated in this manner, can be reused 10–20 times without significant loss of antigen-binding capacity.

Acknowledgments

We thank Dr. Leesa J. Deterding for helpful discussions. This work was supported by Intramural Research program, National Institute of Environmental Health Sciences, NIH.

References

1. Kuby, J. (ed.) (1997) *Immunology.* W.H. Freeman and Co., New York, NY.
2. Hager-Braun, C., and Tomer, K. B. (2005) Determination of protein-derived epitopes by mass spectrometry. *Expert Rev. Proteomics* 2, 745–756.
3. Laver, W. G., Air, G. M., Webster, R. G., and Smithgill, S. J. (1990) Epitopes on protein antigens: misconceptions and realities. *Cell* 61, 553–556.
4. Harlow, E., and Lane, D. (eds.) (1988) *Antibodies–a laboratory manual.* Cold Spring Harbor, New York, NY, pp. 23–36.
5. Zhu, C. S., Liu, X. S., Feng, J. N., Zhang, W., Shen, B. F., Ou'yang, W., Cao, Y. X., and Jin, B. Q. (2006) Characterization of the neutralizing activity of three anti-human TNF monoclonal antibodies and prediction of their TNF epitopes by molecular modeling and mutant protein approach. *Immunol. Lett.* 102, 177–183.
6. Cerutti, M. L., Ferreiro, D. U., Sanguineti, S., Goldbaum, F. A., and de Prat-Gay, G. (2006) Antibody recognition of a flexible epitope at the DNA binding site of the human papillomavirus transcriptional regulator E2. *Biochemistry* 45, 15520–15528.
7. Bentley, G. A., Bhat, T. N., Boulot, G., Fischmann, T., Navaza, J., Poljak, R. J., Riottot, M. M., and Tello, D. (1989) Immunochemical and crystallographic studies of antibody D1.3 in its free, antigen-liganded, and idiotope-bound states. *Cold Spring Harb. Sym.* 54, 239–245.
8. Lescar, J., Stouracova, R., Riottot, M. M., Chitarra, V., Brynda, J., Fabry, M., Horejsi, M., Sedlacek, J., and Bentley, G. A. (1997) Three-dimensional structure of an Fab–peptide complex: structural basis of HIV-1 protease inhibition by a monoclonal antibody. *J. Mol. Biol.* 267, 1207–1222.
9. Padlan, E. A. (1994) Anatomy of the antibody molecule. *Mol. Immunol.* 31, 169–217.
10. Padlan, E. A., Abergel, C., and Tipper, J. P. (1995) Identification of specificity-determining residues in antibodies. *FASEB J.* 9, 133–139.
11. Tormo, J., Blaas, D., Parry, N. R., Rowlands, D., Stuart, D., and Fita, I. (1994) Crystal structure of a human rhinovirus neutralizing antibody complexed with a peptide derived from viral capsid protein VP2. *EMBO J.* 13, 2247–2256.
12. Hager-Braun, C., Katinger, H., and Tomer, K. B. (2006) The HIV-neutralizing monoclonal antibody 4E10 recognizes N-terminal sequences on the native antigen. *J. Immunol.* 176, 7471–7481.
13. Papac, D. I., Hoyes, J., and Tomer, K. B. (1994) Epitope mapping of the gastrin-releasing peptide/antibombesin monoclonal antibody complex by proteolysis followed by matrix-assisted laser desorption ionization mass spectrometry. *Protein Sci.* 3, 1485–1492.
14. Parker, C. E., Papac, D. I., Trojak, S. K., and Tomer, K. B. (1996) Epitope mapping by mass spectrometry: determination of an epitope on HIV-1 IIIB p26 recognized by a monoclonal antibody. *J. Immunol.* 157, 198–206.
15. Parker, C. E., and Tomer, K. B. (2002) MALDI/MS-based epitope mapping of antigens bound to immobilized antibodies. *Mol. Biotechnol.* 20, 49–62.
16. Peter, J. F., and Tomer, K. B. (2001) A general strategy for epitope mapping by direct MALDI-TOF mass spectrometry using secondary antibodies and cross-linking. *Anal. Chem.* 73, 4012–4019.
17. Purcell, A. W., and Gorman, J. J. (2001) The use of post-source decay in matrix-assisted laser desorption/ionisation mass spectrometry to delineate T cell determinants. *J. Immunol. Methods* 249, 17–31.
18. Purcell, A. W., and Gorman, J. J. (2004) Immunoproteomics: mass spectrometry-based methods to study the targets of the immune response. *Mol. Cell Proteomics* 3, 193–208.
19. Purcell, A. W., Zeng, W. G., Mifsud, N. A., Ely, L. K., MacDonald, W. A., and Jackson, D. C. (2003) Dissecting the role of peptides in the immune response: theory, practice and the application to vaccine design. *J. Pept. Sci.* 9, 255–281.

20. Anglister, J., Scherf, T., Zilber, B., Levy, R., Zvi, A., Hiller, R., and Feigelson, D. (1993) Two-dimensional NMR investigations of the interactions of antibodies with peptide antigens. *FASEB J.* 7, 1154–1162.
21. Scherf, T., and Anglister, J. (1993) A T1 rho-filtered two-dimensional transferred NOE spectrum for studying antibody interactions with peptide antigens. *Biophys. J.* 64, 754–761.
22. Butler, J. E. (2000) Solid supports in enzyme-linked immunosorbent assay and other solid-phase immunoassays. *Methods* 22, 4–23.
23. Mullett, W. M., Lai, E. P. C., and Yeung, J. M. (2000) Surface plasmon resonance-based immunoassays. *Methods* 22, 77–91.
24. Wang, L. F., and Yu, M. (2004) Epitope identification and discovery using phage display libraries: applications in vaccine development and diagnostics. *Curr. Drug Targets* 5, 1–15.
25. Williams, J. G., Tomer, K. B., Hioe, C. E., Zolla-Pazner, S., and Norris, P. J. (2006) The antigenic determinants on HIV p24 for CD4+ T cell inhibiting antibodies as determined by limited proteolysis, chemical modification, and mass spectrometry. *J. Am. Soc. Mass Spectrom.* 17, 1560–1569.
26. Jemmerson, R., and Blankenfeld, R. (1989) Affinity consideration in the design of synthetic vaccines intended to elicit antibodies. *Mol. Immunol.* 26, 301–307.
27. Parham, P. (1983) On the fragmentation of monoclonal IgG1, IgG2a, and IgG2b from BALB/c mice. *J. Immunol.* 131, 2895–2902.
28. Borchers, C., and Tomer, K. B. (1999) Characterization of the noncovalent complex of human immunodeficiency virus glycoprotein 120 with its cellular receptor CD4 by matrix-assisted laser desorption/ionization mass spectrometry. *Biochemistry* 38, 11734–11740.
29. Fenn, J. B., Mann, M., Meng, C. K., Wong, S. F., and Whitehouse, C. M. (1989) Electrospray ionization for mass spectrometry of large biomolecules. *Science* 246, 64–71.
30. Loo, J. A. (2000) Electrospray ionization mass spectrometry: a technology for studying noncovalent macromolecular complexes. *Int. J. Mass Spectrom.* 200, 175–186.
31. Sobott, F., and Robinson, C. V. (2002) Protein complexes gain momentum. *Curr. Opin. Struct. Biol.* 12, 729–734.
32. Tanaka, K. (2003) The origin of macromolecule ionization by laser irradiation. *Angew. Chem. Int. Edit.* 42, 3860–3870.
33. Suckau, D., Kohl, J., Karwath, G., Schneider, K., Casaretto, M., Bittersuermann, D., and Przybylski, M. (1990) Molecular epitope identification by limited proteolysis of an immobilized antigen–antibody complex and mass spectrometric peptide mapping. *Proc. Natl. Acad. Sci. USA* 87, 9848–9852.
34. Macht, M., Marquardt, A., Deininger, S. O., Damoc, E., Kohlmann, M., and Przybylski, M. (2004) "Affinity-proteomics": direct protein identification from biological material using mass spectrometric epitope mapping. *Anal. Bioanal. Chem.* 378, 1102–1111.
35. Suckau, D., Mak, M., and Przybylski, M. (1992) Protein surface topology-probing by selective chemical modification and mass spectrometric peptide mapping. *Proc. Natl. Acad. Sci. USA* 89, 5630–5634.

Chapter 8

Identifying Residues in Antigenic Determinants by Chemical Modification

Howard M. Reisner and Roger L. Lundblad

Summary

Chemical modification of the side chains of amino acid residues was one of the first methods developed to investigate epitopes in protein antigens. The principle of the method is that alteration of the structure of a key residue of an epitope by a chemical modification will alter reactivity with antibody by affecting either specificity or avidity or both. Chemical modification has the advantage that it can be applied to discontinuous as well as continuous epitopes and may be of value in identifying cryptic epitopes. We consider here the several recent studies that have applied site-specific chemical modification to the identification of epitopes on antigens, including the use of formaldehyde, glutaraldehyde, and acid anhydrides, to produce allergoids where determinants important to reaction with IgE are modified but the ability to elicit an IgG response is retained. It is noteworthy that modification of amino groups with charge reversal appears to be the most useful approach. The approach to the use of site-specific chemical modification as a tool for the study of protein function is discussed, and emphasis is placed on the necessity to (1) validate the specificity of modification and (2) assess potential conformational change that may occur secondary to modification. Finally, a list of chemical reagents used for protein modification is presented, together with properties and references to use.

Key words: Chemical modification, Side chain modification, Antibody, Epitope mapping, Conformational epitope, Discontinuous epitope, Linear epitope.

1. Introduction

An antigenic determinant is that region of an antigen molecule, termed an epitope, that interacts with the specific antigen-binding site in the variable region of an antibody molecule which is known as a paratope *(1, 2)*. There are a number of methods for identifying epitopes described throughout this volume, including

Ulrich Reineke and Mike Schutkowski (eds.), *Methods in Molecular Biology, Epitope Mapping Protocols, vol. 524*

DOI: 10.1007/978-1-59745-450-6_8

X-ray crystallography *(3–5)* (*see* Chapter "Structural Basis of Antibody–Antigen Interactions") and phage display *(6–8)* (*see* Chapters "Epitope Mapping Using Phage Display Peptide Libraries," "Humoral Response to Immunoselected Phage-Displayed Peptides by Microarray-Based Method," and "Epitope Mapping Using Phage-Display Random Fragment Libraries"), within the developing field of structural immunology *(9, 10)*.

Site-specific chemical modification has been useful in defining structure–function relationships in proteins *(11, 12)*. Chemical modification of the side chains of residues in protein antigens was one of the first methods developed to investigate epitopes. Together with proteolytic fragmentation (*see* also Chapters "Proteolytic Fragmentation for Epitope Mapping" and "Epitope Mapping by Proteolysis of Antigen–Antibody Complexes"), it played a major role in the pioneering efforts of Atassi and others to assign antigenic determinants on the surfaces of lysozyme and myoglobin *(13, 14)*. The principle of the method is that alteration of the structure of a key residue of an epitope by a chemical modification reagent will greatly change its reactivity with an antibody to that epitope. Chemical modification has the advantage that it can be applied to discontinuous as well as continuous epitopes *(15, 16)*. Chemical modification can also be of value in identifying cryptic epitopes *(17)*. Moreover, continuous epitopes can be conformationally constrained in the context of the folded protein and may not always adopt a recognizable conformation when removed from that context. It is usually necessary to combine chemical modification with other technologies to identify a particular residue antigen, rather than a residue type, as being a key epitope constituent. Hence, it is necessarily an adjunct to other tools that can locate the individual residue of the type being modified. The exception is when a particular type of residue occurs infrequently, i.e., only once, perhaps twice, in the protein. This is more common in smaller proteins and peptides or with residues that are infrequently expressed in proteins. Examples include an analysis of the roles of histidines in the expression of Gm allotypes on IgG *(18)* and investigation of a cysteine residue in HLA-B27 *(19)*. Generally, identification of the exact residue will not be achieved. Nevertheless, the range of chemical reactivities of residues has made it possible to prepare partially or even singly modified derivatives of some antigens *(20, 21)*, which thus permits direct identification of key residues.

Neurath and others *(22)* used a variety of chemical reagents, including 1,2-cyclohexanedione (arginine), citraconic anhydride (amino groups including lysine), tetranitromethane (tyrosine), and carbodiimide (carboxyl groups), to identify a hepatitis B surface antigenic determinant. These investigators also used cyanogen bromide cleavage at methionine residues and 2-nitro-5-thiocyanobenzoic acid cleavage at cysteine residues. Subsequent studies by Peterson and colleagues demonstrated that the modification of lysine residues with reductive methylation did not affect monoclonal antibody binding by the surface antigen *(23)*. Studies by Neurath and others *(24)* used a different set of monoclonal antibodies to study the effect of reductive methylation and modification by citraconylation on the hepatitis B surface antigen. This study showed that reductive methylation had an effect on the surface antigenic determinant with this set of antibodies but the effect was less than that observed with citraconylation. It is important to note that modification with citraconic anhydride results in charge reversal (positive to negative) while the positive charge is maintained on reductive methylation. Neurath and co-workers have also used this technology to study the binding of a porphyrin, *meso*-teta(4-carboxyphenylporphine), to HIV-1 glycoprotein, gp120 *(25, 26)*.

Although not directed specifically at the issue of epitope identification by chemical modification, we would be remiss in not mentioning the use of chemical modification in the production of allergoids. Allergoids are modified allergens that have decreased allergenicity but unchanged immunogenicity. Thus, the administration of allergoids would result in the production of IgG "blocking" antibodies preventing the type I hypersensitive (allergic/anaphylactoid) response *(27)*, characterized by a humoral IgE response *(28)*. Marsh and co-workers introduced the concept of allergoids in 1970 *(29)*, when they demonstrated that the modification of allergens with formaldehyde resulted in the formation of allergoids. Subsequent studies have used carbamylation and modification with organic acid anhydrides *(30)*. While recombinant allergens have been developed for this use, there is still interest in the classical chemical approach for the synthesis of allergoids *(31, 32)*.

Site-specific chemical modification can be a powerful tool to study structure–function relationships in proteins. It is, however, necessary to determine whether there are gross structural changes that occur concomitantly with the chemical modification. This is of particular importance when there is more than a single residue modified. It is possible to make some independent assessment of the structural integrity of the protein after modification by various analytical methods, including size-exclusion chromatography, reverse-phase high-performance liquid chromatography, analytical ultracentrifugation, differential scanning calorimetry, and circular dichroism. The development of surface plasmon resonance technology (*see* Chapter "Epitope Mapping by Surface Plasmon Resonance") is of value for the study of the kinetics of antigen–antibody reactions *(33–41)* and is useful for evaluating the effect of epitope modification on antigen–antibody interactions in a quantitative manner.

A list of reagents for chemical modification is given in **Table 1**.

Table 1
Reagents for the chemical modification of proteins

Reagent	Specificity/Conditions[a]	MW	References
Acetic anhydride	Lysine, α-amino groups, tyrosine hydroxyl; preferred reaction is at lysine; pH 8 or greater; reaction can be "driven" to α-amino groups at pH less than 6.5. Avoid nucleophilic buffers such as Tris; hydrolysis of the reagent is an issue above pH 9.5. Acetic anhydride has been used for trace labeling in the study of protein conformation, and more recently the deuterated derivative has been used in proteomics for differential isotope tagging.	102.1	*(42–46)*
N-Acetylimidazole (1-acetylimidazole)	Tyrosine hydroxyl groups, lysine ε-amino groups, transient reaction at histidine; neutral pH.	110.1	*(47–51)*
Bromoacetamide (2-bromoacetamide)	Cysteine, reaction with active site histidine residues, also reaction with lysine, methionine, and possibly carboxylic acids. Reaction at pH 5–9, but reaction with methionine at pH 3.0. Reaction rate below pH 7.5 is usually slow as the modification of cysteine requires thiolate anion (pK_a^2 for cysteine is 8.7. Reaction is slower than iodoacetamide. A neutral reagent.	138	*(52–56)*
Bromoacetic acid (2-bromoacetic acid)	Reaction parameters similar to bromoacetamide except bromoacetic acid is a charged reagent at pH greater than 4 (pK_a is 2.7 at 25°C). Amide and acid derivatives can show different reaction patterns.	139	*(57–61)*
Bromoethylamine	Modification of sulfhydryl groups; conversion of cysteine to lysine analogue (*S*-2-aminoethylcysteine); reaction with cysteine at alkaline pH (see bromoacetamide). Reaction is reasonably specific for cysteine with possible modification at the amino-terminal α-amino group and histidine.	204.9 as HBr salt	*(62–66)*
N-Bromosuccinimide	Modification of tryptophan with some oxidative side reactions; pH 4–6.	178	*(66–71)*
2,3-Butanedione (diacetyl)	Modification of arginine residues; reversible reaction with the product stabilized by the presence of borate; reaction at alkaline pH.	86.1	*(72–76)*

Cyanate	Carbamylation of α-amino groups in proteins at alkaline pH with some preference toward the modification of N-terminal α-amino groups. Reaction also occurs at cysteine residues using 2-nitro-5-thiocyanatobenzoic acid.	NA	*(77–81)*
1,2-Cyclohexanedione	Modification of arginine in the presence of borate. At pH 7–9, the reaction is reversible; above pH 9, the reaction is irreversible with the formation of several products.	112.1	*(82–86)*
1,3-Dicyclohexyl-carbodiimide (DCC)	Modification of carboxyl groups in proteins (activates carboxyl groups which are then modified with a nucleophile such as glycine methyl ester; solubility issues have made EDC a more attractive reagent); also used for synthesis of phosphate ester bonds and peptide bonds; pH less than 5; modification requires protonated carboxyl group. The reagent has been used at more alkaline pH values. The modification of tyrosine and cysteine has been reported. DCC is an inhibitor of the proton-translocating ATPase in mitochondria and has been extensively used to characterize that activity.	206.3	*(87–91)*
Diethylpyrocarbonate	Modification of histidine residues in proteins with transient modification of tyrosine; possible reaction at amino groups. Disubstitution of histidine results in ring-opening.	162.1	*(92–96)*
1-Ethyl-3-(3-dimethylaminopropyl)-carbodiimide (EDC); *N*-(3-dimethylaminopropyl)-*N'*-ethylcarbodiimide	Modification of carboxyl groups in proteins frequently with *N*-hydroxylsuccinimide. Used for zero-length cross-linking in proteins and for the coupling of proteins to matrices and for the preparation of protein conjugates.	155.2	*(97–101)*
N-Ethylmaleimide	Modification of sulfhydryl groups via Michael addition to the maleimide ring. There are some proteins that are distinguished by their modification with *N*-ethylmaleimide, including *N*-ethylmaleimide-sensitive fusion protein and soluble *N*-ethylmaleimide-sensitive factor attachment protein receptors.	125.1	*(102–106)*
2-Hydroxy-5-nitrobenzyl bromide (Koshland's reagent)	Modification of tryptophan by alkylation of the indole ring. Reaction at acid pH (pH 2–6). Disubstitution can occur. This was one of the first "reporter groups."	232.0	*(107–111)*
N-Hydroxysuccinimide	Functional group for modification of amino groups in proteins, frequently with carboiimide for carboxyl modification and coupling in proteins.	115.1	*(112–116)*

(continued)

Table 1 (continued)

Reagent	Specificity/Conditions[a]	MW	References
2-Iminothiolane (Traut's reagent)	Insertion of a sulfydryl group into a protein via modification of a lysine residue. This reaction can be used for other amino functions and can be useful for matrix building.	137.6 as HCl	*(117–120)*
Iodoacetamide	Reaction is faster than with bromo- or chloro derivatives. Reaction characteristics similar to bromoacetamide. As with bromoacetamide, iodoacetamide is neutral.	185	*(121–126)*
Iodoacetic acid	See bromoacetic acid for reaction conditions. As with bromoacetic acid, reagent is charged at pH greater than 5 (pK_a is 3.12 at 25°C).	186	*(127–131)*
Methyl acetimidate	Modification of amino groups. Imido esters are the functional groups for a number of cross-linking agents such as dimethylsuberimidate. One of the more interesting imido esters is methyl picolinimidate. Reaction at pH 8–10.	109.6 as HCl	*(132–136)*
Methyl methane thiosulfonate	Methyl methanethiosulfonate is one of a group of alkyl methanethiosulfonate derivatives which reversibly modify cysteine residues in proteins. Reaction occurs at slightly alkaline pH (pH 7.8).	126.2	*(137–141)*
2-Nitrophenylsulfenyl chloride (*o*-nitrophenyl-sulfenyl chloride)	Modification of tryptophan residues in proteins; reaction occurs at acid pH; modified tryptophan can be converted to the 2-thioltryptophan derivatives. This modification has been used to purify tryptophan peptides from protein hydrolyzates. An analogue, 2-(trifluoromethyl)-benzenesulfenyl chloride has been developed for use in mass spectrometry.	189.6	*(142–146)*
Phenylglyoxal	Modification of arginine residues in proteins; reaction accelerated in the presence of bicarbonate buffers; reaction at alkaline pH. *p*-Hydroxyphenylglyoxal and *p*-nitrophenylglyoxal are useful derivates.	134.1 as hydrate	*(146–151)*
Sodium sulfite	Oxidative sulfitolysis to cleave disulfide bonds; conversion of cysteine to *S*-sulfocysteine.	126	*(152–156)*

Tetranitromethane	Nitration of tyrosine residues in proteins and possible cross-linking; also reacts with sulfhydryl groups; possible reaction with indole ring of tryptophan. Reaction at alkaline pH; does introduce a "reporter group" in proteins. The nitrotyrosine function can be reduced to aminotyrosine with sodium dithionite (sodium hydrosulfite). The modification with peroxynitrite is a similar reaction.	196	*(157–161)*
Trinitro-benzenesulfonic acid	Modification of amino groups in proteins; used for the determination of amino groups. Reaction at alkaline pH; also reacts with sulfhydryl groups.	293.2	*(162–166)*
Woodward's reagent K (*N*-ethyl-5-phenylisoxazolium-3-sulfonate)	Reagent used for the modification of carboxyl groups in proteins. Reaction at acidic pH.	253.3	*(166–171)*

[a]Absolute specificity of modification cannot be guaranteed. Conditions are most common with the caveat that specific buffer effects are observed

References

1. Berzofsky, J. A., and Berkower, I. J. (2003) Immunogenicity and antigen structure, in *Fundamental Immunology* (Paul, W. E., ed.), Lippincott Williams & Wilkins, Philadelphia, PA, pp. 631–683.
2. Cruse, J. M., Lewis, R. E., and Wang, H. (eds.) (2004) *Immunology Guidebook*. Elsevier, Amsterdam, The Netherlands.
3. Stewart, P. L., and Nemerow, G. R. (1997) Recent structural solutions for antibody neutralization of viruses. *Trends Microbiol.* 5, 229–233.
4. Shreder, K. (2000) Synthetic haptens as probes of antibody response and immunorecognition. *Methods* 20, 372–379.
5. Humphries, M. J., Symonds, E. J., and Mould, A. P. (2003) Mapping functional residues onto integrin crystal structures. *Curr. Opin. Struct. Biol.* 13, 236–243.
6. Mertens, P., Walgraffe, D., Laurent, T., Deschrevel, N., Letesson, J. J., and De Bolle, X. (2001) Selection of phage-displayed peptides recognized by monoclonal antibodies directed against the lipopolysaccharide of *Brucella*. *Int. Rev. Immunol.* 20, 181–199.
7. Wang, L. F., and Yu, M. (2004) Epitope identification and discovery using phage display libraries: applications in vaccine development and diagnostics. *Curr. Drug Targets* 5, 1–15.
8. Rowley, M. J., O'Connor, K., and Wijeyewickrema, L. (2004) Phage display for epitope determination: a paradigm for identifying receptor–ligand interactions. *Biotechnol. Annu. Rev.* 10, 151–188.
9. Downard, K. M. (2000) Contributions of mass spectrometry to structural immunology. *J. Mass Spectrom.* 35, 493–503.
10. Anzel, L. M., and Gaffney, B. J. (1995) Structural immunology: problems in molecular recognition. *FASEB J.* 9, 7–8.
12. Foley, T. L., and Burkart, M. D. (2007) Site-specific protein modification: advances and applications. *Curr. Opin. Chem. Biol.* 11, 12–19.
13. Atassi, M. Z. (1978) Antigenic structure of myoglobin: the complete anatomy of a protein and conclusions relating to antigenic structures of proteins. *Immunochemistry* 15, 909–936.
14. Atassi, M. Z. (1978) Precise determination of the entire antigenic structure of lysozyme. *Immunochemistry* 12, 423–438.
15. Becker, W.-M., and Reese, G. (2001) Immunologic identification and characterization of individual food allergens. *J. Chromatogr. B* 756, 131–140.
16. Restani, P., Ballabio, C., Cattaneo, A., Isoardi, P., Tenacciano, L., and Fioochi, A. (2004) Characterization of bovine serum albumin epitopes and their role in allergic reactions. *Allergy* 59 (Suppl. 78), 21–24.
17. Kalluri, R., Cantley, L. G., Kerjaschki, D., and Nielson, E. G. (2000) Reactive oxygen species expose cryptic epitopes associated with autoimmune goodpasture syndrome. *J. Biol. Chem.* 275, 20027–20032.
18. Williams, R. C., Malone, C. C., and Solomon, A. (1993) Conformational dependency of human IgG heavy chain-associated Gm allotypes. *Mol. Immunol.* 30, 341–351.
19. MacLean, L., Macey, M., Lowdell, M., Badakere, S., Whelan, W., Perret, D., and Archer, J. (1992) Sulfydryl reactivity of the HLA-B27 epitope: accessibility of the free cysteine studies by flow cytometry. *Ann. Rheum. Dis.* 51, 456–460.
20. Yang, C. C., Chang, L. S., Ong, P. L., and Tung, T. H. (1992) Immunochemical properties of *Naja Naja atra* (Taiwan cobra) phospholipase A2 using polyclonal and monoclonal antibodies. *Toxicon* 30, 151–159.
21. Lin, S. R., Chang, K. L., and Chang, C. C. (1993) Chemical modification of amino groups in cardiotoxin III from Taiwan cobra (*Naja naja atra*). *Biochem. Mol. Biol. Int.* 31, 175–184.
22. Neurath, A. R., Strick, N., and Oleszko, W. R. (1981) Localization of a hepatitis B surface antigen determinant deduced from results of chemical modification. *J. Virol. Methods* 3, 115–125.
23. Peterson, D. L., Paul, D. A., Lam, J., Tribby, I. I. E., and Achord, D. T. (1984) Antigenic structure of hepatitis B surface antigen: identification of the "d" subtype determinant by chemical modification and use of monoclonal antibodies. *J. Immunol.* 132, 920–927.
24. Neurath, A. R., Kent, S. B. H., and Strick, N. (1984) Monoclonal antibodies to hepatitis B surface antigen (HBsAg) with anti-α specificity recognize a synthetic peptide analogue (S135–155) with unmobilized lysine (141). *J. Virol. Methods* 9, 341–346.
25. Debnath, A. K., Jiang, S., Strick, N., Lin, K., Haberfield, P., and Neurath, A. R. (1994) Three-dimensional structure–activity analysis of a series of porphyrin derivatives with anti-HIV-1 activity targeted to the V3 loop of the gp120 envelope glycoprotein of the human immunodeficiency virus type 1. *J. Med. Chem.* 37, 1099–1108.
26. Neurath, A. R., Strick, N., and Debnath, A. K. (1995) Structural requirements for and consequences of an antiviral porphyrin binding to

the V3 loop of the human immunodeficiency virus (HIV-1) envelope glycoprotein gp120. *J. Mol. Recognit.* 8, 345–357.

27. Flicker, S., and Valenta, R. (2003) Renaissance of the blocking antibody concept in type I allergy. *Int. Arch. Allergy Immunol.* 132, 13–24.
28. Klaustermeyer, W. B. (ed.) (1983) *Practical Allergy and Immunology*. John Wiley & Sons, New York, NY, p. 14.
29. Marsh, D. C., Lichenstein, L. M., and Campbell, D. H. (1970) Studies on allergoids prepared from naturally occurring allergens. I. Assay of allergenicity and antigenicity of formalized rye group I components. *Immunology* 18, 705–727.
30. irkovi , T., Gavrovi -Jankulovi , M., Priši , S., Jankov, R. M., Burazer, L., Vuckovi , O., Sporci , Z., and Paranos, S. (2002) The influence of residual groups in low-molecular-weight allergens of *Artemiisic vulgaris* pollen in their allergenicity, IgE, and IgG-binding properties. *Allergy* 57, 1013–1020.
31. La Grutta, S., Arena, A., D'Anneo, W. R., Gammeri, E., Leonardi, S., Trimarchi, A., Platania, D., and La Rosa, M. (2007) Evaluation of the antiinflammatory and clinical effects of sublingual immunotherapy with carbamylated allergoids in allergic asthma with or without rhinitis. A 12-month perspective randomized, controlled, trial. *Allerg. Immunol. (Paris)* 39, 40–44.
32. Bussmann, C., Maintz, L., Hart, J., Allam, J.-P., Vrtala, S., Chen, K.-W., Bieber, T., Thomas, W. R., Valenta, R., Zuberbier, T., Sager, A., and Novak, N. (2007) Clinical improvement and immunological changes in atopic dermatitis patients undergoing subcutaneous immunotherapy with a house dust mite allergoid: a pilot study. *Clin. Exp. Allergy* 37, 1277–1285.
33. Malmqvist, M. (1993) Surface plasmon resonance for detection and measurement of antibody–antigen affinity and kinetics. *Curr. Opin. Immunol.* 5, 282–286.
34. Szabo, A., Stolz, L., and Granzow, R. (1995) Surface plasmon and its use in biomolecular interaction analysis (BIA). *Curr. Opin. Struct. Biol.* 5, 699–705.
35. Schuck, P. (1997) Reliable determination of binding affinity and kinetics using surface plasmon resonance biosensors. *Curr. Opin. Biotech.* 8, 498–502.
36. Huber, A., Demartis, S., and Neri, D. (1999) The use of biosensor technology for the engineering of antibodies and enzymes. *J. Mol. Recognit.* 12, 198–216.
37. Mullett, W. M., Lai, E. P., and Yeung, J. M. (2000) Surface plasmon resonance-based immunoassays. *Methods* 22, 77–91.
38. Van Regenmortel, M. H. (2001) Analyzing structure–function relationships with biosensors. *Cell. Mol. Life Sci.* 58, 794–800.
39. McDonnell, J. M. (2001) Surface plasmon resonance: towards an understanding of the mechanism of biological molecular recognition. *Curr. Opin. Chem. Biol.* 5, 572–577.
40. Pendley, C., Schantz, A., and Wagner, C. (2003) Immunogenicity of therapeutic monoclonal antibodies. *Curr. Opin. Mol. Ther.* 5, 172–179.
41. Pattnaik, P. (2005) Surface plasmon resonance: applications in understanding receptor–ligand interaction. *Appl. Biochem. Biotech.* 126, 79–92.
42. Giedroc, D. P., Puett, D., Sinha, S. K., and Brew, K. (1987) Calcium effects on calmodulin lysine reactivation. *Arch. Biochem. Biophys.* 252, 136–144.
43. Illy, C., Thielens, N. M., and Arlaud, G. J. (1993) Chemical characterization and location of ionic interactions involved in the assembly of the C1 complex of human complement. *J. Protein Chem.* 12, 771–781.
44. Che, F. Y., and Fricker, L. D. (2002) Quantitation of neuropeptides in Cpefat/Cpefat mice using differential isotopic tags and mass spectrometry. *Anal. Chem.* 74, 3190–3198.
45. Turner, B. T., Jr., Sabo, T. M., Wilding, D., and Maurer, M. C. (2004) Mapping of factor XIII solvent accessibility as a function of activation state using chemical modification methods. *Biochemistry* 43, 9755–9765.
46. Nam, H. W., Lee, G. Y., and Kim, Y. S. (2006) Mass spectrometric identification of K210 essential for rat malonyl-CoA decarboxylase catalysis. *J. Proteome Res.* 5, 1398–1406.
47. Scherer, H. J., Karthein, R., Strieder, S., and Ruf, H. H. (1992) Chemical modification of prostaglandin endoperoxide synthase by *N*-acetylimidazole. Effect on enzyme activities and EPR spectroscopic properties. *Eur. J. Biochem.* 205, 751–757.
48. Cymes, G. D., Igelesias, M. M., and Wolfenstein-Todel, C. (1993) Chemical modification of ovine prolactin with *N*-acetylimidazole. *Int. J. Pept. Protein Res.* 42, 33–38.
49. Vazeux, G., Iturrioz, X., Corvol, P., and Llorens-Cortes, C. (1997) A tyrosine residue essential for catalytic activity in aminopeptidase A. *Biochem. J.* 327, 883–889.
50. Pal, J. K., Bera, S. K., and Ghosh, S. K. (1999) Acetylation of α-crystallin with *N*-acetylimidazole and its influence upon the native aggregate and subunit reassembly. *Curr. Eye Res.* 19, 358–367.
51. Zhang, F., Gao, J., Weng, J., Tan, C., Ruan, K., Xu, C., and Jiang, D. (2005) Structural and functional differentiation of three groups of

tyrosine residues by acetylation of *N*-acetylimidazole in manganese stabilizing protein. *Biochemistry* 44, 719–725.

52. Horn, A., Vandenberg, C. A., and Lange, K. (1984) Statistical analysis of single sodium channels. Effects of *N*-bromoacetamide. *Biophys. J.* 45, 323–335.
53. Pallotta, B. S. (1985) *N*-Bromosuccinimide removes a calcium-dependent component of channel opening from calcium-activated potassium channels in rat skeletal muscle. *J. Gen. Physiol.* 86, 601–611.
54. Huang, R. C. (1995) Novel pharmacological properties of transient potassium currents in central neurons revealed by *N*-bromosuccinimide and other chemical modifiers. *Mol. Pharmacol.* 48, 451–458.
55. Qi, X., Lee, S. H., and Kwon, J. Y. (2003) Aminobromination of unsaturated phosphonates. *J. Org. Chem.* 68, 9140–9143.
56. Wang, H., Vath, G. M., Gleason, K. J., Hanna, P. E., and Wagner, C. R. (2004) Probing the mechanism of hamster arylamine *N*-acetyltransferase 2 acetylation by active site modification, site-directed mutagenesis, and pre-steady state and steady state kinetic studies. *Biochemistry* 43, 8234–8246.
57. Glick, D. M., Goren, H. J., and Barnard, E. A. (1967) Concurrent bromoacetate reaction at histidine and methionine residues in ribonuclease. *Biochem. J.* 102, 7c–10c.
58. Lennette, E. P., and Plapp, B. V. (1979) Kinetics of carboxymethylation of histidine hydantoin. *Biochemistry* 18, 3933–3988.
59. Shapiro, R., Strydom, D. J., Weremowicz, S., and Vallee, B. L. (1988) Sites of modification of human angiogenin by bromoacetate at pH 5.5. *Biochem. Biophys. Res. Commun.* 156, 530–536.
60. Schelte, P., Boeckler, C., Frisch, B., and Schuber, F. (2000) Differential reactivity of maleimide and bromoacetyl functions with thiols: application to the preparation of liposomal diepitope constructs. *Bioconjug. Chem.* 11, 118–123.
61. Chatani, E., Tanimizu, N., Ueno, H., and Hayashi, R. (2001) Structural and functional changes in bovine pancreatic ribonuclease A by the replacement of Phe120 with other hydrophobic residues. *J. Biochem. (Tokyo)* 129, 917–922.
62. Okazaki, K., Yamada, H., and Imoto, T. (1985) A convenient *S*-2-aminoethylation of cysteinyl residues in reduced proteins. *Anal. Biochem.* 149, 516–520.
63. Planas, A., and Kirsch, J. F. (1990) Sequential protection-modification method for selective sulfhydryl group derivatization in proteins having more than one cysteine. *Protein Eng.* 3, 625–628.
64. Bochar, D. A., Tabernero, L., Stauffacher, C. V., and Rodwell, V. W. (1999) Aminoethylcysteine can replace the function of the essential active site lysine of *Pseudomonas mevalonii* 3-hydroxy-3-methylglutaryl coenzyme A reductase. *Biochemistry* 38, 8879–8883.
65. Thevis, M., Ogorzalek Loo, R. R., and Loo, J. A. (2003) In-gel derivatization of proteins for cysteine-specific cleavages and their analysis by mass spectrometry. *J. Proteome Res.* 2, 163–172.
66. Hopkins, C. E., Hernandez, G., Lee, J. P., and Tolan, D. R. (2005) Aminoethylation in model peptides reveals conditions for maximizing thiol specificity. *Arch. Biochem. Biophys.* 443, 1–10.
67. McAllister, K. A., Marrone, L., and Clarke, A. J. (2000) The role of tryptophan residues in substrate binding to catalytic domains A and B of xylanase C from *Fibrobacter succinogenes* S85. *Biochim. Biophys. Acta* 1400, 342–352.
68. Takita, T., Nakagoshi, M., Inouye, K., and Tonomura, B. (2003) Changes observed in the amino acid activation reaction. *J. Mol. Biol.* 325, 677–685.
69. Sargisova, Y., Pierfederici, F. M., Scirè, A., Bertoli, E., Tanfani, F., Febbraio, F., Briante, R., Karapetyan, Y., and Mardanyan, S. (2004) Computational, spectroscopic, and resonant mirror biosensor analysis of the interaction of adrenodoxin with native and tryptophan-modified NADPH-adrenodoxin reductase. *Proteins* 57, 302–310.
70. Faridmoayer, A., and Scaman, C. H. (2005) Binding residues and catalytic domain of soluble *Saccharomyces cerevisiae* processing α-glucosidase I. *Glycobiology* 15, 1341–1348.
71. Kumar, A., Tyagi, N. K., and Kinne, R. K. (2007) Ligand-mediated and conformational changes and positioning of tryptophans in reconstituted human sodium/*d*-glucose cotransporter (hSGLT1) probed by tryptophan fluorescence. *Biophys. Chem.* 127, 69–77.
72. Leitner, A., and Lindner, W. (2005) Functional probing of arginine residues in proteins using mass spectrometry and an arginine-specific covalent tagging concept. *Anal. Chem.* 77, 4481–4488.
73. Foettinger, A., Leitner, A., and Lindner, W. (2005) Solid-phase capture and release of arginine peptides by selective tagging and boronate affinity chromatography. *J. Chromatogr. A* 1079, 187–196.
74. Saraiva, M. A., Borges, C. M., and Florencio, M. H. (2006) Reactions of a modified lysine with aldehydic and diketonic dicarbonyl

compounds: an electrospray mass spectrometry structure/activity study. *J. Mass Spectrom.* 41, 216–228.

75. Holm, A., Rise, F., Sessler, N., Sollid, L. M., Undheim, K., and Fleckenstein, B. (2006) Specific modification of peptide-bound citrulline residues. *Anal. Biochem.* 352, 68–76.
76. Leitner, A., Amon, S., Rizzi, A., and Lindner, W. (2007) Use of the arginine-specific butanedione/phenylboronic acid tag for analysis of peptides and protein digests using matrix-assisted laser desorption/ionization mass spectrometry. *Rapid Commun. Mass Spectrom.* 21, 1321–1330.
77. Griffey, R. H., Scavini, M., and Eaton, R. P. (1988) Characterization of the carbamino adducts of insulin. *Biophys. J.* 54, 295–300.
78. Kraus, L. M., Miyamura, S., Pecha, B. R., and Kraus, A. F., Jr. (1991) Carbamoylation of hemoglobin in uremic patients determined by antibody specific for homocitrulline (carbamoylated ε-*N*-lysine). *Mol. Immunol.* 28, 459–463.
79. Reyes, A. M., Bravo, M., Ludwig, H., Iriarte, A., and Slebe, J. C. (1993) Modification of Cys-128 of pig kidney fructose 1,6-bisphosphatase with different thiol reagents: size dependent effect on the substrate and fructose-2,6-bisphosphate interaction. *J. Protein Chem.* 12, 159–168.
80. Lapko, V. N., Smith, D. L., and Smith, J. B. (2003) Methylation and carbamylation of human gamma-crystallins. *Protein Sci.* 12, 1762–1774.
81. Jaisson, S., Lorimier, S., Ricard-Blum, S., Sockalingum, G. D., Delevallée-Forte, C., Kegelaer, G., Manfait, M., Garnotel, R., and Gillery, P. (2006) Impact of carbamylation of type I collagen conformational structure and its ability to activate human polymorphonuclear neutrophils. *Chem. Biol.* 13, 149–159.
82. Chang, L. S., Wu, P. F., Liou, J. C., Chiang-Lin, W. H., and Yang, C. C. (2004) Chemical modification of arginine residues of *Notechis scutatus scutatus* notexin. *Toxicon* 44, 491–497.
83. Masuda, T., Ide, N., and Kitabatake, N. (2005) Structure–sweetness relationship in egg white lysozyme: role of lysine and arginine residues on the elication of lysozyme sweetness. *Chem. Senses* 30, 667–681.
84. Herrman, A., Svangard, E., Claeson, P., Gullbo, J., Bohlin, L., and Göransson, U. (2006) Key role of glutamic acid for the cytotoxic activity of the cyclotide cycloviolacin O2. *Cell. Mol. Life Sci.* 63, 235–245.
85. Schwartz, M. P., Barlow, D. E., Russell, J. N., Jr., Weidkamp, K. P., Butler, J. E., D'Evelyn, M. P., and Hamers, R. J. (2006) Semiconductor surface-induced 1,3-hydrogen shift: the role of covalent vs zwitterionic character. *J. Am. Chem. Soc.* 128, 11054–11061.
86. Daniel, J., Oh, T. J., Lee, C. M., and Kolattukudy, P. E. (2007) AccD6, a member of the Fas II locus, is a functional carboxyltransferase subunit of the acyl-coenzyme A carboxylase in *Mycobacterium tuberculosis. J. Bacteriol.* 189, 911–917.
87. Azzi, A., Casey, R. P., and Nalecz, M. J. (1984) The effect of *N,N*-dicyclohexylcarbodiimide on enzymes of bioenergetic relevance. *Biochim. Biophys. Acta* 768, 209–226.
88. Dimroth, P., Matthey, U., and Kaim, G. (2000) Critical evaluation of the one- versus the two-channel model for the operation of the ATP synthase's F(o) motor. *Biochim. Biophys. Acta* 14589, 506–513.
89. Aresta, M., Dibenedetto, A., Fracchiolla, E., Giannoccaro, P., Pastore, C., Pápai, I., and Schubert, G. (2005) Mechanism of formation of organic carbonates from aliphatic alcohols and carbon dioxide under mild conditions promoted by carbodiimides. DFT calculation an experimental study. *J. Org. Chem.* 70, 6177–6186.
90. Vgenopoulou, L., Gemperli, A. C., and Steuber, J. (2006) Specific modification of a Na+ binding site in NADH:quinine oxidoreductase from *Klebsiella pneumonia* with dicyclohexylcarbodiimide. *J. Bacteriol.* 188, 3264–3272.
91. Ogino, S., Sato, Y., Yamamoto, G., Sasamoric, K., Kimurac, H., Hashidad, T., Motomiyaa, K., Jeyadevana, B., and Tohji, K. (2006) Relation of the number of cross-links and mechanical properties of multi-walled carbon nanotube films formed by a dehydration condensation reaction. *J. Phys. Chem. B* 110, 23159–23163.
92. Follmer, C., and Carlini, C. R. (2005) Effect of chemical modification of histidine on the copper-induced oligomerization of jack bean urease. *Arch. Biochem. Biophys.* 435, 15–20.
93. Colleluori, D. M., Reczkowski, R. S., Emig, F. A., Cama, E., Cox, J. D., Scolnick, L. R., Compher, K., Jude, K., Han, S., Viola, R. E., Christianson, D. W., and Ash, D. E. (2005) Probing the role of hyper-reactive histidine residue of arginase. *Arch. Biochem. Biophys.* 444, 15–26.
94. Runquist, J. A., and Miziorko, H. (2006) Functional contribution of a conserved mobile loop histidine of phosphoribulokinase. *Protein Sci.* 15, 837–842.
95. Wang, X. Y., Sun, M. L., Zhao, D. M., and Wang, M. (2006) Kinetics of inactivation of phytase (phy A) during modification of histidine residue by IAA and DEP. *Protein Peptide Lett.* 13, 565–570.

96. Nakanishi, N., Takeuchi, F., Okamoto, H., Tamura, A., Hori, H., and Tsubaki, M. (2006) Characterization of heme-coordinating histidyl residues of cytochrome *b*5 based on the reactivity with diethylpyrocarbonate: a mechanism for the opening of axial imidazole rings. *J. Biochem.* 140, 561–571.
97. Ghosh, M. K., Kildsig, D. O., and Mitra, A. K. (1989) Preparation and characterization of methotrexate–immunoglobulin conjugates. *Drug Des. Deliv.* 4, 13–25.
98. Shen, X., Lagergard, T., Yang, Y., Lindblad, M., Fredriksson, M., and Holmgren, J. (2000) Preparation and preclinical evaluation of experimental group B streptococcus type III polysaccharide-cholera toxin B subunit conjugate vaccine for intranasal immunization. *Vaccine* 19, 850–861.
99. Hafemann, B., Ghofrani, K., Gattner, H. G., Stieve, H., and Pallua, N. (2001) Cross-linking by 1-ethyl-3-(3-dimethylaminopropyl)-carbodiimide (EDC) of a collagen/elastin membrane meant to be used as a dermal substitute: effects on physical, biochemical and biological features *in vitro*. *J. Mater. Sci. Mater. Med.* 12, 437–446.
100. Zhang, R., Tang, M., Bowyer, A., Eisenthal, R., and Hubble, J. (2005) A novel pH- and ionic-strength-sensitive carboxy methyl dextran hydrogel. *Biomaterials* 26, 4677–4683.
101. Li, D., He, Q., Cui, Y., Duan, L., and Li, J. (2007) Immobilization of glucose oxidase onto gold nanoparticles with enhanced thermostability. *Biochem. Biophys. Res. Commun.* 355, 488–493.
102. Mollinedo, F., Calafat, J., Janssen, H., Martín-Martín, B., Canchado, J., Nabokina, S. M., and Gajate, C. (2006) Combinatorial SNARE complexes modulate the secretion of cytoplasmic granules in human neutrophils. *J. Immunol.* 177, 2831–2841.
103. Rogers, L. K., Leinweber, B. L, and Smith, C. V. (2006) Detection of reversible protein thiol modifications in tissue. *Anal. Biochem.* 258, 171–184.
104. Kurono, S., Kurono, T., Komori, N., Niwayama, S., and Matsumoto, H. (2006) Quantitative proteome analysis using D-labeled *N*-ethylmaleimide and 13C-labeled iodoacetamide by matrix-assisted laser desorption/ionization time-of-flight mass spectrometry. *Bioorg. Med. Chem.* 14, 8197–8209.
105. Togneri, J., Cheng, Y. S., Munson, M., Hughson, F. M., and Carr, C. M. (2006) Specific SNARE complex binding mode of the Secl/Munc-18 protein, Seclp. *Proc. Natl. Acad. Sci. USA* 103, 17730–17735.
106. Yan, L. J., Yang, S. H., Shu, H., Prokai, L., and Forster, M. J. (2007) Histochemical staining and quantification of dihydrolipoamide dehydrogenase diaphorase activity using blue native PAGE. *Electrophoresis* 28, 1036–1045.
107. Strohalm, M., Kodicek, M., and Pechar, M. (2003) Tryptophan modification by 2-hydroxy-5-nitrobenzyl bromide studied by MALDI-TOF mass spectrometry. *Biochem. Biophys. Res. Commun.* 312, 811–816.
108. Strohalm, M., Santrucek, J., Hynek, R., and Kodicek, M. (2004) Analysis of tryptophan surface accessibility in proteins by MALDI-TOF mass spectrometry. *Biochem. Biophys. Res. Commun.* 323, 1134–1138.
109. Jung, J. W., Kuk, J. H., Kim, K. Y., Jung, K. C., and Park, R. D. (2006) Purification and characterization of exo-β-D-glucosaminidase from *Aspergillus fumigatus* S-26. *Protein Expr. Purif.* 45, 125–131.
110. Tashima, I., Yoshida, T., Asada, Y., and Ohmachi, T. (2006) Purification and characterization of a novel L-2-amino-$2\Delta_2$-thiazoline-4-carboxylic acid hydrolase from *Pseudomonas* sp. Strain ON-4a expressed in *E. coli*. *Appl. Microbiol. Biotechnol.* 72, 499–507.
111. Ma, S. F., Nishikawa, M., Yabe, Y., Yamashita, F., and Hashida, M. (2006) Role of tyrosine and tryptophan in chemically modified serum albumin on its tissue distribution. *Biol. Pharm. Bull.* 29, 1926–1930.
112. Smith, G. P. (2006) Kinetics of amine modification of proteins. *Bioconjug. Chem.* 17, 501–506.
113. Adden, K., Gamble, L. J., Castner, D. G., Hoffmann, A., Gross, G., and Menzel, H. (2006) Phosphonic acid monolayers for binding of bioactive molecules to titanium surfaces. *Langmuir* 22, 8197–8204.
114. Noti, C., de Paz, J. L., Polito, L., and Seeberger, P. H. (2006) Preparation and use of microarrays containing synthetic heparin oligosaccharides for the rapid analysis of heparin-protein interactions. *Chemistry* 12, 8664–8686.
115. Kenawy, el-R., el-Newehy, M., Abdel-Hay, F., and Ottenbrite, R. M. (2007) A new degradable hydroxamate linkage for pH-controlled drug delivery. *Biomacromolecules* 8, 196–201.
116. Pandey, P., Singh, S. P., Arya, S. K., Gupta, V., Datta, M., Singh, S., and Malhotra, B. D. (2007) Application of thiolated gold nanoparticles for the enhancement of glucose oxidase activity. *Langmuir* 23, 3333–3337.

117. Gauvreau, V., Chevalier, P., Vallieres, K., Petitclerc, E., Gaudreault, R. C., and Laroche, G. (2004) Engineering surfaces for bioconjugation: developing strategies and quantifying the extent of the reactions. *Bioconjug. Chem.* 15, 1146–1156.
118. Balthasar, S., Michaelis, K., Dinauer, N., von Briesen, H., Kreuter, J., and Langer, K. (2005) Preparation and characterization of antibody modified gelatin nanoparticles as drug carrier system for uptake in lymphocytes. *Biomaterials* 26, 2723–2732.
119. Kommareddy, S., and Amiji, M. (2005) Preparation and evaluation of thiol-modified gelatin nanoparticles for intracellular DNA delivery in response to glutathione. *Bioconjug. Chem.* 16, 1423–1432.
120. Langoth, N., Kalbacher, H., Schöffmann, G., Schmerold, I., Schuh, M., Franz, S., Kurka, P., and Bernkop-Schnürch, A. (2006) Thiolated chitosans: design and *in vivo* evaluation of a mucoadhesive buccal peptide drug delivery system. *Pharmacol. Res.* 23, 573–579.
121. Fowler, J. M., Stuart, M. C., and Wong, D. K. (2007) Self-assembled layer of thiolated protein G as an immunosensor scaffold. *Anal. Chem.* 79, 350–354.
122. Jao, S. C., English Ospina, S. M., Berdis, A. J., Starke, D. W., Post, C. B., and Mieyal, J. J. (2006) Computational and mutational analysis of human glutaredoxin (thioltransferase): probing the molecular basis of the low p*K*a of cysteine 22 and its role in catalysis. *Biochemistry* 45, 4785–4796.
123. Talib, J., Beck, J. L., and Ralph, S. F. (2006) A mass spectrometric investigation of the binding of gold antiarthritic agents and the metabolite $[Au(CN)2]^-$ to human serum albumin. *J. Biol. Inorg. Chem.* 11, 559–570.
124. Rogers, L. K., Leinweber, B. L., and Smith, C. V. (2006) Detection of reversible protein thiol modification in tissues. *Anal. Biochem.* 358, 171–184.
125. Kurono, S., Kurono, T., Komori, N., Niwayama, S., and Matsumoto, H. (2006) Quantitative proteome analysis using D-labeled *N*-ethylmaleimide and 13C-labeled iodoacetanilide by matrix-assisted laser desorption/ionization time-of-flight mass spectrometry. *Bioorg. Med. Chem.* 14, 8197–8209.
126. Yang, E., and Attygalle, A. B. (2007) LC/MS characterization of undesired products formed during iodoacetamide derivatization of sulfhydryl groups of peptides. *J. Mass Spectrom.* 42, 233–243.
127. Meng, T. C., Hsu, S. F., and Tonka, N. K. (2005) Development of a modified in-gel assay to identify protein tyrosine phosphatases that are oxidized and inactivated in vivo. *Methods* 35, 28–36.
128. Morty, R. E., Shih, A. Y., Fulop, V., and Andrews, N. W. (2005) Identification of the reactive cysteine residues in oligopeptidase B from *Trypanosoma brucei. FEBS Lett.* 579, 2191–2196.
129. Hasegawa, G., Kikuchi, M., Kobayashi, Y., and Saito, Y. (2005) Synthesis and characterization of a novel reagent containing dansyl group, which specifically alkyates sulfhydryl group: an example of application for protein chemistry. *J. Biochem. Biophys. Methods* 63, 33–42.
130. Atsriku, C., Scott, G. K., Benz, C. C., and Baldwin, M. A. (2005) Reactivity of zinc finger cysteines: chemical modification with labile zinc fingers in estrogen receptors. *J. Am. Soc. Mass Spectrom.* 16, 2017–2026.
131. Chao, C. C., Chelius, D., Zhang, T., Mutumanje, E., and Ching, W. M. (2007) Insight into the virulence of *Rickettsia prowazekii* by proteomic analysis and comparison with an avirulent strain. *Biochim. Biophys. Acta* 1774, 373–381.
132. Liu, X., Alexander, C., Serrano, J., Borg, E., and Dawson, D. C. (2006) Variable reactivity of an engineered cysteine at position 338 in cystic fibrosis transmembrane conductance regulator reflects different chemical states of the thiol. *J. Biol. Chem.* 281, 8275–8285.
133. Audia, J. P., Roberts, R. A., and Winkler, H. H. (2006) Cysteine-scanning mutagenesis and thiol modification of the *Rickettsia prowazerkii* ATP/ADP translocase: characterization of the TMs IV-VII and IX-XII and their accessibility to the aqueous translocation pathway. *Biochemistry* 45, 2648–2656.
134. Tombolato, F., Ferrarini, A., and Freed, J. H. (2006) Modeling the effects of structure and dynamics of the nitroxide side chain on the ESR spectra of spin-labeled proteins. *J. Phys. Chem. B* 110, 26260–26271.
135. Karala, A. R., and Ruddock, L. W. (2007) Does *S*-methyl methanethiosulfonate trap the thiol-disulfide state of proteins? *Antioxid. Redox Signal.* 9, 527–531.
136. Thonon, D., Jacques, V., and Desreux, J. F. (2007) A gadolinium triacetic monoamide DOTA derivative with a methanesulfonate anchor group. Relaxivity properties and conjugation with albumin and thiolated particles. *Contrast Media Mol. Imaging* 2, 24–34.
137. Ishikawa, Y., Yamamoto, Y., Otsubo, M., Theg, S. M., and Tamura, N. (2002) Chemical modification of amine groups on PS II protein(s) retards photoassembly of the

photosynthetic water-oxidizing complex. *Biochemistry* 41, 1972–1980.

138. Shortreed, M. R., Lamos, S. M., and Frey, B. L. (2006) Ionizable isotopic labeling reagent for relative quantification of amine metabolites by mass spectrometry. *Anal. Chem.* 78, 6398–6403.

139. Poon, S. F., Stock, N., Payne, J. E., McGuire, A. R., Stearns, B., Yang, X., Chen, W., Munoz, B., and Smith, N. D. (2005) Novel approach to pro-drugs of lactones: water soluble imidate and ortho-ester derivatives of a furanone-based COX-2 selective inhibitor. *Bioorg. Med. Chem. Lett.* 15, 2259–2263.

140. Xu, J., Degraw, A. J., Duckworth, B. P., Lenevich, S., Tann, C.-M., Jenson, E. C., Gruber, S. J., Barany, G., and Distefano, M. D. (2006) Synthesis and reactivity of 6,7-dihydrogeranylazides: reagents for primary amine incorporation into peptides and subsequent Staudinger ligation. *Chem. Biol. Drug Des.* 68, 85–96.

141. Takaku, H., Sato, J., Ishida, H. K., Inazu, T., Ishida, H., and Kiso, M. (2006) A chemical synthesis of UDP-LacNAc and its regioisomer for finding 'oligosaccharide trans-ferases'. *Glycoconjugate J.* 23, 565–573.

142. Marche, P., Girma, J. P., Morgat, J. L., and Fromageot, P. (1975) Specific tritiation of indole derivatives by catalytic desulfenylation. Application to the labelling of tryptophan-containing peptides. *Eur. J. Biochem.* 50, 375–382.

143. Sasagawa, T., Titani, K., and Walsh, K. A. (1983) Selective isolation of tryptophan-containing peptides by hydrophobicity modulation. *Anal. Biochem.* 134, 224–229.

144. Hassani, O., Mansuella, P., Cestèle, S., Bourdeaux, M., Rochat, H., and Sampieri, F. (1999) Role of lysine and tryptophan residues in the biological activity of toxin VII (Ts γ) from the scorpion *Tityus serrulatus*. *Eur. J. Biochem.* 260, 76–86.

145. Ou, K., Kesuma, D., Ganesan, K., Yu, K., Soon, S. Y., Lee, S. Y., Goh, X. P., Hooi, M., Chen, W., Jikuya, H., Ichikawa, T., Kuyama, H., Matsuo, E., Nishimura, O., and Tan, P. (2006) Quantitative profiling of drug-associated proteomic alterations by combined 2-nitrophenylsulfenyl chloride (NBS) isotope labeling and 2DE/MS identification. *J. Proteome Res.* 5, 2194–2206.

146. Matsunaga, H., and Haginaka, J. (2006) Investigation of chiral recognition mechanism on chicken α(1)-acid glycoprotein using separation system. *J. Chromatogr. A* 1106, 124–130.

147. Johans, M., Milanesi, E., Franck, M., Johans, C., Liobikas, J., Panagiotaki, M., Greci, L., Principato, G., Kinnunen, P. K., Bernardi, P., Costantini, P., and Eriksson, O. (2005) Modification of permeability transition pore arginine(s) by phenylglyoxal derivatives in isolated mitochondria and mammalian cells. Structure–function relationship of arginine ligands. *J. Biol. Chem.* 280, 12130–12136.

148. Saraiva, M. A., Borges, C. M., and Florencio, M. H. (2006) Reactions of a modified lysine with aldehydic and diketonic dicarbonyl compounds: an electrospray mass spectrometry structure/activity study. *J. Mass Spectrom.* 41, 216–228.

149. Takazaki, S., Abe, Y., Kang, D., Li, C., Jin, X., Ueda, T., and Hamasaki, N. (2006) The functional role of arginine 901 at the C-terminus of the human anion transporter band 3 protein. *J. Biochem.* 139, 903–912.

150. Greig, N., Wyllie, S., Vickers, T. J., and Fairlamb, A. N. (2006) Trypanothione-dependent glyoxylase I in *Trypanosoma cruzi*. *Biochem. J.* 400, 217–223.

151. Ye, M., and English, A. M. (2006) Binding of polyaminocarboxylate chelators to the active-site copper inhibits the GSNO-reductase activity but not the superoxide dismutase activity of Cu, Zn-superoxide dismutase. *Biochemistry* 45, 12723–12732.

152. Damodaran, S. (1985) Estimation of disulfide bonds using 2-nitro-5-thiosulfobenzoic acid: limitations. *Anal. Biochem.* 145, 200–204.

153. Martin de Llano, J. J. and Gaviulanes, J. G. (1992) Increased electrophoretic mobility of sodium sulfite-treated jack bean urease. *Electrophoresis* 13, 300–304.

154. Emerson, D. and Ghiorse, W. C. (1993) Role of disulfide bonds in maintaining the structural integrity of the sheath of the *Leptothrix discophora* SP-6. *J. Bacteriol.* 175, 7819–7827.

155. Mukhopadhyay, A. (2000) Reversible protection of disulfide bonds followed by oxidativate folding render recombinant hCGβ highly immunogenic. *Vaccine* 18, 1802–1810.

156. Raftery, M. J. (2005) Selective detection of thiosulfate-containing peptides using tandem mass spectrometry. *Rapid Commun. Mass Spectrom.* 19, 674–682.

157. Santrucek, J., Strohalm, M., Kadlcík, V., Hynek, R., and Kodícek, M. (2004) Tyrosine residue modification studied by MALDI-TOF mass spectrometry. *Biochem. Biophys. Res. Commun.* 323, 1151–1156.

158. Negrerie, M., Martin, J. L., and Njgiem, H. O. (2005) Functionality of nitrated acetylcholine receptor: the two-step formation of nitrotyrosines reveals their differential role in effector binding. *FEBS Lett.* 579, 2643–2647.

159. Carven, G. J. and Stern, L. J. (2005) Probing the ligand-induced conformational change in HLA-DRl by selective chemical modification and mass spectrometric mapping. *Biochemistry* 44, 13625–13637.

160. Gruijthuijsen, Y. K., Grieshuber, I., Stöcklinger, A., Tischler, U., Fehrenbach, T., Weller, M. G., Vogel, L., Vieths, S., Pöschl, U., and Duschl, A. (2006) Nitration enhances the allergic potential of proteins. *Int. Arch. Allergy Immunol.* 141, 265–275.

161. Ghesquiere, B., Goethals, M., van Damme, J., Staes, A., Timmerman, E., Vandekerckhove, J., and Gevaert, K. (2006) Improved tandem mass spectrometric characterization of 3-nitrotyrosine sites in peptides. *Rapid Commun. Mass Spectrom.* 20, 2885–2893.

162. Korn, C., Scholz, S. R., Gimadutdinow, O., Pingoud, A., and Meiss, G. (2002) Involvement of conserved histidine, lysine, and tyrosine residues in the mechanism of DNA cleavage by the caspase-3 activated DNase CAD. *Nucleic Acids Res.* 30, 1325–1332.

163. Metz, B., Jiskoot, W., Hennink, W. E., Crommelin, D. J., and Kersten, G. F. (2003) Physicochemical and immunochemical techniques predict the quality of diphtheria toxoid vaccines. *Vaccine* 22, 156–167.

164. Lin, J. C., Chen, Q. X., Shi, Y., Li, S. W., and Zhao, H. (2003) The chemical modification of the essential groups of β-*N*-acetyl-D-glucosaminidase from *Turbo cornutus* Solander. *IUBMB Life* 55, 547–552.

165. Jagtap, S., and Rao, M. (2006) Conformation and microenvironment of the active site of a low-molecular weight 1,4-β-D-glucan glucanohydrolase from an alkalothermophilic *Thermomonospora* sp.: involvement of lysine and cysteine residues. *Biochem. Biophys. Res. Commun.* 347, 428–432.

166. Chang, L. S., Cheng, Y. C., and Chen, C. P. (2006) Modification of Lys-6 and Lys-65 affects the structural stability of Taiwan cobra phospholipase A2. *Protein J.* 25, 127–134.

167. Salhany, J. M., Sloan, R. L., and Cordes, K. S. (2003) The carboxyl side chain of glutamate 681 interacts with a chloride binding modifier site that allosterically modulates the dimeric conformational state of band 3 (AE1). Implications for the mechanism of anion/proton cotransport. *Biochemistry* 42, 1589–1602.

168. Kosters, H. A., and de Jongh, H. H. (2003) Spectrophotometric tool for the determination of the total carboxylate content in proteins: molar extinction coefficient of the enol ester from Woodward's reagent K reacted with protein carboxylates. *Anal. Chem.* 75, 22512–22516.

169. Carvajal, N., Uribe, E., Lopez, V., and Salas, M. (2004) Inactivation of human liver arginase by Woodward's reagent K: evidence for reaction with His141. *Protein J.* 23, 179–183.

170. Jennings, M. L. (2005) Evidence for a second binding/transport site for chloride in erythrocyte anion transporter AE1 modified at glutamate 681. *Biophys. J.* 88, 2681–2691.

171. SinhaRoy, S., Banerjee, S., Ray, M., and Ray, S. (2005) Possible involvement of glutamic and/or aspartic residue(s) and requirement of mitochondrial integrity for the protective effect of creatine against inhibition of cardiac mitochondrial respiration by methylglyoxal. *Mol. Cell. Biochem.* 271, 167–176.

Chapter 9

Epitope Mapping by Differential Chemical Modification of Antigens

Suraj Dhungana, Michael B. Fessler, and Kenneth B. Tomer

Summary

Matrix-assisted laser desorption ionization or electrospray ionization mass spectrometry combined with differential chemical modification have proven to be versatile tools for epitope mapping as well as for studying diverse protein–protein and protein–ligand interactions. Characterization of a discontinuous or a conformational epitope on an antigen demands the ability to map the three-dimensional protein surface along with the interface of two interacting proteins. Classical methods of differentially derivatizing amino acid residues have been successfully merged with highly sensitive and highly accurate mass spectrometric techniques to rapidly profile the three-dimensional protein surface and determine the surface accessibility of specific amino acid residues. Here we discuss the use of mass spectrometry to characterize discontinuous or conformational epitopes by studying antigen–antibody interactions. The steps involved in epitope mapping approaches using differential chemical modification and H/D exchange on the antigen are discussed in detail, with particular emphasis on the experimental protocols.

Key words: Antibody, Antigen, Epitope mapping, Chemical modification, Conformational and discontinuous epitope, Mass spectrometry, MALDI-MS, LC/ESI-MS, H/D exchange, Acetylation.

1. Introduction

Antigen recognition by a highly specific monoclonal antibody involves a multifaceted spatial interaction where the three-dimensional structures of antigen and antibody fully complement each other. Protein antigens almost always have conformational and/or discontinuous epitopes that are recognized by the paratope on the antibody *(1)*. Conformational and discontinuous epitopes consist of several short linear peptides that are spatially organized to form the antigenic determinant. The structural constraints

Ulrich Reineke and Mike Schutkowski (eds.), *Methods in Molecular Biology, Epitope Mapping Protocols, vol. 524*

DOI: 10.1007/978-1-59745-450-6_9

that define the three-dimensional features of the antigenic determinant on a conformational or discontinuous epitope are often complex and difficult to probe, which makes mapping of such epitopes challenging *(2, 3)*. Linear epitope mapping approaches that utilize limited proteolysis of antigen (*see* Chapters "Proteolytic Fragmentation for Epitope Mapping" and "Epitope Mapping by Proteolysis of Antigen–Antibody Complexes") are not ideal for the characterization of conformational and/or discontinuous epitopes, because the proteolytic cleavage often destroys the spatial organization of the epitope and completely abolishes the interaction between the antigen and the antibody *(3, 4)*. In the absence of a high binding affinity, epitope mapping approaches, such as epitope extraction and excision, fail. Given the shortcomings of limited proteolysis in characterizing conformational and discontinuous epitopes, alternative epitope mapping techniques involving differential chemical modification of free and antibody-bound antigen have been developed *(2, 3, 5–11)*.

Similar to limited proteolysis *(12)*, the underlying principle of the epitope mapping approach using differential chemical modification is the steric protection of amino acid residues at the interface of a monoclonal-antibody–antigen complex. The differential surface-exposure and reagent-accessibility of a protein antigen in the presence and absence of an antibody determines the rate and the extent of chemical modification of amino acid residues on the antigen *(10)*. The location of the epitope on a given antigen can be identified by comparing either the rate of chemical modification or the degree of chemical modification of the amino acid side chains on the antigen in the presence and absence of an antibody. Although it is difficult to measure the absolute rates of chemical modifications, a comparison of the extent of chemical modification provides insight into the spatial organization of the epitope *(13)*. Additionally, the kinetics of chemical modification can be used as a quantitative measure for the degree of surface-accessibility of the reactive amino acid side chain. This further helps to map the protein surface on the antigen, primarily the surface around the epitope *(4, 14, 15)*.

The method of differential chemical modification was first used in combination with radioactive labeling of the antigen *(10, 16, 17)*. Using ^{3}H- and ^{14}C-radioisotope-labeled acetic anhydride, lysine residues and the N-terminal amine on the antigen were converted into N^{ε}-acetyl-lysine and acetylated amino terminus, respectively. The degree of chemical modification is then calculated from the amount of radioactivity release from the digested peptides. Adaptation of this method for the use in mass spectrometric (MS) analysis *(18)* has proven to have multiple advantages *(11, 13, 14, 19–24)*. Most important, the ability of a mass spectrometer to measure the mass difference of 1 atomic mass unit (amu) with high sensitivity has allowed the use of

^{2}H- and/or ^{13}C-labeled acetic anhydride. This avoids the use of radioactive material without any loss in sensitivity of the analytical method of detection. Furthermore, the high mass resolving power of modern mass spectrometers has increased their attraction as platforms for robust sample analysis.

Mass spectrometers have been routinely used to study the chemical modifications on proteins and peptides *(11, 13, 14, 19–24)*. Complex structural biology problems, including mapping the surface topology of a protein, determining the active site of an enzyme, or protein–protein interactions, have been addressed by monitoring differential chemical modifications using MS. The use of MS for the analysis of chemically modified proteins and peptides has gained tremendous popularity, and as a result, protocols that are compatible with MS analysis have been developed for the derivatization of lysine, arginine, tyrosine, tryptophan, aspartic acid, and glutamic acid *(11, 13–15, 19–23)* (*see* also Chapter "Identifying Residues in Antigenic Determinants by Chemical Modification" for an overview of possible chemical modifications). Successful epitope mapping by differential chemical modification in conjunction with MS has several key requirements: (1) the chemical modification must be targeted and specific for a selected amino acid or a group of amino acids; (2) chemical modification should not alter the conformation of the proteins and should not affect the antibody–antigen interaction; (3) modification should only occur at the surface of the proteins; and (4) the chemical modification, which also serves as the label for MS analysis, needs to be stable under conditions of MS analysis (strongly acidic or strongly basic conditions).

Epitope mapping using differential chemical modification and MS involves derivatization of specific amino acid residues on a free antigen and the antibody-bound antigen in parallel and subsequent analysis of the modified amino acid residues by MS *(3, 4, 11, 14, 15, 19)*. Examples of epitope mapping based on this approach include differential chemical modification of primary amines to characterize epitopes on the HIV core protein, p24, *(14, 15)* and systematic chemical modification of arginine, lysine, and tyrosine to map a discontinuous epitope in hen egg white lysozyme *(11)*. Following chemical modification, the excess reagent used for derivatization is quenched and the antigen is purified prior to MS analysis. **Fig. 1** depicts the strategy used for epitope mapping by the chemical modification of primary amines (lysine residues and N-terminal amine) on the antigen using excess acetic anhydride. Derivatization of primary amines using excess acetic anhydride has proven to be a versatile approach to modify surface-exposed lysine residues and the N-terminal amine; however, at high concentrations of acetic anhydride the protein can unfold. Unfolding of a protein will result in changes in the conformation of the protein antigen and antibody and will adversely affect their

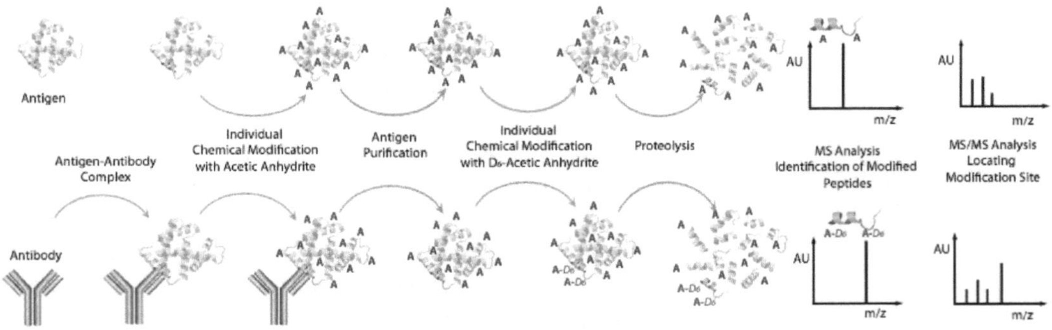

Fig. 1. General approach for characterizing conformational and discontinuous epitopes by differential chemical modification of antigen and mass spectrometry. Free antigen and antigen–antibody complex are individually subjected to chemical modification. Modified antigens (free and antibody-bound) are purified and proteolytically digested. Peptides are analyzed by MS (*see Color Plates*).

interaction, if not completely abolish it. Several studies have been carried out to determine the amount of acetic anhydride that can be used for chemical modification of proteins before the protein begins to unfold *(13, 25, 26)*. Following acetylation, the antigen is purified using high-performance liquid chromatography (HPLC). In the case of affinity-bound antigen, the antigen has to be dissociated from the antibody prior to or during purification. The purified antigens are then subjected to a second acetylation reaction with a high molar excess of the hexadeuteroderivative of acetic anhydride (d_6-acetyl anhydride) (*see* **Note 1**). Following the second purification step, the purified antigens are subjected to proteolytic digestion and the digested peptides are analyzed by matrix-assisted laser desorption ionization mass spectrometry (MALDI-MS) or liquid chromatography/electrospray ionization mass spectrometry (LC/ESI-MS) for the presence of chemically modified label. In addition to detecting the derivatized peptides, MS analyses also provide information on the extent of chemical modification of a particular peptide, while the position of the modified amino acid residue can be pinpointed by a tandem mass spectrometric analysis (MS/MS). The acetylated and trideuteroacetylated peptides are identified based on their mass shift of $42/z$ and $45/z$ amu (where z is the charge on the peptide) compared to the unmodified peptide, respectively *(14, 15)*. The extent of chemical modification on the peptides derived from the affinity-bound antigen and those derived from the free antigen can be determined by calculating the ratio of the monoisotopic peak intensities of H_3/D_3-modified peptides. The calculated H_3/D_3 ratios can also be used to map the relative reactivity of a particular amino acid residue, which is related to its surface accessibility. A reduced reactivity of an amino acid residue in the presence of the antibody suggests a steric hindrance and a reduced accessibility of that antigenic residue. The high association constant between the antigen and antibody is responsible for this shielding and the consequential reduction in reactivity of the amino acid residues that constitute the epitope.

Chemical modification of amino acid residues is highly specific and allows for the derivatization of a surface-accessible amino acid or a group of amino acids on the antigen. This fundamental principle of surface-accessibility and epitope mapping by differential chemical modification can also be extended to the use of hydrogen/deuterium exchange (H/D exchange) to probe conformational and/or discontinuous epitopes *(5)*. Unlike chemical modification, H/D exchange is not limited to a specific type of amino acid residue, e.g. lysine, but probes the entire antigen. During H/D exchange, it is the hydrogen on the protein backbone amides that is primarily exchanged with deuterium. However, some H/D exchange at the side chain amino acids also occurs. Often the exchange at the side chain amino acids are excluded from MS-based structural analysis *(27)*. Although H/D exchange is nondiscriminatory, the rate of exchange is highly dependent on the solvent accessibility of the backbone amide hydrogen. The kinetics of H/D exchange at shielded amides are considerably slower than the rate of exchange of solvent-exposed amide bonds *(28)*. The differences in the rates of H/D exchange at a sterically hindered position vs. solvent-exposed backbone amides have been routinely used in conjunction with MS to study protein–protein or protein–ligand interactions *(29, 30)*. Antigen–antibody interaction is a protein–protein interaction and can be probed using H/D exchange and MS analysis. For most purposes H/D exchange can be thought of as a facile chemical modification that labels the protein amide backbone at different rates depending on the backbone's solvent accessibility.

Epitope mapping using H/D exchange looks at the extent of H/D exchange on the antigen when it is bound to the antibody and compares that with the corresponding extent on the free form of antigen *(5)*. Typically, antigen and antibody are individually incubated in D_2O for a maximal exchange of amide hydrogen by deuterium. The deuterium-labeled antigen and antibody are incubated to generate deuterated antigen–antibody complex. Deuterated immune complex and deuterated free antigen are then individually subjected to back exchange by incubation in H_2O. The back exchange reaction will allow for all the deuterium to be substituted for hydrogen except for the deuterons that are on the epitope and have been shielded by the antibody. The exchange reaction is quenched by dropping the pH to 2–3 and lowering the temperature to 4°C to minimize continuous back exchange. Following the quenching of the exchange, the antigen is proteolytically cleaved by pepsin (a proteolytic enzyme that is compatible with digestion at low pH) for MS analysis. The general approach for epitope mapping by H/D exchange and

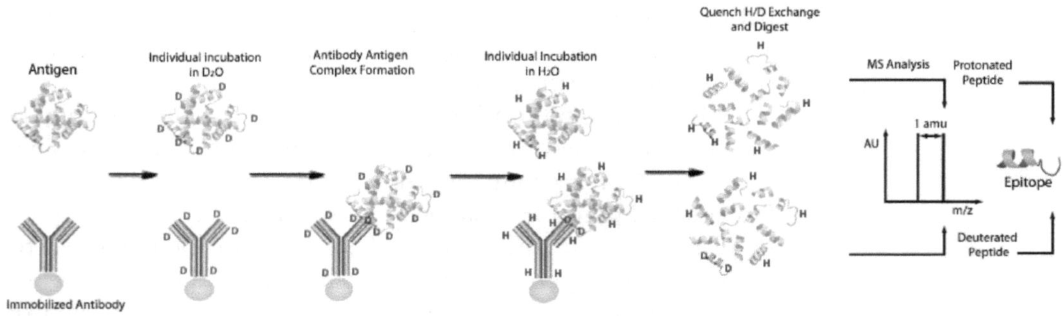

Fig. 2. General approach for characterizing conformational and discontinuous epitopes by H/D exchange and mass spectrometry. Antigen and antibody are individually labeled with deuterium by incubating in D_2O. Immune complex is formed using deuterated antigen and antibody. Back exchange is performed on the free antigen and the immune complex. Following quenching of the exchange reaction, the antigen is proteolytically cleaved using immobilized pepsin at pH 2.25 and the peptides are analyzed by MS (*see Color Plates*).

MS is summarized in **Fig. 2**. In this chapter, the discussion will focus on the use of MALDI-MS and LC/ESI-MS to identify the interaction sites of antigen–antibody complexes involving discontinuous and conformational epitopes.

2. Materials

2.1. Epitope Mapping by Chemical Modification of Antigen

1. Purified antigen and antigen-specific monoclonal antibody.
2. High-purity proteases, e.g., trypsin (Promega, Madison, WI), Glu-C, immobilized pepsin (Roche, Indianapolis, IN).
3. 100 mM Na-phosphate buffer at pH 8.0. Buffer pH adjusted using NaOH.
4. 5 mM acetic anhydride.
5. 2 M d_6-acetic anhydride.
6. Acetylation quenching buffer: 0.1 M Tris-HCl, pH 8.0.
7. Deuteroacetylation quenching buffer: 0.1 M Tris-HCl, pH 8.0.
8. MALDI-MS matrix solution: saturated solution of recrystallized α-cyano-4-hydroxycinnamic acid in ethanol/water/concentrated formic acid (FA) (45/45/10, v/v/v).
9. Acetonitrile for LC separation.

2.2. Proteolysis of Antigen

1. Purified antigen and antigen-specific monoclonal antibody.
2. High-purity proteases, e.g., trypsin (Promega), Glu-C, immobilized pepsin (Roche, Indianapolis, IN).
3. 50 mM NH_4HCO_3 (pH 7.8). The pH of 50 mM NH_4HCO_3 is close to pH 7.8.

4. MALDI-MS matrix solution: saturated solution of recrystallized α-cyano-4-hydroxycinnamic acid in ethanol/water/concentrated FA (45/45/10, v/v/v).
5. Acetonitrile for LC separation.

2.3. Epitope Mapping by H/D Exchange

1. Cyanogen bromide (CNBr)-activated Sepharose 4B beads (Amersham Pharmacia, Uppsala, Sweden).
2. Compact reaction columns (CRCs) with 35-µm column filters (USB, Cleveland, OH).
3. Antibody-coupling buffer: 0.1 M $NaHCO_3$, 150 mM NaCl, pH 8.2. Adjust the pH using NaOH/HCl.
4. Antibody coupling reaction quenching buffer: 0.1 M Tris-HCl, pH 8.0.
5. Unreacted antibody washing buffer: 0.1 M sodium acetate, 0.5 M NaCl, pH 4.0. Adjust the pH using glacial acetic acid.
6. Phosphate-buffered saline (PBS), pH 7.2: 100 mM sodium phosphate buffer, 150 mM NaCl, pH 7.2. Adjust the pH using NaOH.
7. 5 M NaOH.
8. 5 M NaOD (deuterated NaOH).
9. D_2O (deuterated water).
10. Deuterated phosphate buffer: 50 mM KH_2PO_4 at pH 6.6, 50 mM NaCl in D_2O. Adjust the pH using NaOD.
11. 0.1% Formic acid (FA) or trifluroacetic acid (TFA).
12. Immobilized pepsin (Pierce Chemicals, Rockford, IL, or Applied Biosystems, Foster City, CA) and immobilized pepsin columns (Orachrom, Woburn, MA). For custom fabrication of pepsin columns refer to *(31)*.
13. MALDI-MS matrix solution: saturated solution of recrystallized α-cyano-4-hydroxycinnamic acid in ethanol/water/concentrated FA (45/45/10, v/v/v).
14. Acetonitrile for LC separation.
15. 1-Propanol.

3. Methods

3.1. Epitope Mapping by Chemical Modification of Antigen

Different amino acid residues can be chemically modified in order to map the epitope. Lysine residues are generally well represented in a protein sequence, and the chemistry behind acetylation of lysine is well characterized. These two factors make lysine a good target for chemical modification. The experimental protocol to modify the lysine residues using acetic anhydride and deuterated acetic anhydride (d_6-acetyl anhydride) is outlined below (*see* **Note 2**).

3.1.1. Differential Acetylation of Lysine

1. Two vials, each containing a solution of 2 μg of the antigen in 20 μL of 100 mM Na-phosphate buffer, pH 8.0, are prepared.
2. To the first vial add twofold molar excess of antibody relative to the antigen and allow the formation of immune complexes by slowly rotating the sample at room temperature for 2 h. Twofold molar excess ensures complete binding of the antigen by the antibody as long as the concentration is tenfold over the K_d.
3. Verify the ratio of the acetic anhydride such that it is in excess to the reactive groups on the antigen by at least 10,000-fold. Then, individually acetylate the primary amines of antigen (control vial) and the immune complex by treating with 5 mM acetic anhydride for 30 min at room temperature. Continuously monitor the pH of the solution using pH strips or a micro pH electrode and adjust the pH by adding 5 M NaOH as needed to maintain a pH above 7.0 (*see* **Note 3**).
4. Add an equal volume of acetylation quenching buffer to quench any remaining unreacted acetic anhydride.
5. Purify the two acetylated antigen samples by HPLC (*see* **Note 4**).
6. Collect HPLC fractions (1 mL) and lyophilize.
7. Determine the HPLC fractions that contain the antigen using MALDI-MS.
8. Combine the fractions found to contain the acetylated antigen and lyophilize.
9. Resuspend the acetylated and lyophilized antigen in 100 mM Na-phosphate buffer (pH 8.0) and add 2 M d_6-acetic anhydride. Incubate the mixture by slowly rotating at room temperature for 1 h. Note that d_6-acetic anhydride is added in excess such that the ratio of d_6-acetic anhydride to the reactive groups on the antigen is 1:100,000.
10. Continuously monitor the pH of the solution using pH strips or a micro pH electrode and adjust the pH by adding 5 M NaOD as needed to maintain a pH above 7.0.
11. Add an equal volume of deuteroacetylation quenching buffer to quench any remaining unreacted d_6-acetic anhydride.
12. Follow **steps 5–10** to obtain purified antigen that is differentially acetylated by acetic anhydride and d_6-acetic anhydride.
13. Proteolytically digest the acetylated antigen with trypsin and/or Glu-C (*see* **Subheading 3.2** for details on digestion conditions).

14. Lyophilize the peptide digest. The lyophilized samples can be stored at –80°C until MS analysis.
15. The digested and lyophilized peptides can be analyzed using MALDI-MS or LC/ESI-MS and MS/MS (*see* **Subheading 3.3** for the details on MS analyses).

3.2. Proteolysis of Antigen

1. Digestion with trypsin: Resuspend the purified acetylated-antigen (**Subheading 3.1.1, step 11**) in 50 mM NH_4HCO_3 (pH 7.8). Digest the antigen by adding trypsin (enzyme/substrate molar ratio, 1:20) at 37°C with slow rotation for 4 h.
2. Digestion with Glu-C: Resuspend the purified and acetylated antigen (**Subheading 3.1.1, step 11**) in 50 mM NH_4HCO_3 (pH 7.8). Digest the antigen by adding Glu-C (enzyme/substrate molar ratio, 1:20) at room temperature with slow rotation overnight (*see* **Note 5**).
3. In parallel, determine the autodigest products of the proteases by preparing the protease solutions in 50 mM NH_4HCO_3 (pH 7.8) buffer and incubating under identical conditions. The autodigest products can be analyzed and characterized by either MALDI-MS or LC/ESI-MS/MS. Knowing the autodigest products of the protease simplifies the interpretation of the MS data.

3.3. Epitope Mapping by H/D Exchange

3.3.1. Preparation of Immobilized Antibody Column

1. Swell ~0.2 g of dry CNBr-activated Sepharose 4B beads into a Falcon™ tube by adding 10 mL of deionized water (*see* **Note 6**).
2. For each sample, pipet 20 µL of beads slurry into two compact reaction columns (CRC) each.
3. Wash CNBr-activated Sepharose columns six times with 0.8 mL 1 mM HCl and six times with 0.4 mL 0.1 M $NaHCO_3$ (pH 8.2). At the end of the wash, drain all liquid from the beads.
4. To the first CRC add 80 µL of antibody-coupling buffer and 20 µL of antibody (50 µg/mL). To the second CRC, which will serve as the control, add 100 µL of PBS. Incubate at room temperature with slow rotation for 1.5 h.
5. Drain the liquid from the CRCs and block any unreacted sites with antibody coupling quenching buffer. Then, rinse the beads once with 0.4 mL of quenching buffer, followed by the incubation of the beads in a second aliquot of 0.4 mL of quenching buffer for 1 h at room temperature with slow rotation.
6. Use a series of washing steps to remove unbound antibody. First wash the column with 0.4 mL of unreacted antibody washing buffer and then with 0.4 mL of antibody coupling reaction quenching buffer. Repeat the process three times.

7. Wash three times with 0.4 mL PBS.
8. At this point MALDI-MS spectra of the beads from both CRC can be obtained. No ions should be detected if the antibody is fully cross-linked and free antibody is thoroughly washed off the column (*see* **Note 7**).

3.3.2. H/D Exchange on Antigen

1. Resuspend 2 μg of antigen in 5 μL of deuterated phosphate buffer and allow the deuteration of surface amides for 10 min at room temperature.
2. In a second vial, resuspend twofold molar excess of immobilized antibody relative to the antigen and allow the deuteration of surface amides for 10 min at room temperature.
3. Mix the immobilized antibody and the antigen (total volume, 25 μL) and allow the formation of immune complexes by slowly rotating the sample at room temperature for 10 min.
4. Individually carry out the back exchange for the immune complex and the free antigen. For this dilute each of the samples in 275 μL of H_2O.
5. Centrifuge the beads with the immune complex, discard the supernatant and resuspend the beads (~25 μL) in 275 μL H_2O. Repeat this process twice. The second resuspension and decantation will generate 25 μL of decanted beads (*see* **Note 8**).
6. Individually quench the exchange reactions on the free antigen and the decanted beads with the immune complexes by mixing with 25 μL of a chilled (4°C) solution of equal parts of 0.1% TFA or FA (pH 2.25) and 1-propanol.
7. H/D exchange quenching solution elutes the antigen from the immobilized antibody in addition to quenching the H/D exchange. Following this elution the antibody immobilized on the Sepharose beads can be removed by centrifugation. For this, centrifuge the beads and remove the supernatant (~25 μL).
8. Mix the supernatant obtained in **step 7** and the back exchanged free antigen separately with 100 μL of a slurry of immobilized pepsin at 4°C in 0.1% FA or TFA (pH 2.25) using a pepsin/substrate ratio of ~1:1 for 5 min.
9. Following 5 min of digestion, remove the immobilized pepsin by centrifugation. The sample is now ready for MS analysis (*see* **Note 9**). It can also be rapidly frozen in liquid nitrogen and stored at –80°C until analysis within 1–2 days; however, for best results analyze the samples immediately.

3.4. Mass Spectrometric Analysis

3.4.1. MALDI-MS Analysis for Acetylation Studies

1. All MALDI-MS analyses can be carried out on any MALDI mass spectrometer in the positive ion mode.
2. For all MALDI analyses dissolve the lyophilized samples in 50% acetonitrile/0.1% FA.

3. Spot a 0.5-μL aliquot of liquid on a stainless steel target and mix with an equal volume of the MALDI-MS matrix solution.
4. Let the sample on the target air dry at room temperature.
5. For all MALDI-MS analyses use an external calibrant to calibrate the instrument in the mass range of interest for mass accuracy and reliable results.

3.4.2. MALDI-MS Analysis for H/D Exchange Studies

1. All MALDI-MS analyses can be carried out on any MALDI mass spectrometer in positive ion mode.
2. Prepare a MALDI target for spotting by chilling on ice in a plastic case. The plastic case prevents condensation of atmospheric H_2O onto the target *(32)*.
3. Also chill the MALDI-MS matrix at 0°C.
4. If the sample has been frozen at −80°C, quickly defrost the samples to 0°C prior to spotting.
5. Spot a 0.5-μL aliquot of the sample on a chilled stainless steel target and mix with an equal volume of the chilled MALDI-MS matrix solution.
6. Once spotted, immediately place the target in a desiccator under a moderate vacuum such that the spots dry in 1–2 min.
7. For all MALDI-MS analyses use an external calibrant to calibrate the instrument in the mass range of interest for mass accuracy and reliable results.

3.4.3. LC/ESI-MS Analysis for Acetylation Studies

1. For all LC/ESI-MS analyses dissolve the lyophilized samples in H_2O.
2. Load the peptides on the LC column connected to the ESI-MS and elute the peptides using a water/acetonitrile LC gradient.
3. A typical LC gradient (water/acetonitrile gradient starting from 3% acetonitrile to 50% acetonitrile over 1 h) can be used for the LC separation of peptides.
4. Duplicate MS analyses should be carried out on independent acetylation experiments (at least two, preferably three) to determine the ratios of acetylation to d_3-acetylation. Representative ESI-MS spectra for acetylated and d_3-acetylated peptides are shown in **Fig. 3**.

3.4.4. LC/ESI-MS Analysis for H/D Exchange Studies

1. For all LC/ESI-MS analyses dissolve the lyophilized samples in cold acidified H_2O (0.1% FA or TFA) (*see* **Note 10**).
2. To minimize the back exchange and deuterium loss, immerse all LC components (injector, tubing, trap, and column) in an ice bath *(33)*.
3. Load the peptide sample prepared in cold acidified H_2O on the LC column connected to ESI-MS and elute the peptides using a chilled water/acetonitrile LC gradient.

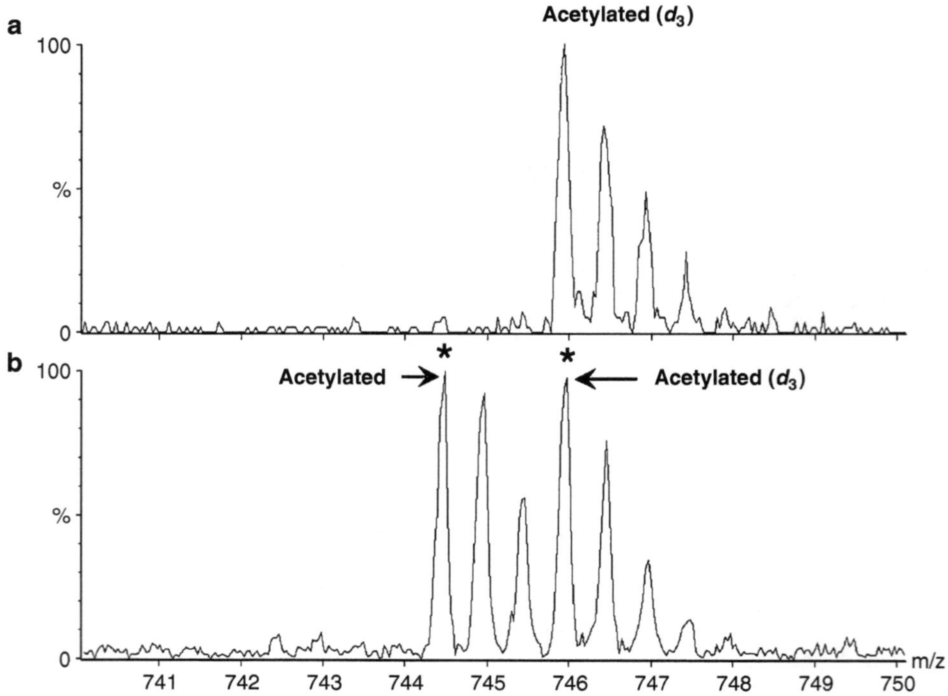

Fig. 3. Mass spectrum of acetylated and deuteroacetylated gp120 trypsin/Glu-C digest showing the doubly protonated peptide GG(AcK)DDTIILP(carboxyamidomethylCys)R, (**a**) in the presence of an anti-HIV gp120 Ab and (B) in the absence the same.

4. Use a fast water/acetonitrile gradient for LC separation of peptides from H/D exchange experiments. Typically, the gradient should include 5% acetonitrile for 3 min, 5–15% acetonitrile over 30 s, and ramp to 50% acetonitrile over 6 min *(31)*. Both mobile phases, water and acetonitrile should contain 0.1% FA or TFA and should be chilled at 4°C (*see* **Note 11**).
5. Online pepsin digestion can also be carried out using an Orachrom StyrosZyme pepsin column. For this, ~300 pmol of protein is loaded into a 20-µL loop and pushed through the pepsin column using 0.1% FA or TFA. After a resident time of 1 min, the peptides are pushed into a trapping column. Following a short desalting period, the peptides are eluted off the peptide trap and separated with a C18 column using the same gradient described in **step 3** *(34, 35)*.

3.5. Interpretation of MS Data

1. Acetylated and deuteroacetylated peptides are identified by their mass shifts of $42/z$ and $45/z$ amu (where z is the charge on the peptide), compared to unmodified peptides.
2. Through MS/MS analysis the presence of acetylated lysine can further be determined by the detection of acetylated lysine innonium ions (143 m/z) as well as the corresponding

b- and y-series ions. Similarly, the deuteroacetylated peptides show the presence of deuteroacetylated lysine innonium ions (146 *m/z*).

3. The monoisotopic peak intensities (peaks marked with asterisk in **Fig. 3b** represent the monoisotopic peaks) are used to calculate ratios of acetylated and deuteroacetylated peptides corresponding to the control antigen and the sample with the antigen bound to the antibody, respectively.
4. The ratios provide quantitative information on surface-accessiblelysine residues as well as the effects antibody binding has on lysine reactivity.
5. H/D ratios in H/D exchange experiments can be determined from the ratio of monoisotopic peak intensities of protonated and deuterated peptides.
6. The ratios provide quantitative information on the surface accessibility of backbone amides on the antigen in the presence and absence of the antibody.

4. Notes

1. During the two successive acetylation steps, d_6-acetyl anhydride is used in the second step. The use of d_6-acetyl anhydride in the second step leads to less problems in determining the d_0/d_6 ratio in the control vs. in the reaction (antigen–antibody complex). This complexity arises for ions whose ^{13}C isotope peak and $^{13}C_1$ and D_1 peaks are of high abundance relative to the monoisotopic peak. Situations where there is low level of acetylation in the first step, acetylation of antigen–antibody complex, are of particular concern because the relative abundance of the isotopic ion containing three ^{13}C atoms to the monoisotopic ion is ~32% *(14)*.
2. Derivatization conditions are dependent on the amino acid residues that are selected for chemical modification. It is important to optimize the conditions for chemical modification prior to attempting antigen modification. Using a commonly available protein such as bovine serum albumin to optimize the conditions for optimal chemical modification is highly recommended.
3. During the acetylation reaction, the pH of the solution drops. It is extremely important to constantly monitor the pH of the solution and maintain it above pH 7 by adding NaOH as needed. Maintaining the pH at 7 helps to retain the native conformation of the proteins.

4. HPLC purification removes salts from the samples and also results in the dissociation of antigen from the antibody (in the case of the immune complex).
5. Tryptic digestion is inhibited when lysines are acetylated. This can result in multiply-acetylated peptides. Thus, digestion with a different endoproteinase is sometimes necessary to generate smaller peptides for MS analyses.
6. Carefully refer to the manufacturer's recommendations for handling and activation of CNBr-activated Sepharose 4B beads. There will be slight variations among protocols across different manufacturers.
7. During MALDI-MS analysis, any unbound antibody will be ionized and the ions will be detected. Absence of detectable ions corresponding to the antibody is a good indication that the antibody is immobilized onto the beads and also that there are no un-cross-linked antibodies nonspecifically interacting with the beads. If ions corresponding to the antibody are detected, then it is necessary to repeat the wash and follow-up with a second MALDI-MS analysis. If any ions are still visible after multiple washes, the CNBr-activated Sepharose 4B beads cross-linking protocol should be re-evaluated according to the manufacture's guidelines.
8. The time required to finish **steps 4** and **5** should take 2–3 min. This is enough time for the back exchange of all the amides that are not shielded by the antibody.
9. MALDI-Time of Flight (MALDI-TOF) or MALDI-TOF/TOF mass spectrometers are best suited for the analysis of H/D exchange samples because the acidic MALDI matrix keeps the H/D exchange quenched. Once the sample is dried on the MALDI target, the samples are stable and do not undergo back exchange. Instruments capable of MALDI-TOF/TOF allow fragmentation and subsequent sequencing of the peptides. LC-ESI/MS and MS/MS can also be used for the peptide analysis. However, during the LC separation there is back exchange taking place in the LC column. A correction for the back exchange is essential prior to data interpretation. Maintaining the temperature of the LC column at 4°C is essential as it slows down the back exchange and the deuterium scrambling events.
10. Keeping the sample acidified and cold is extremely important as it slows down the back exchange.
11. To minimize the back exchange, the peptide separation must be carried out using a fast gradient and the temperature of the LC column and the plumbing must be kept at 4°C as strictly as possible. H/D MS experiments are most successful when the amount of deuterium lost during analysis is relatively small.

Acknowledgments

We thank Dr. Leesa J. Deterding and Dr. Jason G. Williams for helpful discussions. This work was supported by Intramural Research program, National Institute of Environmental Health Sciences, NIH.

References

1. Barlow, D. J., Edwards, M. S., and Thornton, J. M. (1986) Continuous and discontinuous protein antigenic determinants. *Nature* 322, 747–748.
2. Bosshard, H. R. (1996) Epitope mapping by differential chemical modification of antigens. *Methods Mol. Biol.* 66, 85–95.
3. Hager-Braun, C., and Tomer, K. B. (2005) Determination of protein-derived epitopes by mass spectrometry. *Expert Rev. Proteomics* 2, 745–756.
4. Hochleitner, E. O., Gorny, M. K., Zolla-Pazner, S., and Tomer, K. B. (2000) Mass spectrometric characterization of a discontinuous epitope of the HIV envelope protein HIV-gp120 recognized by the human monoclonal antibody 1331A. *J. Immunol.* 164, 4156–4161.
5. Baerga-Ortiz, A., Hughes, C. A., Mandell, J. G., and Komives, E. A. (2002) Epitope mapping of a monoclonal antibody against human thrombin by H/D-exchange mass spectrometry reveals selection of a diverse sequence in a highly conserved protein. *Protein Sci.* 11, 1300–1308.
6. Parker, C. E., Papac, D. I., Trojak, S. K., and Tomer, K. B. (1996) Epitope mapping by mass spectrometry: determination of an epitope on HIV-1 IIIB p26 recognized by a monoclonal antibody. *J. Immunol.* 157, 198–206.
7. Parker, C. E., and Tomer, K. B. (2002) MALDI/MS-based epitope mapping of antigens bound to immobilized antibodies. *Mol. Biotechnol.* 20, 49–62.
8. Peter, J. F., and Tomer, K. B. (2001) A general strategy for epitope mapping by direct MALDI-TOF mass spectrometry using secondary antibodies and cross-linking. *Anal. Chem.* 73, 4012–4019.
9. Purcell, A. W., and Gorman, J. J. (2004) Immunoproteomics: mass spectrometry-based methods to study the targets of the immune response. *Mol. Cell. Proteomics* 3, 193–208.
10. Burnens, A., Demotz, S., Corradin, G., Binz, H., and Bosshard, H. R. (1987) Epitope mapping by chemical modification of free and antibody-bound protein antigen. *Science* 235, 780–783.
11. Fiedler, W., Borchers, C., Macht, M., Deininger, S. O., and Przybylski, M. (1998) Molecular characterization of a conformational epitope of hen egg white lysozyme by differential chemical modification of immune complexes and mass spectrometric peptide mapping. *Bioconjug. Chem.* 9, 236–241.
12. Jemmerson, R., and Paterson, Y. (1986) Mapping epitopes on a protein antigen by the proteolysis of antigen–antibody complexes. *Science* 232, 1001–1004.
13. Glocker, M. O., Borchers, C., Fiedler, W., Suckau, D., and Przybylski, M. (1994) Molecular characterization of surface topology in protein tertiary structures by amino-acylation and mass spectrometric peptide mapping. *Bioconjug. Chem.* 5, 583–590.
14. Hochleitner, E. O., Borchers, C., Parker, C., Bienstock, R. J., and Tomer, K. B. (2000) Characterization of a discontinuous epitope of the human immunodeficiency virus (HIV) core protein p24 by epitope excision and differential chemical modification followed by mass spectrometric peptide mapping analysis. *Protein Sci.* 9, 487–496.
15. Williams, J. G., Tomer, K. B., Hioe, C. E., Zolla-Pazner, S., and Norris, P. J. (2006) The antigenic determinants on HIV p24 for CD4+ T cell inhibiting antibodies as determined by limited proteolysis, chemical modification, and mass spectrometry. *J. Am. Soc. Mass Spectrom.* 17, 1560–1569.
16. Bosshard, H. R. (1979) Mapping of contact areas in protein–nucleic acid and protein–protein complexes by differential chemical modification. *Methods Biochem. Anal.* 25, 273–301.
17. Kaplan, H., Stevenson, K. J., and Hartley, B. S. (1971) Competitive labelling, a method for determining the reactivity of individual groups in proteins. The amino groups of porcine elastase. *Biochem. J.* 124, 289–299.

18. Roesijadi, G., Vestling, M. M., Murphy, C. M., Klerks, P. L., and Fenselau, C. C. (1991) Structure and time-dependent behavior of acetylated and non-acetylated forms of a molluscan metallothionein. *Biochim. Biophys. Acta* 1074, 230–236.

19. Hager-Braun, C., and Tomer, K. B. (2002) Characterization of the tertiary structure of soluble CD4 bound to glycosylated full-length HIVgp120 by chemical odification of arginine residues and mass spectrometric analysis. *Biochemistry* 41, 1759–1766.

20. Santrucek, J., Strohalm, M., Kadlcik, V., Hynek, R., and Kodicek, M. (2004) Tyrosine residues modification studied by MALDI-TOF mass spectrometry. *Biochem. Biophys. Res. Commun.* 323, 1151–1156.

21. Wood, T. D., Guan, Z., Borders, C. L., Jr., Chen, L. H., Kenyon, G. L., and McLafferty, F. W. (1998) Creatine kinase: essential arginine residues at the nucleotide binding site identified by chemical modification and high-resolution tandem mass spectrometry. *Proc. Natl. Acad. Sci. USA* 95, 3362–3365.

22. Alcalde, M., Plou, F. J., Andersen, C., Martin, M. T., Pedersen, S., and Ballesteros, A. (1999) Chemical modification of lysine side chains of cyclodextrin glycosyltransferase from Thermoanaerobacter causes a shift from cyclodextrin glycosyltransferase to alpha-amylase specificity. *FEBS Lett.* 445, 333–337.

23. Strohalm, M., Santrucek, J., Hynek, R., and Kodicek, M. (2004) Analysis of tryptophan surface accessibility in proteins by MALDI-TOF mass spectrometry. *Biochem. Biophys. Res. Commun.* 323, 1134–1138.

24. Steiner, R. F., Albaugh, S., Fenselau, C., Murphy, C., and Vestling, M. (1991) A mass spectrometry method for mapping the interface topography of interacting proteins, illustrated by the melittin-calmodulin system. *Anal. Biochem.* 196, 120–125.

25. Suckau, D., Mak, M., and Przybylski, M. (1992) Protein surface topology-probing by selective chemical modification and mass spectrometric peptide mapping. *Proc. Natl. Acad. Sci. USA* 89, 5630–5634.

26. Glocker, M. O., Nock, S., Sprinzl, M., and Przybylski, M. (1998) Characterization of surface topology and binding area in complexes of the elongation factor proteins EF-Ts and EF-TuGDP from Thermus thermophilus: a study by protein chemical modification and mass spectrometry. *Chem. Eur. J.* 4, 707–715.

27. Yan, X., Watson, J., Ho, P. S., and Deinzer, M. L. (2004) Mass spectrometric approaches using electrospray ionization charge states and hydrogen–deuterium exchange for determining protein structures and their conformational changes. *Mol. Cell. Proteomics* 3, 10–23.

28. Milne, J. S., Mayne, L., Roder, H., Wand, A. J., and Englander, S. W. (1998) Determinants of protein hydrogen exchange studied in equine cytochrome c. *Protein Sci.* 7, 739–745.

29. Busenlehner, L. S., and Armstrong, R. N. (2005) Insights into enzyme structure and dynamics elucidated by amide H/D exchange mass spectrometry. *Arch. Biochem. Biophys.* 433, 34–46.

30. Hoofnagle, A. N., Resing, K. A., and Ahn, N. G. (2003) Protein analysis by hydrogen exchange mass spectrometry. *Annu. Rev. Biophys. Biomol. Struct.* 32, 1–25.

31. Wu, Y., Kaveti, S., and Engen, J. R. (2006) Extensive deuterium back-exchange in certain immobilized pepsin columns used for H/D exchange mass spectrometry. *Anal. Chem.* 78, 1719–1723.

32. Mandell, J. G., Falick, A. M., and Komives, E. A. (1998) Measurement of amide hydrogen exchange by MALDI-TOF mass spectrometry. *Anal. Chem.* 70, 3987–3995.

33. Zhang, Z. Q., and Smith, D. L. (1993) Determination of amide hydrogen exchange by mass spectrometry: a new tool for protein structure elucidation. *Protein Sci.* 2, 522–531.

34. Ehring, H. (1999) Hydrogen exchange/electrospray ionization mass spectrometry studies of structural features of proteins and protein/protein interactions. *Anal. Biochem.* 267, 252–259.

35. Wang, L. T., Pan, H., and Smith, D. L. (2002) Hydrogen exchange-mass spectrometry: optimization of digestion conditions. *Mol. Cell. Proteomics* 1, 132–138.

Subsection B

Peptides Library Approaches

Chapter 10

Linear B-Cell Epitope Mapping Using Enzyme-Linked Immunosorbent Assay for Libraries of Overlapping Synthetic Peptides

Michael W. Heuzenroeder, Mary D. Barton, Thiru Vanniasinkam, and Tongted Phumoonna

Summary

The aim of this chapter is to provide a strategy for mapping linear antibody epitopes of protein antigens in order to discover candidates for vaccines or diagnostic tests. A set of overlapping peptides was designed and synthesised based upon a known amino acid sequence of the target protein, virulence-associated protein A (VapA) of the bacterium *Rhodococcus equi*, an important pulmonary pathogen in foals.

The peptides were biotinylated and used in an ELISA to screen immune sera from foals. These biotinylated peptides were coated directly onto micro titre plates that had been pre-coated with NeutrAvidin™. A linear B-cell epitope was identified by a universal recognition of sera to the synthetic peptides which corresponds to a particular fragment of the VapA protein.

Key words: Epitope mapping, Linear B-cell epitope, Biotinylated peptides, ELISA, VapA, *Rhodococcus equi*.

1. Introduction

B-cell epitope mapping using a series of overlapping synthetic peptides is a very efficient way to identify a linear antigenic determinant(s) recognised by a particular antibody through an immunoassay (*see* also Chapter "Antibody Epitope Mapping Using SPOT- Peptide Arrays"). This approach has successfully been used in several studies *(1–5)* to identify immunogenic epitopes of potential target vaccine proteins.

Mapping epitopes can also be undertaken using X-ray crystallography (*see* Chapter "Structural Basis of Antibody–Antigen

Ulrich Reineke (eds.), *Methods in Molecular Biology, Epitope Mapping Protocols, vol. 524*

DOI: 10.1007/978-1-59745-450-6_10

Interactions") or by using a combination of computer-aided molecular modelling, ligand binding, and spectroscopy to identify the structure of antigen–antibody complexes *(6, 7)*. However, use of these methods, particularly antigen–antibody crystals, are impractical in the vast majority of cases, due to the difficulty in creating the crystals. Thus, the use of synthetic peptides of target proteins has become an alternative method in identifying individual epitopes because synthetic peptide fragments can be sufficiently similar to the native antigen, thus allowing the binding of the antibody produced by B cells. This method provides a rapid, practical, and cost-effective approach in identifying linear epitopes.

In the following sections, details of procedures for detection of linear B-cell epitopes of the highly immunogenic VapA protein of *Rhodococcus equi* that we have developed will be described. Importantly, the amino acid sequence of the protein antigen must be known (either via protein sequencing or Genbank database) before epitopes can be mapped using synthetic peptides. Overlapping peptides of a defined length and homologous with the VapA protein were synthesised and screened with a population of sera from foals that had been diagnosed with *R. equi* disease, using sera from uninfected foals as a negative control (**Fig. 1**).

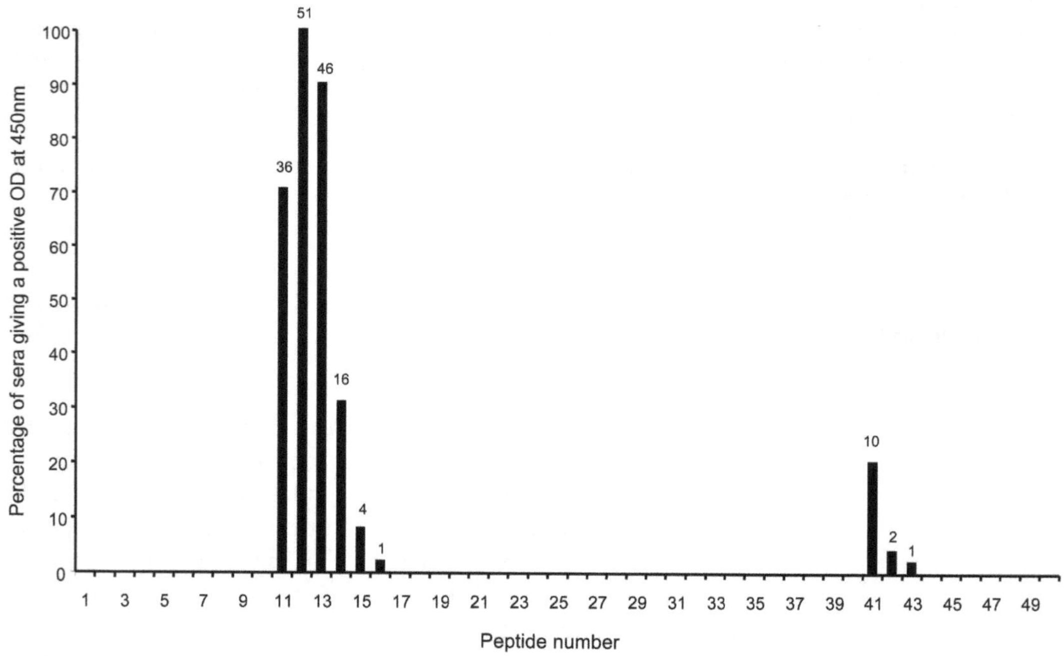

Fig. 1. VapA protein of *R. equi*-derived overlapping peptides reacting with 51 positive horse sera. A total of 50 overlapping peptides was synthesised based upon the amino acid sequences of the antigenic VapA protein of *Rhodococcus equi* and were screened with 51 positive horse sera. Overlapping peptides Nos. 11, 12, 13, and 14 were universally recognised by the sera with peptide No. 12 being the most reactive. These peptides correspond to the region between amino acids 62–81 of VapA and this is thus identified as a linear B-cell epitope of the target protein (reproduced from *(2)*, with permission from The American Society for Microbiology Inc.).

2. Materials

2.1. Design and Synthesis of Biotinylated Peptides

It is important that the target protein is known to be immunogenic and able to elicit an antibody response in the host.

1. The amino acid sequence of the *R. equi* VapA protein (Genbank accession No. D21236) (**Fig. 1**).
2. Biotinylated peptides (Mimotopes, Clayton, Vic., Australia). The peptides are supplied as lyophilised product. Dissolve the peptides in a suitable solvent (following manufacturer's instructions) at concentrations of 2.86 mg/mL and 28.6 μg/mL and store in small aliquots as stock solutions at –20°C (*see* **Note 1**).
3. Solvent: 0.1% (v/v) acetic acid in deionised water. This solution is used to reconstitute the peptides and can be stored at room temperature (*see* **Note 2**).

2.2. ELISA Assay of Biotinylated Peptides

1. Foal sera: Sera obtained from foals that had been diagnosed with *R. equi* disease and sera from healthy foals to screen the synthetic peptide bank.
2. Micro titre plates: Nunc-Immuno® Maxisorp F96-well micro titre plates (Nalge Nunc International, Roskilde, Denmark).
3. NeutrAvidin™ biotin-binding protein (Pierce Chemical, Rockford, IL). Reconstitute the protein in ultra pure water at a concentration of 1 mg/mL. Aliquots of this stock solution should be stored frozen at –20°C.
4. Phosphate-buffered saline (PBS): 10 mM phosphate, pH 7.4, 150 mM NaCl. Dissolve 8.0 g NaCl, 0.2 g KCl, 1.44 g Na_2HPO_4 anhydrous (or 3.63 g $Na_2HPO_4 \times 12\ H_2O$), and 0.24 g KH_2PO_4 in 900 mL of H_2O. Adjust the pH to 7.4 with either 50% (w/v) NaOH or concentrated 37% (w/w) HCl. Adjust the volume of the solution to 1 L with H_2O and sterilise by autoclaving. Store the sterile PBS at room temperature.
5. PBS/Tween 20 (PBST): 0.05% (v/v) Tween 20 in PBS. This solution is used as the washing solution and the diluent of the primary antibody and peptides. It can be stored at room temperature for several weeks. Discard it if there is any sign of contamination.
6. Blocking buffer: 1% (w/v) sodium caseinate in PBST. This solution is the diluent for the secondary antibody and is used to block non-specific binding. It should be refrigerated and discarded if there is any sign of contamination.
7. Secondary antibody: Goat anti-horse IgG horseradish peroxidase-conjugated antibody (Bethyl Laboratories, Montgomery, TX). Keep the reagent in the refrigerator or as per the manufacturer's recommendations.

8. Substrate buffer: Dissolve one phosphate-citrate buffer tablet (Sigma, St. Louis, MO) in 100 mL of deionised water with stirring to obtain 50 mM phosphate-citrate buffer, pH 5.0. Alternatively, dissolve 1.46 g Na_2HPO_4 and 1.02 g citric acid in deionised water to a final volume of 200 mL. Store this solution in the refrigerator and bring the amount required to room temperature before each use. Substrate buffer should be used within 2 weeks of preparation and checked for signs of contamination immediately before use.
9. Chromogenic substrate solution: Dissolve one tablet (1 mg) of 3,3′,5,5′-tetramethylbenzidine (Sigma) in 10 mL substrate buffer. Prepare the substrate solution immediately before use. This chromogenic substrate is used to detect horseradish-peroxidase-conjugated antibodies.
10. Stop solution (0.5 N H_2SO_4): Add 15 mL of concentrated 35.6 N H_2SO_4 slowly to a final volume of 1 L deionised water and with adequate protection from splashes. Store the solution in Pyrex glass bottles at room temperature. *Caution:* Sulfuric acid is extremely corrosive to skin, metals, and clothing. It must always be added slowly to water when making dilutions. Avoid contact with liquid and vapour.

3. Methods

3.1. Design and Synthesis of Biotinylated Peptides

A total of 50 overlapping peptides were designed based upon the 189 amino acid sequence in length of the *R. equi* VapA protein (Genbank accession No. D21236) *(8)*, and were synthesised by Mimotopes (http//:www.mimotopes.com). Every peptide was 11 amino acid residues in length, offset by 3 and overlapping by 8 residues (*see* **Note 3**). Each peptide was coupled with biotin at the N-terminus, followed by a tetrapeptide spacer sequence SGSG prior to the peptide sequence of interest, and a free acid (–OH) at the C-terminus. Please contact Mimotopes if you need help in designing peptide length and endings that are more appropriate to your project. The purity and identity of peptides were quantified by the manufacturer, using reverse-phase high-performance liquid chromatography and confirmed by ion-spray mass spectrometry methods, respectively.

3.2. ELISA Assay of Biotinylated Peptides

It is strongly recommended to optimise concentrations of reagents such as peptides, NeutrAvidin™ biotin-binding protein, and secondary antibodies for each ELISA system or when a new batch of the reagent is purchased.

1. Prepare 10 mL of 1:300 dilution of the stock solution (1 mg/mL) of NeutrAvidin biotin-binding protein in sterile deionised water per plate to be tested.

2. Coat Nunc-Immuno Maxisorp F96-well micro titre plates with 100 μL (0.33 μg) per well of the diluted NeutrAvidin biotin-binding protein. Incubate plates overnight at 37°C or at 4°C if it is not possible to continue the test next day.
3. Add 200 μL of blocking buffer to each well to block non-binding sites and incubate at room temperature for 1 h or for 2 h at 4°C.
4. Flip out the solution from the plates. Wash the plates five times with PBST (400 μL/well) using an automatic ELISA washer (Ultrawash Plus™, Dynex Technologies, Chantilly, VA).
5. Remove any excess buffer by slapping the plates (well side down) on a clean towel or absorbent paper.
6. Just before use, prepare a working strength (0.286 μg/mL) of biotinylated peptides by making 1:100 dilutions of the stock solution (28.6 μg/mL) in PBST. Then, transfer 100 μL of each of the diluted peptide solutions into the corresponding NeutrAvidin-coated wells (*see* **Note 4**).
7. Place the plates on a platform shaker at a low speed (~125 rpm) and allow the reaction to proceed for 1 h at room temperature. After incubation, flick out solution and repeat the washing procedure described in **steps 4** and **5** (*see* **Note 5**).
8. Dilute the horse serum 1:250 in PBST. However, this dilution may not be ideal if other types of sera are being tested, because the optimum dilution of serum depends on the source and the amount of antibodies present in the sample.
9. Add 100 μL of the diluted serum to each of the wells containing captured peptides. The plates are then incubated overnight at 4°C for better sensitivity. Positive and negative control sera should be included in every assay.
10. Remove the incubation mixture by flicking the plate and repeat the washing procedure as described in **steps 4** and **5**.
11. Immediately before use, dilute the secondary antibody (goat anti-horse IgG horseradish peroxidase-conjugated antibody) 1:25,000 in blocking buffer (*see* **Note 6**).
12. Dispense 100 μL of the diluted secondary antibody into each well and incubate at room temperature for 1 h.
13. Repeat **steps 4** and **5** plus three additional washes with PBS containing no Tween 20 to remove traces of Tween.
14. Detect the presence of peroxidase by adding 100 μL of freshly prepared chromogenic substrate solution to each well. Place plates in a dark place for 15 min at room temperature to protect from light.
15. Stop reaction by adding 100 μL of 0.5 N H_2SO_4 into each well.

16. Read plates on an MR7000 ELISA plate reader (Dynatech laboratories, USA) or on an equivalent instrument using a test wavelength of 450 nm and a reference wavelength of 630 nm. Any positives in this assay would be due to conjugated antibodies binding directly to the peptides (*see* **Note 7**).

4. Notes

1. After lyophilisation, peptides retain significant amounts of water. Peptides are oxidised over time at –20°C and slowly degrade. Thus, the peptide stock solution should be stored in small aliquots upon arrival to prevent degradation caused by repeated freezing and thawing.
2. A common problem with synthetic peptides (particularly peptides with a predominance of hydrophobic residues) is insolubility in aqueous solution. Other solvents recommended by Mimotopes are 30% (v/v) dimethylformamide, dimethyl sulfoxide, or 40% (v/v) acetonitrile in water.
3. Often the terms *offset* and *overlap* are confused. Overlapping residues are those amino acids common to two peptides, while the offset is the distance (in amino acid residues) between the N-terminal ends of the two overlapping peptides. For example, the amino acid sequences TSLNLQKDEPN, NLQKDEPNGRA, and KDEPNGRASDT overlap by eight residues and are offset by three residues.
4. The peptide stock solution (28.6 µg/mL) can be diluted further down to 1:200; however, we found that ELISA sensitivity was reduced. The amount required of each of the diluted peptides depends on how many sera are used in screening. For example, if 20 sera are tested, 40 wells are required (test in duplicate) for each peptide. Therefore, a minimum 4 mL of diluted peptide (0.286 µg/mL) should be prepared.
5. If the plates are not to be used immediately, they should be dried at 37°C before storing in the dry state at 4°C. We have tested shelf-life stability of peptide-coated ELISA plates and found that sensitivity was excellent, and well-to-well and plate-to-plate variation was minimal after 12 months of storage.
6. Note that the diluent of horseradish-peroxidase-conjugated antibodies must not contain sodium azide because this would destroy the activity of the peroxidase. In addition, the dilution of the secondary antibody must be determined for each newly purchased batch.
7. A linear B-cell epitope was identified by a universal recognition of sera to the synthetic peptides corresponding to

```
        10         20         30         40         50         60
MKTLHKTVSK AIAATAVAAA AAMIPAGCAN ATVLDSGSSS AILNSGAGSG IVGSGSYDSS

        70         80         90        100        110        120
TTSLNLQKDE PNGRASDTAG QEQQYDVHGD VISAVVYQRF HVFGPEGKVF DGDAGGLTLP

       130        140        150        160        170        180
GAGAFWGTLF TNDLQRLYKD TVSFQYNAVG PYLNINFFDS SGSFLGHIQS GGVSTVVGVG

       190
GGSGSWHNA
```

Fig. 2. Amino acid sequences of the *R. equi* VapA protein (Genbank accession No. D21236) with a linear B-cell epitope (*underlined*).

a particular fragment of the target protein. An example of results is shown in **Fig. 1**. The 51 positive sera screened against the total 50 overlapping peptides recognised four overlapping peptides Nos. 11, 12, 13, and 14. The amino acid sequences of these peptides are TSLNLQKDEPN, NLQKDEPNGRA, KDEPNGRASDT, and PNGRASDTAGQ, respectively. These peptides are equivalent to the N-terminal 20mer-epitope TSLNLQKDEPNGRASDTAGQ (amino acids 62–81) of the *R. equi* VapA protein *(9)* (**Fig. 2**). The region TSLNLQKDEPNGRASDTAGQ is thus identified as an immunodominant region for the VapA protein. B-cell epitopes are contained in this region.

Acknowledgements

We thank Stuart Rodda of Mimotopes for his advice and helpful discussion. We are grateful to Glenn Browning from University of Melbourne for kindly providing the sera. We also thank Tuck Weng Kok and the staff of the Serology Unit of the Institute of Medical and Veterinary Science for the use of equipment and facilities. This work was supported by the Rural Industries Research and Development Corporation (RIRDC) – Horse Programme and Vet Biotechnology Ltd, Adelaide, South Australia.

References

1. Phumoonna, T., Barton, M. D., and Heuzenroeder, M. W. (2005) Recognition of a B-cell epitope of the VapA protein of *Rhodococcus equi* in newborn and experimentally infected foals. *J. Vet. Med. B Infect. Dis. Vet. Public Health* 52, 291–295.
2. Vanniasinkam, T., Barton, M. D., and Heuzenroeder, M. W. (2001) B-cell epitope mapping of the VapA protein of *Rhodococcus equi*: implications for early detection of *R. equi* disease in foals. *J. Clin. Microbiol.* 39, 1633–1637.

3. Panchanathan, V., Naidu, B. R., Devi, S., Pasqule, A. D., Mason, T., and Pang, T. (1998) Immunogenic epitopes of *Salmonella typhi* GroEL heat shock protein reactive with both monoclonal antibody and patients sera. *Immunol. Lett.* 62, 105–109.
4. Norton, R. E., Heuzenroeder, M., and Manning, P. A. (1996) Antigenic epitope mapping of the M24 protein of *Streptococcus pyogenes*: implications for serodiagnosis of rheumatic fever. *FEMS Immunol. Med. Microbiol.* 16, 267–271.
5. Torres, D., and Espino, A. M. (2006) Mapping of B-cell epitopes on a novel 11.5-kilodalton F*asciola hepatica–Schistosoma mansoni* cross-reactive antigen belonging to a member of the *F. hepatica* saposin-like protein family. *Infect. Immun.* 74, 4932–4938.
6. Viswanathan, M., Anchin, J. M., Droupadi, P. R., Mandal, C., Linthicum, D. S., and Subramaniam, S. (1995) Structural predictions of the binding site architecture for monoclonal antibody NC6.8 using computer-aided molecular modeling, ligand binding, and spectroscopy. *Biophys. J.* 69, 741–753.
7. Sheriff, S., Silverton, E. W., Padlan, E. A., Cohen, G. H., Smith-Gill, S. J., Finzel, B. C., and Davies, D. R. (1987) Three-dimensional structure of an antibody–antigen complex. *Proc. Natl. Acad. Sci. USA* 84, 8075–8079.
8. Sekizaki, T., Takai, S., Egawa, Y., Ikeda, T., Ito, H., and Tsubaki, S. (1995) Sequence of the *Rhodococcus equi* gene encoding the virulence-associated 15–17-kDa antigens. *Gene* 155, 135–136.
9. Takai, S., Hines, S. A., Sekizaki, T., Nicholson, V. M., Alperin, D. A., Osaki, M., Takamatsu, D., Nakamura, M., Suzuki, K., Ogino, N., Kakuda, T., Dan, H., and Prescott, J. F. (2000) DNA sequence and comparison of virulence plasmids from *Rhodococcus equi* ATCC 33701 and 103. *Infect. Immun.* 68, 6840–6847.

Chapter 11

Antibody Epitope Mapping Using SPOT™ Peptide Arrays

Ulrich Reineke and Robert Sabat

Summary

Information at the amino acid level about the epitopes of proteins recognized by antibodies or antibody fragments is important for their use as biological and diagnostic tools, therapeutic molecules, and for understanding molecular recognition events in general. The use of chemically prepared arrays of short peptides has emerged as a powerful tool to identify and characterize antibody epitopes. In this chapter the SPOT™ synthesis technique is described in detail. In addition, three different types of peptide libraries and their applications are described: protein sequence-derived scans of overlapping peptides (peptide scans) used to locate epitopes within the protein sequence, truncation libraries used to identify the minimal peptide length required for antibody binding, and complete substitutional analyses to identify the key residues important for antibody binding.

Key words: SPOT synthesis, Peptide array, Peptide library, Epitope mapping, Antibody, Antigen, Truncation anaylsis, Peptide scan, Substitutional analysis.

1. Introduction

The identification of antibody epitopes and their characterization at the amino acid level is extremely important. Understanding antibody specificity at the molecular level provides the key to optimizing their use as research or diagnostic tools as well as their application as therapeutic agents. Among other techniques, the use of chemically prepared arrays of protein sequence-derived short peptides has emerged as a powerful tool to identify and characterize antibody epitopes (*see* also Chapter "Linear B-Cell Epitope Mapping Using Enzyme-Linked Immunosorbent Assay for Libraries of Overlapping Synthetic Peptides"). In particular, the SPOT™ synthesis technique *(1, 2)*, which is described in

Ulrich Reineke and Mike Schutkowski (eds.), *Methods of Molecular Biology, Epitope Mapping Protocols, vol. 524*

DOI: 10.1007/978-1-59745-450-6_11

detail in this chapter, is extremely well suited for this purpose, and has been widely used for epitope mapping and a variety of other applications. Three different types of peptide libraries are described: (1) Protein sequence-derived scans of overlapping peptides (peptide scans) *(3, 4)* used to locate and identify antibody epitopes (**Subheading 3.4.1**). (2) Truncation libraries used to identify the minimal peptide length required for antibody binding. In these libraries, amino acids from the termini of a peptide previously identified by a peptide scan are systematically omitted (**Subheading 3.4.2** (3) Finally, complete substitutional analyses to identify the key residues important for antibody binding (**Subheading 3.4.3**) *(5)*.

1.1. The Principle of the SPOT™ Synthesis Technique

The SPOT™ synthesis concept, a highly parallel and technically simple approach, was developed by Ronald Frank *(1, 2)*. This method is very flexible and economical in comparison with other multiple solid-phase procedures, particularly with regard to miniaturization and array geometries. The basic principle involves the positionally addressed delivery of small volumes of activated amino acid solutions directly onto a coherent membrane sheet. The areas wet by the resulting droplets can be considered as microreactors provided that a nonvolatile solvent is used. One standard membrane support material is cellulose, but other membranes are also used (**Subheadings 2.2** and **3.2**). The SPOT™ synthesis of peptide arrays employs the Fmoc strategy following the general scheme outlined in **Fig. 1**. The peptide libraries can be synthesized manually without great effort in nonspecialized laboratories simply by pipetting activated amino acids onto predefined spots on the functionalized membrane. Depending on the viscosity of the amino acid solution and the type of membrane used, a drop of 1 µL results in a spot of approximately 0.7 cm in diameter. Washing, capping, and deprotection steps, which are the same for all the peptides on the membrane, are carried out in a stainless steel dish by rinsing and shaking with the appropriate solvents and reagents. In order to facilitate the numerous pipetting steps required for SPOT™ synthesis, an automated SPOT™ synthesizer was developed and is briefly described in **Subheadings 2.8** and **3.8.**

Usually peptides between 4 and 15 amino acids in length are synthesized. Peptides of this length are certainly sufficient to identify linear, although not usually discontinuous, antibody epitopes, and have purities similar to peptides synthesized by solid-phase methods in reactors. However, longer peptides can also be synthesized, and give reliable screening results *(6)*.

Since its introduction, the SPOT™ method has become a widespread approach, mainly because of the following advantages: (1) The highly parallel synthesis format permits rapid simultaneous synthesis of many different peptides. Several hundred peptides can be prepared manually within 2–3 days. (2) The technique is very economical compared with other solid-phase synthesis methods

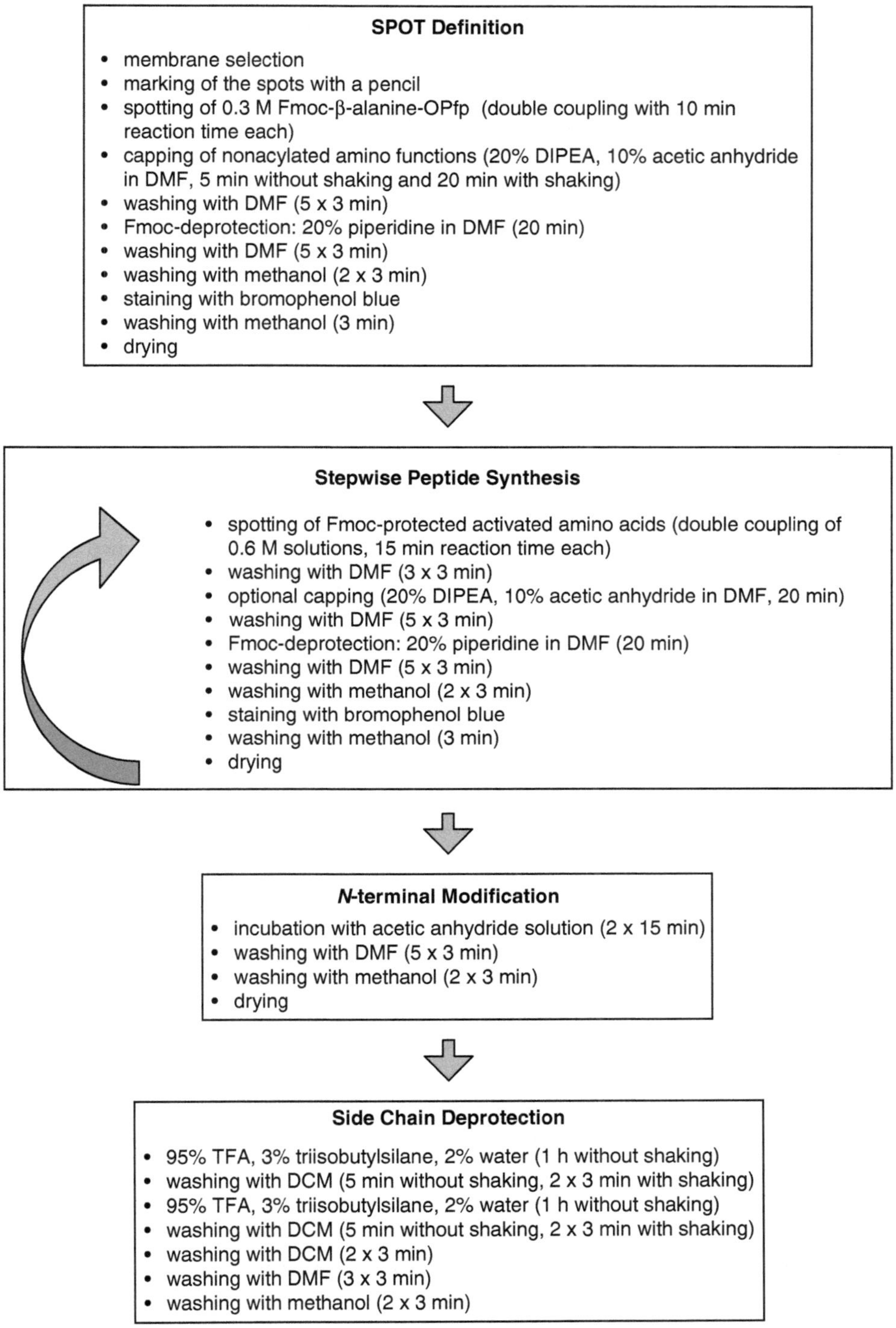

Fig. 1. General scheme summarizing the SPOT™ synthesis process.

because only very small amounts of reagents are used. Depending on the peptide length and the spot size, each spot represents between 5 and 100 nmol of the respective peptide, an amount sufficient to detect antibody binding. (3) The peptide arrays are synthesized on cellulose membranes that are compatible with many biological screening systems, for example, conventional enzyme-linked immunosorbent assay (ELISA) or Western blot procedures. All the array peptides are probed directly and simultaneously, e.g., for antibody binding, on the synthesis support. The sequence of any active peptide is already recorded on the sequence list for the array synthesis, so no coding or decoding procedures are necessary. (4) Usually, the peptide membrane can be reused several times. (5) In addition to the proteinogenic amino acids, a variety of unnatural building blocks such as D-amino acids or β-amino acids can be used. Furthermore, arrays of cyclic as well as branched peptides and peptidomimetics have been described.

1.2. Linear vs. Discontinuous Epitopes

Two different types of epitopes have to be considered. In linear (continuous) binding sites, the key amino acids that mediate the contacts to the antibody are located within one part of the primary structure, usually not exceeding 15 amino acids in length. Peptides that span these sequences have affinities to the antibody that are within the range shown by the entire protein antigen. In discontinuous (conformational) binding sites, the key residues are distributed over two or more binding regions that are separated in the primary structure. Upon folding, these binding regions are brought together on the protein surface to form a composite epitope. Even if the complete epitope forms a high-affinity interaction, peptides covering only one binding region, as synthesized in a scan of overlapping peptides, have very low affinities, which often cannot be measured in normal ELISA or surface plasmon resonance (SPR) experiments (*see* Chapter "Epitope Mapping by Surface Plasmon Resonance"). The mapping of linear epitopes using the protocols described here is an easy and straight forward approach. The incubation of a peptide scan results in the identification of one or a few consecutive overlapping peptides that are able to bind the antibody under investigation. On the other hand, the identification of discontinuous epitopes can fail due to the low affinities described earlier. However, several publications have described the mapping of these epitopes *(7)*, in which two or more binding regions are identified in the peptide scan. In general, the mapping of discontinuous epitopes is facilitated by employing an enzyme-labeled primary antibody (**Subheadings 3.9–3.11**).

1.3. Polyclonal vs. Monoclonal Antibodies (MAbs)

The techniques described here are used for epitope mapping of monoclonal as well as polyclonal antibodies. Two or more antibody binding regions may be observed for polyclonal antibodies, but in this case it is not possible to determine whether the active peptides represent linear epitopes or different binding regions of discontinuous epitopes (*see* also Chapters "Epitope Mapping Using Randomly Generated Peptide Libraries," "Probing the Epitope Signatures of IgG Antibodies in Human Serum from Patients with Autoimmune Disease," and "Microarrayed Allergen Molecules for Diagnostics of Allergy").

1.4. Other Applications

This chapter only addresses the synthesis and application of peptide arrays prepared by SPOT™ synthesis for the mapping of antibody epitopes. Over the last decade many other applications have emerged, including paratope mapping; mapping of protein-protein interactions in general; identification of kinase and protease substrates and inhibitors; identification of metal ion, DNA or cofactor binding peptides; and many others. This was accompanied by the development of novel types of peptide libraries and array-based assay techniques. These diverse applications are reviewed in detail elsewhere *(8–11)*. The most comprehensive review including a bibliography of 617 publications is *(12)*.

2. Materials

2.1. General Equipment

1. Stainless steel dish with a good closing lid that is slightly bigger than the peptide array membrane.
2. A rocker table.
3. A pipet adjustable from 0.5 to 10 µL with suitable plastic tips. A multistep version is highly recommended.
4. A fume cupboard for handling solvents and reagents. The rocker table must fit into the fume cupboard.
5. A pH-meter for adjusting the buffer solutions.

2.1.1. General Remarks for Organic Solvents

The purity of *N,N*-dimethylformamide (DMF) and 1-methyl-2-pyrrolidone (NMP) is critical for the peptide synthesis process, as their degradation can result in free amines that lead to premature deprotection of the Fmoc group or decomposition of amino acid active esters. This will reduce the yield of full-length peptide, and can cause byproduct formation. In order ensure that DMF or NMP free of amines is used, add 10 µL of 1% bromophenol blue solution in DMF to 1 mL of NMP or DMF in a 1.5-mL tube and mix thoroughly. Let this stand for 5 min and then observe the color: yellow indicates it is satisfactory to use, and yellow/green or blue/green means it should not be used.

2.2. Membranes for Peptide Arrays

Amino-functionalized membranes are commercially available from AIMS Scientific Products GmbH, Germany (http://www.aims-scientific-products.de).

2.3. Array Formatting

1. Fmoc-β-alanine-OPfp solution: 0.3 M Fmoc-β-alanine-OPfp in dimethyl sulfoxide (DMSO).
2. DMF, peptide synthesis grade. *Caution:* DMF is toxic. It may cause harm to an unborn child, is harmful through inhalation and through contact with skin, and is irritating to eyes.

3. Acetic anhydride solution: 10% acetic anhydride, 20% *N,N*-diisopropylethylamin (DIPEA) in DMF. Freshly prepare this solution prior to use.
4. Piperidine solution: 20% piperidine in DMF.
5. Methanol.
6. Staining solution: 0.01% (w/v) bromophenol blue in methanol. This solution should be yellow or orange, and must be discarded if it becomes green or green/blue.

2.4. Library Design

Standard word processing software can be used.

2.5. Stepwise Peptide Synthesis

1. All amino acids are applied as 9-fluorenylmethoxycarbonyl (Fmoc)-protected active esters (*see* **Table 1**). Prepare 0.6 M stock solutions of all amino acids except for arginine. Use NMP as solvent except for serine and threonine. These amino acids should be dissolved in DMF. Use 0.5-mL or 1.5-mL reaction tubes. Test NMP and DMF regularly for free amines as described in **Subheading 2.1.1.** The solutions are stable at –20°C for several days. The arginine solution must be prepared freshly each working day. Calculate the consumption for each peptide array carefully because the reagents are quite expensive. If necessary, vortex or sonicate to dissolve the amino acids derivatives. Store at –20°C. Allow to warm up to room temperature prior to use. Discard the stock solutions if precipitates are observed.
2. DMF.
3. Piperidine solution: 20% piperidine in DMF.
4. Methanol.
5. Staining solution: 0.01% (w/v) bromophenol blue in methanol. This solution should be yellow or orange, and must be discarded if it becomes green or green/blue.

2.6. Acetylation of the N-terminus

1. Acetic anhydride solution: 10% acetic anhydride, 20% DIPEA in DMF. Prepare this solution freshly prior to use.
2. DMF.
3. Methanol.

2.7. Side-Chain Deprotection

1. Deprotection solution: 95% trifluoroacetic acid (TFA), 3% tri-isobutylsilane, and 2% water. TFA is toxic and very corrosive, and should be handled with the greatest caution. Do not mix TFA and DMF waste, as it can undergo an exothermic and explosive reaction. Consult your safety officer for approved handling and disposal procedures.
2. Dichloromethane (DCM).
3. DMF.
4. Methanol.

Table 1
L-amino acid derivatives for SPOT™ synthesis

	Amino acid	MW [g/mol]	0.6 M 200 μL	0.6 M 400 μL	0.6 M 1.0 mL
A	Fmoc-Ala-OPfp	477.4	57.3	114.6	286.4
C	Fmoc-Cys(Trt)-OPfp	751.8	90.2	180.4	451.1
D	Fmoc-Asp(OtBu)-OPfp	577.5	69.3	138.6	346.5
E	Fmoc-Glu(OtBu)-OPfp	591.5	71.0	142.0	354.9
F	Fmoc-Phe-OPfp	553.5	66.4	132.8	332.1
G	Fmoc-Gly-OPfp	463.2	55.6	111.2	277.9
H	Fmoc-His(Boc)-OPfp	643.6	77.2	154.5	386.2
I	Fmoc-Ile-OPfp	519.5	62.3	124.7	311.7
K	Fmoc-Lys-(tBoc)-OPfp	634.6	76.1	152.3	380.8
L	Fmoc-Leu-OPfp	519.5	62.3	124.7	311.7
M	Fmoc-Met-OPfp	537.5	64.5	129.0	322.5
N	Fmoc-Asn(Trt)-OPfp	762.8	91.5	183.1	457.7
P	Fmoc-Pro-OPfp	503.4	60.4	120.8	302.0
Q	Fmoc-Gln(Trt)-OPfp	776.8	93.2	186.4	466.1
R	Fmoc-Arg(Pbf)-OPfp	814,9	97.8	195.6	488.9
S	Fmoc-Ser(tBu)-OPfp	549.5	65.9	131.9	329.7
T	Fmoc-Thr(tBu)-OPfp	563.5	67.6	135.2	338.1
V	Fmoc-Val-OPfp	505.4	60.6	121.3	303.2
W	Fmoc-Trp(tBoc)-OPfp	692.4	83.1	166.2	415.4
Y	Fmoc-Tyr(tBu)-OPfp	625.6	75.1	150.1	375.4

The amounts for the preparation of 200 μL, 400 μL, and 1 mL of the 0.6 M stock solutions are given

2.8. Automated Spot Synthesis

An automated SPOT™ synthesizer is commercially available from INTAVIS Bioanalytical Instruments AG, Germany (http://www.intavis.com/en/index.php).

2.9. Screening of Peptide Arrays

1. Methanol.
2. Tris-buffered saline (TBS): 50 mM Tris-HCl, pH to 8.0, 137 mM NaCl, 2.7 mM KCl.
3. T-TBS: TBS containing 0.05% Tween 20.
4. Blocking buffer: e.g., 1 equivalent of the blocking reagent delivered together with the BM Chemiluminescence Blotting

Substrate (POD) (Roche Diagnostics, Mannheim, Germany; Cat. No. 11500708001) to 9 equivalents of TBS.

5. Primary antibody solution: dilute the monoclonal antibody of interest to 0.1–3.0 µg/mL in blocking buffer. Polyclonal sera should be diluted between 1:100 and 1:10,000 in blocking buffer.

2.10. Detection of Antibody Binding by Chromogenic Substrates

1. T-TBS (*see* **Subheading 2.9**).
2. Blocking buffer (*see* **Subheading 2.9**).
3. Secondary antibody solution (alkaline phosphatase conjugated): dilute the antibody to 1.0 µg/mL in blocking buffer or follow the instructions of the supplier for Western blotting protocols. Make sure that the secondary antibody corresponds to the species of the primary antibody under investigation.
4. Nitroblue tetrazolium (NBT) stock solution: dissolve 500 mg NBT in 10 mL of 70% DMF in water. The stock solution is stable for at least 1 year if stored at 4°C.
5. Bromochloroindolyl phosphate (BCIP) stock solution: dissolve 500 mg BCIP disodium salt in 10 mL of DMF. The stock solution is stable for at least 1 year if stored at 4°C.
6. Alkaline phosphatase buffer: 100 mM NaCl, 5 mM $MgCl_2$, 100 mM Tris-HCl, pH 9.5.
7. Enzyme substrate solution: add 330 µL of NBT stock solution to 50 mL of alkaline phosphatase buffer. Mix well and add 165 µL of BCIP stock solution. The enzyme substrate solution must be used within 1 h.
8. Stop solution: 20 mM EDTA in phosphate buffered saline (PBS: 9.2 mM Na_2HPO_4, 1.6 mM NaH_2PO_4, 150 mM NaCl, pH7.4).

2.11. Detection of Antibody Binding by Chemiluminescence

1. T-TBS (*see* **Subheading 2.9** and **Note 1**).
2. Blocking buffer (*see* **Subheading 2.9**).
3. Secondary antibody solution (peroxidase conjugated): dilute the antibody to 1.0 µg/mL in blocking buffer or follow the instructions of the supplier for Western blotting protocols. Make sure that the secondary antibody corresponds to the species of the primary antibody under investigation (*see* **Note 1**).
4. Chemiluminescence substrate (e.g., BM Chemiluminescence Blotting Substrate (POD), Roche Diagnostics; Cat. No. 11500708001): mix luminescence substrate solution A and starting solution B 100:1 just before developing the peptide array. Other chemiluminescence substrates can also be used.
5. X-ray films, film cassette, and developing equipment or a chemiluminescence imager.

2.12. Reutilization and Storage of Peptide Arrays

1. Regeneration buffer I: 62.5 mM Tris-HCl, pH 6.7, 2% SDS in water. Add 70 μL of 2-mercaptoethanol per 10 mL SDS buffer prior to regeneration. Use this buffer only in an operating fume cupboard because of the pungent smell of 2-mercaptoethanol.
2. 10× T-TBS buffer: 500 mM Tris-HCl, pH 8.0, 1.37 M NaCl, 27 mM KCl, 0.5% Tween 20.
3. Regeneration buffer IIA: 8 M urea, 1% SDS, 0.1% 2-mercaptoethanol in water. Use this buffer only in an operating fume cupboard because of the pungent smell of 2-mercaptoethanol.
4. Regeneration buffer IIB: for 1 L, mix 400 mL of water, 500 mL of ethanol, and 100 mL of acetic acid.

3. Methods

3.1. General Overview

The protocols described in this chapter fall into two categories. In **Subheadings 3.2–3.8**, the synthesis of different protein sequence-derived peptide library arrays are described. In **Subheadings 3.9–3.12**, all processes for screening, regeneration, and storage of the peptide arrays are addressed. Solvents or solutions used in washing or incubation steps are always gently agitated on a rocker table at room temperature unless otherwise stated. During incubations and washings, the dishes are closed with a lid. Calculate the consumption of solvents and reagents carefully before starting the synthesis.

3.2. Membranes for Peptide Arrays

Membranes for the synthesis of peptide arrays are commercially available from AIMS Scientific Products GmbH, Germany (http://www.aims-scientific-products.de). Amino-functionalized cellulose APEG-membranes with a peptide loading of 400 nmol/cm^2 are recommended and are supplied in different sizes, e.g., 8 × 12 cm^2 or 10 × 15 cm^2. As an alternative, self-prepared synthesis membranes based on Whatman 50 filter paper can be used. Protocols for membrane preparation have already been described *(1, 8, 12, 13)*. The membranes must be handled with gloves during the entire processes described here. In addition, tweezers are recommended for handling wet membranes because they simplify the handling and avoid tearing.

3.3. Array Formatting

The SPOT™ method is particularly flexible with respect to spot numbers and layout of the array. Any desired array can be designed to fit the individual needs of an experiment. If necessary, the membrane can easily be cut with scissors to obtain a suitable membrane size. A standard format for the arrays is the 8 × 12 array that corresponds to 96-well microtiter plates.

All washing steps must be carried out on a rocker table that must be set up in an operating fume cupboard for safety reasons.

1. Mark as many spots as needed for the peptide library in a suitable configuration on the membrane support using a soft pencil. Graphite is stable against all the chemicals and solvents used during the synthesis and screening steps. The spots should be spaced at a distance of 1.5 cm in order to avoid cross contamination of neighboring spots during synthesis. To facilitate the stepwise synthesis process (**Subheading 3.5**) the spots can also be numbered using the pencil.
2. Use a 0.3 M Fmoc-β-alanine-OPfp solution in DMSO (*see* **Note 2**) for the first coupling step. A volume of 1.0 μL has to be spotted on each predefined position of the peptide array (*see* **Note 3**). A multistep pipet facilitates this procedure. Repeat this spotting step once after a 10 min reaction time to ensure a complete coupling. Go to **step 3** after a second 10 min reaction time.
3. Slowly place the membrane in a stainless steel dish with acetic anhydride solution (*see* **Note 4**). Avoid shaking and air bubbles. After 5 min pour off the solution and incubate the membrane with a sufficient volume of fresh acetic anhydride solution for 20 min. Shake the membrane on a rocker table. This step is carried out to acetylate the remaining amino functions that did not react with the Fmoc-β-alanine-OPfp solution in **step 2**. Thus, defined spots for the peptide library are achieved.
4. Wash the membrane 5× with DMF for 3 min each.
5. Cleave the Fmoc protecting groups by treatment of the membrane with piperidine solution for 20 min.
6. Wash the membrane 5× with DMF for 3 min each.
7. Wash the membrane twice with methanol for 3 min.
8. Rinse the membrane with staining solution (bromophenol blue is an indicator for free amino functions). The bromophenol blue solution should remain yellow and the spots should become blue leaving the surrounding membrane white. Treat the membrane until an equal blue staining of the spots is achieved.
9. Wash the membrane with methanol for 3 min to remove the remaining staining solution.
10. Air-dry the membrane. The process can be accelerated by carefully using a hair-dryer operating with cold air.

3.4. Library Design

Three different types of protein sequence-derived peptide libraries for antibody epitope identification and characterization are described: scans of overlapping peptides *(3, 4)*, truncation analyses *(12)*, and complete substitutional analyses *(5)*.

3.4.1. Scans of Overlapping Peptides

A protein sequence-derived scan of overlapping peptide is used to identify the epitope region recognized by an antibody raised against a certain protein. Therefore, the entire protein sequence or a region of special interest is dissected into short linear peptides. 15-mer peptides are recommended, as most linear binding sites do not exceed this range *(14)*. Consecutive peptides should overlap by at least 12 amino acids. Shorter overlaps can indicate that important peptides are overlooked. In peptide scans derived from proteins containing disulfide bonds or free cysteine residues, these residues should be exchanged by a similar amino acid such as serine to avoid dimerization and oligomerization of the peptides or covalent linkage to thiols of the screening antibody. As an example, a sequence list for a human interferon-γ (hIFNγ)-derived scan of overlapping peptides as well as the screening results for the anti-hIFNγ mab CB/RS/F are shown in **Fig. 2**. The manual synthesis of a peptide scan requires a high degree of concentration due to the irregular order of pipetting steps in each synthesis cycle. It is recommended to prepare a pipetting scheme for each synthesis cycle and to check off those spotting steps that have been carried out.

3.4.2. Truncation Analyses

Peptide epitopes identified from scans of overlapping peptides often comprise dispensable positions resulting from the predefined peptide length (e.g. 15-mers). Thus, truncation libraries are used to narrow down the epitope to the key interaction residues. An example of a sequence list for a truncation library designed for the peptide epitopes of two other hIFNγ antibodies, CB/RS/A and CB/RS/D, is shown in **Fig. 3**. The truncation library comprises peptides that omit *N*-, *C*-, or *N*- and *C*-terminal amino acids.

3.4.3. Complete Substitutional Analyses

Complete substitutional analysis is carried out in order to identify the key residues of a peptide epitope, i.e., those amino acids that mediate the contact with the antibody that cannot be substituted without loss of binding. This type of library consists of all possible single-site substitution analogs. A substitutional analysis library requires an array of 21 columns and as many rows as there are amino acids in the peptide of interest (**Fig. 4**). The starting peptide is synthesized in each row of the first column. In the other columns, each position is sequentially substituted by all 20 genetically encoded amino acids. In **Fig. 4**, the substitutions are arranged in alphabetical order following their one-letter-code. Alternatively, the amino acids can be grouped according to their physicochemical properties (e.g., [DE], [KRH], [NQST], [FYW], [ILVM], [APG]). Carry out the pipetting steps as described under **Subheading 3.5**, and follow these recommendations for the pipetting order:

Human Interferon-γ Sequence

CY**C**QDPYVKEAENLKKYFNAGHSDVADNGTLFLGILKNWKEESDRKIMQSQIVSFYFKLFKN
FKDDQSIQKSVETIKEDMNVKFFNSN**KKKRD**DFEKLTNYSVTDLN**VQRKAIHELIQ**VMAELS
CB/RS/F CB/RS/D

PAAKTGK**RKRSQMLFRG**RRASQ
CB/RS/A

Sequence list for a human INFγ-derived scan of overlapping peptides (15-mers, 14 amino acids overlap). Cysteine residues in the wild-type sequence (bold and underlined) are substituted by serine in the sequence list of overlapping peptides. Epitopes for mabs CB/RS/A, CB/RS/D, and CB/RS/F are highlighted.

1. SYSQDPYVKEAENLK	34. GILKNWKEESDRKIM	67. QSIQKSVETIKEDMN	100. NYSVTDLNVQRKAIH
2. YSQDPYVKEAENLKK	35. ILKNWKEESDRKIMQ	68. SIQKSVETIKEDMNV	101. YSVTDLNVQRKAIHE
3. SQDPYVKEAENLKKY	36. LKNWKEESDRKIMQS	69. IQKSVETIKEDMNVK	102. SVTDLNVQRKAIHEL
4. QDPYVKEAENLKKYF	37. KNWKEESDRKIMQSQ	70. QKSVETIKEDMNVKF	103. VTDLNVQRKAIHELI
5. DPYVKEAENLKKYFN	38. NWKEESDRKIMQSQI	71. KSVETIKEDMNVKFF	104. TDLNVQRKAIHELIQ
6. PYVKEAENLKKYFNA	39. WKEESDRKIMQSQIV	72. SVETIKEDMNVKFFN	105. DLNVQRKAIHELIQV
7. YVKEAENLKKYFNAG	40. KEESDRKIMQSQIVS	73. VETIKEDMNVKFFNS	106. LNVQRKAIHELIQVM
8. VKEAENLKKYFNAGH	41. EESDRKIMQSQIVSF	74. ETIKEDMNVKFFNSN	107. NVQRKAIHELIQVMA
9. KEAENLKKYFNAGHS	42. ESDRKIMQSQIVSFY	75. TIKEDMNVKFFNSNK	108. VQRKAIHELIQVMAE
10. EAENLKKYFNAGHSD	43. SDRKIMQSQIVSFYF	76. IKEDMNVKFFNSNKK	109. QRKAIHELIQVMAEL
11. AENLKKYFNAGHSDV	44. DRKIMQSQIVSFYFK	77. KEDMNVKFFNSNKKK	110. RKAIHELIQVMAELS
12. ENLKKYFNAGHSDVA	45. RKIMQSQIVSFYFKL	78. EDMNVKFFNSNKKKR	111. KAIHELIQVMAELSP
13. NLKKYFNAGHSDVAD	46. KIMQSQIVSFYFKLF	79. DMNVKFFNSNKKKRD	112. AIHELIQVMAELSPA
14. LKKYFNAGHSDVADN	47. IMQSQIVSFYFKLFK	80. MNVKFFNSNKKKRDD	113. IHELIQVMAELSPAA
15. KKYFNAGHSDVADNG	48. MQSQIVSFYFKLFKN	81. NVKFFNSNKKKRDDF	114. HELIQVMAELSPAAK
16. KYFNAGHSDVADNGT	49. QSQIVSFYFKLFKNF	82. VKFFNSNKKKRDDFE	115. ELIQVMAELSPAAKT
17. YFNAGHSDVADNGTL	50. SQIVSFYFKLFKNFK	83. KFFNSNKKKRDDFEK	116. LIQVMAELSPAAKTG
18. FNAGHSDVADNGTLF	51. QIVSFYFKLFKNFKD	84. FFNSNKKKRDDFEKL	117. IQVMAELSPAAKTGK
19. NAGHSDVADNGTLFL	52. IVSFYFKLFKNFKDD	85. FNSNKKKRDDFEKLT	118. QVMAELSPAAKTGKR
20. AGHSDVADNGTLFLG	53. VSFYFKLFKNFKDDQ	86. NSNKKKRDDFEKLTN	119. VMAELSPAAKTGKRK
21. GHSDVADNGTLFLGI	54. SFYFKLFKNFKDDQS	87. SNKKKRDDFEKLTNY	120. MAELSPAAKTGKRKR
22. HSDVADNGTLFLGIL	55. FYFKLFKNFKDDQSI	88. NKKKRDDFEKLTNYS	121. AELSPAAKTGKRKRS
23. SDVADNGTLFLGILK	56. YFKLFKNFKDDQSIQ	89. KKKRDDFEKLTNYSV	122. ELSPAAKTGKRKRSQ
24. DVADNGTLFLGILKN	57. FKLFKNFKDDQSIQK	90. KKRDDFEKLTNYSVT	123. LSPAAKTGKRKRSQM
25. VADNGTLFLGILKNW	58. KLFKNFKDDQSIQKS	91. KRDDFEKLTNYSVTD	124. SPAAKTGKRKRSQML
26. ADNGTLFLGILKNWK	59. LFKNFKDDQSIQKSV	92. RDDFEKLTNYSVTDL	125. PAAKTGKRKRSQMLF
27. DNGTLFLGILKNWKE	60. FKNFKDDQSIQKSVE	93. DDFEKLTNYSVTDLN	126. AAKTGKRKRSQMLFR
28. NGTLFLGILKNWKEE	61. KNFKDDQSIQKSVET	94. DFEKLTNYSVTDLNV	127. AKTGKRKRSQMLFRG
29. GTLFLGILKNWKEES	62. NFKDDQSIQKSVETI	95. FEKLTNYSVTDLNVQ	128. KTGKRKRSQMLFRGR
30. TLFLGILKNWKEESD	63. FKDDQSIQKSVETIK	96. EKLTNYSVTDLNVQR	129. TGKRKRSQMLFRGRR
31. LFLGILKNWKEESDR	64. KDDQSIQKSVETIKE	97. KLTNYSVTDLNVQRK	130. GKRKRSQMLFRGRRA
32. FLGILKNWKEESDRK	65. DDQSIQKSVETIKED	98. LTNYSVTDLNVQRKA	131. KRKRSQMLFRGRRAS
33. LGILKNWKEESDRKI	66. DQSIQKSVETIKEDM	99. TNYSVTDLNVQRKAI	132. RKRSQMLFRGRRASQ

Incubation of an Interferon-γ-Derived Peptide Scan with mab CB/RS/F

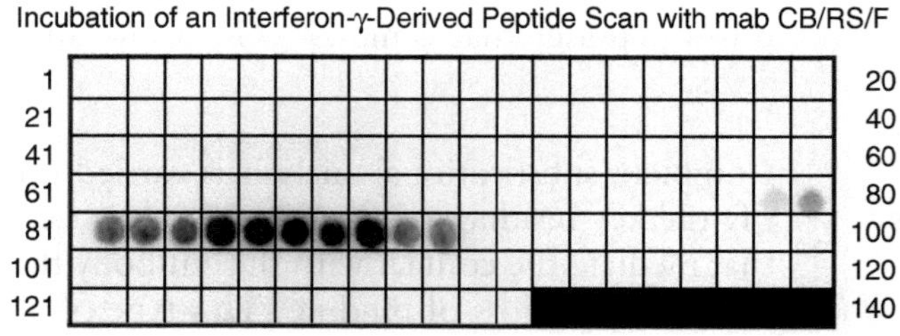

79 DMNVKFFNSN**KKKRD**
80 MNVKFFNSN**KKKRD**D
81 NVKFFNSN**KKKRD**DF
82 VKFFNSN**KKKRD**DFE
83 KFFNSN**KKKRD**DFEK
84 FFNSN**KKKRD**DFEKL
85 FNSN**KKKRD**DFEKLT
86 NSN**KKKRD**DFEKLTN
87 SN**KKKRD**DFEKLTNY
88 N**KKKRD**DFEKLTNYS
89 **KKKRD**DFEKLTNYSV

Fig. 2. Scan of overlapping peptides. The entire human interferon-γ (hIFNγ) sequence *(top)* is dissected into 15-mer overlapping peptides (14 amino acids overlap), resulting in a sequence list for the SPOT™ synthesis process of a total of 132 peptides *(middle)*. The resulting peptide library array was incubated with the anti-hIFNγ mab CB/RS/F (mouse IgG). Subsequently, peptide-bound antibody was detected using a peroxidase-labeled secondary antibody in combination with a chemiluminescence substrate and an imaging system (*lower panel*). Several consecutive peptides were identified that bind mab CB/RS/F. The sequence common to these peptides is framed.

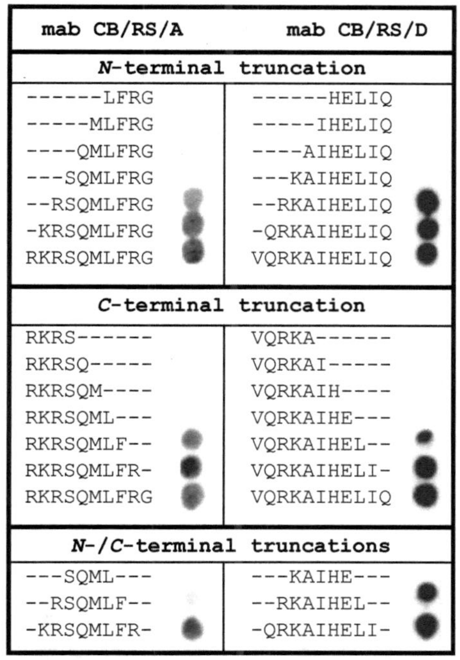

Fig. 3. Truncation analysis peptide library. *N*-, *C*-, and bi-directional truncations of the peptide epitopes of anti-hIFNγ mabs CB/RS/A and CB/RS/D (both mouse IgG) are shown with the sequences and the incubation results using a peroxidase-labeled secondary antibody in combination with a chemiluminescence substrate and an imaging system.

1. In the first coupling step, pipet the *C*-terminal amino acid of the starting peptides onto all rows except the last one.
2. Pipet the same amino acid onto the first spot of the bottom row.
3. Spot all 20 amino acids successively onto the remaining 20 spots in alphabetical order.
4. Use an analogous procedure for the remaining synthesis steps.

3.5. Stepwise Peptide Synthesis

The synthesis of membrane-bound peptides is carried out in an iterative process (**Fig. 1**). This cycle includes double-coupling of activated amino acid solutions, washing with DMF to remove excess amino acids, Fmoc deprotection with piperidine solution, washing with DMF and methanol, staining, washing with methanol, and drying. Subsequently, the next coupling step is carried out. All peptide sequences are written from the N-terminus (left) to the *C*-terminus (right). However, the peptides are chemically synthesized from *C*- to the N-terminus, i.e., for the peptide RKRSQMLFRG (**Fig. 4**) the synthesis steps are: G→R→F→L→M→Q→S→R→K→R.

a

wt	Ala (A)	Cys (C)	Asp (D)	G H I ...
RKRSQMLFRG	**A**KRSQMLFRG	**C**KRSQMLFRG	**D**KRSQMLFRG	..
RKRSQMLFRG	R**A**RSQMLFRG	R**C**RSQMLFRG	R**D**RSQMLFRG	..
RKRSQMLFRG	RK**A**SQMLFRG	RK**C**SQMLFRG	RK**D**SQMLFRG	..
RKRSQMLFRG	RKR**A**QMLFRG	RKR**C**QMLFRG	RKR**D**QMLFRG	..
RKRSQMLFRG	RKRS**A**MLFRG	RKRS**C**MLFRG	RKRS**D**MLFRG	..
RKRSQMLFRG	RKRSQ**A**LFRG	RKRSQ**C**LFRG	RKRSQ**D**LFRG	..
RKRSQMLFRG	RKRSQM**A**FRG	RKRSQM**C**FRG	RKRSQM**D**FRG	..
RKRSQMLFRG	RKRSQML**A**RG	RKRSQML**C**RG	RKRSQML**D**RG	..
RKRSQMLFRG	RKRSQMLF**A**G	RKRSQMLF**C**G	RKRSQMLF**D**G	..
RKRSQMLFRG	RKRSQMLFR**A**	RKRSQMLFR**C**	RKRSQMLFR**D**	..

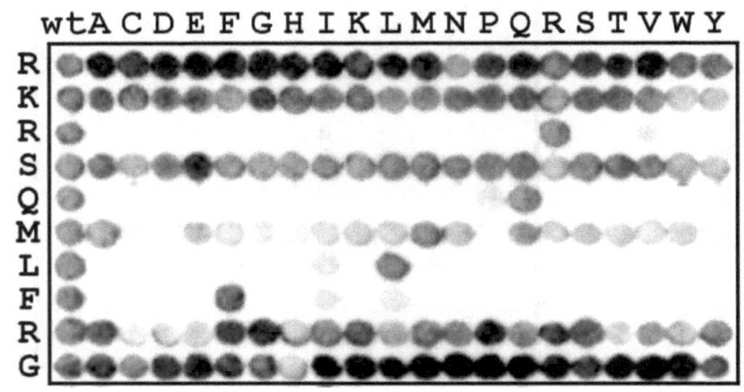

b

wt	Ala (A)	Cys (C)	Asp (D)	G H I ...
VQRKAIHELIQ	**A**QRKAIHELIQ	**C**QRKAIHELIQ	**D**QRKAIHELIQ	..
VQRKAIHELIQ	V**A**RKAIHELIQ	V**C**RKAIHELIQ	V**D**RKAIHELIQ	..
VQRKAIHELIQ	VQ**A**KAIHELIQ	VQ**C**KAIHELIQ	VQ**D**KAIHELIQ	..
VQRKAIHELIQ	VQR**A**AIHELIQ	VQR**C**AIHELIQ	VQR**D**AIHELIQ	..
VQRKAIHELIQ	VQRK**A**IHELIQ	VQRK**C**IHELIQ	VQRK**D**IHELIQ	..
VQRKAIHELIQ	VQRKA**A**HELIQ	VQRKA**C**HELIQ	VQRKA**D**HELIQ	..
VQRKAIHELIQ	VQRKAI**A**ELIQ	VQRKAI**C**ELIQ	VQRKAI**D**ELIQ	..
VQRKAIHELIQ	VQRKAIH**A**LIQ	VQRKAIH**C**LIQ	VQRKAIH**D**LIQ	..
VQRKAIHELIQ	VQRKAIHE**A**IQ	VQRKAIHE**C**IQ	VQRKAIHE**D**IQ	..
VQRKAIHELIQ	VQRKAIHEL**A**Q	VQRKAIHEL**C**Q	VQRKAIHEL**D**Q	..
VQRKAIHELIQ	VQRKAIHELI**A**	VQRKAIHELI**C**	VQRKAIHELI**D**	..

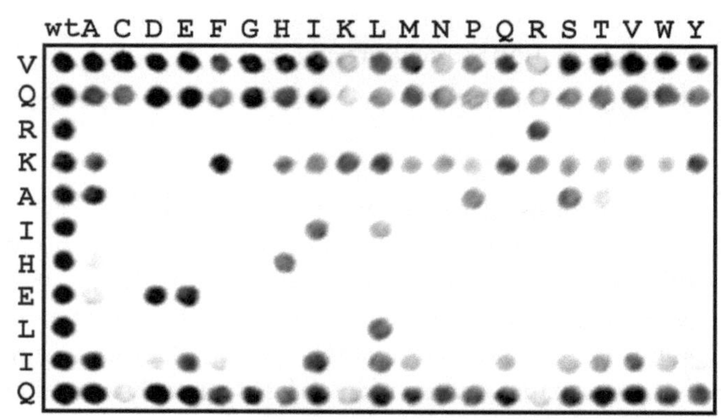

Fig. 4. Complete substitutional analysis libraries of the mab CB/RS/A and CB/RS/D epitopes. In this experiment all possible single-site substitution analogs were synthesized and screened for mab CB/RS/A (**a**) or CB/RS/D (**b**) binding as described in **Fig. 2**. As an example, the peptide sequences of the four left hand columns of each substitutional analysis library are shown at upper of each panel. The results after screening for antibody binding are shown below. Four key residues were identified for mab CB/RS/A, namely arginine at position 3, glutamine at position 5, leucine at position 7, and phenylalanine at positions 8. Critical residues for mab CB/RS/D are arginine 3, alanine 5, isoleucine 6, histidine 7, glutamic acid 8, and leucine 9.

All washing steps must be carried out on a rocker table that must be set up in an operating fume cupboard for safety reasons.

1. Spot 1.5 μL of the activated amino acids onto the respective spots. Use a clean pipet tip for each amino acid. As soon as the droplets of activated amino acid solutions are added to the spots, coupling proceeds with a conversion of free N-terminal amino groups to amide bonds. During elongation of the peptides, a larger volume is spotted compared with array formatting (*see* **Subheading 3.3**, **step 2**). Thus, incomplete coupling at the edges is avoided. Repeat this step after a reaction time of 15 min. Go to the next step after an additional reaction time of 15 min. The coupling reaction can be followed by a color change from blue to green/blue or even yellow. Because the solvent within the spots is slowly evaporating over the reaction time, additional droplets may be added onto the same position without enlarging the spots and risking overlap with neighboring spots. Thus, difficult coupling reactions may be brought to completion by double or triple couplings.
2. Wash the membrane 3× with DMF for 3 min each (*see* **Note 4**).
3. Cleave the Fmoc protecting groups by treatment of the membrane with piperidine solution for 20 min.
4. Wash the membrane 5× with DMF for 3 min each.
5. Wash the membrane twice with methanol for 3 min.
6. Rinse the membrane with staining solution. The bromophenol blue solution should remain yellow, and the spots should become blue. Treat the membrane until an equal blue staining of the spots is achieved (*see* **Note 5**).
7. Wash the membrane with methanol for 3 min to remove the remaining staining solution.
8. Air-dry the membrane. The process can be accelerated by carefully using a hair-dryer, operating with cold air.
9. Go to **step 1** for coupling the next amino acid (*see* **Note 6**). The synthesis can also be stopped at this step to proceed another day. In this case, the membrane should be placed in a plastic bag, sealed, and stored at –20°C.
10. After coupling the last amino acid of the longest peptide of the array follow the procedure without the **step 6** staining with bromophenol blue.

3.6. Acetylation of the N-terminus

This capping step is carried out because acetylation at the N-terminus stabilizes the peptide against proteolytic degradation. Furthermore, this counteracts the N-terminal positive charge, which is an artifact caused by dissecting the protein sequence into short peptides. All steps of this subheading must be carried out on a rocker table that must be set up in an operating fume cupboard for safety reasons.

1. Incubate the membrane twice with acetic anhydride solution for 15 min each.
2. Wash the membrane 5× with DMF for 3 min.
3. Wash the membrane twice with methanol for 3 min.
4. Dry the membrane.

3.7. Side-Chain Deprotection

In the final synthesis step, all side-chain protecting groups are cleaved off. All steps of this subheading must be carried out in an operating fume cupboard for safety reasons.

1. Place the membrane carefully in a stainless steel dish with deprotection solution (*see* **Note 4**). Follow the safety instructions (**Subheading 2.7**). Close the dish tightly with an appropriate lid, put the entire dish into a plastic bag, and seal it. Do not use a rocker table in **steps 1–3**. Incubate the membrane for 1 h and remove the deprotection solution. Now, the membrane is mechanically extremely unstable and has to be handled carefully to avoid tearing.
2. Wash the membrane once with DCM without agitation for 5 min. Pour the solvent into the dish at an edge very slowly to avoid subjecting the membrane to mechanical stress. Wash the membrane twice with DCM under gentle agitation for 3 min.
3. Repeat **step 1** and **2** once.
4. Wash the membrane two more times with DCM under gentle agitation for 3 min.
5. Wash the membrane 3× with DMF for 3 min.
6. Wash the membrane twice with methanol for 3 min.
7. Dry the membrane. Proceed with the screening (**Subheading 3.9**), or store the membrane at –20°C.

3.8. Automated SPOT™ Synthesis

Although the SPOT™ synthesis technique is a robust and easy-to-use technique that can be carried out even in nonspecialized laboratories, the disadvantages of manual synthesis are obvious: the numerous pipetting and synthesis steps are rather time consuming and it is difficult to ensure the required precision during the manual pipetting steps. It was particularly the laborious pipetting steps that led to the development of an automated SPOT™ synthesizer, providing a maximum amount of precision and reliability for the synthesis. This AutoSpot synthesizer is available from INTAVIS Bioanalytical Instruments AG (Köln, Germany). Spot diameters are between 1.5 mm (50 nL dispensing volume) and 1.0 cm (1–2 μL dispensing volume). This unit facilitates consecutive delivery of peptide building blocks from a rack of 44 reservoirs onto the membrane surface. However, washing, capping, and deprotection steps still must be performed manually. In order to manage the increasing amount of data required to control the automated synthetic process, a software package that allows the development of appropriate sequence and chemistry files for a variety of library types has been devised together with the AutoSpot robot. For detailed protocols, please refer to the manufacturers instructions.

3.9. Screening of Peptide Arrays

The screening strategy depends on whether the antibody under investigation is available in an enzyme-labeled form or if an enzyme-labeled secondary antibody must be used. The protocols described in **Subheadings 3.9–3.11** are applicable in both cases. It is helpful to know if the primary antibody is active in Western blot experiments. If it is, this is a strong indication that it recognizes a linear epitope (*see* **Subheading 1.2**), and implies a relatively high-affinity antibody-peptide interaction. Thus, a low concentration of the primary antibody can be used in **step 4**.

If a secondary antibody is used, it is essential to ensure that there is no detectable binding of this antibody to the peptides. Thus, for the detection by chromogenic substrates, the membrane must be incubated according to the protocols in **Subheadings 3.9** and **3.10**. The only difference is that **step 4** of **Subheading 3.9** is carried out with no primary antibody. The membrane should be incubated with the substrate solution (**Subheading 3.10, step 3**) for 30 min. Similarly, for the detection by chemiluminescence, the membrane must be incubated according to the protocols in **Subheadings 3.9** and **3.11**, except that **step 4** of **Subheading 3.9** is carried out with no primary antibody. Expose the X-ray film for at least 1 h (**Subheading 3.11, step 5**). A second exposure can be omitted. If a chemiluminescence imager is used, record an image for 30 min. Spots that give signals in these control experiments must be excluded as false-positives in the final assessment. Perform the control experiments before the main incubation.

All washing and incubation steps of the protocols described in **Subheadings 3.9–3.11** should be carried out on a rocker table at room temperature.

1. Rinse the peptide array membrane with a small volume of ethanol for 1 min. This is done to facilitate the wetting of hydrophobic peptides for the following aqueous washing and incubation steps.
2. Wash the membrane 3× with an appropriate volume of TBS for 10 min each. The membrane should be sufficiently covered by the solution.

3. Block the membrane with a sufficient volume of blocking buffer for 3 h at room temperature (*see* **Note 4**).
4. Incubate the membrane with the same volume of 0.1–3.0 µg/mL primary antibody solution (*see* **Notes** 7 and **8**) in blocking buffer for 3 h. If the primary antibody is directly labeled with alkaline phosphatase (*see* **Note 9**) go to **step 2** in **Subheading 3.10** or with peroxidase to **step 2** in **Subheading 3.11**. If not, proceed with the following steps:
5. Wash the membrane 3× with T-TBS for 10 min each.
6. Go to **Subheadings 3.10** or **3.11**.

3.10. Detection of Antibody Binding by Chromogenic Substrates

If a chemiluminescence imager or a dark room is available, the protocol for detecting antibody binding by chemiluminescence is recommended because of the higher sensitivity (**Subheading 3.11**). Otherwise, the following alkaline phosphatase protocol should be used. The BCIP/NBT substrate generates an intense black–purple precipitate where the enzyme-conjugated antibody has been bound.

1. Incubate the membrane with a sufficient volume of alkaline phosphatase-labeled secondary antibody solution (specific for the primary antibody that was used for the screening) at a concentration of 1 µg/mL in blocking buffer for 2 h, or follow the supplier's instructions for Western blotting (*see* **Note** 7).
2. Wash the membrane 3× with T-TBS for 10 min each.
3. Incubate the membrane with the enzyme substrate solution. Go to **step 4** when the spots have turned suitably dark. This depends on the membrane-bound enzyme activity, and takes between 1 and 30 min.
4. Stop the reaction by rinsing the membrane with stop solution 3× for 3 min each. The Mg^{2+} ions that are essential for alkaline phosphatase activity are complexed by ethylenediaminetetraacetic acid (EDTA) in the stop solution. For long-term documentation of the results, the membrane should be photographed or scanned. If the peptide array is not to be used further the membrane itself can stored as permanent record: wash the membrane 3× with water for 5 min, twice with methanol for 3 min, and dry. If the membrane is to be used again, wash the membrane 3× with T-TBS to remove excess stop solution and proceed with the regeneration procedures described in **Subheading 3.12**.

3.11. Detection of Antibody Binding by Chemiluminescence

Detection by chemiluminescence is highly sensitive, and has short imaging times ranging from a few seconds to 1 h. The reaction reaches its maximum after 1–2 min, and is relatively constant for 20–30 min. After 1 h, the signal intensity decreases to approx 60–70% of the maximum (*see* **Note 1**).

1. Incubate the membrane with a sufficient volume of peroxidase-labeled secondary antibody solution (specific for the primary antibody that was used for the screening) at a concentration of 1 μg/mL in blocking buffer for 2 h or follow the supplier's instructions for Western blotting (*see* **Notes 7** and **8**).
2. Wash the membrane 3× with T-TBS for 10 min each.
3. The following steps must be carried out in a dark room. If a chemiluminescence imager is available, go to **step 9**. Rinse the membrane with the chemiluminescence substrate for about 1 min. The membrane has to be completely covered with the detection solution, which roughly corresponds to 50–100 μL/cm^2. Use a pipet to spread the solution repeatedly over the membrane.
4. Place the membrane in an X-ray film cassette that is lined with plastic film and cover it with film, which must be transparent.
5. Turn dark room light to red and place an X-ray film onto the membrane. Close the cassette and expose for 60 s.
6. Upon opening the cassette, immediately replace the exposed film with a new one, then close and set aside the cassette. Develop the first exposed film at once.
7. Expose the second film for a suitable time (up to 45 min) estimated from the signal intensity on the first film (*see* **Notes 7**, **8**, and **10**).
8. The films can be digitized and processed using a standard scanner.
9. Procedure if a chemiluminescence imager is available: rinse the membrane with the chemiluminescence substrate for about 1 min. The membrane must be completely covered with the detection solution, which roughly corresponds to 50–100 μL/cm^2. Use a pipet to spread the solution repeatedly over the membrane.
10. Place the membrane on the imager and record an image for 30 s. Make additional images with recording times adjusted according to the signal intensities and the signal-to-noise ratio of the first image.
11. Wash the membrane 3× with T-TBS to remove excess substrate solution. Proceed with the regeneration procedures described in **Subheading 3.12**.

3.12. Reutilization and Storage of Peptide Arrays

To reuse the peptide arrays, peptide-bound antibody must be completely removed. Usually, the peptide membranes can be used several times, but in a few cases regeneration fails as a result of strong binding of mature or denatured antibodies to the peptides or the cellulose. Two different protocols described in **Subheadings 3.12.1** and **3.12.2** are applicable. If the chromogenic substrate precipitate is not removed by these protocols, wash the membrane with DMF overnight, then 3× with methanol, and dry.

3.12.1. Regeneration Protocol I

1. Wash the membrane 3× with water for 10 min.
2. Wash the membrane 3× for 10 min with regeneration buffer I at 50°C. Temperatures above 50°C can harm the membrane and/or the peptides.
3. Wash the membranes 3× for 10 min with 10× T-TBS at room temperature.
4. Wash the membranes 3× for 10 min with T-TBS at room temperature.
5. Proceed with the next incubation or store the membrane after washing 3× with water, twice with methanol, and drying.

3.12.2. Regeneration Protocol II

1. Wash the membrane twice with water for 10 min.
2. Incubate the membrane 3× with regeneration buffer IIA for 10 min.
3. Incubate the membrane 3× with regeneration buffer IIB for 10 min.
4. Wash the membrane with water for 10 min.
5. Wash the membrane 3× with T-TBS.
6. Proceed with the next incubation or store the membrane after washing 3× with water, twice with methanol, and drying.

3.12.3. Regeneration Controls

1. If the membrane was incubated with a directly labeled antibody, check the success of the regeneration by rinsing the membrane in substrate solution and then exposing it at least as long as the original exposure if the chemiluminescence method was used, or develop with the chromogenic substrates as in the main experiment. If spots are still detected repeat regeneration protocol I (**step 2** can be prolonged), or go to regeneration protocol II.
2. If the membrane was incubated with a primary antibody in combination with an enzyme-labeled secondary antibody, re-incubate the membrane with the secondary antibody and substrate solution and make an exposure at least as long as the original exposure to show that the primary antibody is completely removed. Perform an analogous procedure if the binding was detected by chromogenic substrates. If spots are still detectable repeat regeneration protocol I (**step 2** can be prolonged) or go to regeneration protocol II.

3.12.4. Storage of Peptide Membranes

1. New membranes should be stored at –20°C until use, where the membranes are placed in a plastic bag and sealed.

2. Incubated membranes that will be used again within a few days should be washed 3× with T-TBS for 10 min and kept with a small volume of T-TBS in a petri dish at 4°C. Drying out of the membrane sometimes leads to poor results in subsequent experiments.
3. Incubated membranes that will be stored for a longer period should be regenerated according to **Subheading 3.12**, washed with methanol twice, air dried, and kept at −20°C.

4. Notes

1. Do not use sodium azide as a preservative in buffers with peroxidase as it is an inhibitor of the enzyme. The presence of azide will greatly reduce or eliminate the signal.
2. If spots with a white center and a dark ring ("ring spots") are obtained after incubation and development of the array, a peptide membrane with a lower peptide loading is recommended *(15)*. This can be achieved by mixing 0.3 M Fmoc-β-alanine-OPfp and 0.3 M Ac-β-alanine-OPfp 1:9 and using this solution for array formatting (**Subheading 3.3, step 2**).
3. The spotting process is most conveniently done by first dispensing the volume to be spotted and then gently touching the pipet tip to the center of each marked spot. This allows most accurate liquid handling.
4. Do not overlay two or more peptide membranes in one incubation dish.
5. Bromophenol blue staining (**Subheading 3.5**, **step 6**): the intensity of staining varies depending on the last spotted amino acid. Some amino acids, such as cysteine, aspartic acid, glutamic acid, and asparagine stain weakly. Alanine, glycine, and proline stain more strongly than others. These differences may serve as an internal control for correct pipetting. During later cycles, the intensity of staining diminishes.
6. Peptides of different length: if a peptide library comprises peptides of various length, the shorter peptides can accidentally be elongated at subsequent cycles by transferring excess Fmoc amino acid active esters from other spots during **step 2** from **Subheading 3.5.** This effect is very unlikely and usually ignored. However, in order to absolutely exclude this peptide elongation, **steps 1** and **2** from **Subheading 3.6** can be carried out in addition between **steps 2** and **3** from

Subheading 3.5. Acetylated spots can no longer be stained by bromophenol blue. This procedure also leads to capping of peptides that did not properly react with the last amino acid.

7. If no signals are observed using the protocols described in **Subheadings 3.9–3.11**, change the protocols as follows (one or more modifications can be adopted simultaneously):
   - Increase the antibody concentrations of the primary and/or secondary antibody.
   - Prolong the incubation time with the primary antibody to overnight. This step should be performed at 4°C.
   - Shorten the washing times. Use washing buffer without Tween 20.
   - Use a supersensitive chemiluminescence substrate (e.g. SuperSignal®Ultra, Pierce, Rockford, IL).
   - Perform a simultaneous incubation of primary and secondary antibody.
   - Check the antibodies and enzymes in an alternative system.
8. Clear spots on dark background can be observed with the chemiluminescence protocol (**Subheading 3.11**). In this case, the primary and/or secondary antibody concentrations may be too high. A high amount of antibody conjugate on the spots results in all the substrate being used up before the X-ray film can be exposed on the membrane or the imaging system can be started. To avoid this, change the protocols as follows:
   - Wash extensively with T-TBS and re-detect.
   - If the problem persists, regenerate the membrane and incubate with lower concentrations of antibodies.
9. Results with directly labeled antibodies are often much better, especially for low-affinity binding antibodies, and may even be essential for the mapping of discontinuous epitopes. Conjugation of antibodies with peroxidase is easy to perform, and protocols are available *(16)*.
10. If the background is too high, change the protocols as follows:
    - Increase the detergent concentration in the washing buffer.
    - Increase the washing times and/or the washing volumes.

References

1. Frank, R. (1992) Spot synthesis: an easy technique for the positionally addressable, parallel chemical synthesis on a membrane support. *Tetrahedron* 48, 9217–9232.
2. Frank, R. (Date of filing: 31.08.1990) Verfahren zur parallelen Herstellung von trägergebundenen oder freien Peptiden, trägergebundene Peptide and ihre Verwendung. DE 4027675.
3. Geysen, H. M., Rodda, S. J., Mason, T. J., Tribbick, G., and Schofs, P. G. (1987) Strategies for epitope analysis using peptide synthesis. *J. Immunol. Methods* 102, 259–274.
4. Geysen, H. M. (Date of filing: 08. 03. 1984) Method for determining antigenically active amino acids sequences. EP 0138855.
5. Pinilla, C., Appel, J. R., and Houghten, R. (1993) Functional importance of amino acid residues making up antigenic determinants. *Mol. Immunol.* 30, 577–585.
6. Molina, F., Laune, D., Gougat, C., Pau, B., and Granier, C. (1996) Improved performances of spot multiple peptide synthesis. *Peptide Res.* 9, 151–155.
7. Reineke, U., Kramer, A., and Schneider-Mergener, J. (1999) Antigen sequence- and library-based mapping of linear and discontinuous protein-protein-interaction sites by spot synthesis. *Curr. Top. Microbiol. Immunol.* 243, 23–36.
8. Wenschuh, H., Volkmer-Engert, R., Schmidt, M., Schulz, M., Schneider-Mergener J., and Reineke U. (2000) Coherent membrane supports for parallel microsynthesis and screening of bioactive peptides. *Biopolymers (Peptide Science)* 55, 188–206.
9. Reineke, U., Volkmer-Engert R., and Schneider-Mergener, J. (2001) Applications of peptide arrays prepared by the SPOT-technology. *Curr. Opin. Biotechnol.* 12, 59–64.
10. Frank, R. and Schneider-Mergener, J. (2002) SPOT-synthesis: scope and applications, in *Peptide Arrays on Membrane Supports – Synthesis and Applications, Springer Laboratory Manual* (Koch, J., and Mahler, M., eds.), Springer-Verlag, Heidelberg, Germany, pp. 1–22.
11. Reimer, U., Reineke, U., and Schneider-Mergener, J. (2002) Peptide arrays: from macro to micro. *Curr. Opin. Biotechnol.* 13, 315–320.
12. Reineke, U., Schneider-Mergener, J., and Schutkowski, M. (2006) Peptide arrays in proteomics and drug discovery, in *BioMEMS and Biomedical Nanotechnology, Volume II Micro/Nano Technology for Genomics and Proteomics* (Ozkan, M., and Heller, M. J., eds.), Springer-Verlag, Berlin, pp. 161–282.
13. Ast, T., Heine, N., Germeroth, L., Schneider-Mergener, J., and Wenschuh, H. (1999) Efficient assembly of peptomers on continuous surfaces. *Tetrahedron Lett.* 40, 4317–4318.
14. Van Regenmortel, M. H. V. (1994) The recognition of proteins and peptides by antibodies, in *Immunochemistry* (van Oss, C. J., and van Regenmortel, M. H. V., eds.), Marcel Dekker, New York, NY, pp. 277–300.
15. Kramer, A., Reineke, U., Dong, L., Hoffmann, B., Hoffmüller, U., Winkler, D., Volkmer-Engert, R., and Schneider-Mergener, J. (1999) Spot synthesis: observations and optimizations. *J. Peptide Res.* 54, 319–327.
16. Wilson, M. B. and Nakane, P. K. (1978) Recent developments in the periodate method of conjugating horseradish peroxidase (HRPO) to antibodies, in *Immunofluorescence Related Staining Techniques* (Knapp, W., Holubar, K., and Wick, G., eds.), Elsevier, Amsterdam, pp. 215–224.

Chapter 12

Peptide Microarrays for Profiling of Modification State-Specific Antibodies

Johannes Zerweck, Antonia Masch, and Mike Schutkowski

Summary

The reversible phosphorylation of serine, threonine, and tyrosine residues is one of the most important intracellular post-translational modifications regulating enzymatic activities and protein/protein interaction in eukaryotic cells. Tools for determining phosphorylation status of proteins and peptides play a prominent role in signal transduction research and proteomics. Pan-specific antibodies claimed to recognize modified amino acid residues independent on the nature of surrounding residues in peptides and proteins are widely used. We used high-content phosphopeptide microarrays and microarrays displaying acetyllysine-containing peptides for comprehensive characterization of commercially available generic anti-phosphopeptide and anti-acetyllysine antibodies. We were able to demonstrate distinct subsite specificity and cross-reactivity for such antibodies.

Key words: Peptide microarray, Phospho-specific antibody, Cross-reactivity, Anti-acetyllysine antibody, Subsite specificity.

1. Introduction

The amount of active protein in cells critically depends on control of translation, assisted folding, trafficking, binding to cofactors and interaction partners, protein turnover, as well as post-translational modification. For example, the actual activity of an enzyme can be altered by a factor of several thousands by kinase-mediated phosphorylation which is one of the key post-translational modifications. Recent estimates suggest that there are more than 500 genes encoding for protein kinases *(1)*. However, only a fraction of these enzymes have thus far been characterized and still little is understood regarding the physiological functions,

Ulrich Reineke (eds.), *Methods in Molecular Biology, Epitope Mapping Protocols, vol. 524*

DOI: 10.1007/978-1-59745-450-6_12

substrate specificity, and downstream targets of the majority of this class of enzymes. The development of phosphorylation-state specific antibodies enabled studies of protein phosphorylation in situ allowing insights into dynamics of this post-translational modification in the spatially complex structures and compartments of cells.

Another important post-translational modification of various proteins is acetylation of lysine residues. Recently two system-wide proteomic analyses using pan-specific anti-acetyllysine antibodies were performed *(2, 3)*, demonstrating that this modification is more abundant in vivo than expected.

Although high-quality pan-specific antibodies are available for phosphotyrosine detection, antibodies specifically recognizing phosphoserine and phosphothreonine residues show subsite specificities for surrounding residues. Limited knowledge about this subsite specificity and cross-reactivity of pan-specific anti-phosphopeptide antibodies requires careful interpretation of experimental results *(4–7)*.

Here we describe the use of high-content phosphopeptide microarrays for the efficient profiling of pan-specific anti-phosphopeptide antibodies. Additionally, we show that peptide microarrays displaying peptides which were chemically acetylated are useful tools to prove quality of pan-specific anti-acetyllysine antibodies.

We generated two different high-content phosphopeptide microarrays according to published procedure combining SPOT synthesis with chemoselective re-immobilization of specifically tagged phosphopeptides *(8–10)*. One microarray displayed 6,262 phosphotyrosine-containing peptides derived from human phosphorylation sites in triplicates (pTyr-chip) and the other microarray displayed 2,034 phosphoserine-containing peptides together with 2,159 phosphothreonine-containing peptides in triplicates (pSer/pThr-chip). Each of these two microarrays was incubated with either pan-specific anti-phosphotyrosine antibody (*see* **Fig. 1**) or pan-specific anti-phosphothreonine antibody (*see* **Fig. 2**) followed by fluorescently labeled appropriate secondary antibody. Anti-phosphotyrosine antibody anti-pTyr-100 from Cell Signaling recognized 90% of all presented phosphotyrosine-containing peptides. Nevertheless, same antibody shows some distinct cross-reactivity if probed with the pSer/pThr-chip (*see* **Fig. 1**). Subsite specificity map of these cross-reactive peptides indicated that phosphothreonine residue was preferentially followed by either tyrosine or phenylalanine (*see* **Fig. 1**). This finding suggests that cross-reactivity maybe caused by mimicking phosphotyrosine by the phosphate group of the phosphothreonine and the aromatic side-chain of the amino acids C-terminal to the phosphoamino acid residue.

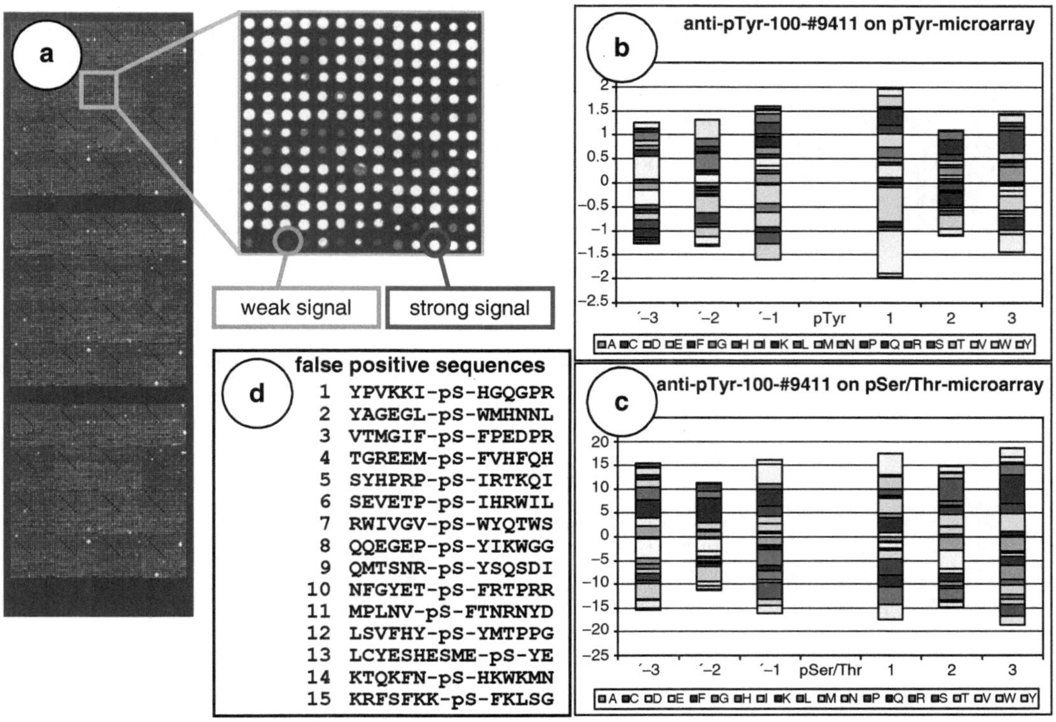

Fig. 1. Profiling of anti-phosphotyrosine antibody. (**a**) Fluorescence image of peptide microarray displaying 6,262 phosphotyrosine-containing peptides derived from annotated human phosphorylation sites together with background controls and fluorescent landmarks. Three identical subarrays result in 3 × 6,262 = 18,786 data points. Microarray was incubated with anti-phosphotyrosine antibody followed by fluorescently labeled secondary antibody. *Enlarged image* shows strong signals (*white color* equal to saturated signal at used laser settings) and weak signals (*red color*). (**b**) Statistical analysis of detected strong binding phosphotyrosine-epitopes is shown. Probability of each amino acid residue relative to the phosphotyrosine moiety is compared with the probability of each residue relative to phosphotyrosine in all displayed peptides. If residue is preferred in the 5,665 strong binding phosphopeptide value is positive and proportional to preference. If residue is disfavored at given position values will be negative. Color code for amino acid residues is given in the legend of the graph. (**c**) Statistical analysis of detected strong binding phosphothreonine-epitopes is shown. Probability of each amino acid residue relative to the phosphothreonine moiety is compared with the probability of each residue relative to phosphothreonine in all displayed peptides. If residue is preferred in the 93 strong binders value is positive and proportional to preference. If residue is disfavored at given position values will be negative. Color code for amino acid resides is given in the legend of the graph. (**d**) Sequences of top 15 false-positive binders are given. Aromatic residues in +1 position relative to the phosphoserine are a hallmark of detected cross-reactivity (*see Color Plates*).

Anti-phosphothreonine antibody anti-pThr from Cell Signaling recognized 42% of presented phosphothreonine/phosphoserine peptides but also 14% of all presented phosphotyrosine-containing peptides (*see* **Fig. 2**). This finding was surprising and shows that results from experiments performed with such antibodies should be interpreted carefully. We found no significant binding for commercially available pan-specific anti-phosphoserine antibodies using described phosphopeptide microarrays.

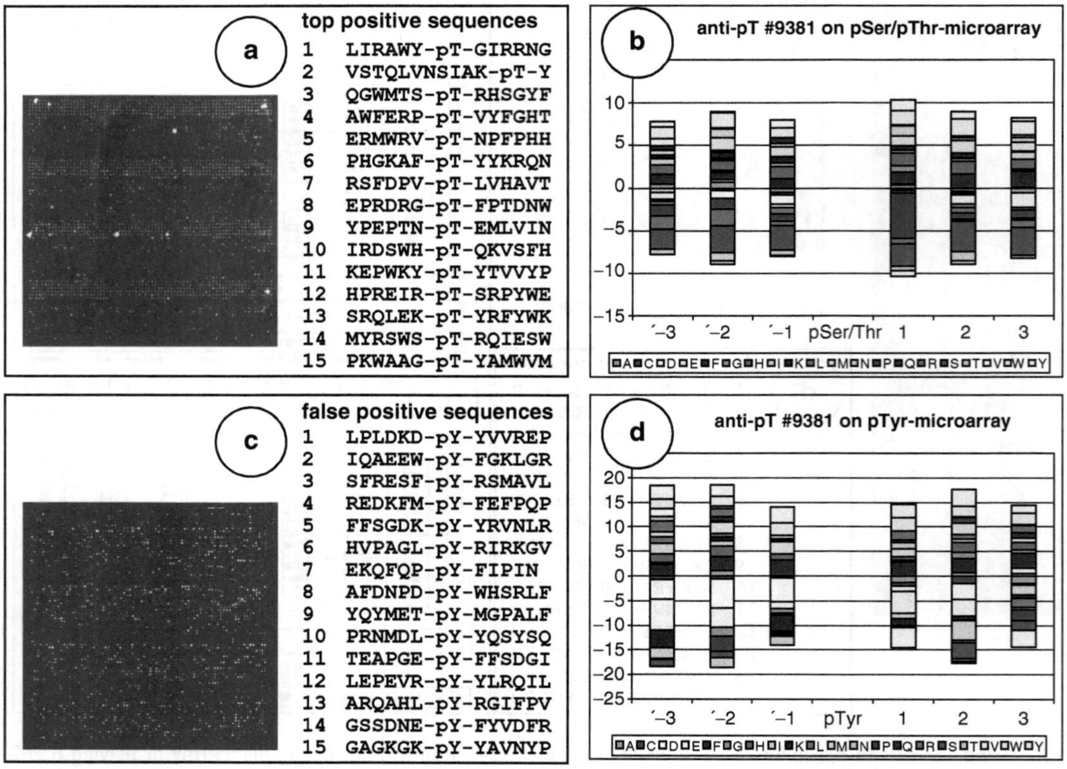

Fig. 2. Profiling of anti-phosphothreonine antibody. (**a**) Fluorescence image of peptide microarray displaying 2,034 phosphoserine-containing peptides and 2,159 phosphothreonine-containing peptides both derived from annotated human phosphorylation sites together with background controls and fluorescent landmarks. One of the three identical subarrays is shown. Microarray was incubated with anti-phosphothreonine antibody followed by fluorescently labeled secondary antibody. *White color* indicates saturated signal at used laser settings and *red color* represents epitopes with weaker affinity. Sequences of top 15 identified phosphothreonine-epitopes are given. (**b**) Statistical analysis of detected strong binding phosphothreonine-epitopes is shown. Probability of each amino acid residue relative to the phosphothreonine/phosphoserine moiety in the 1,739 binders is compared with the probability of each residue relative to phosphothreonine/phosphoserine in all displayed peptides. If residue is preferred value is positive and proportional to preference. If residue is disfavored at given position values will be negative. Color code for amino acid residues is given in the legend of the graph. (**c**) Fluorescence image of peptide microarray displaying 6,262 phosphotyrosine-containing peptides derived from annotated human phosphorylation sites together with background controls and fluorescent landmarks. One of the three identical subarrays is shown. Microarray was incubated with anti-phosphothreonine antibody followed by fluorescently labeled secondary antibody. *White color* indicates saturated signal at used laser settings and *red color* represents epitopes with weaker affinity. Sequences of top 15 cross-reactive phosphotyrosine-epitopes are given. (**d**) Statistical analysis of detected strong binding cross-reactive phosphotyrosine-epitopes is shown. Probability of each amino acid residue relative to the phosphotyrosine moiety is compared with the probability of each residue relative to phosphotyrosine in all displayed peptides. If residue is preferred in the 1,844 cross-reactive binders value is positive and proportional to preference. If residue is disfavored at given position values will be negative. Color code for amino acid residues is given in the legend of the graph (*see Color Plates*).

To analyze quality and subsite specificity of pan-specific anti-acetyllysine antibodies we generated acetyllysine-containing microarrays either by immobilizing 1,536 acetyllysine-containing peptides in triplicates (*see* **Fig. 3**) or by chemical “on-chip” acetylation of 18,397 microarray bound peptides using nitrophenyl-acetate

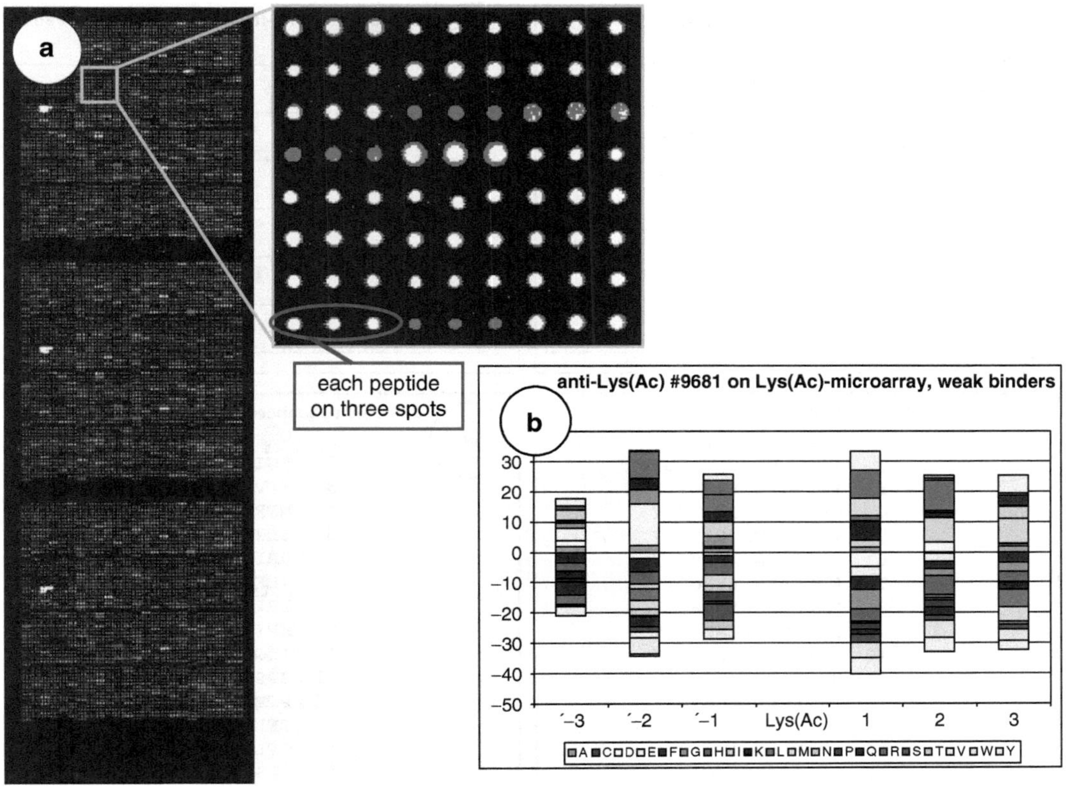

Fig. 3. Profiling of anti-acetyllysine antibody. (**a**) Fluorescence image of peptide microarray displaying 1,536 acetyllysine-containing peptides of general structure hexapeptide-acetyllysine-hexapeptide. Peptide sequences were generated randomly using a software package allowing similar distribution of all amino acid residues in all positions. Each peptide represents a defined single peptide and is printed three times within each of the three identical subarrays resulting in $3 \times 3 \times 1{,}536 = 13{,}824$ data points. Microarray was incubated with anti-acetyllysine antibody followed by fluorescently labeled secondary antibody. *Enlarged image* shows strong signals (*white color* equals to saturated signal at used laser settings) and weak signals (*red color*). (**b**) Statistical analysis of detected weak binding acetyllysine-containing peptides is shown. Probability of each amino acid residue relative to the acetyllysine moiety in the 68 weak binders is compared with the probability of each residue relative to acetyllysine in all displayed peptides. If residue is preferred in the weak binders value is positive and proportional to preference. If residue is disfavored at given position values will be negative. Color code for amino acid residues is given in the legend of the graph (*see Color Plates*).

(*see* **Fig. 3**). It could be demonstrated that anti-acetyllysine antibody recognizes several non-acetylated peptides (*see* **Fig. 4a** and **4c**), but is relative generic with respect to acetylated lysine side chains. The used selective "on-chip" modification approach enables more sophisticated investigation of antibody specificities without the need to synthesize the complete libraries in the modified form. It could be expected that extension of this method to other selective modification reactions like lysine/arginine methylation or tyrosine sulfation allows more comprehensive characterization of subsite specificities of antibodies specific for post-translational modified proteins.

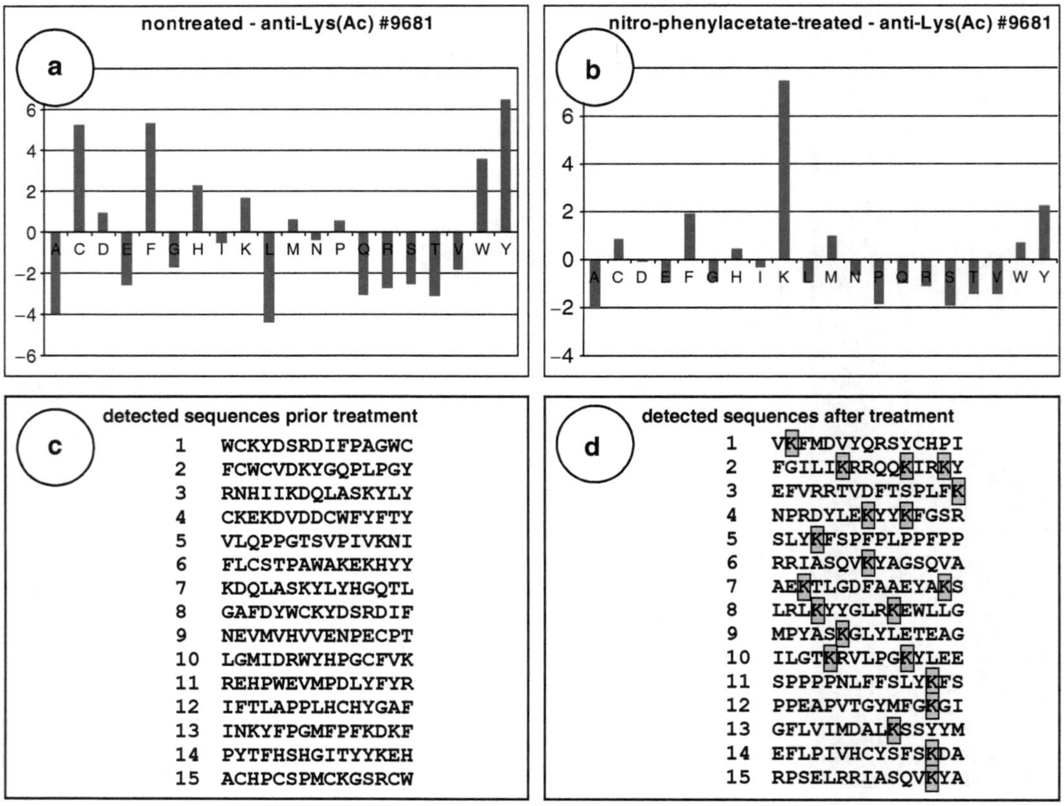

Fig. 4. Profiling of anti-acetyllysine antibody. (**a**) Results of experiment with peptide microarray displaying 18,397 human peptides. Microarray was incubated with anti-acetyllysine antibody followed by fluorescently labeled secondary antibody. Several cross-reactive binders could be detected. Statistical analysis of detected 28 cross-reactive peptides is shown. Presence of each amino acid residue within cross-reactive binders is compared with the average presence of each amino acid residue in all displayed peptides. If residue is preferred in cross reactive binders value is positive and proportional to preference. If residue is disfavored at given position values will be negative. (**b**) Results of experiment with peptide microarray displaying 18,397 human peptides which were treated "on-chip" with reagent acetylating side chains of lysine residues chemoselectively. Microarray was incubated with anti-acetyllysine antibody followed by fluorescently labeled secondary antibody. Statistical analysis of detected 995 strong binders is shown. Presence of each amino acid residue within the binders is compared with the average presence of each amino acid residue in all displayed peptides. If a residue is preferred in the 995 strong binders respective value is positive and proportional to preference. Thus, lysine residues are six times more frequently found in strong binders. If residue is disfavored at given position values will be negative. (**c**) Sequence information for the detected top 15 cross-reactive binders is given. Clusters of hydrophobic amino acid residues seems to be responsible for the cross-reactivity of microarray bound peptides. (**d**) Sequence information for the detected top 15 strong binders is given. All identified peptides have at least one lysine residue and there is a clear preference of the used anti-acetyllysine antibody for hydrophobic amino acid residues in the surrounding of the acetylated lysine residue visible.

2. Materials

2.1. Manual Incubation

2.1.1. Hardware

1. Micropipettes adjustable from 0.5 to 10 μL and from 100 to 1,000 μL (Eppendorf or Gilson) with corresponding plastic tips.
2. Petri dishes made from glass.
3. 2-mL micro tubes (Sarstedt, Nuembrecht) with corresponding screw caps.

4. Shaking system (Labortechnik Fröbel GmbH, Lindau; Rocky 1000).
5. Takara spaced cover glass XL (Matsunami Glass Ind., Ltd).
6. Microarray high speed centrifuge (ArrayIt, MHC220V).
7. Microarray fluorescence scanner or imaging system that is able to perform microarray fluorescence scans with at least 10-μm pixel size.

2.1.2. Buffers and Reagents

1. TBS-buffer: (50 mM Tris-HCl, 137 mM NaCl, 2.7 mM KCl, pH 8.0).
2. TBS-based blocking buffer (Superblock T20, Pierce, Rockford, IL).
3. Primary phosphotyrosine-specific antibodies (anti-pTyr-100, No. 9411, Cell Signaling, Danvers, MA).
4. Primary phosphothreonine-specific antibodies (anti-pThr, #9381, Cell Signaling, Danvers, MA).
5. Fluorescence labelled secondary antibody (goat anti-mouse IgG DyLight 649 conjugated, Pierce, Rockford, IL).
6. Peptide-, phosphopeptide-, and acetyllysine-containing peptide-displaying microarrays (JPT Peptide Technologies GmbH, Berlin, Germany).

2.1.3. Software

1. Spot-recognition software like GenePix Pro 6.0 (Molecular Devices, Sunnyvale, CA).
2. Microarray evaluation database Acuity 4.0 (Molecular Devices, Sunnyvale, CA).

2.2. Semi-Automatic Incubation

2.2.1. Hardware

1. HS400 microarray processing station including incubation chambers suitable for industry glass slide format (1 × 3 in.).
2. Gilson Microman Pipetting system (250 μL).
3. Microarray fluorescence scanner or imaging system which is able to perform microarray fluorescence scans with at least 10 μm/pixel size.
4. Nitrogen supply connected to HS400 microarray processing station for slide drying.

2.2.2. Buffers and Reagents

1. TBS-buffer plus 0.1% Tween 20: (50 mM Tris-HCl, 137 mM NaCl, 2.7 mM KCl, 0.1% Tween 20, pH 8.0).
2. 50 mM sodium phosphate buffer, pH 7.5.
3. SSC-buffer (1.5 mM sodium citrate, 15 mM NaCl, pH 7.0).
4. TBS-based blocking buffer (Superblock T20, Pierce, Rockford, IL).
5. Primary anti-acetyllysine antibody (#9681, Cell Signaling, Danvers, MA).

6. Fluorescence labelled secondary antibody (goat anti-mouse IgG DyLight 649 conjugated, Pierce, Rockford, IL).
7. Nitro-phenylacetate (Sigma-Aldrich, Munich, Germany).
8. Dimethylformamide.
9. Di-*iso*-propyl-ethyl amine.
10. Peptide microarrays displaying either human peptides or acetylated, randomly generated peptides (JPT Peptide Technologies GmbH, Berlin, Germany).

2.2.3. Software

1. Spot-recognition software like GenePix Pro 6.0 (Molecular Devices, Sunnyvale, CA).
2. Microarray evaluation database Acuity 4.0 (Molecular Devices, Sunnyvale, CA).

3. Methods

3.1. Chemoselective Acetylation of Lysine-Containing Peptides Bound to Glass Surfaces

1. Cover microarray displaying lysine-containing peptides with 8% (w/vol) solution of nitro-phenylacetate in dimethylformamide per 50 mM sodium phosphate buffer 1:1 (by vol.) containing 15% di-*iso*-propyl-ethyl amine for 1 h at room temperature.
2. Wash the slides with dimethylformamide (three times, 1-min each).
3. Wash the slides with TBS-buffer (three times, 1-min each).
4. Dry the slides using microarray centrifuge.

3.2. Manual Incubation

1. Pre-treat the microarrays with blocking buffer. Make sure that the microarray slides are totally submerged in the solution and shake the slides for 1 h using shaking system (*see* **Notes 1–8**).
2. Wash the slides with TBS-buffer (three times, 1 min each) (*see* **Notes 13 and 14**).
3. Wash the slides with double distilled water (three times, 1 min each).
4. Dry the slides using microarray centrifuge.
5. Prepare a humidifying chamber by placing a wet wipe in a vessel, which can be tightly sealed and putting a petri dish upside down onto it. Place the microarray slide on top of the petri dish. If the vessel is closed a constant humidity within the vessel will be created during incubation.
6. Prepare solution of the primary antibodies in blocking buffer. The concentration to be applied to the microarray ranges from 1 to 50 μg/mL. For antibodies with high binding affinities

the recommended concentration is between 1 and 5 μg/mL. If a Takara spaced cover glass is used, the total assay volume is about 200 μL.

7. Pipette the primary antibody on the surface of the peptide microarray (*see* **Note 5**). Subsequent to pipetting, cover the solution with the spaced cover glass. Make sure there are no air-bubbles within the created incubation chamber. If the cover glass is applied correctly, excess solution will be visible at the top and lower end of the cover glass.
8. Seal the humidifying chamber and leave it for 2 h at room temperature.
9. Open the humidifying chamber, take the peptide microarray and remove all antibody solution as fast as possible by dipping microarrays in TBS-buffer.
10. Wash the microarray with TBS-buffer using shaking system (three times, 5-min each). Make sure that the microarray is totally submerged in the solution everytime.
11. Prepare a solution of blocking buffer sufficient to wash the microarray in it. Pipette an adequate amount of antibody to it to achieve a final concentration of 1 μg/mL (*see* **Note 12**). Submerge the microarray in solution of fluorescence labeled secondary antibody using a petri dish and incubate for 45 min using the shaking system (*see* **Note 9**). The total volume for the incubation with the secondary antibody should be about 10 mL. Do not use the Takara cover glasses for the secondary antibody but instead incubate the microarray directly in secondary antibody-containing solution (*see* **Note 15**).
12. Wash the microarray thoroughly with TBS-buffer (three times, 5-min each) and double distilled water (three times, 5-min each).
13. Rinse front and back of the microarray with double distilled water for 10-s each side.
14. Spin-dry the microarray using a microarray centrifuge.
15. Scan the dried microarray using a microarray scanner with appropriate laser settings (*see* **Notes 10 and 11**).
16. Apply ".gal-files" to the scanned images using spot-recognition software. Save the resulting values for the signal intensity and background.
17. Import the result files to a microarray evaluation database for detailed evaluation.

3.3. Semi-Automatic Incubation

1. Clip the microarray into the microarray slide holder provided by the HS400 microarray processing station and close the incubation chambers.
2. Make sure that enough liquid is available (2 L of TBS Tween 20 (channels 1 and 2, Ch.: 1 and 2) and 1 L of SSC-buffer (channel 5, Ch.: 5)).

3. Prepare the program steps according to the steps shown below (TECAN programming language):
   a. Step 1: WASH Temp. °C: 30.0, First: Yes, Ch.: 1, Runs: 3, Wash time: 0:02:00, Soak time: 0:02:00
   b. Step 2: PROBE INJECTION Temp. °C: 30.0
   c. Step 3: HYBRIDIZATION Temp. °C: 30.0, Agitation Frequency: Medium, Time: 0:30:00
   d. Step 4: WASH Temp. °C: 30.0, First: No, Ch.: 1, Runs: 2, Wash time: 0:02:30, Soak time: 0:02:30
   e. Step 5: PROBE INJECTION Temp. °C: 30.0
   f. Step 6: HYBRIDIZATION Temp. °C: 30.0, Agitation Frequency: High, Time: 2:00:00
   g. Step 7: WASH Temp. °C: 30.0, First: No, Ch.: 2, Runs: 5, Wash time: 0:02:00, Soak time: 0:02:00
   h. Step 8: PROBE INJECTION Temp. °C: 30.0
   i. Step 9: HYBRIDIZATION Temp. °C: 30.0, Agitation Frequency: Medium, Time: 0:45:00
   j. Step 10: WASH Temp. °C: 30.0, First: No, Ch.: 2, Runs: 4, Wash time: 0:02:30, Soak time: 0:02:30
   k. Step 11: WASH Temp. °C: 30.0, First: No, Ch.: 5, Runs: 2, Wash time: 0:02:30, Soak time: 0:02:30
   l. Step 12: SLIDE DRYING Temp. °C: 30.0, Time: 0:04:00, Final Manifold Cleaning: No, Ch.: No
4. Make sure that the nitrogen supply is connected to the HS400 station and is opened.
5. Start the program and follow the instructions according to the program steps (*see* **Note 16**).
6. When finished, open the incubation chamber (*see* **Note 17**) and scan the dried microarray using a microarray scanner with appropriate laser settings (*see* **Notes 10 and 11**).
7. Apply ".gal-files" to the scanned images using spot-recognition software. Save the resulting values for the signal intensity and background.
8. Import the result files to a microarray evaluation database for detailed evaluation.

4. Notes

4.1. General

1. Always wear laboratory gloves when handling peptide microarray slides.
2. Never touch the peptide microarray surface.

3. Always handle peptide microarray slides with care. They are made of glass and can damage your fingers.
4. Hold peptide microarray slides at the end, which carries the engraved data label. This label provides for unique identification of the micro array.
5. Apply the sample to the peptide displaying side of the glass slide.
6. Never whisk the surface of the peptide microarray slide with a paper cloth.
7. Never use other chemicals as described. Inappropriate chemicals may destroy the chemical bonding of the peptides to the glass surface.
8. Avoid dust or other particles during each step of the experiment. Dust, particles, and resulting scratches will cause artefacts during the final signal readout.
9. Control incubations using labeled secondary antibody alone should be performed in parallel to the epitope mapping experiment to ensure that found signals are not caused by unspecific binding of the secondary antibody to the immobilized peptides.
10. Fluorescence scanning could be very sensitive depending on the used scanner. Avoid any fluorescent impurities/contaminations inside your assay solution or washing solutions. You can easily check for such impurities by incubating and washing a dummy slide with the same solutions followed by fluorescence imaging.
11. Make sure to scan the front face of the microarray slide carrying the peptides.
12. Carefully adjust the final dilution of your labeled secondary antibody. Microarray technology is very sensitive and therefore it could be possible to use the secondary antibody in a higher dilution as proposed by the manufacturer. Generally, 1:1,000 dilutions of a 1 mg/mL stock solution are working very well. Nevertheless, depending on the nature of the secondary antibody, such concentrations may yield high background signals caused by unspecific binding to the coated glass surface. If the signals within the peptide spots are high you could test 1:5,000 or 1:30,000 dilutions of a 1 mg/mL stock as well.

4.2. Manual Incubation

13. Please take care when dispensing solutions onto the slide surface. Make sure not to touch the surface with pipette-tips or dispensers.
14. Filter all solutions for the washing steps through 2-μm, preferably 0.4-μm particle filters before use.
15. For incubation with the fluorescently labeled secondary antibody it is important to use metal trays with a cover or plastic

trays completely covered with aluminum foil as these antibodies are sensitive towards light.

4.3. Semi-Automatic Incubation

16. Make sure to use excess volumes for all pipetting steps to make sure that the incubation chamber is completely filled.
17. Take care when opening and closing the incubation chamber. In case the microarray sticks to the sealing of the chamber it might be damaged.

References

1. Manning, G., Whyte, D. B., Martinez, R., Hunter, T., and Sudarsanam, S. (2002). The protein kinase complement of the human genome. *Science* 298, 1912–1934.
2. Kim, S. C., Sprung, R., Chen, Y., Xu, Y., Ball, H., Pei, J., Cheng, T., Kho, Y., Xiao, L., Grishin, N. V., White, M., Yang, X.-J., and Zhao, Y. (2006). Substrate and functional diversity of lysine acetylation revealed by a proteomics survey. *Mol. Cell* 23, 607–618.
3. Iwabata, H., Yoshida, M., and Komatsu, Y. (2005). Proteomic analysis of organ-specific post-translational lysine-acetylation and –methylation in mice by use of anti-acetyllysine and –methyllysine mouse monoclonal antibodies. *Proteomics* 5, 4653–4664.
4. Kaufmann, H., Bailey, J. E., and Fussenegger, M. (2001). Use of antibodies for detection of phosphorylated proteins separated by two-dimensional gel electrophoresis. *Proteomics* 1, 194–199.
5. Salomon, A. R., Ficarro, S. B., Brill, L. M., Brinker, A., Phung, Q. T., Ericson, C., Sauer, K., Brock, A., Horn, D. M., Schultz, P. G., and Peters, E. C. (2003). Profiling of tyrosine phosphorylation pathways in human cells using mass spectrometry. *Proc. Natl. Acad. Sci. U S A* 21, 443–448.
6. Steinberg, T. H., Agnew, B. J., Gee, K. R., Leung, W. Y., Goodman, T., Schulenberg, B., Hendrickson, J., Beechem, J. M., Haugland, R. P., and Patton, W. F. (2003). Global quantitative phosphoprotein analysis using Multiplexed Proteomics technology. *Proteomics* 3, 1128–1144.
7. Craig, A. L., Bray, S. E., Finlan, L. E., Kernohan, N. M., and Hupp, T. R. (2003). Signaling to p53: The use of phospho-specific antibo-dies to probe for in vivo kinase activation. *Methods Mol. Biol.* 234, 171–202.
8. Wenschuh, H., Volkmer-Engert, R., Schmidt, M., Schulz, M., Schneider-Mergener, J., and Reineke, U. (2000). Coherent membrane supports for parallel microsynthesis and screening of bioactive peptides. *Biopolymers (Peptide Science)* 55, 188–206.
9. Panse, S., Dong, L., Burian, A., Carus, R., Schutkowski, M., Reimer, U., and Schneider-Mergener, J. (2004). Profiling of generic anti-phosphopeptide antibodies and kinases with peptide microarrays using radioactive and fluorescence-based assays. *Mol. Divers.* 8, 291–299.
10. Reineke, U., and Schutkowski, M. (2006). Peptide arrays in proteomics and drug discovery, in *BioMEMS and Biomedical Nanotechnology*, Volume II: Micro and Nano-Technologies for Genomics and Proteomics, (Ozkan, M., and Heller, M. J., eds), Springer Science + Business Media, LLC, New York, pp. 161–282.

Chapter 13

Epitope Mapping Using Phage Display Peptide Libraries

Volker Böttger and Angelika Böttger

Summary

Phage libraries displaying millions of peptides with randomized sequences are extremely useful tools for mapping antibody epitopes. In many cases, antibodies are able to select peptides with reasonable affinity for their combining sites (paratopes) from these libraries. Ideally, consensus motives can be deduced from multiple peptide sequences and matched to areas of the antigen against which the antibody was raised. That way, critical components of the antibody epitope can be defined. This chapter focuses on technical details of epitope mapping employing pre-made filamentous phage peptide display libraries. Examples are given for illustration.

Key words: Phage display library, Filamentous phage, Biopanning, Mimotope, Continuous epitope, Discontinuous epitope, Critical binding residue, Consensus motif.

1. Introduction

Phage display of foreign proteins and peptides originates from the pioneering work of George P. Smith performed about two decades ago *(1, 2)*. Being a revolution for epitope mapping in the early 1990s *(3–5)* it has now become a classical method used by researchers around the world for target selection. What was only a handful of papers in 1992 has now reached almost one publication per day.

Areas of application for phage display technologies have expanded enormously and so has the understanding for their potential merits and limitations. Many variations in library design and selection strategies have been introduced over the years. However, the basic principles are still the same.

Most commonly, filamentous phage (M13, f1, fd) are used for phage display techniques *(6, 7)*. They are rod-shaped structures

Ulrich Reineke (eds.), *Methods in Molecular Biology, Epitope Mapping Protocols, vol. 524*

DOI: 10.1007/978-1-59745-450-6_13

of about 1-μm length and a diameter of 6 nm (**Fig. 1**). For protein and peptide display mainly two phage genes, gIII and gVIII, have been genetically modified *(8)*. Random peptide libraries have been generated by inserting short synthetic oligonucleotides with a completely randomized sequence usually at the 5′ end of these genes. That way, huge libraries of about 10^8–10^{10} individual phage, each containing a unique nucleotide sequence, have been generated. A single phage then provides a unique peptide sequence displayed as fusion to the coat protein (**Fig. 1**). This is the basis for affinity selection and subsequent amplification of specific phage clones as ligands for target molecules like antibodies.

Originally, phage display peptide libraries were mostly used for antibody epitope mapping. Now they are employed in applications comprising in vitro and in vivo selection for numerous organic and inorganic targets *(9, 10)*. In addition to filamentous phage, other biological vector systems have been harnessed for random peptide and protein display, e.g., lytic bacteriophage λ *(11)*, T4 *(12)*, T7 *(13)* (T7 Select System, Novagen), P4 *(14)* or bacteria *(15)* (FliTrx™– Random Peptide Display Library, Invitrogen) to name but a few. However, reflections on their potentials and limitations for epitope mapping of antibodies are not within the scope of this book chapter.

In the following sections we will describe the use of pre-made phage display peptide libraries for mapping of antibody epitopes. Protocols are based on peptide libraries inserted into fd-tet phage vectors *(16)*. These nonlytic phage contain a Tn10 fragment in their genome conferring tetracycline resistance to their bacterial hosts (**Fig. 1**). Infective phage can be detected as transducing units (TU) rather than as plaque forming units (pfu) on tetracycline containing agar plates and amplified in growth media with tetracycline. Affinity selection of phage, coined as biopanning, comprises iterative steps of (1) incubation of antibody with library

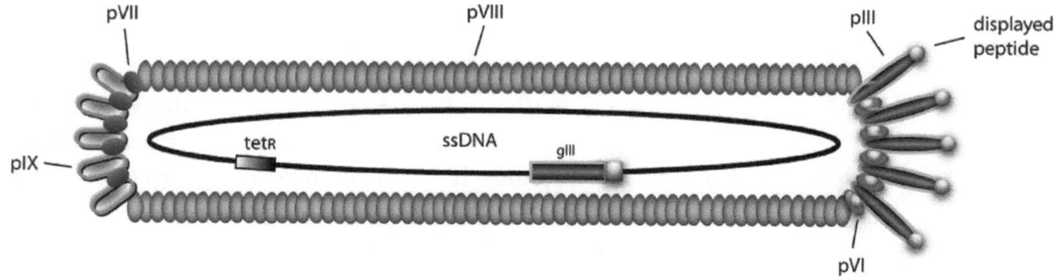

Fig. 1. Schematic view of a filamentous phage (fd-tet). The protein capsid encasing a circular single stranded DNA consists of a major coat protein (pVIII, > 2,500 copies) and several minor coat proteins (five copies each). They comprise pIII and pVI at one end and pVII and pIX at the opposite end of the virion. Filamentous phage infect *E. coli* bacteria carrying the F-plasmid. The infection is initiated by docking of pIII to the F-pilus of these bacteria

phage, (2) removal of nonbinding phage, and (3) amplification of binding phage. Several protocols for this process are provided. In addition, data (sequence) generation and analysis are discussed and illustrated by examples. The methods described in this chapter have been used for many years by the authors to map sequential (continuous) and conformational (discontinuous) epitopes of antibodies directed against a number of different proteins *(17–21)*.

2. Materials

2.1. Amplification of Phage Display Library

2.1.1. Phage Display Libraries and E. coli Host Strains

1. Phage display libraries displaying 6-mer or 15-mer peptides on fUSE5 (fd-tet derived phage vector) as well as *E. coli* host strains were kindly provided by George P. Smith (University of Missouri, Columbia). He continues to pursue a policy of distributing kits (vector, *E. coli* host strains and amplified random peptide libraries) free to anyone who asks, including commercial enterprises, *www.biosci.missouri.edu/smithgp/PhageDisplayWebsite/PhageDisplayWebsiteIndex.html* (*see* **Note 1**).
2. Library f3–6-mer (fUSE/6-mer; GenBank Accession AF246446): Vector fUSE5 (foreign 6-mer displayed on all five copies of pIII); number of primary clones: 2×10^8; size of phage DNA: 9,225 bases; buffer: TBS.
3. Library f3–15-mer (fUSE/15-mer; GenBank Accession AF246445): Vector fUSE5 (foreign 15-mer displayed on all five copies of pIII); number of primary clones: 2.5×10^8; size of phage DNA: 9,252 bases; buffer: TBS.
4. *E. coli* K91: Chromosome: *thi*; sex: Hfr-C (Cavalli); phenotype: stable Hfr, even without selection; excellent for propagating filamentous phage; gives large plaques with wt phage and small but visible plaques with infective members of fd-tet family.
5. *E. coli* K91Kan: Chromosome: same as K91, except that the mini-kan hopper element is inserted into the *lacZ* gene; sex: same as K91 (Hfr-C); phenotype: same as K91 except: $LacZ^-$; kanamycin resistant.

2.1.2. Handling of Phage and Bacteria

1. Sterile polypropylene tubes: 15 mL, 50 mL.
2. 1.5-mL sterile Eppendorf tubes, 1-L and 2-L sterile Erlenmeyer flasks.
3. U96 MicroWell™ Plates, polypropylene, sterile (Nunc, Roskilde, Denmark) for preparing phage dilutions.

4. Sterile petri dishes 100 × 15 mm (Nunc).
5. LB medium: 10 g bacto-tryptone (BD Biosciences, Oxford, UK), 5 g bacto-yeast extract (BD Biosciences), 10 g NaCl, in 1,000 mL distilled water, adjust pH to 7.5 with NaOH, autoclave and store at room temperature.
6. 2× YT medium: 16 g bacto-tryptone (BD Biosciences), 10 g bacto-yeast extract (BD Biosciences), 5 g NaCl in 1,000 mL distilled water, adjust pH to 7.0 with NaOH, autoclave and store at room temperature.
7. Kanamycin stock solution: Prepare kanamycin in distilled water at 100 mg/mL, filter-sterilize (0.2 μm), aliquot, and store at −20°C in a light tight container.
8. Tetracycline stock solution (20 mg/mL): Autoclave 20 mL (25.2 g) glycerol in a 50-mL tube, filter-sterilize 20 mL (20 g) of a 40 mg/mL tetracycline solution in distilled water into the tube once glycerol has cooled down, mix thoroughly and store at −20°C. Wrap the tube in aluminum foil to protect solution from light.
9. 2× YT, Kan 100: add 10 μL kanamycin stock solution to 10 mL 2× YT medium.
10. 2× YT, Tet 20: add 10 μL tetracycline stock solution to 10 mL 2× YT medium.
11. 2× YT, Tet 0.2: add 5 μL tetracycline stock solution to 500 mL 2× YT medium.
12. LB agar, Tet 40, Kan 100 plates: 1,000 mL LB medium, 15 g bacto-agar (BD Biosciences), autoclave, cool to 55°C, add 2 mL tetracycline and 1 mL kanamycin stock solution, swirl up cautiously and pour into sterile petri dishes (approximately 15 mL/dish).
13. LB agar, Tet 20 plates: As described before, but add 1 mL tetracycline stock to 1,000 mL LB medium.
14. Polyethylene glycol 8000: PEG 8000 Powder (Promega, Madison, WI).
15. PEG/NaCl solution: 475 mL distilled water, 100 g PEG 8000, 117 g NaCl, autoclave, store at room temperature. Make sure that the solution looks homogenous before use and heat-up again if necessary.
16. TBS: 6.05 g Tris-base, 8.76 g NaCl in 1,000 mL distilled water, adjust pH to 7.5 with HCl, autoclave and store at room temperature.

2.2. Affinity Selection of Antibody-Binding Phage

1. Target antibodies: Monoclonal antibodies in PBS/TBS, preferably purified with protein A/G.

2.2.1. Biopanning with Magnetic Beads Separation

1. Dynabeads Protein A: (Dynal Biotech, Smestad, Norway), alternatives: Dynabeads Protein G or Dynabeads M-280 sheep anti-mouse IgG (both Dynal Biotech).
2. Dynal magnetic particle concentrator MPC-S: for 1.5-mL conical microcentrifuge tubes (e.g., Eppendorf tubes) (Dynal Biotech).
3. BSA: Dissolve 50 mg/mL dialyzed BSA in water, filter-sterilize (0.2 μm), store at –20°C.
4. Blocking buffer I: PBS, 2% BSA.
5. Wash buffer: PBS, 0.5% Tween 20.
6. Elution buffer: Add 1.5 mL sterile 0.2 N glycine–HCl, pH 2.2 (stored at room temperature) and 60 μL sterile BSA (50 mg/mL) to 1.5 mL sterile distilled water, and 0.1 mg/mL phenol red (optional to check pH after neutralization).
7. Neutralization buffer: 2 M Tris-Base, not pH adjusted.
8. pH paper: pH-universal-indicator sticks pH 0–14.
9. PBS: 8 g NaCl, 0.2 g KCl, 1.44 g Na_2HPO_4, 0.24 g KH_2PO_4, in 1,000 mL distilled water, adjust pH to 7.4, autoclave and store at room temperature.

2.2.2. Biopanning with Separation on Streptavidin Plates

1. Biotinylated goat anti-mouse IgG (Sigma, St Louis, MO).
2. Streptavidin: (Zymed/Invitrogen, Paisley, UK) reconstituted with distilled water at a final concentration of 1 mg/mL, store at 4°C.
3. Coating buffer: 10 μL streptavidin at a concentration of 1 mg/mL, 100 μL 1 M $NaHCO_3$, 900 μL distilled water per 35 × 10 mm petri dish.
4. Petri dish: Falcon™– cell culture 35 × 10 mm (BD Biosciences, Oxford, UK).
5. Blocking buffer II: 5 mg/mL dialyzed BSA in 0.1 M $NaHCO_3$, filter-sterilize (0.2 μm), store at 4°C.
6. Wash buffer: *see* **item 5** (**Subheading 2.2.1**).

2.2.3. Biopanning on Solid-Phase Bound Antibodies

1. Immunotube: MaxiSorp (Nunc).
2. Blocking buffer III: PBS, 5% (w/v) skim milk powder (Merck, Darmstadt, Germany).
3. Dilution buffer: Blocking buffer III with 0.1% Tween 20.
4. Wash buffer: PBS, 0.1% Tween 20.

2.3. Phage ELISA

1. Immunoplate 96-well: MaxiSorp (Nunc).
2. HRP-labeled anti-M13 antibody: Mouse monoclonal antibody B62-FE2 (Abcam, Cambridge, MA).

3. Blocking buffer IV: PBS, 5% (w/v) skim milk powder, 0.1% Tween 20.
4. TMB stock solution (10 mg/mL): Add 100 mg TMB (3,3′,5,5′-tetramethylbenzidine to 10 mL dimethyl sulfoxide (DMSO) in a 15-mL tube, dissolve and wrap in aluminum foil, and store at room temperature.
5. Hydrogen peroxide (H_2O_2) solution: 30% (w/w).
6. Substrate: 0.1 mg/mL TMB, 0.03% H_2O_2 in 0.1 M sodium acetate pH 6.0.
7. Stop solution: 1 M H_2SO_4.
8. Microtiterplate: 96-well, polypropylene, flat bottom (Greiner, Stonehouse, UK), autoclave or U96 MicroWell™ plates, polypropylene, sterile (Nunc).

2.4. Phage Sequencing

1. Minipreps: Wizard Plus SV DNA Purification System (Promega, Madison, WI).
2. Primer: fUSE 35S: 5′-CCC TCA TAG TTA GCG TAA CG-3′ (– strand), M13 gene III ForSeq: 5′-ATT CAC CTC GAA AGC AAG CTG-3′ (+ strand).

3. Methods

3.1. Amplification of Phage Display Library

In order to provide sufficient starting material for the selection procedures, a sample of the phage display library needs to be amplified. Library amplification is always accompanied by loosing some of the original sequence diversity. The loss is minimized in the following procedure by infecting much more bacteria than independent clones are present in the primary library.

1. Start an overnight culture by inoculating 10 mL 2× YT, Kan 100 medium in a 50 mL tube with a scrape of K91Kan bacteria from a –80°C glycerol stock.
2. Shake culture overnight at 200 rpm, 30°C.
3. Transfer an 1 mL aliquot into 100 mL 2× YT in an 1-L Erlenmeyer flask.
4. Shake at 250 rpm at 37°C until the OD at 600 nm reaches ca. 0.2 for a 1/10 diluted bacterial suspension.
5. Slow down shaking speed to 150 rpm for 10 min for F-pili regeneration (necessary for the following phage infection).
6. Add ca. 10 μL of phage display library (10^{12} phage particles), which is about 5×10^{10} TU (infective phage) for a 6-mer library (5% infectivity). Since the original library has a diversity of ca. 2×10^8 each clone (sequence) should be represented more than 100 times in the infected bacteria pool.

7. Shake for another 20 min at 150 rpm.
8. Transfer the culture to a 2-L Erlenmeyer flask containing 500 mL of pre-warmed 2× YT, Tet 0.2.
9. Shake for 30 min at 220 rpm at 37°C.
10. Add 500 μL tetracycline stock solution (ca. 17 μg/mL final concentration).
11. Mix properly and remove 20 μL suspension for titration (**step 13**).
12. Shake overnight with 220 rpm at 37°C.
13. For titration add 180 μL 2× YT medium to each of six Eppendorf tubes and transfer 20 μL bacterial suspension (from **step 11**) to the first tube.
14. Mix properly by aspirating and dispensing ten times and transfer 20 μL to the second tube.
15. Continue until six dilutions have been prepared. Change tips for each dilution to avoid carry-over.
16. Spread 50 μL of each dilution on a quarter of a 10 cm LB, Tet 40, Kan 100 agar plate and incubate inverted plates overnight at 37°C.
17. Count on plate sectors with well-separated colonies. Example for the 6-mer library: 47 colonies at 10^{-5} dilution = 9.4×10^{7}/mL ($n \times 20 \times df$, where n is the number of colonies and df the dilution factor). The total number of infected bacteria in a 600 mL culture is then 5.64×10^{10}, which corresponds very well with approximately 5×10^{10} TU in the phage library sample added for infection.
18. Spin overnight culture in several sterile centrifuge tubes at 3,400 × *g* for 50 min at 4°C and transfer the supernatant to fresh sterile centrifuge tubes.
19. Add 1/5 volume of pre-cooled PEG/NaCl solution to precipitate phage particles, mix properly by inverting tubes several times, and place the tubes on ice for ca. 5 h.
20. Spin at 3,400 × *g* for 30 min at 4°C and discard the supernatant.
21. Resuspend pellets in sterile distilled water and combine them in one 15-mL tube (total volume of 12 mL).
22. Spin at 3,400 × *g* for 20 min at 4°C to remove bacterial debris and other insoluble matter.
23. Transfer the cleared supernatant to a fresh 15 mL tube and add 2 mL of pre-cooled PEG/NaCl solution, mix properly and keep overnight at 4°C.
24. Spin at 3,400 × *g* for 30 min at 4°C, remove supernatant, spin again shortly and remove remaining liquid.
25. Resuspend the pellet in 2.5 mL TBS and filter-sterilize (0.2 μm).

26. Determine the phage concentration by titration (**step 28–36**).
27. Store the phage library at 4°C or after adding an equal amount of sterile glycerol at –20°C (*see* **Note 2**).
28. Prepare LB agar, Tet 20 plates and let the plates dry for several hours with the lid open. Draw a 4 × 4 grid on the back of each plate.
29. Fill 45 µL TBS into ten wells of a sterile 96-well polypropylene round bottom plate and add 5 µL phage sample to first well.
30. Mix properly by aspirating and dispensing ten times and transfer 5 µL to the second well.
31. Continue until ten dilutions have been prepared (change tips for each dilution to avoid phage carry-over).
32. Add 155 µL log-phase K91 to each well, mix and incubate at 37°C for 30 min.
33. Resuspend bacteria and spot 20 µL of each dilution onto one square of the 4 × 4 grid of the dried agar plate.
34. Wait until the liquid of each spot has completely disappeared.
35. Close the lid and incubate the plate face down overnight at 37°C.
36. Count on squares with well-separated colonies and calculate the phage concentration. Example for 6-mer library: 23 colonies at 10^{-8} dilution = 4.6×10^{11} TU/mL ($n \times 200 \times df$, where n is the number of colonies and df the dilution factor).

3.2. Affinity Selection of Antibody-Binding Phage

The methods for biopanning described in the following subheadings were frequently used by the authors and other groups. Many more techniques have been explored and are referenced in the literature *(9)*. It has been shown that different procedures can result in completely different phage sequences being selected *(22)*, which makes it advisable to try variable conditions. However, one should be guided by certain principles:

Choose conditions for biopanning, which have been found to be most suitable for the interaction of the antibody with its antigen. For instance, some antibodies show impaired binding when they are passively adsorbed to solid supports. In such cases phage selection should be performed in solution followed by a suitable capture procedure for phage–antibody complexes.

Choose a capture system that binds your antibody efficiently. Remember, protein A and G have species and immunoglobulin class preferences. Biotinylated antibodies have been frequently used as targets for biopanning with a subsequent capture step on streptavidin-coated plates. However, after the labeling procedure they need to be tested for their uncompromized immunoreactivity.

Start the first round with a high antibody concentration. This increases the chance for low affinity or underrepresented sequences to be selected. The phage input in the first round of biopanning contains at least 100 infective phage of each sequence in the naive library, a number sufficiently high to allow catching any phage species that has reasonable affinity for the antibody paratope. However, certain sequences interfere with phage infectivity, bacterial translocation and/or assembly events. In addition, different phage compete for binding, and nonspecific sterical hindrance takes place. Therefore, not every suitable phage is selected. This notion is supported by the observation that you usually do not select the same phage sequences in "identical" biopanning procedures.

If the outcome of biopanning is unexpected, one can go back to the amplified libraries of previous round(s) and start over again with a varied protocol. For instance, lower antibody concentrations in context with a stringent wash protocol will favor higher affinity ligands to be selected. Thus, by experimenting with the starting conditions the whole sequence space for the target can be explored.

3.2.1. Biopanning with Magnetic Beads Separation

First Round

1. Transfer 100 µL of the amplified library (ca. 4.6×10^{10} phage for the aforementioned 6-mer library) to an 1.5-mL Eppendorf tube.
2. Add 10 µL BSA (50 mg/mL) and 10 µL purified monoclonal antibody (1 mg/mL) (*see* **Notes 3** and **4**).
3. Mix and incubate overnight at 4°C.
4. Start a 10 mL culture of *E. coli* K91 in LB medium (from a single colony picked from a plate or from a scrape of a –80°C glycerol culture), grow overnight at 30°C, and shake at 200 rpm.
5. Next day, inoculate 10 mL LB medium with 10–100 µL overnight culture and grow to late log phase ($OD_{600nm} \geq 1.0$).
6. Transfer 250 µL of Dynabeads Protein A slurry into an 1.5 mL Eppendorf tube (*see* **Note 5**).
7. Equilibrate and block with 1 mL blocking buffer I.
8. Mix and separate in the MPC-S device for 1 min.
9. Discard the liquid and repeat blocking step two times.
10. Resuspend beads in 1 mL blocking buffer I and aliquot 200 µL per target antibody into 1.5-mL Eppendorf tubes. The binding capacity is ≥10 µg IgG, which is equimolar to the amount of antibody added to the library.
11. Add 100 µL blocking buffer I to overnight reaction of phage/antibody mix and transfer the content to the tube with 200 µL Protein A beads.

12. Shake at 130 rpm for 1 h at room temperature.
13. Wash nine times with 1 mL wash buffer and two times with 1 mL TBS. Separate the beads between the washing steps in the MPC-S device.
14. Add 400 µL elution buffer for 20 min and shake occasionally.
15. Separate the beads in the MPC-S device and transfer the eluate into an Eppendorf tube containing 24 µL of neutralization buffer.
16. Check pH of elution buffer and neutralized eluate with pH paper.
17. Optional: Determine the phage concentration in neutralized eluate by titration as described in **step 28–36** of **Subheading 3.1** (*see* **Note 6**).
18. Add neutralized eluate to 2.5 mL log-phase K91 ($OD_{600nm} \geq 1.0$) in a 50-mL tube and incubate for 1 h at 37°C.
19. Add 10 mL 2× YT, Tet 20 and shake ca. 20 h at 250 rpm at 37°C. Cover the tube loosely with the screw cap to allow aeration.
20. Spin at 3,400 × *g* for 30 min at 4°C.
21. Transfer 12 mL supernatant to a fresh 15-mL tube (keep 50 µL for titration) and add 2.5 mL PEG/NaCl for phage precipitation.
22. Mix properly by inverting the tube several times and keep it on ice for 2.5 h.
23. Spin as before, discard the supernatant, and spin again shortly to remove remaining liquid.
24. Add 1 mL of distilled water, resuspend the pellet properly and transfer to an Eppendorf tube. Spin at 13,000 × *g* for 5 min at 4°C.
25. Transfer the supernatant into a fresh Eppendorf tube and add 200 µL PEG/NaCl. Mix properly and keep the tube on ice for 1 h.
26. Spin at 13,000 × *g* for 10 min at 4°C, remove the supernatant, and spin again shortly to remove remaining liquid.
27. Resuspend the phage pellet in 1 mL TBS and store the tube at 4°C (optional: filter-sterilize through 0.2-µm syringe filter). This is the "phage library after the first round" of biopanning. It is used as starting material for the second round.
28. Determine the phage concentration by titration as described in **step 28–36** of **Subheading 3.1** (optional).

Second Round

29. Transfer 100 µL of the amplified library after the first round (usually 10^{10}–10^{11} phage) to an 1.5-mL Eppendorf tube.

30. Add 10 μL BSA (50 mg/mL) and 4 μL purified monoclonal antibody (1 mg/mL).
31. A negative control can be added by setting up a parallel incubation omitting the monoclonal antibody and/or adding an irrelevant antibody (same species and isotype).
32. Mix and incubate overnight at 4°C.
33. Proceed as described for the first round from **step 4** in **Subheading 3.2.1**, but transfer 100 μL Protein A slurry instead of 250 μL in a 1.5-mL Eppendorf tube (i.e., 200 μL of conditioned Protein A beads will have a binding capacity of ca. 4 μg IgG, which again is equimolar to the amount of antibody added to the library).
34. It is worthwhile to determine the phage concentration in the eluates of the second round since a selective phage enrichment can already show up here.

Further Round(s)

35. Additional cycle(s) of biopanning can be performed if no selective phage enrichment has been obtained after the second round.

3.2.2. Biopanning with Separation on Streptavidin Plates

1. Mix 100 μL of the primary phage library (ca. 5×10^{10} phage particle) with 10 μL BSA (50 mg/mL) and monoclonal antibody (0.1 mg/mL final concentration) and incubate overnight at 4°C.
2. Coat a polystyrene petri dish (e.g., Falcon 3001) with 1 mL streptavidin coating buffer overnight at 4°C in a humidified box.
3. Start overnight culture of *E. coli* K91 (*see* **steps 4** and **5** of **Subheading 3.2.1**).
4. Add biotinylated goat anti-mouse IgG to the phage–antibody mixture equimolar to the monoclonal antibody (ca. 10 μg here). This will minimize the risk of selecting phage clones specific for goat-anti-mouse IgG.
5. Incubate for 1 h at room temperature and mix occasionally.
6. In the meantime, remove the streptavidin solution from the petri dish (*see* **step 2**), fill it with blocking buffer II, and incubate for 1 h at room temperature.
7. Aspirate and wash three times with wash buffer.
8. Add 1 mL wash buffer to the phage–antibody reaction suspension from **step 5** and transfer the suspension to the streptavidin plate (*see* **step 6**).
9. Incubate for 30 min at room temperature on an orbital shaker.
10. Remove unbound phage and antibodies and add 2.5 mL wash buffer, swirl and pour out the solution immediately.

11. Repeat the wash cycle 10–20 times.
12. Add 400 μL elution buffer and incubate for 20 min with occasional shaking to elute bound phage.
13. Transfer the eluate to an 1.5-mL Eppendorf tube and neutralize with 24 μL neutralization buffer.
14. Continue with **step 16** of **Subheading 3.2.1**.
15. The second and any further rounds of phage selection are carried out essentially as described for the first round. However, in order to select phage clones with higher affinity to the monoclonal antibody, the concentration of the antibody can be decreased to a final concentration of 1 and 0.1 μg/mL in the second and third round of biopanning. Remember to add only equimolar amounts of the secondary antibody.

3.2.3. Biopanning on Solid-Phase Bound Antibodies

1. Coat an immunotube with 4 mL of purified monoclonal antibody at a concentration of 5–20 μg/mL in PBS.
2. Seal the tube with parafilm and incubate at 4°C overnight.
3. Remove the solution and wash three times with PBS. Perform all washing and blocking steps by completely filling the tube with solution.
4. Add blocking buffer III for 1 h at room temperature.
5. Wash the tube three times with PBS and add 10^{10}–10^{11} phage in 4 mL dilution buffer.
6. Incubate for 1 h at room temperature on a rotation incubator.
7. Remove the phage suspension and wash 10–20 times with wash buffer.
8. Add 800 μL elution buffer and incubate for 20 min at room temperature with constant shaking (*see* **Note 7**).
9. Transfer the suspension to an Eppendorf tube and neutralize with 48 μL neutralization buffer.
10. Continue with **step 16** of **Subheading 3.2.1**.
11. Subsequent rounds can essentially be performed as described for the first round. Again, modifying antibody concentration, wash stringency or switching to a different biopanning strategy all together can be advisable.

3.3. Phage ELISA

A phage ELISA should be performed to determine whether phage have been specifically selected by the target antibody. Amplified libraries from all rounds of biopanning can be used for this purpose.

1. Coat the wells of a 96-well immunoplate with 100 μL monoclonal antibody at 1–20 μg/mL in PBS overnight at 4°C (*see* **Note 8**).
2. Wash the plate with water or PBS and add 300 μL blocking buffer IV and incubate for 1 h at room temperature.

3. In the meantime dilute phage stocks 1:1 to 1:100 in blocking buffer IV.
4. Remove the blocking buffer IV from the immunoplate, transfer 100 μL diluted phage to the wells, and incubate for 2 h at room temperature.
5. Wash the plate as before, add 100 μL of HRP-labeled anti-M13 antibody (1:5,000 in blocking buffer IV), and incubate for 1 h at room temperature.
6. Wash as before, add 100 μL HRP substrate, and incubate for 15 min.
7. Stop color development by adding 100 μL of stop solution.
8. Measure the OD at 450 nm.
9. Proceed with ELISA-positive phage stocks (*see* **Note 9**): Add 1 μL phage suspension from several serial tenfold dilutions of amplified libraries to 1 mL log-phase *E. coli* K91.
10. Incubate for 30 min at 37°C and spread 50 μL of infected bacteria on LB, Tet 20 agar plates.
11. Grow colonies overnight at 37°C.
12. Transfer single well-separated colonies into each well of a sterile 96-well polypropylene microtiterplate filled with 200 μL 2× YT, Tet 20 medium.
13. Close the lid and incubate plates in a humidified box for 24 h at 37°C. Shake at 250 rpm.
14. Spin plates for 15 min at 1,000 × *g* at 4°C.
15. Transfer 100 μL phage supernatant into a fresh 96-well plate filled with 100 μL blocking buffer IV and use for ELISA on antibody-coated plates, as described in **steps 4–8** of **Subheading 3.3**.
16. Seal original plate with parafilm and keep as master plate at 4°C.

3.4. Phage Sequencing

1. Transfer 20 μL of bacterial suspension of ELISA-positive single clones from the master plate into 5 mL 2× YT, Tet 20 and grow overnight at 37°C. Shake at 250 rpm.
2. Spin at 3,400 × *g* at 4°C for 30 min, transfer the phage containing supernatant to a fresh tube and store in a fridge as backup.
3. Prepare the double stranded replicative form of phage DNA from the bacterial pellets using standard DNA miniprep procedures.
4. The yield of 10–20 μg is sufficient for several automated sequencing reactions using fUSE 35S and/or M13 gene III ForSeq primer(s).

3.5. Data Analysis

Peptide insert sequences obtained in **Subheading 3.4** are considered epitope mimetics for the antibody and can fall into different categories.

1. Peptide sequences show some sequence resemblance to the antigen sequence. The antibody recognizes a linear (continuous) epitope, i.e., critical binding residues of the antibody are within the sequence space provided by the length of the phage peptide.
2. Peptide sequences have no obvious sequence similarity to the antigen sequence. As antibodies usually bind to hydrophilic surface patches of proteins, critical binding residues can be widely spaced in the primary sequence but brought together by folding of the polypeptide chain (discontinuous epitope). Here, mapping of the binding region is not straightforward. However, epitope information may be gained from such peptide sequences in combination with 3D structural data of the antigen and sophisticated algorithms for phage peptide sequence analysis *(23)* or/and comprehensive mutational analysis *(24)*.
3. Peptide sequences are true mimotopes, i.e., they mimic the antibody epitope structurally without having any sequence resemblance to the latter. To test this, whole phage or synthetic peptides coupled to a carrier protein can be used as immunogens. If they elicit an immune response cross-reactive with the antigen they are considered structural surrogates of the antibody epitope *(25)* (*see* **Note 10**).

3.5.1. Example 1

Monoclonal antibody IIIF10 directed against the human urokinase receptor, uPAR *(26)* was used to select phage from the 6-mer and 15-mer phage random peptide libraries. Two rounds of biopanning were carried out essentially as described in **Subheading 3.2.1** using magnetic Protein A beads for separation. Eluates from the second round contained 8.5×10^8 TU/mL for the 6-mer library selection and 2×10^8 TU/mL for the 15-mer library selection. A negative control selection (with an antibody known to fail in phage selection) yielded only 1×10^6 and 3×10^5 TU/mL, respectively. Since this result indicated a specific phage selection with IIIF10, no further round of biopanning was carried out. Instead, a phage ELISA was performed on amplified libraries after the second round and subsequently on supernatants of single clones as described in **Subheading 3.3**. Amplified libraries and all single clones tested clearly positive. Nucleotide and deduced amino acid sequences for one 6-mer phage are shown in **Fig. 2**

Six unique peptide sequences were obtained from a limited sequencing approach (only seven clones were tested from each phage pool). Aligning these sequences with each other and with a segment of the human uPAR sequence which is known to contain the IIIF10 epitope *(26)* revealed obvious similarities (**Fig. 3**). The presence of an SYR motif in the 6-mer phages indicates the importance of the corresponding region in human

```
IIIF10-6mer        M13 gene III ForSeq primer >>>

V  K  K    L  L  F  A    I  P  L    V  V  P    F  Y  S  H
GTGAAAAAAT TATTATTCGC AATTCCTTTA GTTGTTCCTT TCTATTCTCA
CACTTTTTTA ATAATAAGCG TTAAGGAAAT CAACAAGGAA AGATAAGAGT

  S  A  D    G  A  T    S  S  Y  R    L  G  A    A  G  A
CTCGGCCGAC GGGGCTACGA GTTCTTATCG TCTTGGGGCC GCTGGGGCCG
GAGCCGGCTG CCCCGATGCT CAAGAATAGC AGAACCCCGG CGACCCCGGC

E  T  V  E    S  C  L    A  K  P    H  T  E  N    S  F  T
AAACTGTTGA AAGTTGTTTA GCAAAACCTC ATACAGAAAA TTCATTTACT
TTTGACAACT TTCAACAAAT CGTTTTGGAG TATGTCTTTT AAGTAAATGA

 N  V  W    K  D  D
AACGTCTGGA AAGACGAC
TTGCAGACCT TTCTGCTG
                                  <<< fUSE 35S primer
```

Fig. 2. Nucleotide and deduced amino acid sequence for a phage clone obtained by selection with monoclonal antibody IIIF10. The phage 6-mer insert is underlined and the pIII signal peptide is shown in italic. The sequencing direction of fUSE 35S and M13 gene III ForSeq primers is indicated.

```
uPAR  49  E K T N R T L S Y R T G L K I   63
                      P M S Y R Y
                      T S S Y R L
                      V E S Y R V
                  T A R G L S A R T A F A G L V
              P I S G R H L S S S F R F G G
        G E P S V V I S R S S R L S P
```

Fig. 3. Phage sequences selected from a 6-mer and 12-mer phage display library by monoclonal antibody IIIF10, specific for human uPAR. Sequences are aligned and show similarity to the IIIF10 epitope region on human uPAR (the region labeled with a double line comprises the epitope as defined by overlapping peptide analysis). Amino acid identity with uPAR in at least 50% of the phage sequences is labeled in bold, other sequence matches are underlined (*see* Note 11).

uPAR for antibody binding. The 15-mer sequences show additional amino acids probably engaged in binding of the antibody to uPAR. When these amino acids are present, the requirement for the complete SYR motif is diminished.

3.5.2. Example 2

In another experiment the 6-mer phage library was probed with a panel of monoclonal antibodies against human proenkephalin (huPENK). The antibodies used for the analysis were PE14, PE15, PE16, PE17, PE18, PE19, PE23, and PE25 *(27)*. Previous inhibition experiments and testing of overlapping PENK peptides had revealed that they all recognize the same epitope on PENK *(18)*. The employed biopanning procedure was a combination of methods described in **Subheadings 3.2.2** and **3.2.3**. Two rounds of solution phase interaction with decreasing antibody concentrations

(**Subheading 3.2.2**) were followed by a final round of solid-phase capture (**Subheading 3.2.3**). Most of the 6-mer phage clones selected by the PENK antibodies showed clear sequence similarities (**Fig. 4**). A consensus sequence was defined as D-L-L-W-X-L. The phage sequences and their consensus motif could

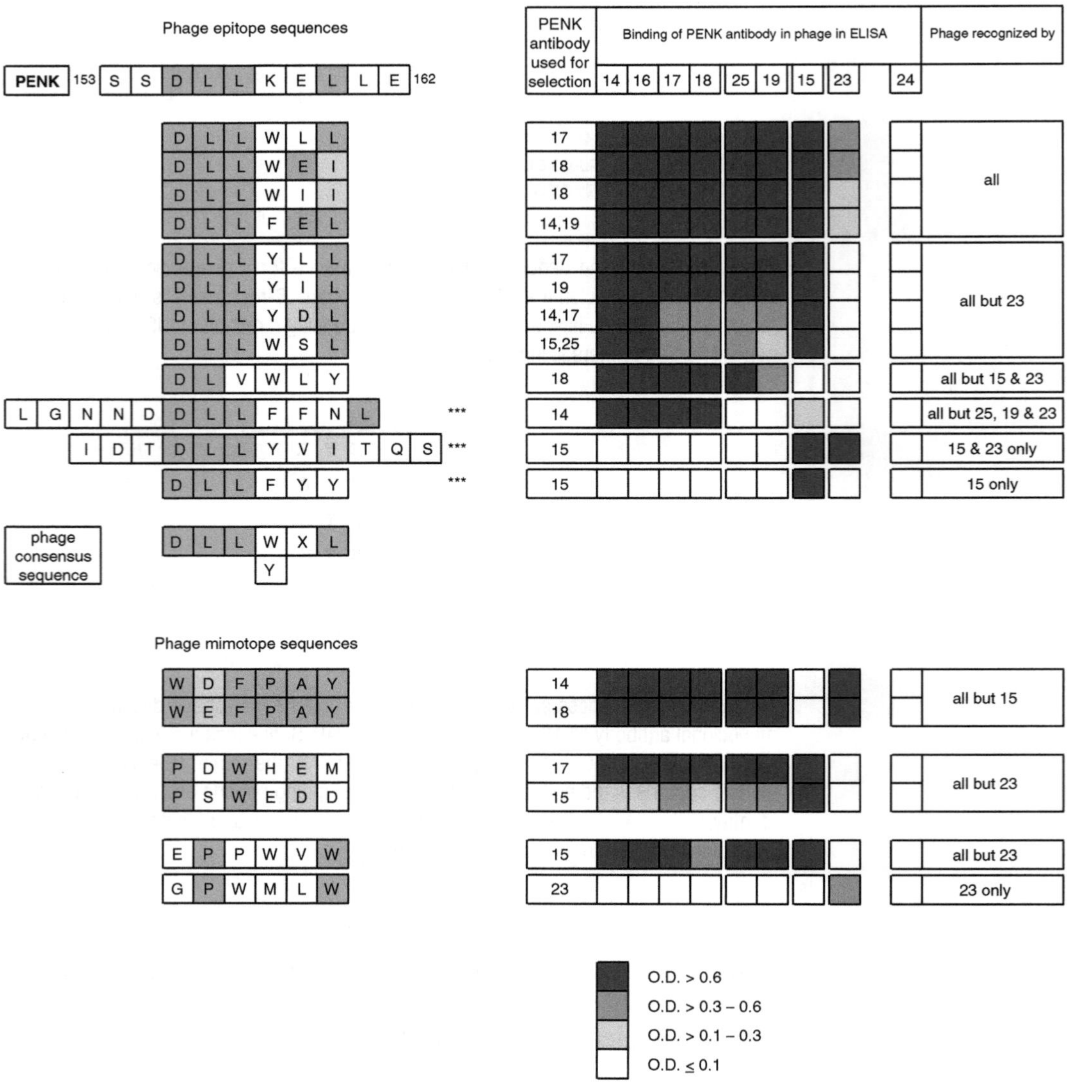

Fig. 4. *Left*: Phage epitope and mimotope sequences selected with several monoclonal antibodies specific for the same epitope on human proenkephalin. Sequences are aligned to each other and to the epitope region on human proenkephalin (phage epitope sequences only). Amino acids shared by at least 50% of the phage sequences (phage consensus sequence) and the proenkephalin sequence are shaded in *grey*, similar amino acids are indicated in *light grey*. Identical amino acids between mimotopes are shaded in *grey*, similar amino acids in *light grey*. *Asterisks* (***) indicate sequences obtained in earlier biopanning experiments *(18)*. *Right*: Proenkephalin (PENK) antibodies used to select particular phage sequences (*left*) are listed alongside with the magnitude of reaction (O.D. range in shades of *grey*) of each antibody with these phage in ELISA. PE24, which binds to a different region on human proenkephalin, was used as negative control. Epitope and mimotope phage are sorted according to their reactivity with the PE antibodies (summarized on the *right*). Likewise, PENK antibodies are arranged and grouped according to their similarity in phage (sequence) binding.

be easily aligned with the PENK sequence defined by overlapping peptides to contain the common epitope for these antibodies. In contrast, some of the selected phage express sequences with no obvious similarity to any region of the human PENK amino acid sequence. They are considered to be mimotopes.

Epitope and mimotope phage were tested in ELISA for binding to individual PENK antibodies. Interestingly, nonuniform binding was observed. Phage sequences were grouped according to their binding spectrum as shown in **Fig. 4**. The nature of these binding differences was further investigated on the antibody side by sequencing the variable regions of heavy and light chain genes, VH and VL (*(18)* and unpublished data). This sequence analysis led to the following findings:

- All antibodies have similar variable regions belonging to the same VH and VL subgroups, respectively.
- All antibodies share nearly identical light chain CDRs and a very similar CDR3 of the heavy chain.
- The variable regions of the heavy chains (and to a lesser degree of the light chains) are distinguishable by a different number of nonsilent nucleotide substitutions.

Numerical differences calculated from sequence alignments of the VH and VL regions are displayed as distance trees in **Fig. 5**. Here, a close sequence similarity is revealed for antibodies PE14, PE16, PE17, and PE18. Antibodies PE25 and PE19 are

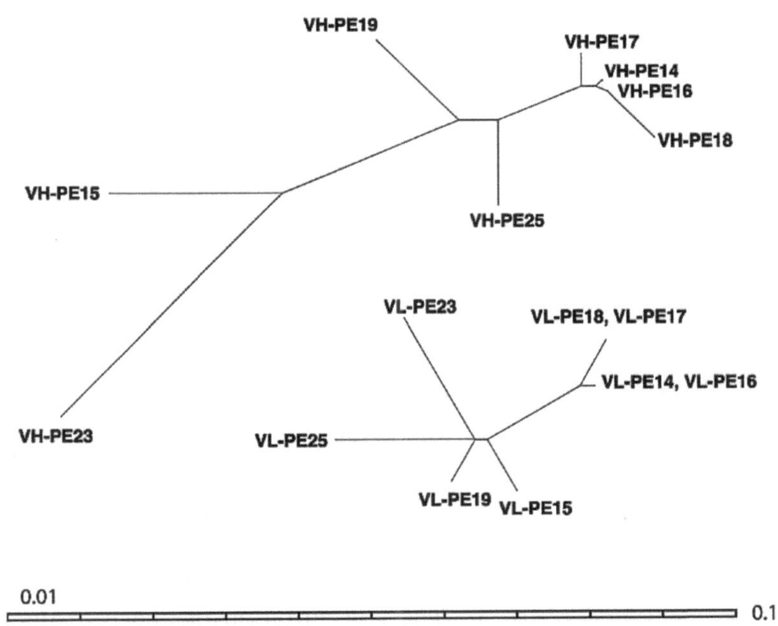

Fig. 5. Distance trees calculated from sequence alignments of the variable regions of heavy and light chain genes (VH and VL) for all PENK antibodies. Sequence distances between antibodies are drawn to scale (a distance of 0.01 represents a sequence difference of 1%). Data were generated using ClustalW and SplitsTree v.4.

placed separately from this group because of moderate sequence differences. However, these are minor compared with PE15 and PE23, which feature the biggest sequence distance to all other antibodies. PE15 and PE23 are also clearly different from each other. Strikingly, sequence distances are almost perfectly mirrored by the differences seen in the binding of individual antibodies and antibody groups to epitope and mimotope phage (**Fig. 4**).

This example clearly demonstrates the power of the phage peptide library approach for discerning fine specificities of antibodies not revealed by their binding to the antigen or antigen-derived overlapping synthetic peptides.

4. Notes

1. Other filamentous phage display libraries are available from academic (f88–4/15-mer, f88–4/Cys0 to Cys6 in pVIII, G. P. Smith, www.biosci.missouri.edu/smithgp/PhageDisplayWebsite/PhageDisplayWebsiteIndex.html) and commercial sources (Ph.D^{TM}-12, 7 and C7C Phage Display Peptide Libraries in pIII, M13, New England Biolabs, www.neb.com), which can be employed for epitope mapping. Please consult respective websites and published papers *(28, 29)* for details.
2. This is the starting material for antibody biopanning described in the following protocols. Scale up the amplification procedure and/or resuspend phage pellets in less TBS if a greater phage output and/or a higher stock concentration are required.
3. To save valuable library stocks several antibodies can be used as cocktail in the first round. In subsequent rounds antibodies should be used separately. It is good practice to run a negative control alongside with your target, especially to monitor specific selection in later rounds. In addition, a positive control (an antibody that has shown its ability to select peptide phage before) is good for troubleshooting in case your antibody fails to select anything.
4. Biopanning can be done with antibodies pre-loaded on protein A beads. Here, it is possible to use hybridoma supernatant instead of purified antibody. Make sure to saturate all IgG binding sites on Protein A before adding phage.
5. Dynabeads Protein G, Dynabeads M-280 sheep anti-mouse IgG or others can be used alternatively.
6. The concentration of phage in the eluates may vary over a wide range (e.g., 10^5–10^8 TU/mL). Phage titers of target and negative control are usually comparable. This is due to the fact that a specific selection is obscured by nonspecific

high background binding in the first round. Even if 50 different phage clones were specifically selected by the target this would translate to ca. 5,000–20,000 TU in total only. In contrast, in the second round a significant difference in phage titers for target and control can be expected if the antibody has found specific ligands in the primary library. Since each specific phage is amplified 10^5–10^6 times before entering the next round it will exceed background binding. Selective enrichment may further increase in round three or be first detected here. However, even if phage numbers indicate that specific selection has not happened it can be worthwhile to continue. Nonspecific phage might still have a growth advantage and so obscure specific selection. Test your amplified libraries from different rounds and/or single phage clones after the last round in phage ELISA.

7. Alternatively to elution, bound phage can be used directly in the tube to infect bacteria. Add 4 mL log phase K91 bacteria to the immunotube and incubate at 37°C for 30–60 min.
8. Make sure to include appropriate controls. Negative controls: no coating antibody and irrelevant antibody of the same species and Ig-class/-subclass. Positive controls: antibody-phage combinations that are already available; coat with anti-phage pVIII antibody. It is very important to test several concentrations of coating antibody against several dilutions of phage suspension to recognize specific (sometimes only weak) signals. This step is very important, otherwise valuable source material might be discarded.
9. If you fail to get ELISA-positive clones you should reconsider your selection strategy. If possible, start over again using a different library format (different length of inserts, loop or constraint inserts) or use different libraries in parallel for biopanning.
10. Researchers are strongly advised to thoroughly investigate the binding specificity of any selected phage sequence for the paratope (antibody combining site) of the antibody. A straightforward experiment is to test for competition of the selected phage and the antigen for antibody binding. This is especially important for mimotope sequences, i.e., motifs, which cannot be clearly assigned to any known sequence of the protein used to raise the antibody. Make sure to avoid or recognize any phage sequence selected by assay components other than the target antibody (protein A, streptavidin, biotin, plastic, species-specific antibodies etc.) used during biopanning. We recommend to read "A very detailed analysis on the nature of target-unrelated peptides recovered in the screening of phage-displayed random peptide libraries with antibodies" published by Menendez and Scott *(30)*.

11. Even if you get clear consensus sequences from a library with longer inserts it may be still worthwhile to screen a library with shorter inserts. This allows defining the critical binding residues much clearer, as seen from the IIIF10 example. On the other hand, antibodies binding to short, linear epitopes can fail to select phage with meaningful peptide sequences, if the library inserts are too long. We have tested human proenkephalin specific antibodies with phage peptide libraries of different insert lengths and repeatedly pulled out epitope phage from 12-mer and especially 6-mer libraries (example shown in **Subheading 3.5.2**). On the other hand, from a 20-mer library we almost exclusively obtained mimotopes.

References

1. Parmley, S. F. and Smith, G. P. (1988) Antibody-selectable filamentous fd phage vectors: affinity purification of target genes. *Gene* 73, 305–318.
2. Smith, G. P. (1985) Filamentous fusion phage: novel expression vectors that display cloned antigens on the virion surface. *Science* 228, 1315–1317.
3. Cwirla, S. E., Peters, E. A., Barrett, R. W., and Dower, W. J. (1990) Peptides on phage: a vast library of peptides for identifying ligands. *Proc. Natl. Acad. Sci. USA* 87, 6378–6382.
4. Devlin, J. J., Panganiban, L. C., and Devlin, P. E. (1990) Random peptide libraries: a source of specific protein binding molecules. *Science* 249, 404–406.
5. Scott, J. K. and Smith, G. P. (1990) Searching for peptide ligands with an epitope library. *Science* 249, 386–390.
6. Kehoe, J. W. and Kay, B. K. (2005) Filamentous phage display in the new millennium. *Chem. Rev.* 105, 4056–4072.
7. Smith, G. P. and Petrenko, V. A. (1997) Phage Display. *Chem. Rev.* 97, 391–410.
8. Pini, A., Giuliani, A., Ricci, C., Runci, Y., and Bracci, L. (2004) Strategies for the construction and use of peptide and antibody libraries displayed on phages. *Curr. Protein Pept. Sci.* 5, 487–496.
9. Rowley, M. J., O'Connor, K., and Wijeyewickrema, L. (2004) Phage display for epitope determination: a paradigm for identifying receptor-ligand interactions. *Biotechnol. Annu. Rev.* 10, 151–188.
10. Szardenings, M. (2003) Phage display of random peptide libraries: applications, limits, and potential. *J. Recept. Signal Transduct. Res.* 23, 307–349.
11. Sternberg, N. and Hoess, R. H. (1995) Display of peptides and proteins on the surface of bacteriophage lambda. *Proc. Natl. Acad. Sci. USA* 92, 1609–1613.
12. Efimov, V. P., Nepluev, I. V., and Mesyanzhinov, V. V. (1995) Bacteriophage T4 as a surface display vector. *Virus Genes* 10, 173–177.
13. Krumpe, L. R., Atkinson, A. J., Smythers, G. W., Kandel, A., Schumacher, K. M., McMahon, J. B., Makowski, L., and Mori, T. (2006) T7 lytic phage-displayed peptide libraries exhibit less sequence bias than M13 filamentous phage-displayed peptide libraries. *Proteomics* 6, 4210–4222.
14. Lindqvist, B. H. and Naderi, S. (1995) Peptide presentation by bacteriophage P4. *FEMS Microbiol. Rev.* 17, 33–39.
15. Stahl, S. and Uhlen, M. (1997) Bacterial surface display: trends and progress. *Trends Biotechnol.* 15, 185–192.
16. Smith, G. P. and Scott, J. K. (1993) Libraries of peptides and proteins displayed on filamentous phage. *Methods Enzymol.* 217, 228–257.
17. Bottger, V. (2001) Epitope mapping with random peptide libraries, in *Antibody engineering* (Kontermann, R. and Dubel, S., eds.), Springer Verlag, Heidelberg, Germany, pp. 460–472.
18. Bottger, V., Bottger, A., Lane, E. B., and Spruce, B. A. (1995) Comprehensive epitope analysis of monoclonal anti-proenkephalin antibodies using phage display libraries and synthetic peptides: revelation of antibody fine specificities caused by somatic mutations in the variable region genes. *J. Mol. Biol.* 247, 932–946.
19. Bottger, V. and Lane, E. B. (1994) A monoclonal antibody epitope on keratin 8 identified using a phage peptide library. *J. Mol. Biol.* 235, 61–67.
20. Bottger, V., Stasiak, P. C., Harrison, D. L., Mellerick, D. M., and Lane, E. B. (1995)

Epitope mapping of monoclonal antibodies to keratin 19 using keratin fragments, synthetic peptides and phage peptide libraries. *Eur. J. Biochem.* 231, 475–485.

21. Kirschenhofer, A., Magdolen, V., Schmitt, M., Albrecht, S., Krol, J., Farthmann, J., Kopitz, C., Prezas, P., Kruger, A., Luther, T., and Bottger, V. (2003) Recombinant single chain antibody scFv-IIIF10 directed to human urokinase receptor. *Recent Res. Devel. Cancer* 5, 9–25.
22. D'Mello, F. and Howard, C. R. (2001) An improved selection procedure for the screening of phage display peptide libraries. *J. Immunol. Methods* 247, 191–203.
23. Bublil, E. M., Freund, N. T., Mayrose, I., Penn, O., Roitburd-Berman, A., Rubinstein, N. D., Pupko, T., and Gershoni, J. M. (2007) Stepwise prediction of conformational discontinuous B-cell epitopes using the Mapitope algorithm. *Proteins* 68, 294–304.
24. Gershoni, J. M., Roitburd-Berman, A., Siman-Tov, D. D., Tarnovitski Freund, N., and Weiss, Y. (2007) Epitope mapping: the first step in developing epitope-based vaccines. *Biodrugs* 21, 145–156.
25. Irving, M. B., Pan, O., and Scott, J. K. (2001) Random-peptide libraries and antigen-fragment libraries for epitope mapping and the development of vaccines and diagnostics. *Curr. Opin. Chem. Biol.* 5, 314–324.
26. Luther, T., Magdolen, V., Albrecht, S., Kasper, M., Riemer, C., Kessler, H., Graeff, H., Muller, M., and Schmitt, M. (1997) Epitope-mapped monoclonal antibodies as tools for functional and morphological analyses of the human urokinase receptor in tumor tissue. *Am. J. Pathol.* 150, 1231–1244.
27. Spruce, B. A., Curtis, R., Wilkin, G. P., and Glover, D. M. (1990) A neuropeptide precursor in cerebellum: proenkephalin exists in subpopulations of both neurons and astrocytes. *EMBO J.* 9, 1787–1795.
28. Kay, B. K., Kasanov, J., and Yamabhai, M. (2001) Screening phage-displayed combinatorial peptide libraries. *Methods* 24, 240–246.
29. Noren, K. A. and Noren, C. J. (2001) Construction of high-complexity combinatorial phage display peptide libraries. *Methods* 23, 169–178.
30. Menendez, A. and Scott, J. K. (2005) The nature of target-unrelated peptides recovered in the screening of phage-displayed random peptide libraries with antibodies. *Anal. Biochem.* 336, 145–157.

Chapter 14

Antibody Epitope Mapping Using *De Novo* Generated Synthetic Peptide Libraries

Ulrich Reineke

Summary

Identification of antibody binding peptides may be based on the primary structure of the protein antigens used to raise the antibodies (knowledge- or sequence-based approach). This involves scanning the entire sequence of the antigen with overlapping peptides (peptide scan), which are then probed for binding to the respective antibody. If a natural protein binding partner is not known, one has to use combinatorial synthetic libraries with peptide mixtures, randomly generated chemically synthesized libraries of single individual sequences, or biologically produced libraries (e.g., phage display libraries, *see* Chapter "Epitope Mapping Using Phage Display Peptide Libraries"). This chapter describes chemically synthesized combinatorial, as well as randomly generated peptide libraries, collectively called *de novo* approaches, and their application for antibody epitope mapping.

Key words: Peptide library, Combinatorial peptide library, Randomly generated peptide library, Peptide synthesis, SPOT™ synthesis, Peptide array, One-bead-one-peptide, Positional scanning library, Interleukin-10, Epitope mapping.

1. Introduction

Identifying peptides that bind to antibodies is an important step in characterizing antibody specificity in order to study molecular recognition occurring during humoral immune responses as well as to investigate cross-reactivity potentially implicated in autoimmune diseases. In addition, many processes using antibodies as research tools, diagnostics, reagents, or therapeutics require more detailed information about their interaction with peptide antigens. Identification of antibody binding peptides may be based on the primary structure of

Ulrich Reineke and Mike Schutkowski (eds.), *Methods in Molecular Biology, Epitope Mapping Protocols, vol. 524*

DOI: 10.1007/978-1-59745-450-6_14

the protein antigens used to raise the antibodies (knowledge- or sequence-based approach). This involves scanning the entire sequence of the antigen with overlapping peptides (peptide scan), and these are then probed for binding to the respective antibody. The sequence common to the interacting peptides is the epitope (*see* Chapters "Linear B-Cell Epitope Mapping Using Enzyme-Linked Immunosorbent Assay for Libraries of Overlapping Synthetic Peptides" and "Antibody Epitope Mapping Using SPOT™ Peptide Arrays"). If a natural protein binding partner is not known, or if peptide ligands have to be identified without any previous knowledge, for example to study the origin of an autoimmune disease, one has to use combinatorial libraries with peptide mixtures or randomly generated libraries of single individual sequences. These strategies, collectively called *de novo* approaches, are described in this chapter, including brief summaries of examples from the scientific literature. Lebl has published a very lively historical review, with personal comments by the authors of "classical" papers form the beginning of combinatorial chemistry *(1)*.

The main problem for the *de novo* identification of peptides is how to handle the immense number of potential peptide sequences, referred to as "combinatorial explosion." Even if using only the genetically encoded amino acids, the number of possible sequences markedly increases with the peptide length:

- Dimers $20^2 = 400$
- Trimers $20^3 = 8,000$
- Tetramers $20^4 = 160,000$
- Pentamers $20^5 = 3,200,000$
- Hexamers $20^6 = 64,000,000$
- Heptamers $20^7 = 1,280,000,000$
- Octamers $20^8 = 25,600,000,000$

Unfortunately, no technology is yet available to synthesize and handle billions of different compounds individually. There are two solutions to this dilemma. (1) Using combinatorial libraries the aim is to completely cover the potential sequence space (**Subheading 2**). The idea is to synthesize peptide mixtures with degenerated or randomized positions by statistically incorporating amino acids of a certain set (**Fig. 1**). Defined amino acids are used only at a limited number of positions. This results in a manageable number of peptide pools to be screened. The randomized positions of active pools must then be deconvoluted iteratively using deconvolution libraries individually designed for the project, ultimately selecting the active compounds. (2) Since deconvolution is a time-consuming process, arrays of randomly generated peptides (**Subheading 3**) have also been applied. Since such libraries only cover a small percentage of the potential

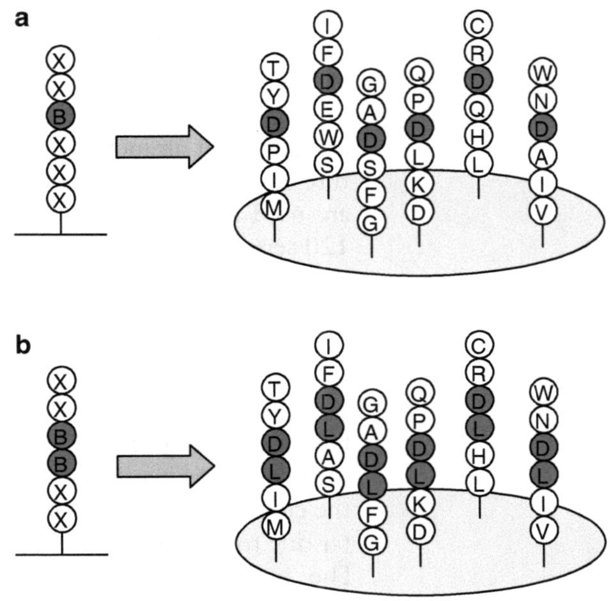

Fig. 1. Peptide mixtures with defined positions B and randomized positions X. (**a**) one defined position and (**b**) two defined positions within a hexapeptide.

sequence space, initially selected peptides often have low affinities to the binding partner and must subsequently be optimized, for example using substitutional analyses (*see* Chapter "Antibody Epitope Mapping Using SPOT™ Peptide Arrays").

2. Combinatorial Peptide Libraries

Chemically prepared combinatorial peptide libraries can be classified into two different types:

(1) Combinatorial library techniques generating mixtures of beads with one individual peptide each *(2, 3)* are prepared by a process called portion mixing *(4)* or the "one-bead-one-peptide" approach *(5)*. Depending on bead size and reactor volume, up to 10^7 or even 10^8 peptides with natural as well as nonnatural building blocks can be generated *(3)*. The disadvantage of this technique is the need to identify the structure of the active compounds after screening using sophisticated but rather tedious coding and decoding or sequencing processes *(6)*.

(2) To circumvent the sequence identification step, combinatorial peptide libraries with randomized as well as defined positions

can be used *(7)*. Here, the entire library is subdivided into a small number of peptide mixtures that have single amino acids at certain positions: O_1XXXXX, XO_2XXXX, XXO_3XXX, XXX-O_4XX, XXXXO_5X and XXXXXO_6 (O = position defined with an individual amino acid, X = position composed of a mixture of amino acids). If the 20 naturally encoded amino acids are used for the defined positions (O) this library comprises 120 separated mixtures that are screened for binding, e.g., to an antibody *(8)*. Subsequently, individual peptides representing all possible combinations of the most active amino acids at each position (positional scanning approach) are synthesized and screened (**Fig. 2a**). Alternatively, two (dual positional approach) or even more positions are defined in the first library (**Fig. 2b**) *(7, 9)*. Although two defined positions involve greater synthesis efforts (20^2 = 400 peptide mixtures) the chance of successful primary screening is significantly better due to interactions with higher affinity and specificity.

The positional and dual positional scanning approaches assume that the contributions of preferred amino acids at each position are additive, or at least not interfering. However, this cannot be taken for granted in every system. To circumvent this limitation, the randomized positions can be deconvoluted by an iterative process (**Fig. 3**). Here, each deconvolution library is designed based on screening results from the starting or precursor library *(10–13)*. Whereas, the initial library is not predefined

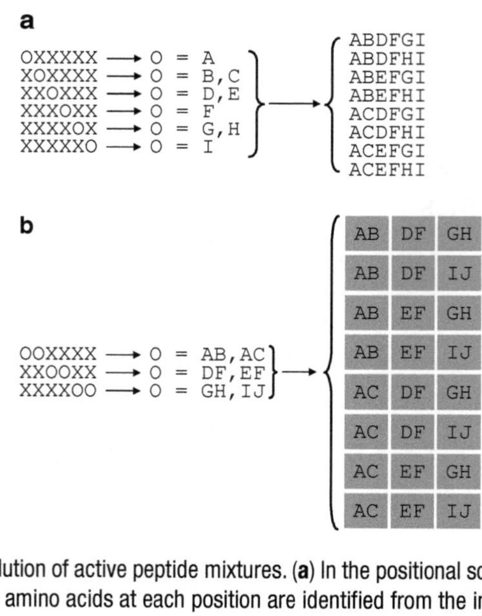

Fig. 2. Deconvolution of active peptide mixtures. (**a**) In the positional scanning approach the most active amino acids at each position are identified from the initial library comprising peptide mixtures (X = randomized position; O = defined position with an individual amino acid). The deconvolution library consists of individual peptides representing all possible combinations of the most active amino acids. (**b**) In the dual positional scanning approach two positions are defined interdependently in the starting library.

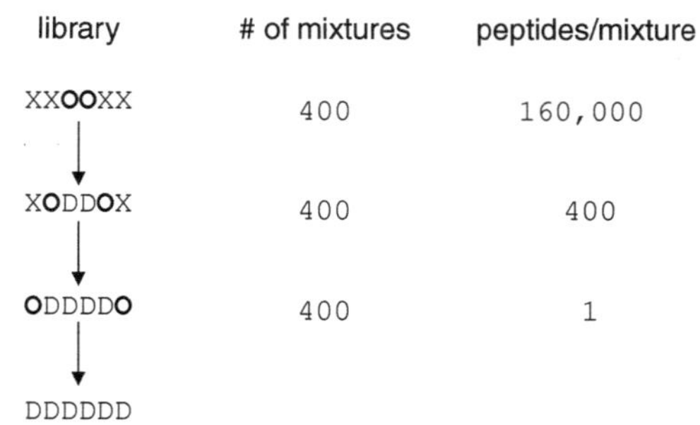

Fig. 3. Iterative deconvolution process of active peptide mixtures. A starting hexamer library of the type XXOOXX (X = randomized position; O = defined position) is screened and the best dipeptide combination OO is selected for the first deconvolution library (XODDOX; D = defined position identified from the preceding library). Subsequently, the second deconvolution library ODDDDO is based on the best tetrapeptide motif ODDO from the preceding library and leads to a single peptide.

for a given screening molecule and can be applied universally, the follow-up libraries are tailor-made for specific purposes. Finally, a re-evaluation is recommended since there might be other amino acids at positions defined early in the process that have a more positive effect on those defined later in the deconvolution.

A marked increase in the effectiveness of peptide and peptide mixture multiple automated synthesis paved the way for more complex libraries of the type $XXXXO_1O_2O_3XXXX$ with 8,000 peptide mixtures *(14)*. The most complex library described so far is one of the type XXXX[3O3X]XXXX prepared by SPOT™ synthesis (*see* Chapter "Antibody Epitope Mapping Using SPOT™ Peptide Arrays"). The internal core [3O3X] is an abbreviation for three defined and three randomized positions arranged in all possible combinations, e.g., $XXXX[O_1O_2O_3XXX]XXXX$; $XXXX[O_1O_2XO_3XX]$ XXXX, and so on *(15)*. This library comprised 68,000 spots and has not only been used to identify antibody epitopes but other peptides binding to the paratope of the antibody in a completely different mode, referred to as mimotopes. This complex library was essential for identifying the epitope of the anti-p24 (human immunodeficiency virus (HIV)-1) monoclonal antibody CB4–1, whereas libraries with one, two, or even three defined neighboring positions failed *(14)*. In many cases such complex libraries are essential to identify peptide epitopes that may require a certain number of key residues in a distinct pattern.

An alternative way to reduce the number of peptide mixtures that need to be prepared, yet match as many defined positions as practicable uses so-called combinatorial clustered amino acid peptide libraries *(16, 17)*. Each cluster contains physicochemically similar amino acids. The rationale of this approach is based on the assumption that physicochemically related amino acids contribute similarly to binding. For instance, grouping the amino acids into six clusters would lower the number of peptide mixtures in a combinatorial library containing four nonrandom positions from 20^4 (160,000, with four defined positions) to 6^4 (1,296, with four cluster positions). Kramer et al. described the epitope mapping of anti-transforming growth factor α (TGFα) mab Tab2 using a library of the type $XC_1C_2C_3C_4X$ (C = one of six amino acid clusters [APG], [DE], [HKR], [NQST], [FYW], [ILVM]) in comparison with phage display techniques *(17)* (*see* Chapter "Epitope Mapping Using Phage Display Peptide Libraries"). The synthetic peptide library array identified several motifs unrelated to the known TGFα-derived linear epitope sequence, whereas the phage display technique only revealed peptide ligands closely related to the wild-type epitope.

Several other combinatorial library techniques have been introduced either as combinations or modifications of the principles described above or with unrelated design strategies. A very interesting technique worth mentioning here is the so-called orthogonal library concept *(18, 19)*. The principle is that the same peptide (or compound in general) is represented in two different mixtures. Comparative activities of different mixtures observed after screening enables identification of the compound responsible for activity.

In summary, the five most critical parameters for identifying peptide ligands, e.g., antibody epitopes from combinatorial libraries with randomized as well as defined positions are: (1) the number of peptide mixtures tested, (2) the number of defined positions, (3) the ratio between defined and randomized positions, (4) the appropriate spacing of defined positions within the entire sequence length, and (5) the overall length of the peptides. These parameters determine the ratio between active and inactive compounds in the peptide mixtures and consequently the signal to noise ratio and likelihood of identifying bioactive peptides.

3. Randomly Generated Peptide Libraries

An alternative to protein sequence-derived or combinatorial peptide array libraries is to use sets of randomly generated peptide sequences. Recently, we described a peptide array approach using a library of 5,520 randomly generated individual 15-mer peptide sequences prepared by SPOT™ synthesis (*see* Chapter "Antibody Epitope Mapping Using SPOT™ Peptide Arrays") that incorporated all genetically encoded amino acids except cysteine *(20)*. Of course, this only covers an extremely small fraction of

a

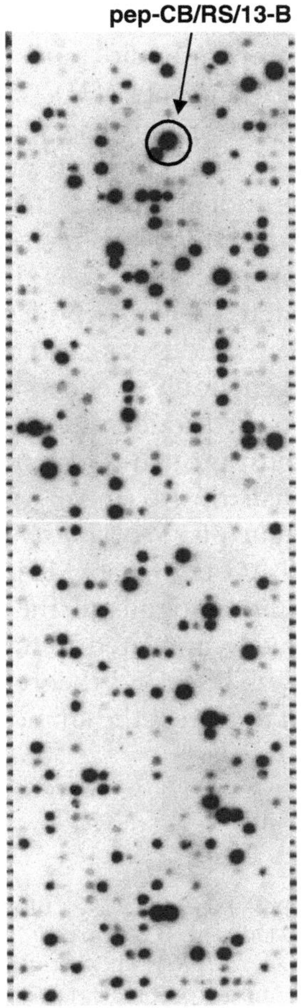

b

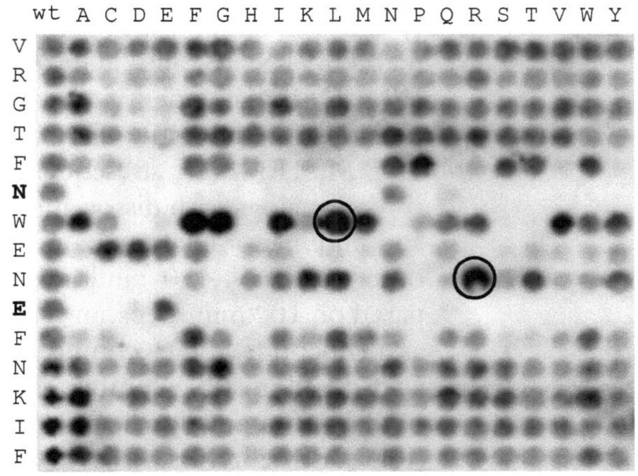

c

Peptide	Sequence	K_D [μM]
wild-type epitope	GNLP**N**ML**RD**LR**D**AFS	0.8
pep-CB/RS/13-B	VRGTF**N**WEN**E**FNKIF	500
pep-CB/RS/13-B1	VRGTF**NL**E**RE**FNKIF	0.46

Fig. 4. Randomly generated peptide library for identifying a peptide epitope. (**a**) Screening of a library of randomly generated 15-mer L-peptides prepared by SPOT™ synthesis with the anti-IL-10 monoclonal antibody CB/RS/13. 1,400 peptide spots are shown. Detection was carried out by chemiluminescence using an anti-mouse IgG peroxidase-labeled polyclonal antibody in combination with an imaging system. The peptide pep-CB/RS/13-B (*circled*) was selected for further analysis. (**b**) Substitutional analysis (*see* Chapter "Antibody Epitope Mapping Using SPOT™ Peptide Arrays") of pep-CB/RS/13-B. Each residue of the peptide was substituted (*rows*) by all other L-amino acids and analyzed for binding of the anti-IL-10 monoclonal antibody CB/RS/13. The sequences corresponding to the left column are identical and represent the starting peptide. Other spots are single-site substitution analogs. The spot intensities correlate qualitatively with the binding affinities. Key residues for the interaction cannot be exchanged by any other, or only by physicochemically similar amino acids, without loss of binding: these are marked in *bold*. Substitution analogs that were selected for affinity measurements and competition are circled. (**c**) Characterization of CB/RS13 binding peptides by affinity measurements using surface plasmon resonance (antibody CB/RS13 coupled to a CM5 sensor chip). The sequences are aligned with the known wild-type epitope by the key residues for interaction (*bold*). Pep-CB/RS/13-B was selected from the randomly generated peptide library. Pep-CB/RS/13-B1 was designed from the substitutional analysis. Substitutions that were selected for affinity optimization are marked in *bold* and *underlined*. Although the sequence of pep-CB/RS/13-B1 differs significantly from the IL-10-derived epitope sequence, the key residues for binding resemble each other very well. Reprinted with permission from **ref.** (*20*).

the potential sequence repertoire. However, the peptide library array was successfully used to identify specifically binding peptide epitopes and mimotopes of three different antibodies (anti-IL-10 mab CB/RS/13, anti-TGFα mab Tab2, anti-p24 (HIV-1) mab CB4–1). Initially identified peptide ligands mostly had very low affinities for the antibodies, with dissociation constants around 10^{-4} M. However, subsequent substitutional analyses revealed several analogs with dissociation constants in the low micromolar and high nanomolar range in a one step process (**Fig. 4**). In two other studies 4,450 randomly generated 12-mer peptides prepared on 10 "mini-pepscan cards" (455 peptides per card) as well as a tripeptide library comprising the genetically encoded amino acids in all possible combinations were used to identify peptides binding to monoclonal antibodies against protein-S of transmissible gastroenteritis virus (TGEV) (mab 6A.A6 and 57.9), an EGF-like domain of the surface protein pfs25 of *Plasmodium falciparum* (mab 32F81), and the FLAG-tag (mab M2). Several peptides were identified as either homologous to the wild-type epitope sequence *(21)* or completely unrelated mimotope sequences *(22)*. Later, this approach was discussed theoretically and the authors described an algorithm to extract the amino acids required for binding *(23)*.

References

1. Lebl, M. (1999) Parallel personal comments on "classical" papers in combinatorial chemistry. *J. Comb. Chem.* 1, 3–24.
2. Beck-Sickinger, A. G. and Jung, G. (1996) From multiple peptide synthesis to peptide libraries, in *Combinatorial Peptide and Nonpeptide Libraries* (Jung, G., ed.), VCH Verlagsgesellschaft, Weinheim, Germany, pp. 79–109.
3. Furka, A. (1996) Chemical synthesis of peptide libraries, in *Combinatorial Peptide and Nonpeptide Libraries* (Jung, G., ed.), VCH Verlagsgesellschaft, Weinheim, Germany, pp. 111–137.
4. Furka, A., Sebestyen, F., Asgedom, M., and Dibo, G. (1991) General method for rapid synthesis of multicomponent peptide mixtures. *Int. J. Pept. Protein Res.* 37, 487–493.
5. Lam, K. S., Salmon, S. E., Hersh, E. M., Hruby, V. J., Kazmierski, W. M., and Knapp, R. J. (1991) A new type of synthetic peptide library for identifying ligand-binding activity. *Nature* 354, 82–84.
6. Lebl, M., Krchnak, V., Sepetov, N. F., Seligmann, B., Strop, P., and Felder, S. (1995) One-bead-one-structure combinatorial libraries. *Biopolymers* 37, 177–198.
7. Pinilla, C., Appel, J., Dooley, C., Blondelle, S., Eichler, J., Dörner, B., Ostresh, J., and Houghten, R. A. (1996) The versatility of non-support bound combinatorial libraries, in *Combinatorial Peptide and Nonpeptide Libraries* (Jung, G., ed.), VCH Verlagsgesellschaft, Weinheim, Germany, pp. 139–172.
8. Pinilla, C., Appel, J. R., Blanc, P., and Houghten, R. A. (1992) Rapid identification of high affinity peptide ligands using positional scanning synthetic peptide combinatorial libraries. *Biotechniques* 13, 901–905.
9. Frank, R. and Overwin, H. (1996) SPOT Synthesis: epitope analysis with arrays of synthetic peptides prepared on cellulose membranes, in *Methods in Molecular Biology Epitope Mapping Protocols* (Morris, G. E., ed.), Humana Press, Totowa, NJ, vol. 66, pp. 149–169.
10. Houghten, R. A., Pinilla, C., Blondelle, S. E., Appel, J. R., Dooley, C. T., and Cuervo, J. H. (1991). Generation and use of synthetic peptide combinatorial libraries for basic research and drug discovery. *Nature* 354, 84–86.
11. Houghten, R. A., Appel, J. R., Blondelle, S. E., Cuervo, J. H., Dooley, C. T., and Pinilla, C. (1992). The use of synthetic peptide libraries for the identification of bioactive peptides. *Biotechniques* 13, 412–421.

12. Kramer, A., Volkmer-Engert, R., Malin, R., Reineke, U., and Schneider-Mergener, J. (1993) Simultaneous synthesis of peptide libraries on single resin and continuous cellulose membrane supports: examples for the identification of protein, metal and DNA binding peptide mixtures. *Peptide Res.* 6, 314–319.
13. Kramer, A., Schuster, A., Reineke, U., Malin, R., Volkmer-Engert, R., Landgraf, C., and Schneider-Mergener, J. (1994) Combinatorial cellulose-bound peptide libraries: screening tools for the identification of peptides that bind ligands with predefned specifcity. *Methods: A Companion to Methods Enzymol.* 6, 912–921.
14. Schneider-Mergener, J., Kramer, A., and Reineke, U. (1996) Peptide libraries bound to continuous cellulose membranes: tools to study molecular recognition, in *Combinatorial Libraries: Synthesis, Screening and Application Potential* (Cortese, R., ed.), Walter de Gruyter, Berlin, Germany, pp. 53–68.
15. Kramer, A., Keitel, T., Winkler, K., Stöcklein, W., Höhne, W., and Schneider-Mergener, J. (1997) Molecular basis for the binding promiscuity of an anti-p24 (HIV-1) monoclonal antibody. *Cell* 91, 799–809.
16. Blake, J. and Litzi-Davis, L. (1992) Evaluation of peptide libraries: an iterative strategy to analyze the reactivity of peptide mixtures with antibodies. *Bioconjugate Chem.* 3, 510–513.
17. Kramer, A., Vakalopoulou, E., Schleuning, W. D., and Schneider-Mergener, J. (1995) A general route to fingerprint analyses of peptide-antibody interactions using a clustered amino acid peptide library: comparison with a phage display library. *Mol. Immunol.* 32, 459–465.
18. Pirrung, M. C. and Chen, J. (1995) Preparation and screening against acetylcholinesterase of a non-peptide "indexed" combinatorial library. *J. Am. Chem. Soc.* 117, 1240–1245.
19. Déprez, B., Willard, X., Bourel, L., Coste, H., Hyafil, F., and Tartar, A. (1995) Orthogonal combinatorial chemical libraries. *J. Am. Chem. Soc.* 117, 5405–5406.
20. Reineke, U., Ivascu, C., Schlief, M., Landgraf, C., Gericke, S., Zahn, G., Herzel, H., Volkmer-Engert, R., and Schneider-Mergener, J. (2002) Identification of distinct antibody epitopes and mimotopes from a peptide array of 5520 randomly generated sequences. *J. Immunol. Methods* 267, 37–51.
21. Slootstra, J. W., Puijk, W. C., Ligtvoet, G. J., Langeveld, J. P. M., and Meloen, R. H. (1995) Structural aspects of antibody-antigen interaction revealed through small random peptide libraries. *Mol. Divers.* 1, 87–96.
22. Slootstra, J. W., Puijk, W. C., Ligtvoet, G. J., Kuperus, D., Schaaper, W. M. M., and Meloen, R. H. (1997) Screening of a small set of random peptides: a new strategy to identify synthetic peptides that mimic epitopes. *J. Mol. Recognit.* 10, 217–224.
23. van der Veen, P. J., Wessels, L. F. A., Slootstra, J. W., Meloen, R. H., Reinders, M. J. T., and Hellendoorn, J. (2001) Determination of binding amino acids based on random peptide array screening data, in *Lecture Notes in Computer Science, Workshop on Algorithms in Bioinformatics (WABI 2001)*, (Gascuel, O. and Moret, B. M. E., eds.), Springer-Verlag, Berlin, Heidelberg, Germany, pp. 264–277.

Chapter 15

Antibody Specificity Profiling on Functional Protein Microarrays

Dawn R. Mattoon and Barry Schweitzer

Summary

Antibodies represent the end product of an exquisitely complex biological process including recombination, somatic hypermutation, affinity maturation, and self-tolerance, culminating in binding reagents directed against a vast repertoire of antigens. The resultant high affinity and diversity of specificity of these biomolecules has been exploited through the development of immunoassays and biotherapeutics that inaugurated a new era in experimental molecular biology and pharmaceutical drug development. Despite the utility of antibodies for research applications and in disease treatment, they must be employed in the context of an accurate understanding of their binding profile. High-content microarrays comprised of thousands of native, full length human proteins are an important tool in the assessment of antibody specificity.

Key words: Protein microarray, Antibody specificity, Off-target binding, Microarray, Antibody, Proteomics, Protein array.

1. Introduction

Antibodies represent invaluable tools that are critical to a vast array of research applications, and comprise a rapidly expanding segment of the pharmaceutical industry. In research, antibodies are commonly used for affinity purification, protein quantitation, and localization of proteins in tissues or cells. Antibodies are also widely used in diagnostic laboratories for protein quantification. The therapeutic antibody market is expected to nearly triple between 2004 and 2010, with more than 370 programs currently in the pipeline *(1)*. To sustain this rate of growth, antibody developers must take advantage of emerging technologies that

Ulrich Reineke and Mike Schutkowski (eds.), *Methods in Molecular Biology, Epitope Mapping Protocols, vol. 524*

DOI: 10.1007/978-1-59745-450-6_15

both refine the pipeline and accelerate the development process. Despite the broad utility of antibodies, many methods for determining antibody specificity remain costly and time consuming. However, formal or established methods to rigorously investigate antibody specificity are not currently available, and exceptions to the "one antibody one antigen" rule are often reported only when experimental evidence suggests that an antibody is leading to an unanticipated result due to off-target binding *(2–10)*. While specificity is the hallmark of antibodies, the widely acknowledged potential for antibody cross-reactivity makes interpretation of results in basic research applications more difficult, and increases the potential for unanticipated adverse side effects in clinical applications *(4–7, 9–12)*. Indeed, the importance of precise and thorough characterization of antibody specificity during the development of antibody therapeutics has recently been underscored by the FDA *(13)*.

Clinical trials of therapeutic monoclonal antibodies often reveal side effects that cannot be immediately explained by dose toxicity or by the known protein binding partners. Through the use of protein microarray technology, it is now possible to simultaneously profile thousands of functional proteins to quickly and accurately assess the specificity of any antibody *(14–17)*. Utilizing this technology at an early stage in the therapeutic antibody development process may allow enhanced selection of antibodies that are likely to have improved performance in clinical trials. Protein microarray content comes in many forms, including micro-scale samples of tissue, cells, cellular lysates, or purified antigens *(18)*. While these formats may all have utility in profiling antibody specificity, optimal value is derived from protein microarrays comprised of full length, native proteins with appropriate post-translational modifications. The use of protein microarray technology carries a number of advantages over currently used methods for assessing antibody specificity. The low concentration of many proteins present in cell lysates make western blotting an ineffective approach for comprehensive screening. In addition, the results obtained from western blotting are likely to under-represent antibody cross-reactivity because the denaturing conditions employed in this assay prevent conformational epitope recognition. Therapeutic antibodies in development are commonly surveyed for cross-reactivity through immunocytochemical or immunohistochemical (IHC) methods, and this approach has been extended recently through the use of tissue microarrays. Although tissue microarrays are amenable to higher throughput analysis through the use of automated image processing, the identification of potential cross-reactivity by any IHC-based approach may be limited because of poor antigen availability or by low antigen concentration in a specific cell or tissue sample *(19–21)*. More recently, investigators have utilized mass-spectrometry to identify cross-reactive antigens following gel electrophoresis or antigen capture by bead-immobilized antibodies *(22–24)*, but these methods remain technically challenging and costly to execute. The sensitivity, reproducibility, and ease-of-use all support the widespread adoption of protein microarray technology for profiling antibody specificity. High-density human protein microarray technology has enormous potential as a tool to markedly accelerate development of optimized antibody research tools and therapeutics.

2. Materials

1. ProtoArray® Human Protein Microarray (Invitrogen, No. PAH052402).
2. Quadri-PERM 4-chamber incubation tray (Greiner, Germany).
3. Alexa Fluor® 647 goat anti-human IgG (H + L), 2 mg/mL (Invitrogen, Carlsbad, CA).
4. Gene Pix Pro Software (recommended) (Molecular Probes).
5. GenePix 4000B Microarray Scanner (Molecular Probes).
6. Eppendorf centrifuge (5810) (Lab-Line Instruments).
7. Polyacetal slide rack (RA Lamb).
8. Protein Microarrays: Human clones used to produce proteins for ProtoArray® Human Protein Microarrays (Invitrogen, Carlsbad, CA) were obtained from Invitrogen's Ultimate™ ORF (open reading frame) collection or from a Gateway® collection of kinase clones developed by Protometrix. The nucleotide sequence of each clone was verified by full length sequencing. All clones were transferred into a system for expressing recombinant proteins in insect cells via baculovirus infection. Using a proprietary high-throughput insect cell expression system, thousands of recombinant human proteins were produced in parallel. Each protein is tagged with Glutathione-S-Transferase (GST), which enables high-throughput affinity purification under conditions that retain activity. After purification, a sample of every purified protein is checked to ensure that the protein is present at the predicted molecular weight. ProtoArray® Protein Microarrays are manufactured using a contact printer equipped with 48 matched quill-type pins. Each protein is deposited along with a set of control proteins in duplicate spots on 1 × 3 inch glass slides that have been coated with a thin layer of nitrocellulose. APiX™ slides are manufactured by GenTel BioSciences, Inc. The printing of these arrays is carried out in a cold room

under dust-free conditions in order to preserve the integrity of both samples and printed microarrays. Before releasing protein microarrays for use, each lot of slides is subjected to a rigorous quality control (QC) procedure, including a gross visual inspection of all the printed slides to check for scratches, fibers and smearing. Since each of the proteins on the array contains an N-terminal GST tag, a GST-directed antibody detects human proteins in a second QC assay. The procedure measures the variability in spot morphology, the number of missing spots, the presence of control spots, and the amount of protein deposited in each spot. The arrays are designed to accommodate 12,288 spots. For the ProtoArray® Human Protein Microarray, samples are printed in 150-μm spots arrayed in 48 subarrays (4,400-μm^2 each) and are equally spaced in vertical and horizontal directions

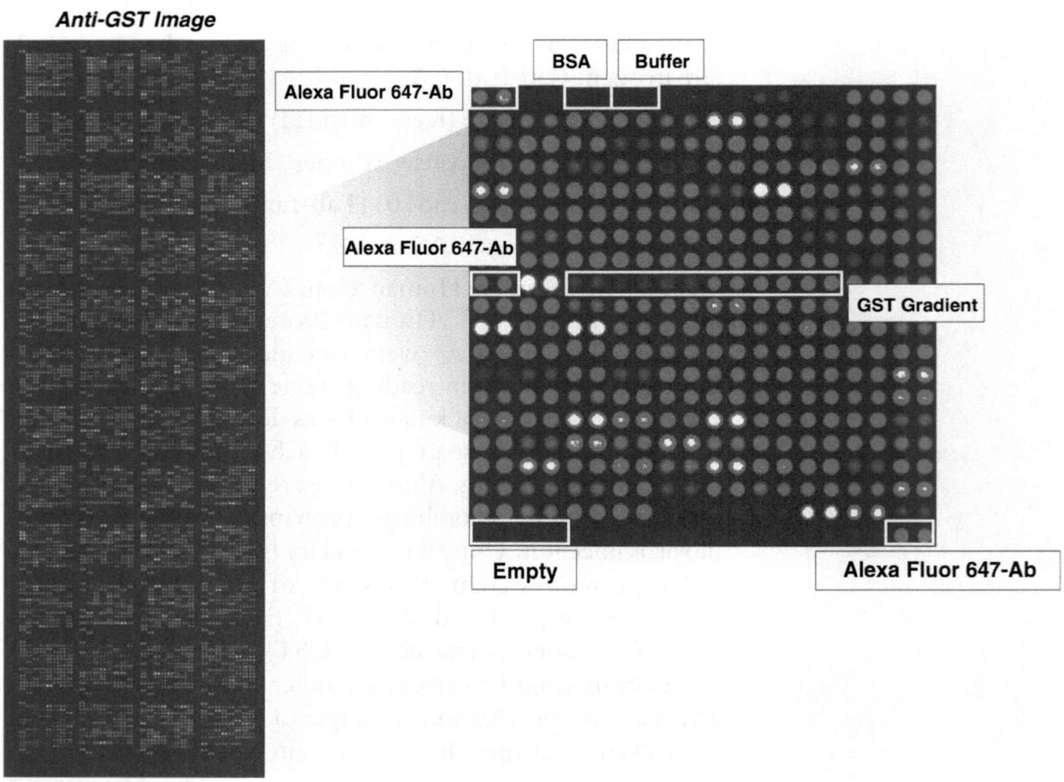

Fig. 1. A ProtoArray® Human Protein Microarray was probed with an anti-GST antibody conjugated to Alexa Fluor 647. The microarray was dried and scanned at 635 nm on an Axon 4000B scanner. The ProtoArray® is divided up into 48 individual subarrays, each comprised of an identical set of negative and positive control elements, and variable human protein content. An enlarged image of a single subarray is shown on the right, with fluorescent positional mapping markers and a subset of the control elements highlighted. Note the fluorescence pattern associated with the spotting of features as adjacent duplicates (*see Color Plates*).

with 20 columns and 20 rows per subarray. Spots are printed with a 220-μm spot-to-spot spacing. An extra 100-μm gap between adjacent subarrays allows quick identification of subarrays (**Fig. 1**).

9. Protein microarray blocking buffer: 5 mM $MgCl_2$, 0.5 mM DTT, 0.05% Triton X-100, 5% glycerol, in PBS, with 1% BSA added fresh prior to assays.
10. Protein microarray PBST probing buffer: 0.1% Tween 20 (w/v) in PBS (Gibco), with 1% BSA (Sigma, protease-free) added fresh prior to blocking.
11. LifterSlip raised edge coverslip (Thermo Fisher Scientific).
12. Secondary Detection Reagent: Alexa Fluor® 647-conjugated anti-species IgG (H + L) at 1.0 μg/mL diluted in 5-mL probing buffer.

3. Methods

Protein microarrays represent an important new tool in proteomic systems biology. This technology is ideally suited to reveal off-target binding events, and as such provides an approach to facilitate selection of antibody candidates for further development. Additionally, researchers utilizing specific antibodies as reagents in high-throughput assays can employ this technology as a method to rapidly and easily evaluate the quality of these high affinity tools. The method described here assumes the use of ProtoArray® Human Protein Microarrays. The workflow for antibody specificity profiling using protein microarray technology is outlined in **Fig. 2**.

3.1. Protein Microarray Antibody Specificity Profiling Assay

1. Prior to initiating the antibody specificity profiling assay, obtain an appropriate number of protein microarrays for the experimental design (ProtoArray® Human Protein Microarrays from Invitrogen are recommended). Assays are generally performed with two concentrations of antibody over at least a tenfold concentration range. The protein microarrays are stored at –20°C, and must be allowed to equilibrate to 4°C for 10 min prior to initiating the blocking step (*see* **Note 1**).
2. The blocking buffer is prepared ahead and stored at 4°C, and BSA is added to a final concentration of 1% just prior to use in blocking the protein microarrays. For blocking, protein microarrays are placed protein-side up in a quadriPERM tray, one array per chamber. If the protein microarrays are manufactured on barcoded slides, ensure that the barcode end of

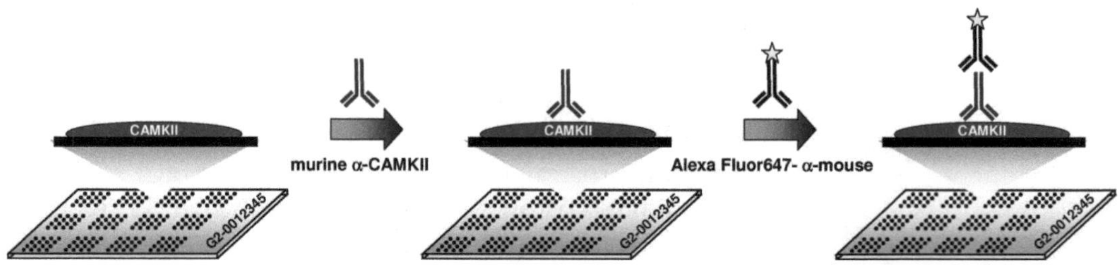

Fig. 2. Protein microarray antibody specificity profiling workflow: Following blocking, the protein microarrays are probed with a dilute solution of the antibody of interest. The antibody binds to its cognate protein on the microarray, and potentially to additional cross-reactive proteins. Following incubation with primary antibody, the microarrays are washed, and incubated with a dye-labelled anti-species antibody. Following incubation with the secondary detection reagent, arrays are washed, dried, and imaged in a fluorescent microarray scanner (*see Color Plates*).

the slide is near the end of the tray with the indented numeral. The indent in the bottom of the tray will be used as the site of buffer exchange. The 4-well trays are gently rocked to ensure that each slide is completely immersed in the blocking buffer. Arrays are blocked for 1 h at 4°C on an orbital shaker in 5-mL blocking buffer per array.

3. During the protein microarray blocking step, the primary antibody solutions are prepared. Antibodies are diluted to suitable concentrations (*see* **Note 2**) in PBST probing buffer. The probing buffer is prepared ahead and stored at 4°C, and BSA added to a final concentration of 1% just prior to use in probing the protein microarrays. Antibodies can be probed in a total volume of 5 mL. However, if antibody sample is limited, antibody probes can be incubated on the microarray under a LifterSlip in a total volume of 100 μL.
4. Following blocking, the blocking buffer is aspirated by vacuum or by the use of a pipettor. The tip of the aspirator or pipettor is positioned in the indented numeral in order to remove as much of the liquid as possible. When each well is dry, the indented numeral end of the tray is lifted to facilitate removal of the liquid which pools at the base of the well (*see* **Notes 3 and 4**).
5. Once the blocking buffer has been completely removed, 5 mL of the dilute antibody solution is added to the quadriPERM tray by pipetting, taking care to avoid pipetting directly onto the array surface. Arrays are then incubated in the primary antibody probing solution for 90 min at 4°C with gentle orbital shaking (*see* **Note 5**).
6. After probing, the diluted antibody solution is aspirated by vacuum or by the use of a pipettor. The tip of the aspirator or pipettor is again positioned in the indented numeral in order to remove as much of the liquid as possible. When

each well is dry, the indented numeral end of the tray is lifted to facilitate removal of the liquid which pools at the base of the well.

7. The protein microarrays with bound primary antibody are then washed with a series of five 5-min washes in probing buffer. For each wash, 5 mL of probing buffer is added to each well of the tray and allowed to incubate for 5 min with gentle agitation on an orbital shaker before removal by aspiration or pipetting and addition of the next wash.
8. Once the final wash has been completely removed, 5 mL of the dilute secondary detection solution is added to the tray by pipetting, taking care to avoid pipetting directly onto the array surface. Alexa Fluor-conjugated secondary detection reagents diluted to 1 μg/mL are recommended to maximize signal stability (*see* **Note 6**). Arrays are then incubated in the secondary detection solution for 90 min at 4°C with gentle orbital shaking.
9. Following the 90 min incubation with the detection reagent, the solution is aspirated by vacuum or by the use of a pipettor. The tip of the aspirator or pipettor is again positioned in the indented numeral in order to remove as much of the liquid as possible. When each well is dry, the indented numeral end of the tray is lifted to facilitate removal of the liquid which pools at the base of the well.
10. The protein microarrays with bound primary and secondary antibodies are then washed with a series of five, five-minute washes in probing buffer. For each wash, 5 mL of probing buffer is added to each well of the tray and allowed to incubate for five minutes with gentle agitation on an orbital shaker before removal by aspiration or pipetting and addition of the next wash.
11. Once the final wash has been completely aspirated, the protein microarrays are removed from the tray and placed in a slide drying rack. To facilitate removal, forceps can be inserted into the indented numeral and used to gently pry the edge of the slide upward. The arrays are then transferred to a slide drying rack with a gloved hand, taking care to only touch the slide by its edges. The arrays are then centrifuged at low speed (800 × *g*) for one min in a centrifuge equipped with a plate rotor.
12. Following complete drying of the protein microarrays, they are scanned with a fluorescent microarray scanner. The GenePix 4000B scanner set at 635 nm with a laser power of 100% and a focus point of 0 μm is recommended for generating high resolution images (*see* **Notes 7 and 8**). These images are saved in Tagged Image File Format (.tif-files)

a

Database ID	Signal - Background		Z-Score		Replicate Spot CV		Protein Description
	α-CAMKII	neg	α-CAMKII	neg	α-CAMKII	neg	
NM_012920.1	16818	296	37.6	0.6	10.7%	1.2%	calcium/calmodulin-dependent protein kinase (CaM kinase) II alpha, transcript variant 1
BC040457.1	5995	130	11.8	0.2	5.6%	5.7%	calcium/calmodulin-dependent protein kinase (CaM kinase) II alpha, transcript variant 2
NM_024316.1	3390	545	5.6	1.1	31.2%	15.7%	leukocyte receptor cluster (LRC) member 1 (LENG1)
NM_002738.5	1696	159	3.3	0.2	2.5%	19.2%	protein kinase C, beta 1 (PRKCB1), transcript variant 2

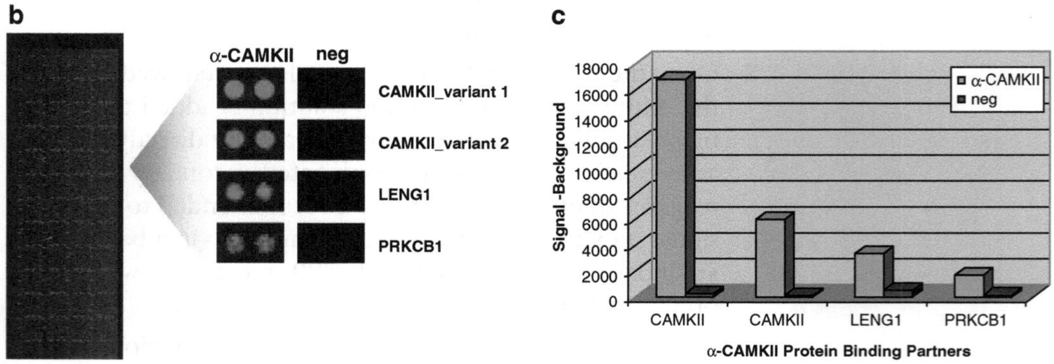

Fig. 3. Protein microarray specificity profiling for anti-CAMKII antibody. (**a**) Murine monoclonal anti-calmodulin kinase II (α-CAMKII, Invitrogen) was probed on a ProtoArray at 1.0 μg/mL. Following completion of the antibody specificity profiling assay protocol, the arrays were scanned on a fluorescent microarray scanner, and pixel intensity data was used to calculate background subtracted signal values, *Z*-scores, and coefficients of variation (CV) for adjacent duplicate spots (**b**) The >8,000 human proteins were evaluated for features giving rise to a *Z*-score >3.0 in the α-CAMKII assay, with a corresponding *Z*-score <1.5 in the negative control (Alexa Fluor 647-anti-mouse antibody, detection reagent only) assay. Four proteins met these threshold criteria, with the two splice variants of CAMKII identified as the most robust binding partners. Duplicate spots from the α-CAMKII and negative control assays are shown. (**c**) background subtracted signal values are plotted (*see Color Plates*).

and are used in extracting pixel intensity information. As an example, the specificity profile for anti-calmodulin kinase II (CAMKII) generated on a high content protein microarray is shown in **Fig.** 3.

13. Once the protein microarrays are completely dry, store the slides in a light-tight slide box. Prolonged exposure to light will diminish signal intensities. After the arrays have been probed and dried, they can be stored either vertically or horizontally.
14. The array list file (.gal-file) is then uploaded to the image analysis software (GenePix 6.1 from Molecular Devices is recommended). This text file describes the layout of the protein microarray and contains the details of the microarray content, including relevant control elements. The .gal-file is used to map the location of each array features, initially with a fixed feature size based on the diameter of the spotted proteins. To maximize accuracy, a pixel-based segmentation algorithm

is recommended for pixel intensity data extraction (irregular feature finding setting, located under the *Alignment* tab in the *Options* menu of GenePix 6.1) (*see* **Note 9**). After aligning all features using fluorescent positional mapping markers, pixel intensities for each spot on the array are calculated by the software and saved to a .txt-file formatted for use in GenePix, the GenePix Result file (.gpr-filename extension). These files are subsequently opened in other text editing or spreadsheet programs for analysis.

15. Assuming ProtoArray® Human Protein Microarrays are used in the assays, quantified spot files are processed using the ProtoArray® Prospector freeware to determine which proteins interact with the antibody probes. The software incorporates background correction, *Z*-factor and *Z*-score calculations and replica CV filtering.

4. Notes

1. Use a shaker that keeps the arrays in one plane during rotation. Nutating or rocking shakers are not to be used because of increased risk of cross-well contamination.
2. The choice of antibody concentration should be made in the context of the final working concentration for the antibody. For example, monoclonal or polyclonal research antibodies are commonly probed at concentrations ranging from 0.1 to 1.0 ng/μL. Therapeutic antibodies, particularly those administered by subcutaneous injection, are commonly probed at much higher concentrations, in the range of 1–10 mg/mL.
3. Do not aspirate buffers from the surface of the slide in order to reduce the risk of surface scratches.
4. Do not allow any part of the array surface to dry before adding the next solution as this can cause high and/or uneven background.
5. The probe buffer may require modification to increase the stringency of washes in order to minimize background and enhance the detection of high affinity antibody–antigen interactions. Examples of buffer modifications designed to increasing washing stringency include (a) increasing salt (b) increasing amount of detergent during washes (c) increasing temperature during washes or (d) increasing time of washes.
6. The recommended concentration of anti-rabbit/mouse antibody Alexa Fluor conjugate may require optimization to maximize signal to background.

7. During microarray scanning, the PMT should be adjusted such that the feature signals on the array are not saturated. This is done to keep signals within the linear range.
8. For the GenePix 4000B, the best signal-to-noise ratio can be achieved when PMT gain is between 500 and 900. In general, using a PMT gain below 400 is not recommended. If the fluorescence intensity is saturating at full laser power, the laser power should be lowered rather than using an extremely low PMT voltage.
9. In general, the use of pixel-based segmentation (irregular feature finding) results in more reproducible signal used values. The spot diameter and background signal intensities are generally similar between circular and irregular feature finding in antibody-specificity profiling assays.

References

1. Kim, S. J., Park, Y., and Hong, H. J. (2005) Antibody engineering for the development of therapeutic antibodies. *Mol. Cells* 20, 17–29.
2. Bonsing, B. A., Corver, W. E., Gorsira, M. C., van Vliet, M., Oud, P. S., Cornelisse, C. J., and Fleuren, G. J. (1997) Specificity of seven monoclonal antibodies against p53 evaluated with Western blotting, immunohistochemistry, confocal laser scanning microscopy, and flow cytometry. *Cytometry* 28, 11–24.
3. Coers, W., Timens, W., Kempinga, C., Klok, P. A., and Moshage, H. (1998) Specificity of antibodies to nitric oxide synthase isoforms in human, guinea pig, rat, and mouse tissues. *J. Histochem. Cytochem.* 46, 1385–1392.
4. Liu, B., Sun, D., Xia, W., Hung, M. C., and Yu, D. (1997) Cross-reactivity of C219 anti-p170(mdr-1) antibody with p185(c-erbB2) in breast cancer cells: cautions on evaluating p170(mdr-1). *J. Natl. Cancer Inst.* 89, 1524–1529.
5. Otte, L., Knaute, T., Schneider-Mergener, J., and Kramer, A. (2006) Molecular basis for the binding polyspecificity of an anti-cholera toxin peptide 3 monoclonal antibody. *J. Mol. Recognit.* 19, 49–59.
6. Schuermann, J. P., Henzl, M. T., Deutscher, S. L., and Tanner, J. J. (2004) Structure of an anti-DNA fab complexed with a non-DNA ligand provides insights into cross-reactivity and molecular mimicry. *Proteins* 57, 269–278.
7. Spellerberg, M., Chapman, C., Hamblin, T., and Stevenson, F. (1995) Dual recognition of lipid A and DNA by human antibodies encoded by the VH4–21 gene. A possible link between infection and lupus. *Ann. N Y Acad. Sci.* 764, 427–432.
8. Suka, N., Suka, Y., Carmen, A. A., Wu, J., and Grunstein, M. (2001) Highly specific antibodies determine histone acetylation site usage in yeast heterochromatin and euchromatin. *Mol. Cell.* 8, 473–479.
9. Tamaoka, A., Endoh, R., Shoji, S., Takahashi, H., Hirokawa, K., Teplow, D. B., Selkoe, D. J., and Mori, H. (1996) Antibodies to amyloid beta protein (A beta) crossreact with glyceraldehyde-3-phosphate dehydrogenase (GAPDH). *Neurobiol. Aging* 17, 405–14.
10. Xu, Y., Ramsland, P. A., Davies, J. M., Scealy, M., Nandakumar, K. S., Holmdahl, R., and Rowley, M. J. (2004) Two monoclonal antibodies to precisely the same epitope of type II collagen select non-crossreactive phage clones by phage display: implications for autoimmunity and molecular mimicry. *Mol. Immunol.* 41, 411–419.
11. Zhang, B., Zheng, R., Wang, J., Bu, D., and Zhu, X. (2006) Epitopes in the linker subdomain region of envoplakin recognized by autoantibodies in paraneoplastic pemphigus patients. *J. Invest. Dermatol.* 126, 832–840.
12. Zhu, X., Wentworth, P., Jr., Kyle, R. A., Lerner, R. A., and Wilson, I. A. (2006) Cofactor-containing antibodies: crystal structure of the original yellow antibody. *Proc. Natl. Acad. Sci. U S A* 103, 3581–3585.
13. U.S. Food and Drug Administration Center for Biologics Evaluation and Research (1997) Points to consider in the manufacture and testing of monoclonal antibody products for human use. *J. Immunother.* 20, 214–243.
14. Michaud, G. A. and Snyder, M. (2002) Proteomic approaches for the global analysis of proteins. *Biotechniques* 33, 1308–1316.

15. Michaud, G. A., Salcius, M., Zhou, F.,Bangham, R., Bonin, J., Guo, H., Snyder, M., Predki, P. F., and Schweitzer, B. I. (2003) Analyzing antibody specificity with whole proteome microarrays. *Nat. Biotechnol.* 21, 1509–1512.

16. Michaud, G. A., Samuels, M. L., and Schweitzer, B. (2006) Functional protein arrays to facilitate drug discovery and development. *IDrugs* 9, 266–272.

17. Predki, P. F., Mattoon, D., Bangham, R., Schweitzer, B., and Michaud, G. (2005) Protein microarrays: a new tool for profiling antibody cross-reactivity. *Hum. Antibodies* 14, 7–15.

18. Mattoon, D., Michaud, G., Merkel, J., and Schweitzer, B. (2005) Biomarker discovery using protein microarray technology platforms: antibody-antigen complex profiling. *Expert Rev. Proteomics* 2, 879–889.

19. Rogers, A. B., Cormier, K. S., and Fox, J. G. (2006) Thiol-reactive compounds prevent nonspecific antibody binding in immunohistochemistry. *Lab. Invest.* 86, 526–533.

20. Sobek, J., Bartscherer, K., Jacob, A., Hoheisel, J. D., and Angenendt, P. (2006) Microarray technology as a universal tool for high-throughput analysis of biological systems. *Comb. Chem. High Throughput Screen.* 9, 365–380.

21. Tabrizi, M. A., Tseng, C. M., and Roskos, L. K. (2006) Elimination mechanisms of therapeutic monoclonal antibodies. *Drug Discov. Today* 11, 81–88.

22. Cha, H., Hancock, C., Dangi, S., Maiguel, D., Carrier, F., and Shapiro, P. (2004) Phosphorylation regulates nucleophosmin targeting to the centrosome during mitosis as detected by cross-reactive phosphorylation-specific MKK1/MKK2 antibodies. *Biochem. J.* 378, 857–865.

23. Potgens, A. J., Schmitz, U., Kaufmann, P., and Frank, H. G. (2002) Monoclonal antibody CD133-2 (AC141) against hematopoietic stem cell antigen CD133 shows crossreactivity with cytokeratin 18. *J. Histochem. Cytochem.* 50, 1131–1134.

24. Weiler, T., Sauder, P., Cheng, K., Ens, W., Standing, K., and Wilkins, J. A. (2003) A proteomics-based approach for monoclonal antibody characterization. *Anal. Biochem.* 321, 217–225.

Chapter 16

Peptide Microarrays for Determination of Cross-Reactivity

Alexandra Thiele

Summary

Polyclonal antibodies raised against full-length antigens are often used for localization experiments. Exact knowledge of epitopes in the antigen recognized by the antiserum is important if the target antigen belongs to a large family of proteins which are highly conserved. We have shown that epitope mapping using peptide microarrays represents a powerful tool for determination of immunodominat regions in a proteome-wide manner. As examples we show results of epitope mapping using peptide microarrays displaying overlapping peptide scans through either all human cyclophilins or all human FK506-binding proteins.

Key words: Peptide microarray, Linear epitope, Polyclonal antibody, Overlapping peptides, Peptidyl-prolyl cis/*trans* isomerase, Cyclophilin, FK506-binding protein.

1. Introduction

Overlapping peptides immobilized on planar surfaces represent a powerful tool for epitope mapping of polyclonal antibodies *(1)*. Originally developed by Geysen using peptides synthesized on plastic pins *(2–4)*, overlapping peptides immobilized on cellulose membranes *(5-8)*, polymer membranes *(5)* or plastic cards *(9)* are widely used for determination of immunodominant regions in protein antigens. Epitope discovery and epitope mapping using peptides synthesized *(10–12)* or re-immobilized and purified *(13–15)* on modified glass surfaces was demonstrated. The advantages of the microarray method are that it allows comprehensive profiling of polyclonal antiserum within just a few hours and that only minute amount of sample is needed.

Antigenic protein sequences can usually be found on the surface of a native protein and preferably consist of long stretches

Ulrich Reineke and Mike Schutkowski (eds.), *Methods in Molecular Biology, Epitope Mapping Protocols, vol. 524*

DOI: 10.1007/978-1-59745-450-6_16

of hydrophilic residues *(16)*. Depending on the size and the 3-dimensional structure of target antigen a polyclonal antiserum can contain a number of antibodies that are directed against several epitopes. Valle et al. identified epitopes of a polyclonal serum raised against the *Bacillus subtilis* bacteriophage l29 connector *(17)* and found 11 immunodominant regions within this antigen. Cellulose-bound peptides were used for affinity purification resulting in fractions of the antiserum specific for only one of the epitopes *(18)*. This represents a simple method to generate epitope-specific polyclonal antibodies with low cross-reactivity, which are extremely useful for the determination of cellular localization of target antigens.

We were interested in a biased proteome-wide characterization of polyclonal antibodies directed against peptidyl-prolyl *cis/trans* isomerases (PPIases). These enzymes are able to catalyze the isomerization about peptidyl–prolyl bonds. This isomerization seems to be important in regulation of protein folding, transport through membranes, cell cycle progression, channel gating, and virus replication. There exist three different classes of PPIases in humans: the cyclophilins, FK506-binding proteins (FKBPs), and parvulins. Hallmark of the first two PPIase classes is their ability to bind to the immunosuppressive drugs cyclosporine and FK506, respectively. The sequences of the appropriate PPIase domains are conserved within different organisms and within different gene products in humans. Therefore, it could be expected that antibodies raised against PPIase domains can have strong cross reactivities to other members of the family of the target antigen. If such antibodies should be used in immunofluorescent staining experiments exact knowledge of the epitopes recognized is necessary. Otherwise, misinterpretation of results will cause false-positive localizations of individual PPIases within cells. We collected all human FKBPs (16 family members, Swissprot IDs: P20071, Q16645, P26885, Q9NYL4, Q9NWM8, Q9Y680, Q00688, O75344, O00170, Q14318, Q9NZN9, Q13451, Q02790, O95302, Q96AY3, Q9Y4DO), cyclophilins (16 family members, Swissprot IDs: P62937, P62937, P62937, Q9Y3C6, O43447, P30405, P23284, P45877, Q9UNP9, Q9UNP9, Q08752, Q8WUA2, Q13356, Q96BP3, Q13427, P30414), and parvulins (three family members: Swissprot IDs: Q52M21, Q9Y237, Q13526) from databases and generated overlapping 15meric peptides from the full-length proteins resulting in 1,450 different peptides for all the human FKBPs and 1,830 peptides for the human cyclophilins and parvulins. All of the 3,280 peptides derived from human PPIases were synthesized by SPOT-™ synthesis with an *N*-terminal amino-oxy-acetylated *N*-(3-{2-[2-(3-Amino-propoxy)-ethoxy]-ethoxy}-propyl)-succinyl moiety. This moiety serves as a spacer and allows chemoselective immobilization of the peptide derivatives onto epoxy modified glass surfaces subsequent to release from the cellulose membranes. This chemoselective immobilization allows both directed immobilization resulting in optimal presentation of the peptide to the antibodies and purification of the peptides by removal of truncated by-products during microarray printing *(1)*. All FKBP derived peptides were printed sixfold to one glass slide resulting in 6 × 1,450 = 8,700 data points per FKBP-Chip. Peptides derived from human cyclophilins and parvulins were immobilized in quadruplicate on one glass slide resulting in 4 × 1,830 = 7,320 spots on the Cyp-Chip. All peptides and peptide microarrays were designed and produced by JPT Peptide Technologies GmbH (Berlin, Germany). Polyclonal antisera from rabbit were incubated with these microarrays and peptide bound antibodies were detected by fluorescence imaging of the chips subsequent to incubation with fluorescently labelled goat anti-rabbit IgG DyLight 649 conjugate. **Fig. 1** shows an image of an FKBP-Chip which was incubated with a polyclonal anti-FKBP24 antiserum. Five immunodominant regions were detected each consisting of at least six amino acid residues. The epitopes E1–E3 are located within the PPIase domain, and epitopes E4 and E5 were found within an EF-hand motif of FKBP24. Additionally, there are also

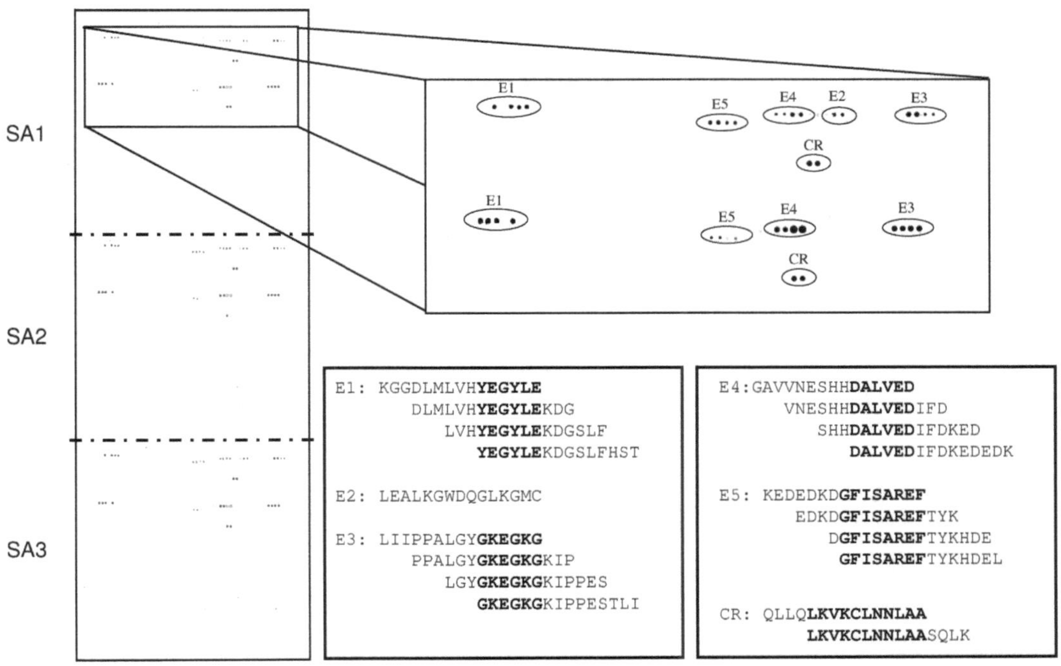

Fig. 1. Image of the FKBP-Chip displaying 1,450 overlapping 15meric peptides covering the primary structure of all human FKBPs. Peptide microarray was incubated with rabbit anti-FKBP24 antiserum (dilution 1:1,000) followed by goat anti-rabbit DyLight labelled antibody (dilution 1:30,000 resulting in 0.03 µg/mL). The *left side* shows the whole chip with the three identical subarrays (SA1–SA3). *Top of right side* shows the enlarged image of SA1. Each of the peptides was immobilized two times per subarray resulting in pairs of identical spots. This arrangement allows to differentiate real signals from fluorescent artefacts. Signals corresponding to the five immunodominat regions of FKBP24 are marked with E1–E5 and detected cross reactivity to peptides derived from FKBP38 was marked with CR. The peptide sequences corresponding to E1–E5 and to CR are given in one letter code. Overlapping sequences within one epitope are marked in *bold*. For better visualization images were inverted. *Black spots* represent high fluorescent signals.

signals found that were caused by cross reactivity to another protein of the FKBP family. The detected cross reacting FKBP38 derived peptides sequences (marked with CR in **Fig. 1**) show no similarity to epitopes found in FBBP24. Nevertheless, the analyzed anti-FKBP24 antiserum shows very limited cross reactivity to other members of the human FKBP family and is therefore suited for immunofluorescent staining experiments.

Fig. 2 summarizes the profiling results of polyclonal rabbit anti-cyclophilin18.1 antiserum. Two immunodominant regions

```
Cyp18|     -FFDIAVDGEPLGRVSFELFADKVPKTAENFRALSTGEK-----GFG--------YKGSCFHRIIPGFMCQG
Cyp18.1|   --MSVTLHTD-VGDIKIEVFCERTPKTCENFLALCASNY----------------YNGCIFHRNIKGFMVQT
Cyp18.2|   -FFDITVDGKPLGRISIKLFADKILKTAENFRALSTGEK-----GFR--------YKGSCFHRIIPGFMCQG
Cyp18.2a|  QPPNVYLETS-MGIIVLELYWKHAPKTCKNFAELARRGY----------------YNGTKFHRIIKDFMIQG
Cyp19.2|   -FFDVSIGGQEVGRMKIELFADVVPKTAENFRQFCTGEFRKDGVPIG--------YKGSTFHRVIKDFMIQG
Cyp22|     -YLDVDANGKPLGRVVLELKADVVPKTAENFRALCTGEK-----GFG--------YKGSTFHRVIPSFMCQA
Cyp23|     -YFDLRIGDEDVGRVIFGLFGKTVPKTVDNFVALATGEK-----GFG--------YKNSKFHRVIKDFMIQG
Cyp23a|    -FFDVRIGDKDVGRIVIGLFGKVVPKTVENFVALATGEK-----GYG--------YKGSKFHRVIKDFMIQG
Cyp33|     -YMDIKIGNKPAGRIQMLLRSDVVPMTAENFRCLCTHEK-----GFG--------FKGSSFHRIIPQFMCQG
Cyp35|     -FLDICIDSSPIGRLIFELYCDVCPKTCKNFQVLCTGKAGFSQRGIR------LHYKNSIFHRIVQNGWIQG
Cyp40|     -FFDVDIGGERVGRIVLELFADIVPKTAENFRALCTGEK-----GIGHTTGKPLHFKGCPFHRIIKKFMIQG
Cyp57|     --MAVLLETT-LGDVVIDLYTEERPRACLNFLKLCKIKY----------------YNYCLIHNVQRDFIIQT
Cyp59|     KKGYVRLHTN-KGDLNLELHCDLTPKTCENFIRLCKKHY----------------YDGTIFHRSIRNFVIQG
Cyp73|     VSDSAIIHTS-MGDIHTKLFPVECPKTVENFCVHSRNGY----------------YNGHTFHRIIKGFMIQT
Cyp89|     -FFDIAINNQPAGRVVFELFSDVCPKTCENFRCLCTGEK-----GTGKSTQKPLHYKSCLFHRVVKDFMVQG
Cyp165|    -HFDIEINREPVGRIMFQLFSDICPKTCKNFLCLCSGEK-----GLGKTTGKKLCYKGSTFHRVVKNFMIQG

Cyp18|     GDFTRHNGTGGKSIYGEK-------FEDE-NFILKHTGPGILSMANAGPNTNGSQFFICTAK-TEWLDGKHV
Cyp18.1|   GDPT-GTGRGGNSIWGKK-------FEDEYSEYLKHNVRGVVSMANNGPNTNGSQFFITYGK-QPHLDMKYT
Cyp18.2|   GDFTRHNGTGDKSIYGEK-------FDDE-NLIRKHTGSGILSMANAGPNTNGSQFFICAAK-TEWLDGKHV
Cyp18.2a|  GDPT-GTGRGGASIYGKQ-------FEDELHPDLKFTGAGILAMANAGPDTNGSQFFVTLAP-TQWLDGKHT
Cyp19.2|   GDFVNGDGTGVASIYRGP-------FADE-NFKLRHSAPGLLSMANSGPSTNGCQFFITCSK-CDWLDGKHV
Cyp22|     GDFTNHNGTGGKSIYGSR-------FPDE-NFTLKHVGPGVLSMANAGPNTNGSQFFICTIK-TDWLDGKHV
Cyp23|     GDFTRGDGTGGKSIYGER-------FPDE-NFKLKHYGPGWVSMANAGKDTNGSQFFITTVK-TAWLDGKHV
Cyp23a|    GDITTGDGTGGVSIYGET-------FPDE-NFKLKHYGIGWVSMANAGPDTNGSQFFITLTK-PTWLDGKHV
Cyp33|     GDFTNHNGTGGKSIYGKK-------FDDE-NFILKHTGPGLLSMANSGPNTNGSQFFLTCDK-TDWLDGKHV
Cyp35|     GDIVYGKGDNGESIYGPT-------FEDE-NFSVPHNKRGVLGMANKGRHSNGSQFYITLQA-TPYLDRKFV
Cyp40|     GDFSNQNGTGGESIYGEK-------FEDE-NFHYKHDREGLLSMANAGRNTNGSQFFITTVP-TPHLDGKHV
Cyp57|     GDPT-GTGRGGESIFGQLYGDQASFFEAEKVPRIKHKKKGTVSMVNNGSDQHGSQFLITTGENLDYLDGVHT
Cyp59|     GDPT-GTGTGGESYWGKP-------FKDEFRPNLSHTGRGILSMANSGPNSNRSQFFITFRS-CAYLDKKHT
Cyp73|     GDPT-GTGMGGESIWGGE-------FEDEFHSTLRHDRPYTLSMANAGSNTNGSQFFITVVP-TPWLDNKHT
Cyp89|     GDFSEGNGRGGESIYGGF-------FEDE-SFAVKHNKEFLLSMANRGKDTNGSQFFITTKP-TPHLDGHHV
Cyp165|    GDFSEGNGKGGESIYGGY-------FKDE-NFILKHDRAFLLSMANRGKHTNGSQFFITTKP-APHLDGVHV

Cyp18|     VFGKVKEGMNIVEAME-RFGS-RNGKTSKKITIADCGQ
Cyp18.1|   VFGKVIDGLETLDELEKLPVNEKTYRPLNDVHIKDITI
Cyp18.2|   AFGKVKERVNIVEAME-HFGY-RNSKTSKKITIADCGQ
Cyp18.2a|  IFGRVCQGIGMVNRVGMVETNSQ-DRPVDDVKIIKAYP
Cyp19.2|   VFGKIIDGLLVMRKIE-NVPTGPNNKPKLPVVISQCGE
Cyp22|     VFGHVKEGMDVVKKIE-SFGS-KSGRTSKKIVITDCGQ
Cyp23|     VFGKVLEGMEVVRKVE-STKTDSRDKPLKDVIIADCGK
Cyp23a|    VFGKVIDGMTVVHSIE-LQATDGHDRPLTNCSIINSGK
Cyp33|     VFGEVTEGLDVLRQIE-AQGS-KDGKPKQKVIIADCGE
Cyp35|     AFGQLIEGTEVLKQLE-LVPT-QNERPIHMCRITDSGD
Cyp40|     VFGQVIKGIGVARILE-NVEV-KGEKPAKLCVIAECGE
Cyp57|     VFGEVTEGMDIIKKINETFVDKD-FVPYQDIRINHTVI
Cyp59|     IFGRVVGGFDVLTAMENVESDPKTDRPKEEIRIDATTV
Cyp73|     VFGRVTKGMEVVQRISNVKVNPKTDKPYEDVSIINITV
Cyp89|     VFGQVISGQEVVREIE-NQKTDAASKPFAEVRILSCGE
Cyp165|    VFGLVISGFEVIEQIE-NLKTDAASRPYADVRVIDCGV
```

Fig. 2. Sequence alignment of PPIase domains of all human cyclophilins. Amino acid sequences corresponding to the peptides found to be recognized by anti-Cyp18.1 antiserum are highlighted in *grey*. Amino acids marked in *bold* represent the residues that are in close contact with the peptide substrate *(15)*.

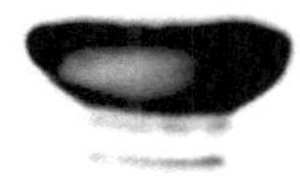

Fig. 3. Western blot analysis of the cross reactivity of anti-cyclophilin18.1 antiserum with recombinant human cyclophilin33. Anti-cyclophilin18.1 antiserum was diluted 1:500 with TBT pH 7.5 containing 3% milk powder. Secondary anti-rabbit peroxidase labelled antibody produced in goat was diluted 1,000-fold using TBT pH 7.5 containing 3% milk powder. Bound antibody was detected using ECL plus Western blotting detection reagents (Amersham Biosciences) and peQLab Chemismart 5000 chemoluminescence reader. 5 µg of Cyp33 were boiled with 5-µL sample buffer and used for separation in 15% polyacrylamide gel electrophoresis. Separated proteins were transferred onto nitrocellulose membranes using semi-dry blotting (Biometra, Germany).

were detected in cyclophilin18.1 (*see* **Fig. 2**). Both regions are located around highly conserved residues which were claimed to be important for catalytic mechanism and substrate binding *(19)*. Therefore, this antiserum shows cross reactivity against peptides derived from all other members of human cyclophilins. This antiserum should never be used for immunofluorescent staining experiments to detect intracellular localization of distinct cyclophilins. On the other hand this polyclonal preparation could be considered as "pan-specific" anti-cyclophilin antiserum. We performed Western blot analysis using full-length cyclophilin33 and anti-cylophilin18.1 antiserum to proof if the results from the peptide microarrays could be transferred to in vitro protein assays. As shown in **Fig. 3** the anti-cyclophilin18.1 antiserum recognizes cyclophilin 33 very strongly.

It was demonstrated that peptide microarrays displaying overlapping peptides covering complete families of human proteins represent a useful tool for comprehensive mapping of linear epitopes recognized by polyclonal antisera. Detected epitopes allow: (a) deduction of peptides valuable for generation of specific anti-peptide antibodies, (b) determination of cross reactivity against other members of the same protein family and (c) estimation of cross reactivity in the complete proteome of the target antigen. Additionally, detected immunodominant regions in target antigens should enable design of peptide sequences for effective affinity purification yielding sequence-specific polyclonal antibody preparations.

2. Materials

1. Micropipettes adjustable from 0.5 to 10 µL and from 100 to 1,000 µL (Eppendorf or Gilson) with corresponding plastic tips.
2. Metall trays with cover (VWR, Germany).

3. 1 mL Eppendorf tubes for the preparation of the antibody solution.
4. A one-dimensional rocking shaker (Heidolph duomax, Schwabach, Germany).
5. A plastic sheet cut in 1 mm × 3 cm pieces. We used sheets with a thickness of 0.8 mm resulting in a final assay volume of 400 µL.
6. A centrifuge for standard industry glass slides.
7. Spot-recognition software like GenePix Pro 6.0.
8. Fluorescence scanner/imager which is capable of excitation of appropriate fluorophore of the labelled secondary antibody and with a pixel size of at least 50 µm. Pixel sizes smaller than 50 µm will result in more accurate data points. We used a GenePix 4000B scanner at 10 µm resolution (Axon Instruments, Sunnyvale, CA).
9. Tris-buffered saline (TBS) buffer: 8.8 g NaCl (Merck, Darmstadt, Germany), 6.1 g Tris-base (AppliChem, Darmstadt, Germany), in 1 L double distilled water, adjust pH to 8.0 with HCl (Roth, Karlsruhe, Germany) and store at 4°C.
10. Blocking solution: 3% bovine serum albumin (BSA) (Roche, Mannheim, Germany) in TBS, pH 8.0.
11. Secondary fluorescence labelled antibody (goat anti-rabbit IgG DyLight 649 conjugated, Pierce, Rockford, IL).
12. Syringe filters for the filtration of all antibody solutions (Sartorius, Göttingen, Germany).

3. Methods

1. The microarrays have to be pre-treated with blocking buffer (3% BSA in TBS, pH 8.0) for 1 h at room temperature on a one-dimensional rocking shaker.
2. Wash the slides with TBS buffer pH 8.0 and water (three times each) in a metal tray (*see* **Notes 1–4**).
3. Dilute the antiserum 1:500 to 1:1,000 in blocking buffer and incubate the arrays with 400 µL of this antibody solution for 1 h at room temperature (*see* **Notes 5–6**). For incubation two slides, one displaying the peptides and another slide without any peptides, have to be assembled according to **Fig. 4** in a sandwich like format. If two peptide microarrays should be screened the top slide could be another peptide-displaying chip. The two slides are separated by two spacers generated from a plastic sheet

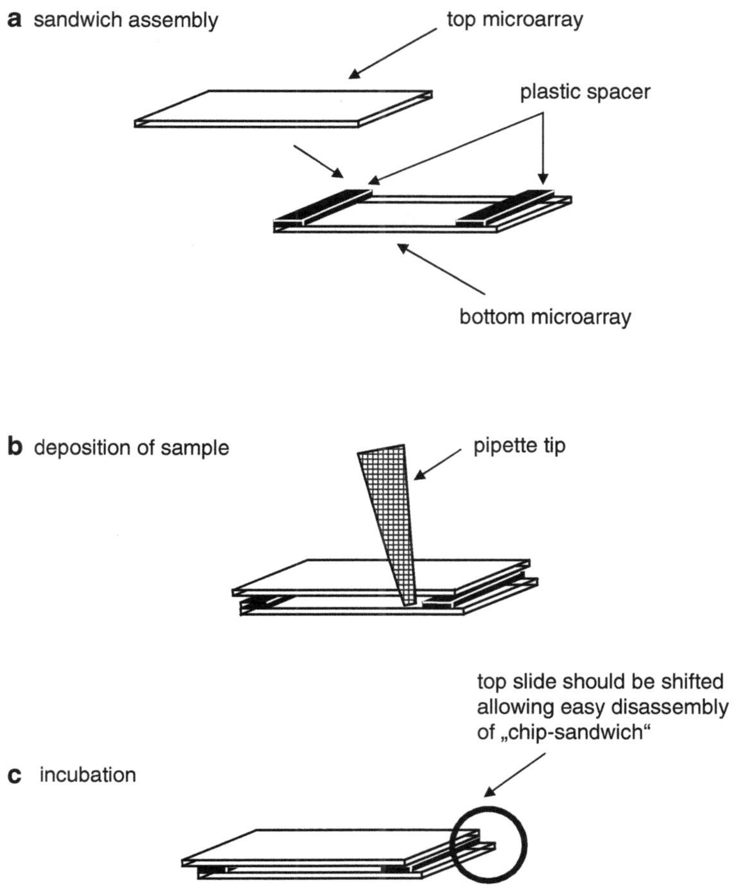

Fig. 4. (**a**) Assembly of "chip sandwich" is shown. Two plastic spacers are placed between the peptide displaying microarray (*bottom microarray*) and the dummy slide or second peptide displaying microarray (*top microarray*) resulting in a defined reaction chamber. (**b**) Assay solution is applied via pipette tip into the reaction chamber formed by the two slides. Capillary forces will soak-in the solution without formation of bubbles. (**c**) Top microarray is shifted resulting in overlaying ends of the glass slides. This arrangement enables convenient disassembly after the incubation step.

(*see* **Fig. 4**). The final assay volume will depend on the thickness of these plastic spacers (0.2-mm thickness will result in 100 µL assay volume, we recommend at least 0.8 mm/400 µL). The sample has to be applied in between the two slides. Therefore, the top slide is shifted about 1 mm to one side. If the pipette tip is adjusted on the position directly over the uncovered bottom slide the capillary forces allow proper distribution of the sample solution without formation of bubbles (*see* **Notes 7**).

4. Remove the plastic spacers and rinse the peptide microarray with distilled water (*see* **Notes 8**).
5. Wash microarray with TBS buffer and distilled water (three times each) (*see* **Notes 8–11**).

6. Dilute the fluorescently labelled secondary antibody (see manufacturers manual) and incubate the microarray with 400 μL of this solution for 1 h at room temperature in the dark. Alternatively, about 20 mL of secondary antibody solution in a metal tray can be used. In this case incubate microarray 1 h using a horizontal shaker. Make sure that the complete microarray surface is permanently covered with solution during shaking (*see* **Notes 12–15**).
7. Wash microarray with TBS buffer and distilled water (three times each).
8. Dry microarray using a chip centrifuge. Alternatively, drying using a stream of oil-free nitrogen or argon can be applied.
9. Scan the dry peptide microarray using a microarray scanner with laser settings corresponding to the fluorescence label of the secondary antibody.
10. Use spot recognition software to get signal intensities for each peptide spot.
11. Calculate the mean intensities of the signals from identical peptides (replicates on the microarray).

4. Notes

1. Always handle peptide microarray slides with care. They are made of glass and can damage your fingers.
2. Never touch the peptide microarray surface.
3. Always wear laboratory gloves when handling peptide microarray slides.
4. Hold peptide microarray slides at the end, which carries the engraved data label. This label provides for unique identification of the array.
5. Please take care when dispensing solutions onto the slide surface. Make sure not to touch the surface with pipette-tips or dispensers.
6. Apply the sample to the peptide displaying side of the glass slide.
7. We strongly recommend arrangement of the two slides during incubation as shown in **Fig. 4** (circled region bottom). The slight shift of the top slide compared with the bottom slide allows easy disassembly of the two glass slides.
8. Never whisk the surface of the peptide microarray slide with a cloth.

9. Never use other chemicals as described. Inappropriate chemicals may destroy the chemical bonding of the peptides to the glass surface.
10. Avoid dust or other particles during each step of the experiment. Dust, particles, and resulting scratches will cause artefacts during the final signal readout.
11. Filter all solutions for the washing steps through 2-μm, preferably 0.4-μm particle filters before use.
12. For incubation with the fluorescently labelled secondary antibody it is important to use metal trays with a cover or plastic trays completely covered with aluminium foil as these antibodies are sensitive towards light.
13. Control incubations using labelled secondary antibody alone should be performed in parallel to the epitope mapping experiment to ensure that found signals are not caused by unspecific binding of the secondary antibody to the immobilized peptides.
14. Fluorescence scanning could be very sensitive depending on the used scanner. Avoid any fluorescent impurities/contaminations inside your assay solution or washing solutions. You can easily check for such impurities by incubating and washing a dummy slide with the same solutions followed by fluorescence imaging.
15. Carefully adjust the final dilution of your labelled secondary antibody. Microarray technology is very sensitive and therefore it could be possible to use the secondary antibody in a higher dilution as proposed by the manufacturer. Generally, 1:1,000 dilutions of a 1 mg/ mL stock solution are working very well. Nevertheless, depending on the nature of the secondary antibody, such concentrations may yield high background signals caused by unspecific binding to the coated glass surface. If the signals within the peptide spots are high you could test 1:5,000 or 1:30,000 dilutions of a 1 mg/mL stock as well.

References

1. Reineke, U. and Schutkowski, M. (2006) Peptide arrays in proteomics and drug discovery, in *BioMEMS and Biomedical Nanotechnology*, Volume II: Micro and Nano-Technologies for Genomics and Proteomics, (Ozkan, M. and Heller, M. J., eds.), Springer Science+Business Media,LLC, New York, NY, pp. 161–282.
2. Geysen, H. M., Meloen, R. H., and Barteling, S. J. (1984) Use of peptide synthesis to probe viral antigens for epitopes to a resolution of a single amino acid. *Proc. Natl. Acad. Sci. U S A* 81, 3998–4002.
3. Geysen, H. M., Rodda, S. J., and Mason, T. J. (1986) A priori delineation of a peptide which mimics a discontinuous antigenic determinant. *Mol. Immunol.* 23, 709–715.
4. Geysen, H. M., Rodda, S. J., Mason, T. J., Tribbick, G., and Schoofs, P. G. (1987) Strategies for epitope analysis using peptide synthesis. *J. Immunol. Methods* 102, 259–274.
5. Wenschuh, H., Volkmer-Engert, R., Schmidt, M., Schulz, M., Schneider-Mergener, J., and Reineke, U. (2000) Coherent membrane

supports for parallel microsynthesis and screening of bioactive peptides. *Biopolymers (Peptide Science)* 55, 188–206.

6. Frank, R. (2002) The SPOT-synthesis technique. Synthetic peptide arrays on membrane supports–principles and applications. *J. Immunol. Methods* 267, 13–26.
7. Reineke, U., Volkmer-Engert, R., and Schneider-Mergener, J. (2001) Applications of peptide arrays prepared by the SPOT-technology. *Curr. Opin. Biotechnol.* 12, 59–64.
8. Reimer, U., Reineke, U., and Schneider-Mergener, J. (2002) Peptide arrays: from macro to micro. *Curr. Opin. Biotechnol.* 13, 315–320.
9. Rigter, A., Langeveld, J. P., Timmers-Parohi, D., Jacobs, J. G., Moonen, P. L., and Bossers, A. (2007) Mapping of possible prion protein self-interaction domains using peptide arrays. *BMC Biochem.* 8, 1–14.
10. Fodor, S. O., Read, J. L., Pirrung, M. C., Stryer, L., Lu, A. T., and Solas, D. (1991) Light-directed, spatially addressable parallel chemical synthesis. *Science* 251, 767–773.
11. Pellois, J. P., Zhou, X., Srivannavit, O., Zhou, T., Gulari, E., and Gao, X. (2002) Individually addressable parallel peptide synthesis on microchips. *Nat. Biotechnol.* 20, 922–926.
12. Pirrung, M. C. (1997) Spatially addressable combinatorial libraries. *Chem. Rev.* 97, 473–488.
13. Falsey, J. R., Renil, M., Park, S., Li, S., and Lam, K. S. (2001) Peptide and small molecule microarray for high throughput cell adhesion and functional assays. *Bioconjugate Chem.* 12, 346–353.
14. Melnyk, O., Duburcq, X., Olivier, C., Urbès, F., Auriault, C., and Gras-Masse, H. (2002) Peptide arrays for highly sensitive and specific antibody-binding fluorescence assays. *Bioconjugate Chem.* 13, 713–720.
15. Panse, S., Dong, L., Burian, A., Carus, R., Schutkowski, M., Reimer, U., and Schneider-Mergener, J. (2004) Profiling of generic anti-phosphopeptide antibodies and kinases with peptide microarrays using radioactive and fluorescence-based assays. *Mol. Divers.* 8, 291–299.
16. Kyte, J. and Doolittle, R. F. (1982) A simple method for displaying the hydropathic character of a protein. *J. Mol. Biol.* 157, 105–132.
17. Valle, M., Kremer, L., Martinez-A. C., Roncal, F., Valpuesta, J. M., Albar, J. P., and Carrascosa, J. L. (1999) Domain architecture of the bacteriophage phi29 connector protein. *J. Mol. Biol.* 288, 899–909.
18. Valle, M., Munoz, N., Kremer, L., Valpuesta, J. M., Martinez-A., C., Carrascosa, J. L., and Albar, J. P. (1999) Selection of antibody probes to correlate protein sequence domains with their structural distribution. *Protein Sci.* 8, 883–889.
19. Fanghänel, J. and Fischer, G. (2004) Insights into the catalytic mechanism of peptidyl prolyl cis/trans isomerases. *Front. Biosci.* 9, 3453–3478.

Subsection C

Profiling of Antibody Signatures in Biological Fluids

Chapter 17

Epitope Mapping Using Randomly Generated Peptide Libraries

Juliane Bongartz, Nicole Bruni, and Michal Or-Guil

Summary

Characterizing the immune response towards a pathogen is of high interest for vaccine development and diagnosis. However, the characterization of disease-related antigen–antibody interactions is of enormous complexity. Here, we present a method comprising binding studies of serum antibody pools to synthetic random peptide libraries, and data analysis of the resulting binding patterns. The analysis can be applied to classify and predict different groups of individuals and to detect the peptides which best discriminate the investigated groups. As an example, the analysis of antibody repertoire binding patterns of different mice strains and of mice infected with helminth parasites is shown. Due to the design of the library and the sophisticated analysis, the method is able to classify and predict the different mice strains and the infection with very high accuracy and with a very small number of peptides, illustrating the potential of random library screenings in determining molecular markers for diagnosis.

Key words: Serum antibody repertoire, Random peptide library, Microarray, Binding pattern analysis, Machine learning, Feature selection, Diagnosis.

1. Introduction

Diseases are often diagnosed by testing serological antibody reactivity. This is the case for several infections, allergies, and autoimmune diseases. Well known examples are HIV infections and Hashimoto's diseases, which are diagnosed by commercially available serum antibody tests.

Antibody reactivity tests are widely used in cases where the antigen eliciting the immune response is known. If the antigenic epitope is linear, a peptide representing the linear epitope might suffice as a diagnostic molecular marker. Several strategies for

Ulrich Reineke and Mike Schutkowski (eds.), *Methods in Molecular Biology, Epitope Mapping Protocols, vol. 524*

DOI: 10.1007/978-1-59745-450-6_17

protein epitope mapping have been developed *(1–6)*. In this context, peptide microarrays have become a class of widely used tools for analyzing antibody binding *(7–10)*. For instance, Quintana et al. *(11)* were able to discriminate between mice resistant and susceptible to diabetes by analyzing the IgG autoantibody repertoire using protein and peptide microarrays.

Despite these methods being successfully used in the diagnosis and even prognosis of many diseases, they obviously fail if no antigen is known. However, the lack of a known antigen is not an impediment for diagnosis via serum antibody profiling. This problem can be circumvented by searching differentially recognized epitopes which do not stem from a pathogenic antigen. Instead, the epitopes can stem from synthetic random sequence peptide libraries. Here, peptides or groups of peptides which are recognized differently by a diseased group in comparison with a control group are chosen as candidate molecular markers.

The rationale behind the assertion that differentially recognized epitopes might be present in random peptide libraries derives from the fact that antibodies are strongly cross-reactive *(2, 12)*. Indeed, studies conducted by Nóbrega et al. on serum binding reactivities to protein mixtures from cell extracts using quantitative immunoblot revealed significant reactivity differences between healthy and infected individuals *(13, 14)*.

Screening for differentially recognized epitopes is best performed by using libraries of peptides printed on glass slides. Here, the binding of serum antibodies to each peptide is detected via fluorescence labelled secondary antibodies, and the resulting signal intensities of each individual peptide are committed to data analysis. The proposed analysis pathways are sketched in **Fig. 1c**. The analysis aims at classification of a diseased group and a control. To gain a first impression on whether the data quality allows for easy classification, linear techniques to reduce data dimensionality, as principal component analysis (PCA), can be used. Linear discriminant analysis (LDA) on the first few principal components allows then for determination of classifier and prediction accuracy. However, only further analysis enables to enhance the classification accuracy and to determine the set of peptides most suitable for prediction. The selection of the most suitable peptides can be performed with a supervised learning, feature selection, and classification tool like potential support vector machines (P-SVM) *(15)*.

To illustrate this method, BALB/c mice were infected with the intestinal helminthic parasite *Heligmosomoides polygyrus* (*H. polygyrus*). Serum samples were collected before and 14 days after infection, and antibody reactivities were measured. Moreover, the antibody reactivities of the healthy BALB/c mice samples were compared with those of a second healthy mouse strain, C57BL/6. The investigated groups and number of samples are summarized in **Fig. 1a**.

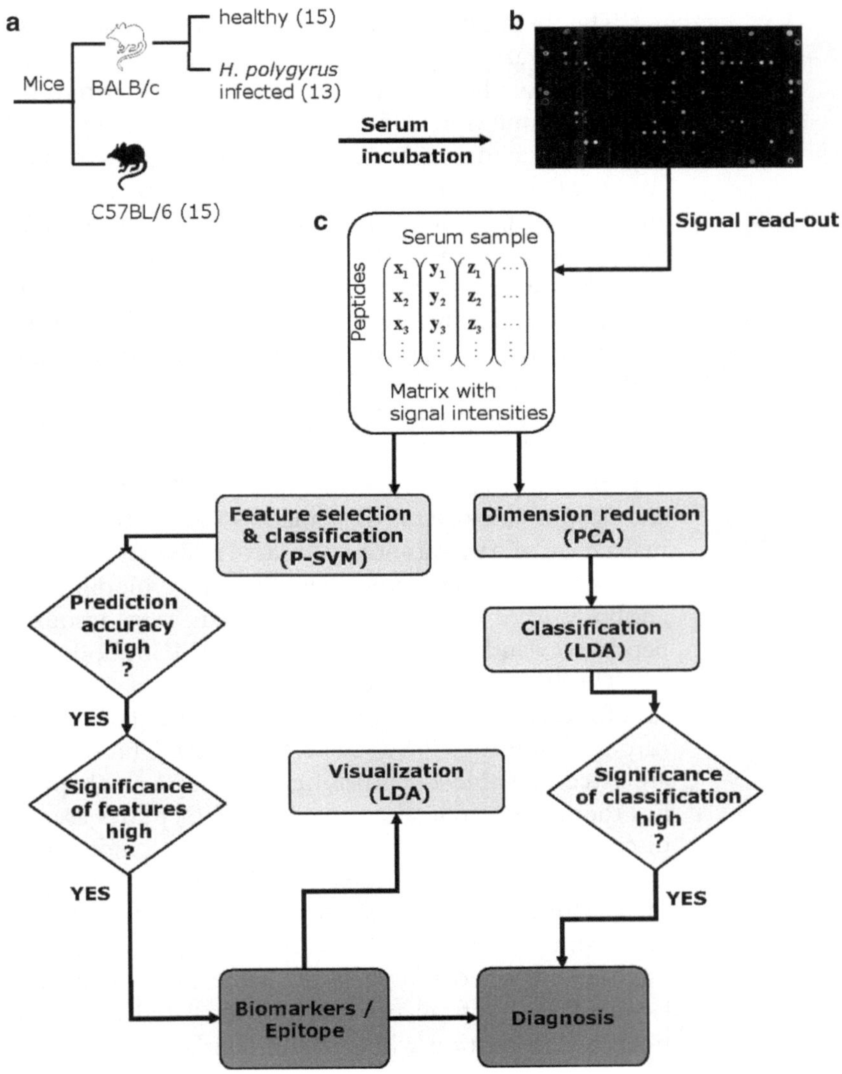

Fig. 1. (**a**) Serum samples of the mice strains BALB/c and C57BL/6 are incubated and the inter-strain differences are used for classification and prediction. The intra-strain differences are analyzed by comparing healthy BALB/c mice with *H. polygyrus* infected BALB/c mice. The numbers in brackets represent the number of different sera in each group tested. (**b**) Binding pattern of serum antibodies to the random peptide library. TAMRA, as internal fluorescence control and murine IgM and IgG as secondary antibody controls are displayed four times on each array (*upper left* and *right* and *lower right corner*). Binding is detected by fluorescently labelled secondary antibodies (anti-IgM-Alexa Fluor 546, anti-IgG-Alexa Fluor 647). (**c**) Statistical analysis pathway. Read-out signal intensities do not need to be normalized. False-positive blank signals that derive from secondary antibody binding are eliminated in all data sets analysed (six out of 255 peptides for IgM). PCA is applied in order to reduce the dimensionality of the dataset by extraction of the highest variances. P-SVM and LDA are applied to classify the different groups. Best classification, though, is not achieved by considering the highest variances stemming from PCA, but from very few peptides selected by the classification and feature selection tool P-SVM. LDA allows a visualization of the classification by again reducing the dimension of the data set.

The binding reactivities were analyzed using a synthetic peptide library consisting of 255 different 14mer peptides. The sequences of the library's peptides were determined randomly, based on amino acid frequencies corresponding to the amino acid's appearance on solvent accessible protein surfaces. No repeat of three consecutive amino acids was allowed. The synthesized peptides were printed on glass slides. TAMRA derived peptide was attached to the glass as internal fluorescence control. Full-lenght mouse-IgM and mouse-IgG were included as secondary antibody controls. The peptide library was displayed in five identical sub-arrays on each slide. The incubation of serum with the peptide library (*see* **Notes 1–8**) and subsequent detection with fluorescence labelled anti-mouse IgM and IgG antibodies resulted in characteristic binding patterns (*see* **Fig. 1b**). The resulting signal intensities were read out with a microarray scanner for subsequent data analysis (*see* **Note 9** and **10**). Normalization of data was not necessary (*see* **Note 11**). False-positive "blank" signals that derive from unspecific binding of secondary antibody were eliminated from all data sets, amounting to six peptides excluded by the anti-mouse IgM antibody.

The minimum set of differentially recognized peptides necessary for classification was selected using the classification and feature selection tool P-SVM (*see* **Note 12**). The selected peptides are listed in **Table 1**. Surprisingly, already a single peptide classifies the mouse strains best, while three peptides are sufficient to discriminate samples of healthy and infected individuals. In both cases, classification results were 100% correct. The significance of the classification results is calculated by shuffling the data group labels and performing classification under the condition that the number of peptides used is the same as with the correct data labels. The significances are then calculated by the number of times a classification with the shuffled labels results in better or

Table 1
Sequences of peptides for best classification of murine IgM binding patterns selected by P-SVM from a randomly generated library of 255 peptides

	Sequences of peptides for best classification
Healthy BALB/c vs. healthy C57BL/6	SGFPDKIEFPTQDC
Healthy BALB/c vs. infected BALB/c	THEDFRYDDVFEGN
	FFDEIIHSCRSQNG
	VRQVQRSKKMHKKG

Table 2
P-SVM classification and leave-one-out prediction accuracies of murine IgM binding patterns for healthy mice of different strains and after infection with the nematode *H. polygyrus*

	Healthy BALB/c vs. healthy C57BL/6	Healthy BALB/c vs. infected BALB/c
Classification accuracy	100%	100%
Significance p	<0.002	0.002
No. of features	1	3
Prediction accuracy	100%	100%
Significance p	<0.002	<0.001
No. of features	1	4

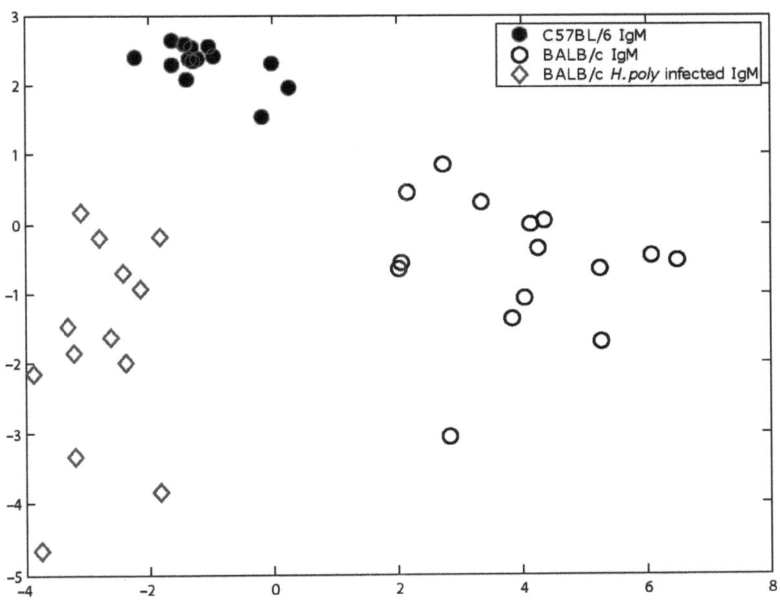

Fig. 2. Linear discriminant analysis of IgM binding patterns for two mouse strains and a group infected with a parasite using four peptides previously selected by P-SVM.

the same classification accuracies. As shown in **Table 2**, the significance lies below $p = 0.002$ in both cases.

Given the small number of samples in each group, the accuracy of prediction was calculated by taking one data point out of the training set ("leave-one-out"), calculating the best classifying features, and using this classifier to predict the test data point. This procedure was repeated for all data points. As shown in **Table 2**, an additional peptide is needed in the case of the healthy vs. infected mice to ensure the best prediction accuracy. Here also, the accuracy of prediction achieved 100% in both cases, again with a significance below p=0.002.

Taking together, the above classification results reveal that four peptides are sufficient to unstitch the three investigated mice groups. For better visualization of the classification, the four-dimensional space defined by these four peptides is reduced to a two-dimensional space using LDA. The resulting representation of all data points is depicted in **Fig. 2**.

2. Materials

1. RepliTope™ Microarrays (JPT Peptide Technologies GmbH, Berlin, Germany), ready to use microarrays.
2. Ethanol.
3. Double distilled water.
4. Working buffer (T-PBS): 9.2mM $Na_2HPO_4.12H_2O$, 1.6mM $NaH_2PO_4.H_2O$, 150mM NaCl, 10% Tween 20, add double distilled water up to 1,000mL (pH 7.4).
5. Multiwell GeneFrame– (Abgene, Epsom, United Kingdom).
6. Serum concentration: microarrays are incubated with a 1:10 dilution of serum in working buffer.
7. Secondary antibody: goat anti-mouse IgG-Alexa Fluor 647 conjugate (20μg/mL) (Invitrogen, Carlsbad, CA), goat anti-mouse IgM-Alexa Fluor 546 conjugate (20μg/mL) (Invitrogen, Carlsbad, CA).
8. Centrifuge (Centrifuge 5403, Eppendorf, Hamburg, Germany).
9. Microarray scanner (Genepix 4200AL, Molecular Devices GmbH, Ismaning, Germany) and the associated GenePix Pro software.
10. Genespotter software (MicroDiscovery GmbH, Berlin, Germany).

3. Methods

3.1. Microarray Incubation and Signal Read-Out

1. Shortly wash microarrays with 100% ethanol.
2. Wash microarrays three times for two min with T-PBS, three times for 2min with deionized water, rinse with flowing

deionized water and then dry by centrifugation (*see* **Notes 1–3**).

3. RepliTope Microarrays are pre-treated to minimize unspecific binding of the target antibodies. Therefore, no blocking step is required prior to incubation.
4. All incubations are performed using a five-well incubation chamber with a total assay volume of 45µL per well. Each well is incubated with 45µL of a 1:10 dilution of serum in T-PBS for 4h at room temperature (*see* **Notes 4–8**).
5. Remove incubation chamber and wash microarrays three times for 2min with T-PBS and three times for 2min with deionized water.
6. Incubate microarrays with 300µL of anti-mouse IgG-Alexa Flour 647 (20µg/mL) and IgM-Alexa Flour 546 (20µg/mL) in T-PBS for 1h at room temperature.
7. Wash microarrays three times for 2min with T-PBS, three times for 2min with deionized water, rinse with flowing deionized water and dry by centrifugation.
8. Fluorescence signals are measured on the GenePix 4200AL microarray scanner (*see* **Note 9**). Both lasers (535nm and 635nm) can be used simultaneously using a red (~650–690nm) and green (~550–600nm) emission filter. An image file is generated at a resolution of 10µm using the associated GenePix Pro software.

3.2. Data Processing and Statistical Analyses

1. Signal intensities are quantified with Genespotter™ software.
2. Statistical analyses are performed using MATLAB 7.0 (The MathWorks Inc.) and R 2.3.1 *(16)*.
3. Principal component analysis (PCA) is a technique used to reduce multidimensional data sets to lower dimensions for analysis. PCA transforms the data to a new coordinate system such that the greatest variance by any projection of the data comes to lie on the first coordinate (called the first principal component), the second greatest variance on the second coordinate, and so on. The function *princomp* from MATLAB 7.0 is used to calculate the principal components *(17)*.
4. Linear discriminant analysis (LDA) is a method used in statistics and machine learning to find the linear combination of features which best separate two or more classes of objects. LDA was performed using the algorithm *lda* of R 2.3.1 *(17,18,20)* (*see* **Note 12**).
5. Potential support vector machines (P-SVM) *(19)* is a new supervised learning method, which can be used with kernels like standard support vector machine (SVM) methods.

SVMs transform the data space into a high dimensional feature space in order to find a hyper-plane that separates the data points with maximum distance to the closest data points from both classes. P-SVM is able to synchronously select appropriate peptides (= features) for classification and prediction.

3.3. Closing

The results reveal that both classification and prediction via serum antibody reactivity can be performed based on an unanticipated small number of peptides carefully selected from a rather small synthetic random peptide library. The resulting high prediction accuracies emphasize the potential of nonantigenic, differentially recognized epitopes as molecular markers for diagnosis.

4. Notes

1. Water, ethanol and PBS should be filtered in order to remove small particles that might grate the microarray surface.
2. Handle the microarrays with care, always wear gloves and never touch the microarray surface.
3. Never add buffers and water directly on the microarray surface during the washing steps.
4. When drying the microarrays, they should intensively be prewashed with double distilled water in order to remove all residual proteins and salts. Salts may corrode the microarray surface resulting in scratches disturbing signal read out.
5. Before incubation with the serum samples, the microarrays need to be dry, otherwise the multiwell frames will not stick to the microarray surface. When using single well incubation chambers this step is not required. We use multiwell incubation chambers because the sample volume is very small, which allows us to repeat an analysis without using large quantities of rare biological material.
6. Carefully drop the serum samples into the incubation chambers without touching the microarray surface with the pipette tip.
7. When using a multiwell incubation chamber, carefully attach the cover slip. Make sure that different samples in nearby chambers do not mix. Apply slight pressure to the cover slip in order to avoid air bubbles.
8. Make sure that the microarrays never dry out during incubation. Use a humidity chamber.

9. Assure yourself that prior to signal read out, the microarrays are dry, clean and that no residual adhesive from the incubation chamber resides on the microarray surface. This can cause problems with the engineering mechanics of the microarray scanner.
10. Store incubated microarrays at 4°C under inert gas. This will prevent fluorescence signals from bleaching for at least 3 months.
11. Always use microarrays from the same batch, printed at the same time, since different batches have shifts in signal readout, which are hard to be corrected.
12. In case of different numbers of individuals per group, make sure to eliminate the group bias in the statistical methods (use '*prior*' in *lda* of R 2.3.1 and '*-b*' in P-SVM).

References

1. Wenschuh, H., Volkmer-Engert, R., Schmidt, M., Schulz, M., Schneider-Mergener, J., and Reineke, U. (2000) Coherent membrane supports for parallel microsynthesis and screening of bioactive peptides. *Biopolymers* 55, 188–206.
2. Frank, R. (2002) The SPOT-synthesis technique. Synthetic peptide arrays on membrane supports–principles and applications. *J. Immunol. Methods* 267, 13–26.
3. Geysen, H. M., Meloen, R. H. , and Barteling, S. J. (1984) Use of peptide synthesis to probe viral antigens for epitopes to a resolution of a single amino acid. *Proc. Natl. Acad. Sci. U S A* 81, 3998–4002.
4. Weiser, A. A., Or-Guil, M., Tapia, V., Leichsenring, A., Schuchhardt, J., Frommel, C. , and Volkmer-Engert, R. (2005) SPOT synthesis: reliability of array-based measurement of peptide binding affinity. *Anal. Biochem.* 342, 300–311.
5. Wenschuh, H., Gausepohl, H., Germeroth, L., Ulbricht, M., Matuschewski, H., Kramer, A., Volkmer-Engert, R., Heine, N., Ast, T., Scharn, D., and Schneider-Mergener, J. (2000) in Combinatorial Chemistry: A Practical Approach (Fenniri, H.), Oxford University Press, Oxford, UK, pp. 95–116.
6. Reineke, U., Volkmer-Engert, R. , and Schneider-Mergener, J. (2001) Applications of peptide arrays prepared by the SPOT-technology. *Curr. Opin. Biotech.* 12, 59–64.
7. Tapia, V., Bongartz, J., Schutkowski, M., Bruni, N., Weiser, A., Ay, B., Volkmer, R. , and Or-Guil, M. (2007) Affinity profiling using the peptide microarray technology: a case study. *Anal. Biochem.* 363, 108–118.
8. Reimer, U., Reineke, U., and Schneider-Mergener, J. (2002) Peptide arrays: from macro to micro. *Curr. Opin. Biotech.* 13, 315–320.
9. Schutkowski, M., Reimer, U., Panse, S., Dong, L., Lizcano, J. M., Alessi, D. R. , and Schneider-Mergener, J. (2004) High-Content Peptide Microarrays for Deciphering Kinase Specificity and Biology. *Angew. Chem.* 116, 2725–2728.
10. Jones, R. B., Gordus, A., Krall, J. A., and MacBeath, G. (2006) A quantitative protein interaction network for the ErbB receptors using protein microarrays. *Nature* 439, 168–174.
11. Quintana, F. J., Hagedorn, P. H., Elizur, G., Merbl, Y., Domany, E. , and Cohen, I. R. (2004) Functional immunomics: microarray analysis of IgG autoantibody repertoires predicts the future response of mice to induced diabetes. *Proc. Natl. Acad. Sci. U S A* 101(Suppl 2), 14615–14621.
12. Frank, S. A. (ed.) (2002) *Immunology and Evolution of Infectious Disease.* Princeton University Press, Princeton, NJ.
13. Reineke, U., Ivascu, C., Schlief, M., Landgraf, C., Gericke, S., Zahn, G., Herzel, H., Volkmer-Engert, R. , and Schneider-Mergener, J. (2002) Identification of distinct antibody epitopes and mimotopes from a peptide array of 5520 randomly generated sequences. *J. Immunol. Methods* 267, 37–51.
14. Nobrega, A., Grandien, A., Haury, M., Hecker, L., Malanchere, E. , and Coutinho, A.

(1998) Functional diversity and clonal frequencies of reactivity in the available antibody repertoire. *Eur. J. Immunol.* 28, 1204–1215.

15. Haury, M., Grandien, A., Sundblad, A., Coutinho, A. , and Nobrega, A. (1994) Global analysis of antibody repertoires. 1. An immunoblot method for the quantitative screening of a large number of reactivities. *Scand. J. Immunol.* 39, 79–87.
16. R Development Core Team (2007) R: A language and environment for statistical computing. R Foundation for Statistical Computing, Vienna, Austria. ISBN 3-90051-07-0, URL http://www.R-project.org.
17. Jackson, J. E. (ed.) (1991) A User's *Guide to Principal Components.* Wiley, Hoboken, NJ.
18. Ripley, B.D. (ed.) (1006) *Pattern Recognition and Neural Networks.* Cambridge University Press, New York, NY.
19. Hochreiter, S. and Obermayer, K. (2006) Support vector machines for dyadic data. *Neural Comput.* 18, 1472–1510.
20. Venables, W. N. and Ripley, B. D. (eds.) (2002) *Modern Applied Statistics with S.* Springer, New York, NY.

Chapter 18

Probing the Epitope Signatures of IgG Antibodies in Human Serum from Patients with Autoimmune Disease

Peter Lorenz, Michael Kreutzer, Johannes Zerweck, Mike Schutkowski, and Hans-Jürgen Thiesen

Summary

High density peptide microarray technologies can be applied in experimental medicine in general and in clinical immunology in particular. Laboratory diagnostics of autoimmune diseases strongly rely on screening human sera for antibodies against known autoantigens. These assays are still difficult to standardize and quantify. Typically, the results are presented as antibody titers within an assay system. Most assays use recombinant or purified autoantigens that are difficult to obtain and require great efforts of quality control. Here we describe a method to obtain patterns of epitope signatures with peptide microarrays from patients suffering from autoimmune diseases in comparison with healthy controls. One of the final aims will be to define subsets of peptides indicative for marker autoantibodies of autoimmune diseases. Finally, informative epitopes can be used for immunopurifying epitope-specific autoantibodies. Eventually, these antibodies can be further characterized on peptide microarrays displaying mutated epitopes obtained by scanning mutagenesis. Any disease or physiological status that affect humoral immune responses such as autoantibodies in oncology or responses to infections or vaccinations can be monitored.

Key words: Peptide microarrays, Autoimmunity, Systemic sclerosis, Systemic lupus erythemotosus.

1. Introduction

Autoimmunity is ultimately a result of the breakdown of tolerance, i.e., the inappropriate targeting of the cells and tissues of the human body by its own immune system *(1)*. This is reflected in the appearance of autoreactive T cells and autoantibodies secreted by autoreactive plasma cells. Autoantibodies are of great importance for diagnostic purposes and may also be connected to the pathophysiology of a particular autoimmune disease *(2)*.

Ulrich Reineke and Mike Schutkowski (eds.), *Methods in Molecular Biology, Epitope Mapping Protocols, vol. 524*

DOI: 10.1007/978-1-59745-450-6_18

However, in many cases it is unclear if the autoantibody pattern, i.e., the reactivity towards specific autoantigens, is indeed causally linked to the disease or just bystander effect or circumstantial. Still, autoantibody patterns are an important cornerstone for diagnostics at least as surrogate markers *(3)*. Our aims are an in-depth characterization of the autoantibody patterns in rare autoimmune diseases by developing, improving, and implementing modern enabling technologies for (1) identification of novel autoantigens (2) determination of specific autoreactive epitopes of autoantigens, and (3) characterization of autoantibody signatures in rare autoimmune diseases. The discovered epitopes have to be compared with epitopes found in healthy controls, and to epitopes presented by other diseases or physiological states such as infections or vaccinations. One technology we are using is based on high density peptide microarrays. In contrast to whole protein microarrays *(4)*, the use of peptides is more robust and provides the following advantages: peptides are easily synthesized in a standardized, reproducible, and cost-effective manner, they can be covalently and specifically attached (including spacers or any chemical modification) with equal number of molecules per spot, and peptide microarrays are rather stable. For autoantibody profiling the peptides are derived from known autoantigens based on translated cDNA sequences in databases. Any whole sequence or part of a given protein can serve as input to synthesize overlapping peptides that might reflect autoantibody epitopes. In the first instance such peptides have to be considered as linear epitopes that lack any post-translational modification. However, common conformational epitopes found for autoantibodies might be reflected in a peptide sequence resembling and mimicking this 3D structure. Such a peptide is often called a mimotope and can even mimic carbohydrate antigens *(5)*. Furthermore, chemical moieties like e.g. phosphate groups can add complexity and reflect post-translational modifications. The signals detected on a peptide microarray used for antibody profiling are a measure of two parameters of this antibody, affinity and concentration.

2. Materials

2.1. Peptide Microarrays

Replitope™ high density microarrays were custom-made at JPT (JPT Peptide Technologies GmbH, Berlin, Germany). Their layout consisted of the whole peptide library spotted in triplicates as three identical subarrays with fields each equal to 2 × 8 blocks and 324 spots/block (*see* **Note 1**). For the image analysis after staining the layout with spot-specific information is supplied as a standard GenePix Array List (GAL) file. For proper alignment fluorescent landmarks containing spots of tetramethyl-rhodamin-labeled peptide (positive controls for Cy3-channel of microarray readers) were co-immobilized onto each peptide microarray.

15-mer overlapping peptides (11 amino acid overlap) were derived from a panel of known autoantigens with particular emphasis on marker antigens for systemic sclerosis and systemic lupus erythematosus (SLE). The autoantigens included: SSA1/Ro52, SSA2/Ro60K, SSB/La, CenpA, CenpB, CenpE, RPC1, RPC2, DNA-topoisomerase I/Scl70, Exosc10 (PM/Scl100), proteinase 3, myeloperoxidase, KIF11/Eg5, apolipoprotein H/beta-2-glycoprotein I, fibrillarin, histone H4 (6 isoforms) and histone H1 (8 isoforms). Furthermore, 9 known auto-epitopes from SSA1/Ro52 (2 peptides), SSA2/Ro60K (3 peptides) and apolipoprotein H (4 peptides) were added and finally the microarray displayed human IgG, and full-length recombinant Ro52K, Ro60K, DNA-topoisomerase I, CenpA, CenpB and apolipoprotein H proteins (all from Diarect AG, Freiburg, Germany).

Amino-oxy-acetylated peptides were synthesized on cellulose membranes in a parallel manner using SPOT synthesis technology according to *(6, 7)*. Following side chain deprotection the solid phase bound peptides were transferred into 96-well microtiter filtration plates (Millipore, Bedford, MA) and treated with 200 μL of aqueous triethylamine (1.5% by vol.) in order to cleave the peptides from the cellulose. Peptide-containing triethylamine solution was filtered off and solvent was removed by evaporation under reduced pressure. Resulting peptide derivatives (50 nmol) were re-dissolved in 25 μL of printing solution (70% DMSO, 25% 0.2 M sodium acetate pH 4.5, 5% glycerol, by vol.) and transferred into 384-well microtiter plates. Printed peptide microarrays were kept at room temperature for 5 h, washed with deionized water, quenched for 1 h with buffered 1% BSA solution at 42°C, washed extensively with water followed by ethanol, and dried using a microarray centrifuge. Resulting peptide microarrays were stored at 4°C and maintained reactivity for more than 18 months.

2.2. Sera of Patients and Immunoglobulin Preparations

All patient sera were obtained according to the rules and with the approval of the local ethics committees and kindly supplied through Dr. Alexandre Voskuyl (Department of Rheumatology, VU University Medical Center, Amsterdam, Netherlands), Dr. Gabriela Riemekasten (Department of Rheumatology and Clinical Immunology, Charité University Hospital, Berlin, Germany), Dr. Susanne Schäd (Department of Dermatology and Venereology, University of Rostock, Rostock, Germany), Dr. Miri Blank (Sheba Medical Center, Ramat Gan, Israel) and Dr. Renate Claus (Diagnostic group, Institute of Immunology, University of Rostock, Rostock, Germany). The intravenous immunoglobulin (IVIG) preparation (OMRIGAM from Omrix Pharmaceutical

Ltd. Nes-Ziona, Israel) was a kind gift of Dr. Miri Blank (Sheba Medical Center, Ramat Gan, Israel) and supplied at 50 mg/mL human IgG in 10% maltose. Sera and IVIG preparations were cleared from complexes by centrifugation at 20,000 × *g* for 10 min at 4°C.

2.3. Staining of Peptide Microarrays

1. TTB buffer for washing and to dilute both, sera and secondary antibody preparations, is composed of TBS (50 mM Tris-HCl, pH 7.4, 150 mM NaCl), 0.05% Tween 20, 0.1% BSA.
2. Optional blocking buffer is Pierce SuperBlock® T20 (TBS) blocking buffer which contains a proprietary protein formulation in Tris-HCl pH 7.4 containing 0.05% Tween-20 detergent and Kathon® anti-microbial additive.
3. Secondary antibodies: Zenon™ goat anti-human-IgG-Alexa-Fluor647 from Invitrogen (0.2 mg/mL Fab fragments).
4. Filter paper Whatman GB002.
5. Standard microscope slides 76 × 26 mm.
6. Glass staining jar with removable rack for standard microscope slides (Carl Roth, Karlsruhe, Germany) filled with up to five slides.
7. Slide carrier with inset for storage and centrifugation, i. e. Mbox (Nunc/Fisher Scientific, Schwerte, Germany).
8. Polypropylene staining box with lid (Nalgene/Fisher Scientific, Schwerte, Germany; 22.5 × 22.5 cm^2).
9. Serological pipettes, 5 mL.

3. Methods

3.1. Handling of the Peptide Microarray

1. Care has to be taken to avoid touching/scratching the surface of the microarray. Grab the microarray at its end that is etched with the lot number.
2. Before use, take the microarray from its container (4°C) and wait for 10–15 min until equilibration to room temperature occurs.
3. Never wipe off fluid with cloth or tissue.
4. All solutions are filtrated through 0.2-μm filters to remove any insoluble matter, dust, and lint.
5. Avoid drying out of the microarray once protein-containing solutions have been applied and avoid the formation of air bubbles.
6. Incubations of antibody solutions were done in a chamber formed by a 2-slide sandwich made by first adding 3-mm wide

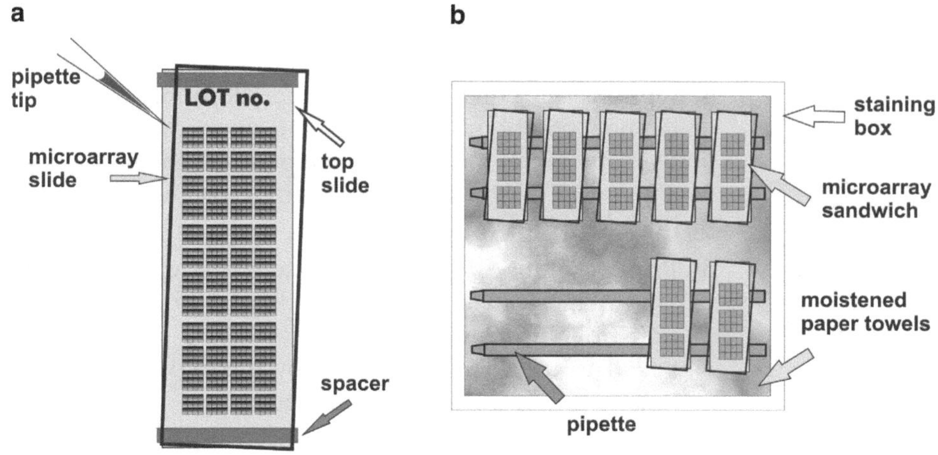

Fig. 1. (**a**) Two-slide staining assembly by placing a second clean slide on top of the microarray slide while generating space for fluid addition by means of two plastic spacers. (**b**) Humid staining chamber for up to ten microarray slide assemblies made of a plastic box with moistened towels and serological pipettes as carriers to keep the microarrays away from the wet towels. The box is closed with a lid.

and 1-mm thick plastic spacers to either end of the microarray slide and then putting a new clean slide on top of the spacers (*see* **Fig. 1a**). A slight misalignment of the top slide creates space to load fluid into the room between the two slides. A volume of 500 μL is more than sufficient to fill the space without any air bubbles.

3.2. Staining of the Peptide Microarray

1. All steps are performed at room temperature.
2. Incubations are done with 500 μL antibody solutions in the chamber formed by the 2-slide sandwich. The slide sandwiches are assembled within a humid chamber consisting of a staining box with well moistened paper towels on the bottom and 5-mL standard plastic serological pipettes for separators from the towels (*see* **Fig. 1b**). Up to ten slides fit into such a humid chamber.
3. Washing is performed as follows: Carefully add TTB buffer to the assembled, fluid-filled 2-slide sandwich until the top slide gently floats on the fluid. Then remove top slide and place microarray slide into the rack of the glass staining jar. Immediately immerse the slides in TTB buffer. Rinse shortly and then transfer rack into new jar with fresh TTB buffer. Incubate for 5 min under slight agitation on a rocker platform. Repeat this incubation three times, always transfer the rack with the microarrays to a jar with fresh buffer. Thereafter wash in the same way in distilled water. Take out microarrays and remove residual fluid by centrifugation for 1 min at 1,000 × *g* and at room temperature in an Mbox slide carrier with underlying two sheets of filter paper. Additional centrifugation steps can

be added if necessary; however, always replace the used filter paper for fresh sheets. Proceed with the next step.

4. Optional blocking step: If desired additional blocking of the peptide microarray slides can be done by loading 500 µL blocking buffer and incubation for 1 h in the humid chamber. Thereafter washing is performed according to **Subheading 3.2, step 3**.
5. Primary antibody incubation: Assemble the 2-slide microarray sandwich. Human sera are pre-diluted (for some considerations *see* **Note 2**) in TTB buffer and 500 µL of this solution added to the chamber formed by the two slides. Incubation was for 4 h in the humid chamber.
6. Proceed with washing according to **Subheading 3.2, step 3**.
7. Secondary antibody incubation: Reassemble the microarray sandwich and add 500 µL of goat anti-human IgG Fab fragments conjugated to AlexaFluor647. Incubate for 2 h in the humid chamber in the dark by covering the chamber with aluminum foil.
8. Proceed with washing according to **Subheading 3.2, step 3**. but add two further rinses in a large beaker of distilled water before drying off the microarray slide by centrifugation. Centrifuge a second time fresh with filter paper added to the centrifuge rack. Acquire images immediately or store slides at 4°C in the dark up to several days (*see* **Note 3**).

3.3. Image Aquisition

Fluorescence signals after staining should be acquired with a 5 µm resolution scanner capable to read standard format slides and equipped with at least a 635 nm laser for excitation. We are using a Molecular Devices Axon Instruments GenePix 4000B scanner with 532 as well as 635 nm laser excitation sources. The scan at 523 nm reads the Cy3 guide dots and the one at 635 nm the signals due to staining. Images have a resolution of 2,540 dpi at 16 bit depth resulting in 2 TIFF format files of around 60 MByte with dimensions height 7,150 × width 2,200 pixels. EXIF interchange extensions to these files are being used to gather all scan parameters. See **Fig. 2** for some examples of acquired images.

3.4. Primary Data Analysis

Primary data analysis is based on the images with their associated GenePix Array List (*see* **Note 4**) GAL file and is performed with the GenePix Pro software (version 4.0, Molecular Devices; Ismaning, Germany). The GAL files describe the individual features on a microarray in a position-specific manner, here in particular peptide spots, in their layout in blocks, columns and rows in connection with their description. In our case information for the respective peptide is given, i.e., peptide sequence and the protein name and database accession the peptide is derived from.

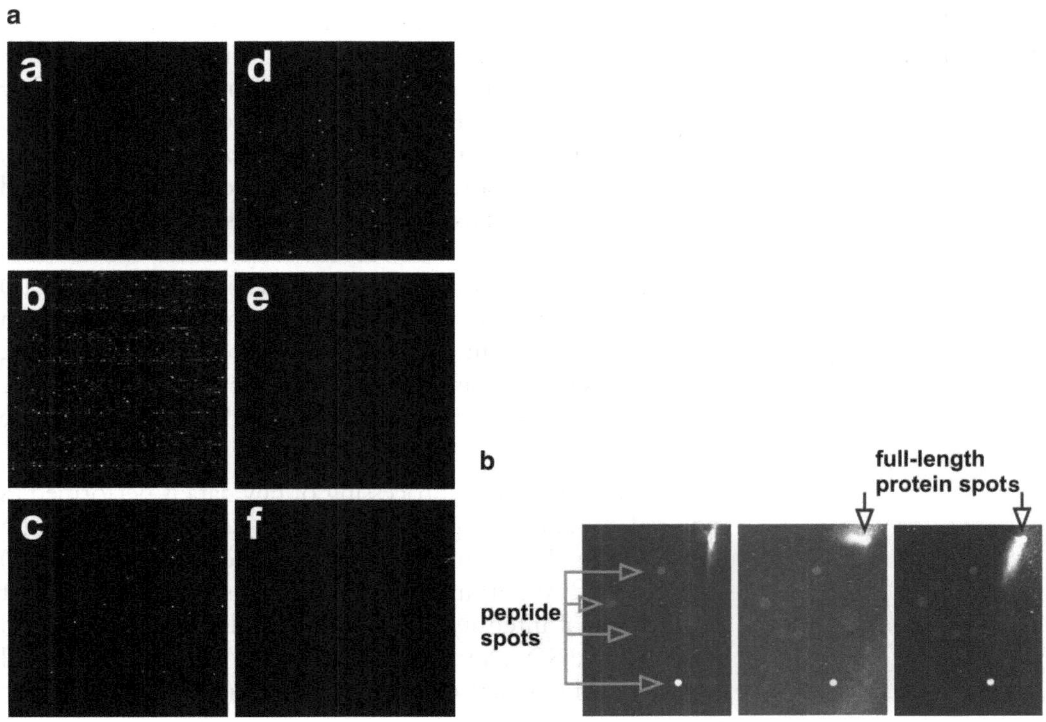

Fig. 2. Examples of acquired peptide microarray images. (**a**) One subarray of peptide microarrays stained with sera from patients with systemic sclerosis (a, b, c), with intravenous immunoglobulin (**d**), and healthy control serum (**e**) or with secondary antibody only (control for false-positives caused by unspecific binding to detection antibody). Note the diversity of the systemic sclerosis patterns and the absence of signals with secondary antibody only. (**b**) Individual high magnification pane showing the corresponding spots from each of the three identical replicate subarrays. Note the good reproducibility in the peptide spots representing the same peptides and the problem with smearing when whole protein was spotted.

Based on the GAL file the GenePix software overlays a mask of circles to define feature-specific signal. To fine-tune this overlay feature alignment is performed with the software package. Besides automated software adjustments the user can manually rearrange the feature alignment. The immediate surrounding area of each feature outside a circle is defined as local background. Flags are being used to mark features with signals lower than the background or obscured features e.g., due to smears and scratches on the peptide microarray. After feature selection the microarray is imaged and the results are exported as GenePix-Results-File (GPR file, *see* **Note 4**), which is a text-based format that can be opened with any spreadsheet program. It contains general information about image acquisition and analysis, as well as the data extracted from each individual feature and user-defined parameters originally contained in the GAL file. The median fluorescence and the respective background at 635 nm excitation of each feature is used to generate a list of signals specific for the peptides on the chip. Since the image depth is 16 bit we have a grey value distribution from 0 up to 65,536 for each pixel.

3.5. Data Management and Visualization

In antibody profiling we are facing the problem that we generate for each tested polyclonal patient serum a diverse list of signals representing peptides from different antigens. In principle we have 3-dimensional data: Each serum shows reactivity of a certain magnitude against a certain peptide derived from a certain protein. For conclusive comparative analysis it will be important to find ways to visualize and statistically evaluate these data.

1. Database: The description of the experiment (GAL file) together with the primary image processing data (GPR file) are imported into a MySQL database. Information includes in particular protein names, accession numbers (e.g. Swissprot, protein Refseq), and amino acid sequence of the proteins the peptides are derived from. Additional descriptions can be added as well.

2. In a second step simple processing of raw data is performed as required. In antibody profiling the mean signal intensity of a feature for its three replicates and the standard deviation (SD) are calculated. We arbitrarily have defined signals as significant if either signal intensity is at least fivefold above background intensities and SD is not greater than 50% of the mean or if signal intensity of the replicates is above 10,000 (*see* **Notes 5–7**).

3. The visualization is carried out through an Apache webserver-based client solution, which takes the data from the backend MySQL database and visualizes them at the front end computer of any user by providing PHP-based scripts. Our web application, dubbed "EpiMap," manages the data tables and translates them into graphical representations (*see* **Fig. 3**). For a chosen experiment each 15mer peptide with signal intensities

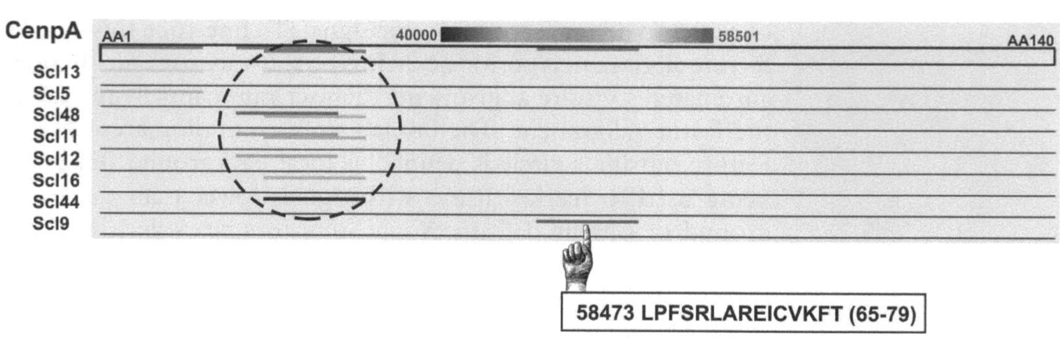

Fig. 3. Pane taken out of the EpiMap dynamic web visualization of peptide microarray data. Example for reactivities of sera from patients with systemic sclerosis (Scl) against centromeric autoantigen CenpA. Here, only sera with signal intensity of at least 40,000 are shown. *Individual bars* in the line with the sample designations represent the peptides which are displayed against the whole protein sequence depicted as a *bar on top* (CenpA from amino acids 1–140). The intensities of the peptides are color coded and the respective look-up table is shown above the depicted protein with the low and high maxima indicated. By moving over a peptide with the mouse ("*hand*" symbol) the user can inspect signal value, peptide sequence and location of the peptide within the whole protein. This visual display quickly identifies immunodominant epitopes within a protein for a selected group of samples (*encircled* peptides representing amino acids 21–39).

above a freely selectable signal threshold is displayed as little bar against the linear representation of the protein the peptides are part of. Thus the information on one microarray is split up to a display of individual peptides against the different proteins they originate from. Each dataset of a particular microarray staining for a given protein takes up one line. By adding the data of other microarrays we create a composite figure for each protein that directly visualizes any accumulation of signals in a particular domain of a protein or even against the same epitope. The peptide sequence plus the mean signal intensity of the triplicates is provided if the user moves the computer mouse over the bar representing the peptide. In addition the peptide bars are color coded according to a look-up table to directly indicate signal intensity. Furthermore, by clicking with the mouse on a peptide bar, the linear sequence of the protein is displayed with the chosen peptide sequence highlighted in color.

4. The webserver also provides the interface through the experiment I.D.s to inspect 8 bit representations of the original 16 bit images. For this, the image of the whole array is subdivided into images of the three subarrays. Furthermore, depending on the chosen threshold for peptide signals, spots included under that condition are displayed in an encircled manner and signal intensities provided if the user moves with the mouse over them.
5. The primary as well as the processed data can be downloaded from the webserver.

3.6. Perspectives

With the establishment of high density peptide microarray technologies in experimental and clinical medicine, epitope repertoires of humoral immune responses can be quantitatively measured. By mutating informative epitopes the clonality of immune responses can be determined as well as changes in antibody titers, specificities and affinities can be monitored over time in conjunction with the nature of antibody subtypes involved. Thus, the etiology of autoantibody generation in autoimmune diseases might be revealed. Immune responses and efficiencies of vaccination programs might be monitored. The contribution of humoral immune responses in fighting cancer might be further elucidated. Though one should keep in mind that only linear epitopes are determined, this information can serve as surrogate information for antibodies that might even preferentially bind conformational epitopes in vivo (mimotope recognition, *see* **ref.** *5*). In the near future, the whole human proteome represented by peptide arrays will be under study driven by world-wide initiatives that are aiming at generating and characterizing antibodies to each human protein (www.proteinatlas.org). Additionally, automated incubations using devices originally developed as hybridization stations for DNA microarrays will improve reproducibility of peptide microarrays experiments.

4. Notes

1. The layout of a peptide microarray is very flexible. For antibody profiling we recommend to use at least triplicates. The maximum number of peptides that can be put on the chip with its fixed dimension is dependent on the spot diameter and the spot to spot distance. Up to 100,000 spots with a diameter of about 65 μm have been deposited onto one standard industry glass slide.
2. The dilution of the sera/antibody solutions is sample-dependent. Usually, the serum of a healthy individual has an IgG concentration range from 7–16 mg/mL. IgG type autoantibodies are the most common type and represent a small fraction of the total serum IgG. In our experience a serum dilution of 1:100 is appropriate to find low affinity and low abundance antibodies without raising the background unduly. For high titer sera a higher dilution is plausible.
3. So far we found a gap of up to 4 days between staining and scanning unproblematic.
4. For details of the file formats see http://www.moleculardevices.com/pages/software/ gn_genepix_file_formats.html.
5. For each peptide microarray we calculate the median of all background signals around the features with spotted peptides. These microarray specific value is usually between 400–800 grey values (out of 65,536 total). Thus, we consider signals higher than 2,000–4,000 grey values (five times over background) as potentially specific ones.
6. In contrast to gene expression microarrays global scaling of signals for normalization is inappropriate for peptide microarrays used for antibody profiling. This is due to the fact that gene expression profiles always show a considerable number of features/genes (40–60%) with significant signals, which provide a broad basis to assess overall signal brightness. In contrast antibody profiles can include from only very few to hundreds of signals above background (*see* **Fig. 4**). Thus, they can be so diverse in the number of signals that global scaling would be highly error-prone.
7. Initially, the human IgG protein spotted on the microarray was meant to be used to normalize peptide microarrays from

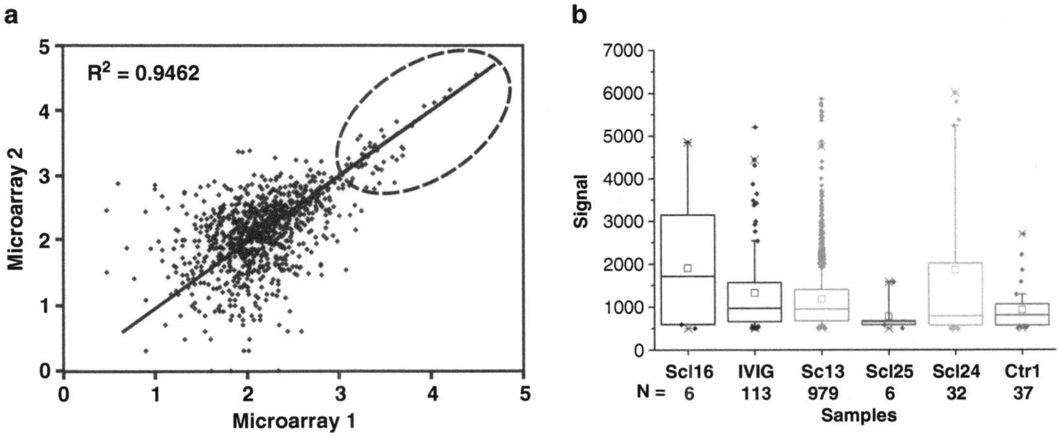

Fig. 4. Reproducibility and diversity of peptide signals. (**a**) Plot of all peptide signals (mean of triplicates after local background subtraction) of microarray one against microarray two, both stained with a serum from the same patient. Both axes are in log10 scale. The regression coefficient R^2 indicates a good reproducibility. The encircled signals can be considered as specific signals whereas the signal "cloud" between one and three (means 10 and 1,000 grey values, respectively) does not fulfill our arbitrary condition to be fivefold over background. (**b**) Box plots of the distribution of the peptide signals (mean of triplicates) of six different microarrays to show the high variance in the number and magnitude of the signals for different samples. Only signals at least fivefold over background are included. *Scl* serum from patients with systemic sclerosis; *IVIG* intravenous immunoglobulin preparation; *Ctr1* serum from healthy control individual. *Line* in box = median (50%); the box includes the values for 0.25 to 0.75 quantile with the middle line representing the median; whisker = 1 standard deviation; little square = mean; *dots* represent outliers. *N* = number of specific signals/microarray.

different experiments as it provided a standard for the secondary dye-conjugated anti-human IgG antibodies. However, we realized that the spotting process of protein resulted in too much deviation and sometimes some smearing (*see* **Fig. 4**) so that this standard was not acceptable. Alternatively, depending on the secondary antibody used a number of false-positive peptides will be on each peptide microarray resulting from unspecific binding of these peptides to the antibody. The signal intensities of these false-positives could be used for standardization efforts.

Acknowledgements

This work was supported by the EU (LSHM-CT-2004–005264; www.autorome.org). We wish to thank all our collaborators in the clinic, and, for excellent technical assistance, Ms. Eva Lorbeer. Further information on the software package can be obtained from H.-J. Thiesen.

References

1. Goodnow, C. C., Sprent, J., Fazekas de St Groth, B., Vinuesa, C. G. (2005) Cellular and genetic mechanisms of self tolerance and autoimmunity. *Nature* 435, 590–597.
2. Shoenfeld, Y., Gershwin, M. E., Meroni, P. L. (2007) *Autoantibodies*. Elsevier, Amsterdam.
3. Gonzalez-Buitrago, J. M., Gonzalez, C. (2006) Present and future of the autoimmunity laboratory. *Clin. Chim. Acta* 365, 50–57.
4. Cretich, M., Damin, F., Pirri, G., Chiari, M. (2006) Protein and peptide arrays: Recent trends and new directions. *Biomol. Eng.* 23, 77–88.
5. Monzavi-Karbassi, B., Shamloo, S., Kieber-Emmons, M., Jousheghany, F., Luo, P., Lin, K. Y., Cunto-Amesty, G., Weiner, D. B., Kieber-Emmons, T. (2003) Priming characteristics of peptide mimotopes of carbohydrate antigens. *Vaccine* 21, 753–760.
6. Frank, R. (1992) Spot-Synthesis: An easy technique for the positionally addressable, parallel chemical synthesis on a membrane support. *Tetrahedron* 48, 9217–9232.
7. Wenschuh, H., Volkmer-Engert, R., Schmidt, M., Schulz, M., Schneider-Mergener, J., Reineke, U. (2000) Coherent membrane supports for parallel microsynthesis and screening of bioactive peptides. *Biopolymers* 55, 188–206.

Chapter 19

Microarrayed Allergen Molecules for Diagnostics of Allergy

Jing Lin, Ludmilla Bardina, and Wayne G. Shreffler

Summary

With the evolution of peptide synthesis techniques and microarray technology, it is now possible to map and characterize allergenic epitopes on a microarray platform. The peptide microarray-based immunoassay allows simultaneous analysis of thousands of target peptides using small volumes of diluted serum samples, and has a promising future to become a superior testing tool for aspects of food allergy diagnosis and prognosis, as well as for designing recombinant allergens for safe immunotherapy. Characterization of allergenic epitopes provides fundamental knowledge for understanding mechanisms of food allergy. This chapter describes in detail the development of a sensitive and reliable peptide microarray-based immunoassay. The information includes a comparison of different slide substrates, effect of buffers on spot morphology, performance of printing pins, immunolabeling detection systems with different levels of sensitivity, and suggested approaches to data analysis.

Key words: Microarray, Peptide array, Epitope mapping, Allergy diagnosis.

1. Introduction

Food allergy has emerged as an important public health problem in the United States *(1)*. Approximately 4% of the American population and 6–8% of children <3-years old are affected by IgE-mediated food allergy, with symptoms ranging from mild oral pruritus to potentially life-threatening anaphylactic shock. Food allergy is currently diagnosed by clinical history and *in vivo* or *in vitro* confirmation of allergen-specific serum IgEs *(2)*.

However, in many cases the history is uncertain and the efficacy of current testing is limited. First of all, positive IgE testing is not highly predictive of clinical allergy. Confirmation of clinical allergy, therefore, often requires oral provocation testing, which

Ulrich Reineke and Mike Schutkowski (eds.), *Methods Molecular in Biology, Epitope Mapping Protocols, vol. 524*

DOI: 10.1007/978-1-59745-450-6_19

carries some risk to the patient. This is not due to technical assay limitations, but follows from the fact that although allergen-specific IgE is necessary for these reactions, additional variables determine clinical reactivity and sensitivity. One of these variables is the quantity of specific IgE or the ratio of specific IgE to the total IgE pool. Numerous studies have documented the positive relationship between IgE quantity and the likelihood of clinical allergy *(2)*, yet there is a large "grey zone" between high levels with good positive predictive values and undetectable levels that have good negative predictive values. In addition, the relationship between current IgE testing and the patients' clinical sensitivity, usually defined as one of or a combination of reaction severity and the dose of allergen required to provoke a reaction, is very weak. Finally, another limitation of current testing is regarding the prognosis of a patient's disease course. The majority of pediatric patients with some forms of food allergy (e.g., milk, egg, wheat, soy), will outgrow them during childhood. For at least some allergens there is again a positive but weak correlation between specific IgE level and duration of clinical allergy.

Recent studies have suggested that these aspects of clinical reactivity to food allergens may correlate better with features of allergen-specific IgE when looking at epitope-specific recognition. For example, patients with persistent allergy or with a history of more severe allergic reactions to milk *(3, 4)*, peanut *(5)* and egg *(6)* were found to recognize a larger number of IgE epitopes. IgE epitope mapping may become an additional tool for allergy diagnosis and prediction. Additional research in larger prospective studies is underway to determine whether some of these correlations are stronger than those seen with current tests.

Characterization of allergenic epitopes is also of fundamental importance for understanding mechanisms of food allergy and for designing safe immunotherapeutics *(7, 8)*. Previous efforts to treat food allergy using traditional allergen injection immunotherapy were suspended due to very high rates of iatrogenic anaphylaxis. Targeted alterations of the IgE epitopes is one strategy for reducing this risk while maintaining the proven efficacy of immunotherapy *(9)*. Human trials using recombinant peanut allergens modified to alter IgE epitopes are beginning now and if successful, will likely usher in a wave of demand for trials of other food allergens.

In the past, spot membrane based immunoassay has been widely applied for epitope mapping *(3, 10)*. In this system, the peptides are synthesized on the membrane and then incubated with the patient's sera. However, synthesis of large numbers of peptides is relatively error prone, time consuming, labor intensive, and expensive. Immunoassay in this format also requires a relatively large volume of serum from each patient and is limited to the number of targeted peptides.

With the development of microarray technology *(11–13)* and evolution in peptide synthesis techniques *(14, 15)* we have been applying peptide microarray-based immunoassays for epitope mapping of several food allergens *(16–19)*. There are several advantages of using microarray-based immumoassay for epitope mapping. It is possible to assay thousands of target peptides in parallel using small volumes of diluted serum samples, greatly reducing the biological sample cost of individual assays and allowing for more robust replication and statistical approaches to analysis and epitope determination. Several immunoglobulin subclasses other than IgE can be tested simultaneously, allowing us to simultaneously investigate potential regulatory responses (e.g., IgG4) that may influence clinical reactivity. In this chapter we will describe the methods for our preparation and processing of the peptide microarray-based immunoassay.

2. Materials

2.1. Microarray Printing Preparation

1. Glass slides coated with epoxide groups are from TeleChem International, Inc. (Sunnyvale, CA) (*see* **Note 1**). The slides should be used within 6 months of the manufacturing date and stored under vacuum before printing. Note the notch on the upper right corner of the slide indicating that samples should be printed on the side facing upward.
2. Peptides consisting of 15 or 20 amino acids length with an offset of three corresponding to the primary sequences of food allergens are commercially synthesized by JPT Peptide Technologies GmbH (Berlin, Germany). The synthesized peptides are either custom-made peptides supplied as lyophilized powder with purity over 70% (analyzed by HPLC and mass spectrometry) or peptides derivatives synthesized using the PepStar technique (*see* **Note 2**).
3. Protein printing buffer (PPB) (TeleChem International, Inc.).
4. Sodium lauryl sarcosinate (Sarkosyl) (Fisher Scientific, Fair Lawn, NJ).

2.2. Microarray Immunolabeling

1. Phosphate buffered saline (PBS) with Tween 20 (0.05%) (PBST): a 10× PBS solution is obtained from Roche Diagnostics Inc. (Indianapolis, IN) containing 0.01 M KH_2PO_4, 0.01 M Na_2HPO_4, 1.37 M NaCl, 0.027 M KCl, pH 7.0. Dilute 200 mL of 10×PBS with 1,800 mL of distilled water and add 1 mL of Tween 20.

2. Blocking buffer: 1% human serum albumin (HSA, Sigma, 85% purity) in PBST.
3. Secondary antibody for 1-step Alexa detection system: Polyclonal goat anti-human IgE (gift from Phadia, Uppsala, Sweden) and monoclonal mouse anti-human IgG4 (BD Pharmingen Inc., San Diego, CA) are conjugated with Alexa 546 and Alexa 647 (Invitrogen, Carlsbad, CA), respectively (*see* **Subheading 3.1**).
4. Secondary antibody for 2-step UltraAmp signal amplification system: Polyclonal biotinylated anti-human IgE (Kirkegaard & Perry Laboratories Inc., Gaithersburg, MD) and monoclonal anti-human IgG4-FITC (SouthernBiotech, Clone HP6025, Birmingham, AL).
5. UltraAmp detection reagents: Anti-biotin oyster 550 (350) and anti-FITC oyster 550 (350) and UltraAmp buffer are obtained from Genisphere Inc., Hatfield, PA. Salmon sperm DNA is from Invitrogen (Carlsbad, CA).
6. 15 mM Tris-HCl buffer, pH 8.0: 1 M Tris-HCl buffer, pH 8.0 is prepared by dissolving 121 g Tris base in 1 L distilled water and adjusting pH to 8.0 with concentrated HCl. Dilute 15 mL 1 M Tris-HCl buffer, pH 8.0 with 985 mL distilled water to make 15 mM Tris-HCl buffer, pH 8.0.

3. Methods

3.1. Preparation of Fluorescent Labeled Secondary Antibody

1. Concentration of secondary antibody: For the optimal conjugation with fluorochrome the concentration of antibody should be >2 mg/mL. Monoclonal antibodies are concentrated to >2 mg/mL using Amicon centricon concentrator YM30 (Amicon, Inc., Beverly, MA) according to manufacturer's instructions. The antibodies must be in a buffer free of primary amines, ammonium ions, sodium azide (>3 mM) and thimerosal (>1 mM), otherwise the conjugation reaction will be interfered. Buffer can be replaced (preferably by PBS) using dialysis.
2. Conjugation: A conjugation mixture is prepared as followed: 200 μL antibodies/proteins (>2 mg/mL) + 20 μL of 1 M $NaHCO_3$ + 20 μL of 10 mg/mL Alexa546/647, and incubated at room temperature for 1 h with gentle stirring (protected from light). The fluorescent labeled antibody/protein is then loaded (120 μL/column) onto the BioSpin 30 column (BioRad, Hercules, CA) and centrifuged at 1,000 × *g* for 4 min. The eluate containing the purified fluorescent labeled

antibody/protein is collected and stored at 4°C protected from light. The conjugate should be stable in the presence of 2 mM sodium azide for several months.

3. Determination of degree of labeling: A 10 µL of the purified fluorescent conjugate is diluted 1/10, followed by serial dilutions 1/20, 1/40 and 1/80 with PBS and the absorbance of the diluted conjugates are measured in a 100 µL quartz cuvette with a 1 cm path length at both 280 nm (A_{280}) and either 650 nm (A_{650}) for Alexa 647 or 558 nm (A_{558}) for Alexa 546. The protein concentration in the diluted sample and the degree of labeling are calculated as follows:

For Alexa 546 conjugate:

$$\text{Protein concentration (M)} = \frac{(A_{280} - (A_{558} \times 0.12)) \times \text{dilution factor}}{203{,}000}$$

$$\text{Moles dye per mole protein} = \frac{A_{558} \times \text{dilution factor}}{104{,}000 \times \text{protein concentration (M)}}$$

For Alexa 647 conjugate:

$$\text{Protein concentration (M)} = \frac{(A_{280} - (A_{650} \times 0.031)) \times \text{dilution factor}}{203{,}000}$$

$$\text{Moles dye per mole protein} = \frac{A_{650} \times \text{dilution factor}}{239{,}000 \times \text{protein concentration (M)}}$$

The fluorescent conjugate usually has fluorochrome:antibody molar ratios of 5–10.

3.2. Printing of Peptide Microarrays

1. Preparation of printing plate: The lyophilized peptides are resuspended in dimethyl sulfoxide (DMSO) at 1 mg/mL, diluted 1/2 in 2×protein printing buffer (PPB) with 0.02% Sarkosyl (*see* **Note 3**) (**Fig. 1**), and transferred (10–15 µL/well) into the printing plate (assay plate 384 well, low volume, nonbinding surface round bottom, white polysterene, Corning). The printing plate is covered with Robolid (Corning), centrifuged at 300 × *g* for 5 min, and stored in a sealed plastic bag at 80°C until use. The centrifuge step is to remove any possible bubbles in the wells and it is very important.
2. Microarray printing instrument: NanoPrint Microarrayer 60 (TeleChem International, Inc.), a contact microarray printer, is used for peptide printing. It is equipped with 2 × 4 ArrayIt Stealth Micro Spotting Pin (SMP3B), which allows pick up of sample solutions (~0.6 µL) from the printing plate and deposition of a specific amount (~0.9 nL) by touching the slide surface. The printing consistency of these eight pins have

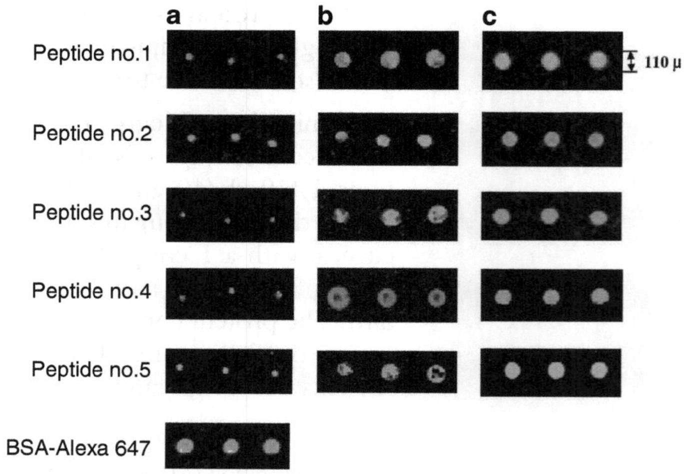

Fig. 1. Effects of buffers on spot morphology. The triplicate spots of five milk peptide samples on the slides were immunolabeled with a 1/5 diluted serum pool (specific IgE to milk: 850 kU_A/L) of five milk allergic patients using the Alexa detection system as described in **Subheading 3.3**. The images taken were scanned with red laser (for Alexa 647). The triplicate spots of BSA-Alexa 647 (1.5 μg/mL in PPB) are also shown as a comparison. (**a**) peptides printed as 0.33 mg/mL in PPB, 33% DMSO, (**b**) peptides printed as 0.33 mg/mL in PPB, 33% DMSO, 0.02% Sarkosyl (**c**) peptides printed as 0.5 mg/mL in PPB, 50% DMSO, 0.02% Sarkosyl.

been tested in the experiment in which the eight different pins showed the ability to deposit the same amount of samples and print at least 250 consistent spots of a fluorescent labeled protein without sample reloading (*see* **Fig. 2**). Therefore the pins are usually used to print not more than 200 spots from each loading. With each loading, the first two printed spots are usually very large and will not be used (set of two dummy slides). If possible, it is better to keep the microarray printing facility in a clean room to avoid any dirt or dust that gets into the printing plate or onto the slide surface during printing.

3. Microarray printing: The whole printing process is controlled by the NanoPrint™ Microarray Manager Software (TeleChem International, Inc.). Peptide samples are usually printed in triplicate and approximately 1/4 of the printed spots are PPB alone (>400 spots/array) which are used as negative control for background normalization. Alexa 546 and Alexa 647-labeled bovine serum albumin elements (prepared following antibody conjugation protocol, *see* **Subheading 3.1**) are used for the purpose of grid alignment. Other proteins, such as different concentrations of purified human IgE, may be printed as well as positive controls. The humidity is maintained at 55–65%

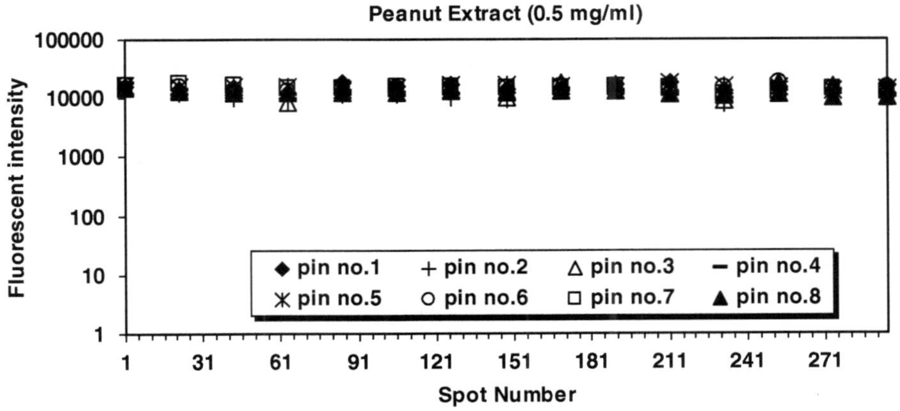

Fig. 2. Pin performance test. 288 spots of 0.5 mg/mL peanut extract were printed from each of the eight spotting pins in a row without reloading. One printed slide was then immunolabeled with the serum of a patient with high specific IgE (>100 kU_A/L) to peanut using the Alexa detection system (*see* **Subheading 3.3**) and scanned with the green laser (*see* **Subheading 3.4**). The fluorescent intensity (dFU) shown here is the median fluorescent intensity of the spot subtracted by local background.

during printing to avoid sample evaporation (*see* **Note 4**). All array elements are printed in duplicate (two sets of triplicates) to improve precision, and to determine intra-assay variation. After the first set of triplicate spots are printed, the plate is turned around so that the second set of triplicate spots are printed using different pins to minimize the possible effect of printing inconsistency of different pins. In order to save slide space and serum volume used for immunolabeling, two arrays are printed on a single slide. Printed slides were dried overnight in the printer (at low humidity) to allow full binding to the slide substrate, stored at room temperature (*see* **Note 5**) and used within 2 months. A.gal-file containing information on sample location and identification from each printing is generated using the Microarray Manager Software.

After printing, it is recommended to perform quality control (QC) scan on dried slides using lower resolution for faster process (*see* **Note 6**). The purpose of QC scan is to monitor the quality and reproducibility of the printing as well as to serve as a reference before labeling.

3.3. Immunolabeling with Patient Serum

The printed arrays are incubated with sera. Sera of at least ten negative controls, either nonatopic people or atopic patients but not allergic to the same food should be run in parallel. Depending on the application, either 1-step Alexa detection system or 2-step UltraAmp could be applied (*see* Note 7) (**Fig. 3**).

1. Slide preparation: A rectangular incubation area around the printed arrays is demarcated using a hydrophobic Dako

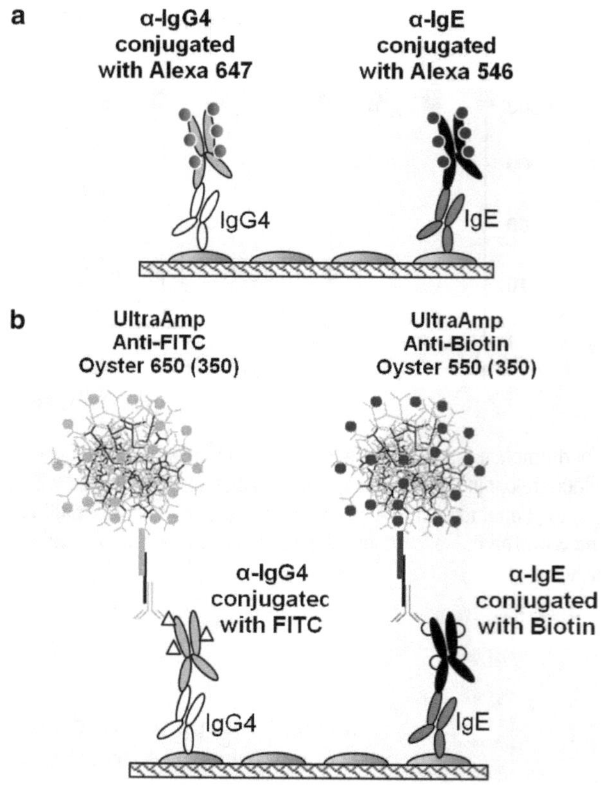

Fig. 3. Immunolabeling procedure. (**a**) 1-step Alexa detection system (**b**) 2-step UltraAmp amplification system.

Cytomation Pen (Glostrup, Denmark) which separates two arrays on the same slide. This step must be done at least 3 h before immunolabeling (better overnight) to ensure that the ink is totally dry. All the incubation steps are performed in a humidity chamber (Binding Site, Birmingham, UK) covered by a black cloth on a rotating platform (Lab-Line Instruments Inc., IL) with gentle agitation. It is very important to keep the slide covered with solution at all times, never let it dry during the whole immunolabeling procedure.

2. Blocking and serum incubation: The slides are placed into the humidity chamber, secured by a magnetic strip, and rinsed with PBST. After removing PBST from the slide surface by aspiration at the array corner, 400 μL of blocking buffer are added onto each array to block the nonspecific binding sites, and incubated for 1 h at room temperature. The volume used for incubation depends on the size of the array area. Usually 200–400 μL is enough to cover one array (half a slide). If the volume cannot spread out and cover the slide surface by itself,

one can try to spread it out by pipette tips held longitudinally but never touched the slide surface. Patients' sera are prepared by diluting 50 μL of serum with 200 μL of blocking buffer (*see* **Note 8**) and applied onto the slide after the blocking buffer is removed by aspiration. As the two arrays are separated by the hydrophobic pen, two different sera can be applied on the same slide. The slides are incubated with the sera for overnight at 4°C (*see* **Note 9**) and washed carefully and thoroughly (5 × 1 min) with PBST. Avoid contamination between different arrays especially arrays on the same slide.

3.3.1. Detection of IgE Binding Using 1-Step Alexa Detection System

The slides are incubated (250 μL/array) for 1 h at room temperature with a cocktail of Alexa 546 conjugated polyclonal goat anti-human IgE and Alexa 647 conjugated monoclonal mouse anti-human IgG4, both of which diluted 1/5,000 (*see* **Note 10**) in blocking buffer. The slides are washed with PBST (3 × 1 min), distilled water (3 × 1 min), and spin dried using a bench top slide spinner (TeleChem International, Inc.).

3.3.2. Detection of IgE Binding Using 2-Step Ultraamp Amplification Detection System

The slides are incubated (250 μL/array) for 1 h at room temperature with a cocktail of polyclonal biotinylated anti-human IgE (*see* **Note 11**) and monoclonal anti-human IgG4-FITC, both diluted 1/500 in the blocking buffer (*see* **Note 11**). Slides are washed with PBST (3 × 1 min), incubated for 4 min with 1 mM ethylene diamine tetraacetic acid (EDTA) in PBST, washed again with PBST, equilibrated for 1 min with UltraAmp buffer and, after aspiration, incubated 3 h at room temperature with a cocktail of anti-Biotin Oyster 550 (350) and anti-FITC oyster 550 (350) in UltraAmp buffer at concentration of 0.6 μg/mL, each with addition of 0.02 μg/mL of salmon sperm DNA. The slides are washed with PBST (3 × 1 min), rinsed with 15 mM Tris-HCl buffer, pH 8.0 (30 s, this step helps to retain signal intensity), spin dried, followed by wash with 0.05×PBS (30 s), and finally spin dried.

3.4. Microarray Scanning and Slide Alignment

Reactions of serum IgE to each peptide sample are represented by the fluorescent intensity of the peptide spot.

1. Scanning: The immunolabeled slides are scanned with both green (laser excitation wavelength at 543 nm for Alexa 546) and red lasers (laser excitation wavelength at 633 nm for Alexa 647) using a ScanArray®Gx (PerkinElmer, Waltham, MA) equipped with the program ScanArray Express (Perkin Elmer). Adjust the laser power and PMT gain as high as possible without saturating the signal of any spot (*see* **Note 12**). For image resolution use a resolution of no more than 20 μm (*see* **Note 6**). Keep the same setting for the slides from the same project. Scanned images at both wavelengths are saved as TIF files.

2. Alignment: The .gal-file is loaded and the scanned images are quantitated with ScanArray Express using the "adaptive threshold" (*see* **Note 13**) quantitation algorithm to generate a spread sheet containing values for median/mean signal intensity, median/mean background intensity, background intensity standard deviation, and signal-to-noise ratio for each spot on the array. The boundary of each spot are carefully checked and re-adjusted if it is not located on the real spot due to possible artefacts. The spread sheets are exported and saved as comma-delimited text files.
3. Data analysis: Data are analyzed with R programming language (http://www.r-project.org/) version 2.6.0 as followed: Briefly, the read out (S) used for each spot, including the replicates for peptides and PPB, is the median fluorescent signal of the spot divided by local background and $\log_2$ transformed. A *Z*-score is calculated for each array element (spot) using PPB values within the same array:

$$Z_i = \frac{S_i - \text{Median}\,(S_{PPB})}{\text{MAD}\,(S_{PPB})}$$

4. Where S_i is the read out for the array element transformed into Z score (Z_i); and the median(S_{PPB}) and MAD(S_{PPB}) are the median and the median absolute deviation (MAD) of all the read outs of PPB spots, respectively.

The total Z value for each peptide is the Median of *Z*-scores of the six replicate spots. As the peptides are overlapped by 12 or 17 amino acids, which means overlapping peptides might share the same epitopes, a weighted average of Z value could be calculated by the formula $Z = 0.25*Z_{-1} + 0.5*Z_0 + 0.25*Z_{+1}$ and subsequent analysis is carried out based on the weighted average index. An individual peptide sample is considered positive if its index exceeds 3. For comparison of *Z*-scores between groups, a Wilcoxon test is applied ($p < 0.01$).

4. Notes

1. Printing slides with various substrates, with either 2- or 3-dimensional surfaces, are commercially available now *(20)*. We have tested slides with aldehyde substrates and found that they have less binding capacity. We have also tested the 3-dimensional nitrocellulose substrate slides such as FAST slide from Whatman and SuperNitro slide from TeleChem International, Inc. They are more difficult to handle and

usually show high background. Moreover, the presence of organic solvents, even as low as 20% DMSO, which is necessary to maintain peptide solubility, damages the nitrocellulose membrane.

2. Depending on the specific project, it might be possible to use the peptide synthesized using the PepStar technique from JPT Peptide Technologies GmbH (Berlin, Germany). The synthesized peptide derivatives carry a reactivity tag and a linker at their N-terminus allowing chemoselective immobilization. It provides much less amount of peptides (50 nmol for each peptide) but at much lower cost for each peptide and faster production time.
3. It is desirable to keep peptides in high concentration of organic solvents, such as dimethylsulfoxide (DMSO) and acetonitrile, which may increase peptide solubility and prevent peptide aggregation. In addition, some synthetic peptides seem to form very small spots after printing. We have solved this problem by adding 0.02% Sarkosyl and increasing DMSO concentration from 33 to 50% (**Fig. 1**).
4. We have found that the signal intensities of the spots of the same fluorescent samples are higher when printed at higher humidity, possibly by increasing the reactivity of the epoxide substrate on the slide or more materials getting deposited. Therefore it is desirable to keep the humidity stable but as high as possible during printing.
5. Although it has been suggested by some researchers to store the printed slides *in vacuum* and store the arrays in refrigerator, we found that storing the printed slides at room temperature may minimize the nonspecific binding (noise) to the slide background, possibly by deactivation of the remaining active epoxide substrate on the slide surface.
6. A quick scan can be performed using lower resolution (high pixel size such as 50 μm) to check the printing quality or locate the area with the printed spots before a second scan with higher resolution is performed.
7. Both of these two detection systems have their own advantages and disadvantages: For the 1-step Alexa detection system, the Alexa fluorochrome and secondary antibody usually form stable conjugate with fluorochrome:antibody molar ratios of 5–10. Our results have shown that it gives very reproducible results ($R > 0.90$) regardless of printing lots and immunolabeling day but the sensitivity is not high enough for allergic patients with less than 5 kU_A/L of specific IgE level. The UltraAmp amplification detection system, on the contrary, contains 5–15 specific antibody molecules and 350 Fluorochromes on each UltraAmp molecule.

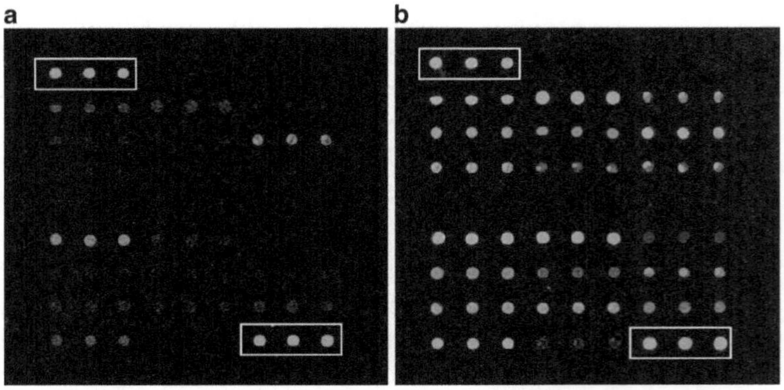

Fig. 4. Comparison between 1-step Alexa detection system and 2-step UltraAmp amplification detection system. In this experiment, replicate milk peptide arrays were immunolabeled with a 1/200 diluted serum pool (specific IgE to milk: 850 kU_A/L) of five milk allergic patients using either the Alexa detection system (**a**) or the UltraAmp detection system (**b**) as described in **Subheading 3.3**. The images shown here are the scanned images (*green laser for Alexa 546*) of one subarray of each array. The peptides were printed as triplicate spots and the spots of positioning control BSA-Alexa 546 (1.5 μg/mL in PPB) are indicated in the white frames.

Therefore it can greatly (*see* **Fig. 4**) increase the fluorescent signal intensity of the bound secondary antibody and be applied for allergic patients with specific IgE level even less than 0.35 kU_A/L. However, the reproducibility so far we can achieve (R = ~0.75) is lower than Alexa detection system due to high sensitivity and high susceptibility to variations in handling. In addition, the UltraAmp reagents cost much more than the Alexa reagents.

8. It is very important to vortex the serum before mixing with the blocking buffer. Keep the serum and the diluted serum on ice during preparation.
9. In order to decide the optimal incubation time, we have tested several serum incubation conditions (1 h at 37°C, 4 h at room temperature, and overnight at 4°C) using replicate arrays and found that overnight at 4°C allows more positive binding while not creating too much nonspecific binding and background noise.
10. The working concentration of the secondary antibody is in the range of 0.4 μg/mL for polyclonal antibody to 2 μg/mL for monoclonal antibody. However, a serial dilution of the secondary antibody should are tested in preliminary experiment to decide the optimal dilution factor which has the best balance between sensitivity and specificity.
11. As an option, monoclonal biotinylated anti-human IgE could be used to increase specificity. In addition, a cocktail of

several clones can be applied to compensate steric hindrance effect.

12. Higher laser power is usually desirable for scanning. If the pixels have been saturated (beyond the detection limit of the scanner) it will appear as white-color pixels.
13. There are three other quantitation methods: adaptive circle, fixed circle, and histogram. The main advantage of the adaptive threshold method is that it can adapt to different spot morphology better than the other methods. Please read "ScanArray Express Microarray Analysis System User Manual" for more details of these four methods.

Acknowledgements

The authors acknowledge the contribution of Yongchao Ge, Ph.D., Assistant Professor Mount Sinai, Department of Neurology, to the statistical analysis methods and authoring of an R analysis script used for some analysis steps. We also thank Todd Martinsky, Telechem International Inc. for technical advice and support.

References

1. Sampson, H. A. (2005) Food allergy – accurately identifying clinical reactivity. *Allergy* 60, 19–24.
2. Sampson, H. A. (2001) Utility of food-specific IgE concentrations in predicting symptomatic food allergy. *J. Allergy Clin. Immunol.* 107, 891–896.
3. Chatchatee, P., Järvinen, K. M., Bardina, L., Vila, L., Beyer, K., and Sampson, H. A. (2001) Identification of IgE and IgG binding epitopes on beta- and kappa-casein in cow's milk allergic patients. *Clin. Exp. Allergy* 31, 1256–1262.
4. Jarvinen, K. M., Beyer, K., Vila, L., Chatchatee, P., Busse, P. J., and Sampson, H. A. (2002) B-cell epitopes as a screening instrument for persistent cow's milk allergy. *J. Allergy Clin. Immunol.* 110, 293–297.
5. Beyer, K., Ellman-Grunther, L., Järvinen, K. M., Wood, R. A., Hourihane, J., and Sampson, H. A. (2003) Measurement of peptide-specific IgE as an additional tool in identifying patients with clinical reactivity to peanuts. *J. Allergy Clin. Immunol.* 112, 202–207.
6. Cooke, S. K. and Sampson, H. A. (1997) Allergenic properties of ovomucoid in man. *J. Immunol.* 159, 2026–2032.
7. Bannon, G. A., Cockrell, G., Connaughton, C., West, C. M., Helm, R., Stanley, J. S., King, N., Rabjohn, P., Sampson, H. A., and Burks, A. W. (2001) Engineering, characterization and in vitro efficacy of the major peanut allergens for use in immunotherapy. *Int. Arch. Allergy Imm.* 124, 70–72.
8. Li, X. M., Srivastava, K., Grishin, A., Huang, C. K., Schofield, B., Burks, W., and Sampson, H. A. (2003) Persistent protective effect of heat-killed *Escherichia coli* producing "engineered," recombinant peanut proteins in a murine model of peanut allergy. *J. Allergy Clin. Immunol.* 112, 159–167.
9. Valenta, R. and Niederberger, V. (2007) Recombinant allergens for immunotherapy. *J. Allergy Clin. Immunol.* 119, 826–830.
10. Frank, R. (2002) The SPOT synthesis technique – Synthetic peptide arrays on membrane supports – principles and applications. *J. Immunol. Methods* 267, 13–26.

11. Hiller, R., Laffer, S., Harwanegg, C., Huber, M., Schmidt, W. M., Twardosz, A., Barletta, B., Becker, W. M., Blaser, K., Breiteneder, H., Chapman, M., Crameri, R., Duchêne, M., Ferreira, F., Fiebig, H., Hoffmann-Sommergruber, K., King, T. P., Kleber-Janke, T., Kurup, V. P., Lehrer, S. B., Lidholm, J., Müller, U., Pini, C., Reese, G., Scheiner, O., Scheynius, A., Shen, H. D., Spitzauer, S., Suck, R., Swoboda, I., Thomas, W., Tinghino, R., Van Hage-Hamsten, M., Virtanen, T., Kraft, D., Müller, M. W., and Valenta, R. (2002) Microarrayed allergen molecules: diagnostic gatekeepers for allergy treatment. *FASEB J.* 16, 414–416.

12. Harwanegg, C., Laffer, S., Hiller, R., Mueller, M. W., Kraft, D., Spitzauer, S., and Valenta, R. (2003) Microarrayed recombinant allergens for diagnosis of allergy. *Clin. Exp. Allergy* 33, 7–13.

13. MacBeath, G. and Schreiber, S. L. (2000) Printing proteins as microarrays for high-throughput function determination. *Science* 289, 1760–1763.

14. Shin, D. S., Kim, D. H., Chung, W. J., and Lee, Y. S. (2005) Combinatorial solid phase peptide synthesis and bioassays. *J. Biochem. Mol. Biol.* 38, 517–525.

15. Panicker, R. C., Huang, X., and Yao, S. Q. (2004) Recent advances in peptide-based microarray technologies. *Comb. Chem. High Throughput Screen.* 7, 547–556.

16. Shreffler, W. G., Beyer, K., Chu, T. H., Burks, A. W., and Sampson, H. A. (2004) Microarray immunoassay: Association of clinical history, in vitro IgE function, and heterogeneity of allergenic peanut epitopes. *J. Allergy Clin. Immunol.* 113, 776–782.

17. Shreffler, W. G., Lencer, D. A., Bardina, L., and Sampson, H. A. (2005) IgE and IgG(4) epitope mapping by microarray immunoassay reveals the diversity of immune response to the peanut allergen, Ara h 2. *J. Allergy Clin. Immunol.* 116, 893–899.

18. Lencer, D. A., Rosenfeld, L., Bardina, L., Shreffler, W. G., Sampson, H. A., and Beyer, K. (2006) Determination of allergenic cross-reactivity between 11S-globulins (seed storage proteins) in peanut, tree-nut and sesame allergic patients using peptide microarray immunoassay. *J. Allergy Clin. Immunol.* 117, S34.

19. Wang, J., Bardina, L., Lencer, D., Shreffler, W. G., and Sampson, H. A. (2006) Determination of epitope diversity in cow's milk hypersensitive using microarray immunassay. *J. Allergy Clin. Immunol.* 117, S39.

20. Kusnezow, W. and Hoheisel, J. D. (2003) Solid supports for microarray immunoassays. *J. Mol. Recognit.* 16, 165–176.

Chapter 20

Monitoring B Cell Response to Immunoselected Phage-Displayed Peptides by Microarrays

Lina Cekaite, Eiving Hovig, and Mouldy Sioud

Summary

Successful adaptation of microarray technology for high-throughput screening of proteins requires a large number of purified recombinant proteins, e.g., antibodies for use as capture molecules. Phage surface display technology has been used for the surface expression of proteins, peptides or cDNA repertoires expressed by tumor cells. It does not require protein purification, as recombinant phages can be spotted on glass slides and used in a high-throughput screening format. Biopanning of phage libraries on patient serum antibodies is expected to enrich for antibody-binding phages for the fabrication of diagnostic and/or prognostic B-cell epitope microarrays. In contrast to other immunological techniques, microarrays can measure the antibody levels against different epitopes in a single test. This chapter highlights the recent advances in phage-based microarray technology to profile humoral immune responses in cancer patients.

Key words: B-cell epitopes, Peptide-phage libraries, Microarray, Cancer, Autoimmunity, Antibody signatures.

1. Introduction

During the last years transcriptional microarrays have been the technology of choice to monitor the abundance of several thousands of mRNA transcripts. However, this technology provides relatively little information regarding the proteins encoded by mRNA transcripts. Because proteins, rather than RNA, carry out the majority of cellular functions, there is a large interest in analyzing the complete repertoire of proteins, the proteome, in a manner comparable with transcriptional microarrays. Similar to DNA microarrays, proteins can be deposited in a predetermined spatial order on a glass or nitrocellulose coated slide allowing them to

Ulrich Reineke and Mike Schutkowski (eds.), *Methods in Molecular Biology, Epitope Mapping Protocols, vol. 524*

DOI: 10.1007/978-1-59745-450-6_20

be made available as probes for binding in a high-throughput, parallel manner. The most common protein microarray type is an antibody microarray, where immobilized antibodies are used as capture molecules to detect proteins from cell lysates or body fluids, such as patient sera and synovial fluids *(1–5)*. Protein microarrays can be used to identify protein–protein interactions, the substrates of protein kinases, or the targets of biologically active small molecules. Also, they could be suitable for clarification and mapping of cellular networks, i.e., signal transduction cascades, phosphorylation networks, and proteins involved in tumor genesis *(6–8)*. However, a bottleneck in fabri-cating protein arrays, especially those for global measurements, is the production of a large number of purified proteins such as antibodies. To overcome this technical problem, other types of capture molecules that are more uniform in their nature including peptides and aptamers have been developed *(9,10)*. In addition, proteins expressed in cell-free in vitro transcription/translation systems *(11,12)*, nucleic acids, and small molecules *(13)* have been spotted onto arrays and then used as capture molecules in order to study protein function.

Although the knowledge on autoantigens and the specificity of autoantibodies increased during the last 20 years, typically the analysis of humoral immune responses in patients requires prior information about the antigens that activated immunity. We have been interested in investigating the possibility of selecting binding epitopes for antibodies in patient sera from random peptide-phage libraries, whether or not the parental antigens are known. Such an approach would allow the understanding of immune responses in patients without having to establish B-cell hybridomas. Furthermore, extending the technique to polyclonal antibodies would permit a comparison of the antibody repertoires of individuals infected by the same or related pathogens. Antibody signatures identified by this technique might facilitate the diagnosis of patients with cancers. In previous studies, we have demonstrated for the first time that B-cell responses in patients with either rheumatoid arthritis *(14)* or cancer *(15)* can be probed by the use of random peptide-phage libraries. The rationale for this technology is that antibodies from patient sera might bind to the phage containing the epitopes responsible for initiating the activation of B-cells. In parallel, we also developed phage-expression cDNA libraries and nitrocellulose arrays for monitoring humoral responses in patients *(16)*.

In the context of developing diagnostic protein microarrays, we have adapted phage-displayed B-cell epitopes as capture molecules to profile humoral immune responses in patients with breast cancer in a high-throughput format *(17)*. In contrast to other protein microarrays, the use of peptides or proteins expressed on the surface of phages or bacteria does not require purification of the expressed peptides or proteins. Although phage-displayed peptide microarrays are quite new, studies showed that they are useful in probing antibody signatures in cancer patients. *(17–23)*.

2. Phage-Displayed Peptides as Novel Serum Antibody Capture Molecules

Phage display technology has emerged as a powerful tool for identifying proteins with specific binding properties. In this technology, amino acid sequences are added to the carboxy terminus of a phage capsid protein, generating a fusion protein displayed on the surface of the phage *(24)*. Phage-displayed peptides or proteins are amenable for binding to antibodies or protein partners. Moreover, the phage display approach is far superior over one-step screening methods because sensitivity and selectivity are extremely high when selecting through iterative and powerful enrichment steps *(14, 25)*. It should be noted that during the biopanning procedure, the antibody selects its binding peptide(s). Thus, the antibody-binding peptides do not necessarily resemble the parental epitopes, but they can mimic their binding properties. Through this strategy it is possible to probe specific humoral immune responses in patients without preconceptions about the parental antigens that initiated and/or perpetuated the immune responses. Profiling immune responses in patients with cancer with peptides or cDNA phage libraries has led to the identification of a series of peptides and proteins, some of which were recognized more frequently by patient serum antibodies when compared with antibodies from control groups *(1, 16, 26)*. These immunoselected phages should constitute an invaluable source of B-cell epitopes as diagnosis and/or prognosis test of patients with cancers.

3. Characteristics of Phage-Displayed Peptide Microarrays

In 2004, we have published the first proof-of-principle of high-throughput analysis of humoral immune response in patients with cancer using phage microarrays *(17)*. Briefly, the workflow of phage-based microarray design was the following. First, a random peptide-phage library was biopanned against patient's serum antibodies resulting in the selection of a large number of antibody-binding phages. Second, some of the immunoselected phage clones were spotted onto glass slides and hybridized with either fluorescence-labeled anti-phage monoclonal antibody or patient serum antibodies. Subsequent to washing, the slides were incubated with fluorescence conjugated anti-human IgG antibodies. Under our experimental conditions, the peptides expressed on phage surface retained their ability to bind serum antibodies after immobilization. The phage microarrays were screened against either cancer or healthy donor sera in order to test whether they can discriminate between nine patients and ten normal individuals. With exception of one patient, all cancer patients were grouped in one single cluster that is different that of normal individuals *(17)*.

Compared with random peptide-phage libraries, phage displaying cDNA repertoires from cells of interest may have an advantage for the identification of the parental antigens because they represent naturally expressed antigens. However, the cloning of full cDNAs as fusions with phage coat proteins was hindered by the presence of stop codons in the 3′ untranslated regions of mRNAs. Using the cloning vectors developed by Jespers et al. *(27)*, we have expressed the full cDNA repertoires from cancer cell lines as fusion proteins with the C terminus of pVI phage protein *(26)*. The biopanning of these libraries on serum antibodies from patients with breast cancer identified several antibody-reacting clones *(26)*. Subsequent to our study, other groups have also developed cDNA expression libraries. It should be noted that various cDNA libraries are now commercially available *(18–20, 22, 28)*.

4. Immunnoselection of Phage-Displayed Peptides or Proteins

Subsequent to the construction of random peptide-phage libraries or phage cDNA libraries, the second step would be the enrichment for phage clones that are recognized by patient serum antibodies. Among the described enrichment steps, two are normally applied *(25)*. First, the phage library was incubated with serum antibody from healthy individuals in order to remove nondisease specific phage clones. Second, the subtracted library is incubated with serum antibodies from patients in order to enrich for disease-specific phage clones. This step thus generates an enriched phage library relevant to a particular disease. After three to five rounds of biopanning on patient serum antibodies, high affinity-binding peptides or proteins are selected. Such immunoselected phage clones can be spotted on a diagnostic phage microarray. However, one should consider that the enrichment steps might reduce the diversity of the phage library by removing phage clones that did not have binding partners in the selected patient serum or serum pool. This potential problem can be overcome by spotting a large number of clones (e.g., 1,000–5,000) from the first round of biopanning on glass slides that will be screened in high-throughput format with a large panel of patients and normal sera. After normalization and statistical analysis, the most relevant phage clones should be included in a diagnostic microarray for further validation with serum from patients that are not included in the first screening steps. To date, only IgG antibody fraction was used for biopanning and screen using phage-displayed peptide microarrays. However, in other human diseases such as allergy one may employ anti-human IgE for enrichment and screening on the arrays.

5. Phage-Displayed Peptide Microspots

Contrary to DNA microarrays where microspots in the array are unique homogeneous oligonucleotides, the microspots of the phage-based microarray display a relatively constant amount of both phage capsid proteins, and the patient B-cell epitopes encoded by the phage genome. This allows two types of hybridizations, first, with antibodies specific for the wild- type phage capsid proteins and second with patient antibodies which will bind to the phage-displayed peptides. The advantage of including two hybridizations is to have accurate signal normalization against the total spotted phage clones. For proof of principle, we have used one array from the array printed batch to normalize against phage concentrations *(17)* (*see* **Fig. 1a**). In the recent studies, however, this was done for each array by using two different fluorescence labeling *(19, 20)* (*see* **Fig. 1b**). In these experiments, it is preferable to use anti-phage monoclonal antibody than polyclonal antibodies in order to reduce cross reactivity. The epitope specific signal readout can be carried out by allowing the microspots onto microarray to react with patient serum antibodies, followed by incubation with fluorescence labeled anti-human IgG (*see* **Fig. 1**).

6. Practical Considerations and Data Analysis

As a consequence of the use of immunoselected phage clones, one would expect positive signals with patient serum antibodies but not with serum antibodies from normal individuals. We have found that even when all analyzed peptides were clustered prior to the significance analysis, two major cluster groups

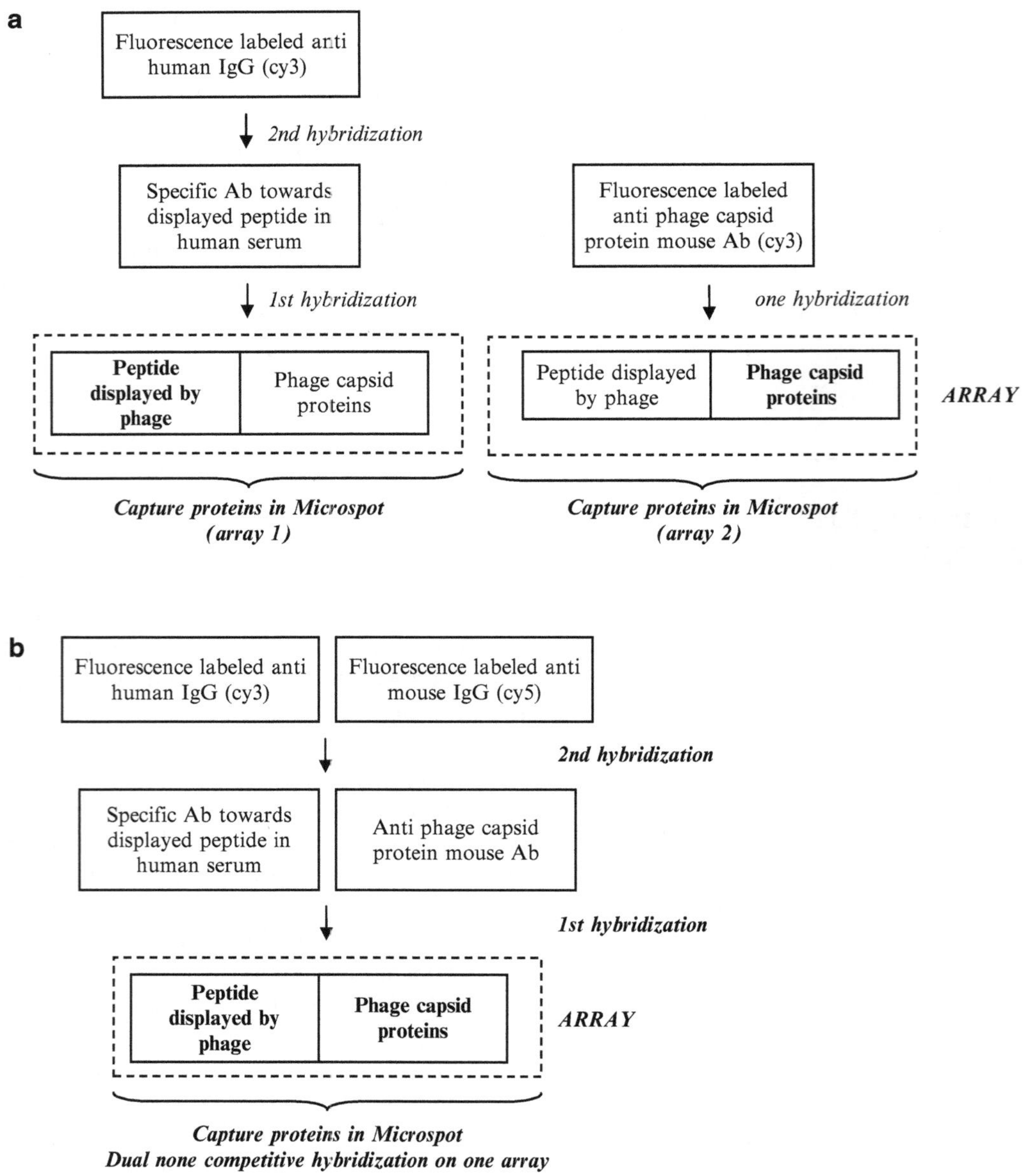

Fig. 1. (**a**) The epitope specific signal readout is carried out by allowing the microspots on microarray to react with patient serum antibodies, followed by incubation with fluorescence labeled anti-human IgG (*array 1*) and one array from the array printed batch is hybridized to specific Ab towards phage capsid protein (*array 2*) and used to normalize against phage concentration. (**b**) Both types of hybridizations are performed on the same array by using two different fluorescence labeled antibodies.

were distinguished: A disease-related cluster, where patients are grouped, and a normal-like cluster, where healthy donors are mainly grouped *(17)*. The accuracy of distinguishing between patients and normal depends on the specificity and diversity of phage-displayed peptides. Controls may include empty phages, additional anti-phage array hybridization or implementation of a dual noncompetitive hybridization where anti-phage array hybridization is performed in parallel to allow standardization and normalization of the raw data. For this purpose, different analytical approaches have been used. For one channel data analysis, a multi variance ANOVA model was applied *(17)*, while for dual channel data ratios of peptide/phage capsid signal were calculated followed by statistical analysis of differences between patients and healthy donor groups using for example, Student t test *(19, 20, 28)*.

To assess whether autoantibody signatures can be used for detection of disease, samples were randomly assigned into two groups, where one group was used for training purposes and the other for validation. Both logistic regression *(29)* and receiver operating characteristics (ROC) *(30)* have been implemented to evaluate the sensitivity and specificity for predictive power *(19, 28)*.

7. Verification Methods of Phage-Displayed B Cell epitopes

As for DNA microarray identified target genes, immunoselected phage clones need to be validated by additional methods. Among the validation techniques, ELISA and immunoblotting techniques are widely used. If the identified B-cell epitopes are not known, sequence analysis of the DNA inserts should provide useful information about their nature, although such information is not critical for use in diagnostic and/or prognostic assays. In the case of random peptide libraries, parental antigens can be predicted from sequence searches against proteins sequences deposed in GenBank database. Given the degenerate nature of the genetic code, homology search with the DNA sequence-encoding peptides are not useful. However, in the case of phage cDNA libraries, BLASTN and BLASTX search tolls based on nucleotide matches (bit score, e-value, and percent sequence match) are expected to reveal an important information about the parental antigens that should be confirmed.

8. Disease Signatures Imprinted in Patient Sera

With the significant advances in proteomics technologies, protein biomarker discovery has become one of the central applications of proteomics. A biomarker could be defined as an identified protein or group of proteins, which change in concentration or structural composition due to a particular disease state. Although tissue biopsies could be seen as the ideal specimen for disease biomarker study, in terms of disease diagnosis and prognosis utility, human body fluids (e.g., blood, urine, or saliva) appear to be more attractive. This is because body fluid testing provides several key advantages, including low invasiveness, minimum cost, and easy sample collection and processing *(31)*. Human plasma proteins originate from a variety of tissues and blood cells as a result of secretion or leakage. Numerous biomedical studies have demonstrated that plasma protein levels reflect human physiological or pathological states and can be used for disease diagnosis and prognosis *(32)*. A critical issue is the complexity of the plasma proteome. Plasma/serum contains a huge number of proteins, differing with the extraordinary dynamic range of at least 9–10 orders of magnitude *(33)*. The IgG fraction is one of the most abundant protein in plasma, and thus is likely to be an ultimate source of diagnostic markers for many human diseases. Detection of particular antibodies is a very common form of medical diagnosis, e.g., a titer of antibodies directed against a particular tumor marker is estimated from the blood. However, immune responses to a single protein or peptide are not expected in the majority of the cancer patients. It is therefore necessary to measure the antibody levels for multiple antigens or B cell epitopes. The phage microarrays can probe serum antibody levels against a large number of B-cell epitopes in a high-throughput format.

Recent studies demonstrated that phage microarrays are useful in monitoring immune responses in various cancer types including breast, non–small-cell lung, prostate, and ovarian cancers. An overview of literature on phage-based microarrays using immunoselected phage-displayed B-cell epitopes is summarized in **Table 1**. As shown, a variety of cDNA libraries also has been constructed from tumor tissues or cell lines. The biopanning of these libraries on serum antibodies derived from cancer patients and/or normal individuals identified cancer-specific B-cell signatures. Interestingly, when advanced-stage non–small-cell lung cancer patients were compared with high-risk control subjects, a diagnostic accuracy of 88.9% was obtained by using selected phage-displayed peptides *(19)*. Notably, this prediction signature was found better than the currently reported prediction values of the clinically available markers for non–small-cell lung cancer patients, such as tissue polypeptide antigen, 80%; CA19–9, 62%; carcinoembryonic antigen, 73%; squamous cell carcinoma antigen, 62%; and neuron-specific enolase, 63% (*28*). In a second study, prostate cancer-specific B-cell epitopes were identified *(18)*. Using a panel of 22 peptides, the data indicate that it is possible to detect prostate cancer with a specificity of 88.2% and a sensitivity of 81.6%. These results were significantly better than PSA test *(18)*

Table 1
Overview of phage-based microarrays using immunoselected phage-displayed B-cell epitopes[a]

Cancer type	Clone number on array	Clone number on diagnostic chip	Phage expression system	Biopanning	Serum samples	Selected cancer-specific peptides	Predictive proteins of phage-displayed peptides	Reference
Breast cancer	36		M13 random nucleotide library	5 rounds using 1 breast cancer patient with stage IV	19 (9 breast cancer patients, 10 healthy female donors)	28		*(14)*
Prostate cancer	2,304	186	T7 cDNA library from 6 prostate cancer tissues	10 normal serum, 19 patients localized prostate cancer	20 patients with prostate cancer	22		*(15, 19)*
					11 healthy male donors			
					Analyzed with diagnostic chip-			
					119 prostate cancer, 138 healthy male donors			
	22				14 patients with prostatectomy, 11 patients hormone refractory prostate cancer, 30 patients with long cancer (specificity validation)			
Non—small-cell lung cancer	4,000	120	T7 NSCLC cDNA library (novagen)	4 rounds using 5 samples, NSCLC, stages 2–4	50 plasma samples (10 used for selection, 40 used for analysis	17	GAGE7, EEF1A, PMS2L15, NOPP140, SEC15L2, RP11–499F19, paxillin	*(16)*

(continued)

Table 1 (continued)

Cancer type	Clone number on array	Clone number on diagnostic chip	Phage expression system	Biopanning	Serum samples	Selected cancer-specific peptides	Predictive proteins of phage-displayed peptides	Reference
Ovarian	480	65	T7 SKOV3 ovarian cell line cDNA library	1 late stage cancer patient (4 rounds)	129 (70 ovarian cancer, 10 bening gynecological diseases, 4 endometrial cancer, 25 healthy donors)	45 stage 1–4	RCAS1, eIF-5A, nibrin	*(26)*
						17 late stage		
Lung adenocarcinoma	2304		T7 cDNA library from 7 lung cancer tissues (stage I-III)	10 lung cancer serum (stage I-III)	250 (150 lung adenocarcinoma, 100 normal donor)	22	Ubiquilin	*(17)*
				3 none cancer serum				
	1,129				122 (62 lung adenocarcinoma, 60 normal donor) (validation set)			

9. Concluding Remarks

Successful adaptation of microarray assays to high-throughput protein screening by array approach requires large panels of purified proteins, e.g., antibodies to be used as capture molecules. The combination of phage display and microarray high-throughput method has helped the analysis of multiple biomarkers. However, the use of phage-expressed proteins has some drawbacks, in that peptides or tumor antigens expressed in a prokaryotic system would not be expected to include post-translationally modified proteins. Modifications such as protein glycosylations that are altered in tumor cells might trigger immune response. This is the main reason that the peptide approach will not generate a comprehensive proteomic profile, although the intrinsic ability of this technology to potentially identify a variety of aberrantly expressed tumor-associated proteins inducing autoimmune responses is promising.

Acknowledgment

The Authors thank Dr. Anne Dybwad for critical reading of the manuscript. This work is supported in part by the Norwegian Cancer Society to M. Sioud.

References

1. Miller, J. C., Zhou, H., Kwekel, J., Cavallo, R., Burke, J., Butler, E. B., Teh, B. S., and Haab, B. B. (2003) Antibody microarray profiling of human prostate cancer sera: antibody screening and identification of potential biomarkers. *Proteomics* 3, 56–63.
2. Belov, L., de la Vega, O., dos Remedios, C. G., Mulligan, S. P., and Christopherson, R. I. (2001) Immunophenotyping of leukemias using a cluster of differentiation antibody microarray. *Cancer Res.* 61, 4483–4489.
3. Usui-Aoki, K., Shimada, K., Nagano, M., Kawai, M., and Koga, H. (2005) A novel approach to protein expression profiling using antibody microarrays combined with surface plasmon resonance technology. *Proteomics* 5, 2396–2401.
4. Wingren, C., Steinhauer, C., Ingvarsson, J., Persson, E., Larsson, K., and Borrebaeck, C. A. (2005) Microarrays based on affinity-tagged single-chain Fv antibodies: sensitive detection of analyte in complex proteomes. *Proteomics* 5, 1281–1291.
5. Ghobrial, I. M., McCormick, D. J., Kaufmann, S. H., Leontovich, A. A., Loegering, D. A., Dai, N. T., Krajnik, K. L., Stenson, M. J., Melhem, M. F., Novak, A. J., Ansell, S. M, and Witzig, T. E (2005) Proteomic analysis of mantle-cell lymphoma by protein microarray. *Blood* 105, 3722–3730.
6. Hesselberth, J. R., Miller, J. P., Golob, A., Stajich, J. E., Michaud, G. A., and Fields, S. (2006) Comparative analysis of Saccharomyces cerevisiae WW domains and their interacting proteins. *Genome Biol.* 7, R30.
7. Rung, J., Schlitt, T., Brazma, A., Freivalds, K., and Vilo, J. (2002) Building and analysing genome-wide gene disruption networks. *Bioinformatics* 18 Suppl 2, S202–S210.
8. Spaller, M. R. (2006) Act globally, think locally: systems biology addresses the PDZ domain. *ACS Chem. Biol.* 1, 207–210.

9. Li, Y., Lee, H. J., and Corn, R. M. (2006) Fabrication and characterization of RNA aptamer microarrays for the study of protein-aptamer interactions with SPR imaging. *Nucleic Acids Res.* 34, 6416–6424.
10. Jones, R. B., Gordus, A., Krall, J. A., and MacBeath, G. (2006) A quantitative protein interaction network for the ErbB receptors using protein microarrays. *Nature* 439, 168–174.
11. Snoek, R., Rennie, P. S., Kasper, S., Matusik, R. J., and Bruchovsky, N. (1996) Induction of cell-free, in vitro transcription by recombinant androgen receptor peptides. *J. Steroid Biochem.* 59, 243–250.
12. Angenendt, P., Kreutzberger, J., Glokler, J., and Hoheisel, J. D. (2006) Generation of high density protein microarrays by cell-free in Situ expression of unpurified PCR products. *Mol. Cell. Proteomics* 5, 1658–1666.
13. Perrin, A., Duracher, D., Perret, M., Cleuziat, P., and Mandrand, B. (2003) A combined oligonucleotide and protein microarray for the codetection of nucleic acids and antibodies associated with human immunodeficiency virus, hepatitis B virus, and hepatitis C virus infections. *Anal. Biochem.* 322, 148–155.
14. Dybwad, A., Forre, O., Kjeldsen-Kragh, J., Natvig, J. B., and Sioud, M. (1993) Identification of new B-cell epitopes in the sera of rheumatoid arthritis patients using a random nanopeptide-phage library. *Eur. J. Immunol.* 23, 3189–3193.
15. Hansen, M. H., Ostenstad, B., and Sioud, M. (2001) Antigen-specific IgG antibodies in stage IV long-time survival breast cancer patients. *Mol. Med.* 7, 230–239.
16. Sioud, M., and Hansen, M. H. (2001) Profiling the immune response in patients with breast cancer by phage-displayed cDNA libraries. *Eur. J. Immunol.* 31, 716–725.
17. Cekaite, L., Haug, O., Myklebost, O., Aldrin, M., Ostenstad, B., Holden, M., Frigessi, A., Hovig, E., and Sioud, M. (2004) Analysis of the humoral immune response to immunoselected phage-displayed peptides by a microarray-based method. *Proteomics* 4, 2572–2582.
18. Bradford, T. J., Wang, X., and Chinnaiyan, A. M. (2006) Cancer immunomics: using autoantibody signatures in the early detection of prostate cancer. *Urol. Oncol.* 24, 237–242.
19. Zhong, L., Hidalgo, G. E., Stromberg, A. J., Khattar, N. H., Jett, J. R., and Hirschowitz, E. A. (2005) Using protein microarray as a diagnostic assay for non–small-cell lung cancer. *Am. J. Respir. Crit. Care* 172, 1308–1314.
20. Chen, G., Wang, X., Yu, J., Varambally, S., Yu, J., Thomas, D. G., Lin, M. Y., Vishnu, P., Wang, Z., Wang, R., Fielhauer, J., Ghosh, D., Giordano, T. J., Giacherio, D., Chang, A. C., Orringer, M. B., El-Hefnawy, T., Bigbee, W. L., Beer, D. G., and Chinnaiyan, A. M. (2007) Autoantibody profiles reveal ubiquilin 1 as a humoral immune response target in lung adenocarcinoma. *Cancer Res.* 67, 3461–3467.
21. Bradford, T. J., Wang, X., and Chinnaiyan, A. M. (2006) Cancer immunomics: using autoantibody signatures in the early detection of prostate cancer. *Urol. Oncol. Sem. Origin. Invest.* 24, 237–242.
22. Wang, X., Yu, J., Sreekumar, A., Varambally, S., Shen, R., Giacherio, D., Mehra, R., Montie, J. E., Pienta, K. J., Sanda, M. G., Kantoff, P. W., Rubin, M. A., Wei, J. T., Ghosh, D., and Chinnaiyan, A. M. (2005) Autoantibody signatures in prostate cancer. *N. Engl. J. Med.* 353, 1224–1235.
23. Nowak, J. E., Chatterjee, M., Mohapatra, S., Dryden, S. C., and Tainsky, M. A. (2006) Direct production and purification of T7 phage display cloned proteins selected and analyzed on microarrays. *Biotechniques* 40, 220–227.
24. Smith, G. P. (1985) Filamentous fusion phage: novel expression vectors that display cloned antigens on the virion surface. *Science* 228, 1315–1317.
25. Sioud, M., Forre, O., and Dybwad, A. (1996) Selection of ligands for polyclonal antibodies from random peptide libraries: potential identification of (auto)antigens that may trigger B and T cell responses in autoimmune diseases. *Clin. Immunol. Immunopathol.* 79, 105–114.
26. Sioud, M., Hansen, M., and Dybwad, A. (2000) Profiling the immune responses in patient sera with peptide and cDNA display libraries. *Int. J. Mol. Med.* 6, 123–128.
27. Jespers, L. S., Messens, J. H., De Keyser, A., Eeckhout, D., Van den Brande, I., Gansemans, Y. G., Lauwereys, M. J., Vlasuk, G. P., and Stanssens, P. E. (1995) Surface expression and ligand-based selection of cDNAs fused to filamentous phage gene VI. *Biotechnology (N Y)* 13, 378–382.
28. Chatterjee, M., Mohapatra, S., Ionan, A., Bawa, G., Ali-Fehmi, R., Wang, X., Nowak, J., Ye, B., Nahhas, F. A., Lu, K., Witkin, S. S., Fishman, D., Munkarah, A., Morris, R., Levin, N. K., Shirley, N. N., Tromp, G., Abrams, J., Draghici, S., and Tainsky, M. A. (2006) Diagnostic markers of ovarian cancer by high-throughput antigen cloning and detection on arrays. *Cancer Res.* 66, 1181–1190.
29. Eilers, P. H., Boer, J. M., Van Ommen, G. J. B., and Van Houwelingen, H. C. (2001) Classification of microarray data with penalized logistic regression. *Proc. Soc. Photo Opt. Instrum. Eng.* 4266, 187–198.

30. Zweig, M. H. and Campbell, G. (1993) Receiver-operating characteristic (ROC) plots: a fundamental evaluation tool in clinical medicine. *Clin. Chem.* 39, 561–577.

31. Hu, S., Loo, J. A., and Wong, D. T. (2006) Human body fluid proteome analysis. *Proteomics* 6, 6326–6353.

32. Thadikkaran, L., Siegenthaler, M. A., Crettaz, D. Queloz, P. A., Schneider, P., and Tissot, J. D. (2005) Recent advances in blood-related proteomics. *Proteomics* 5, 3019–3034.

33. Anderson, L. (2005) Candidate-based proteomics in the search for biomarkers of cardiovascular disease. *J. Physiol.* 563, 23–60.

Subsection D

Antigen Expressed from Recombinant DNA

Chapter 21

Epitope Mapping Using Homolog-Scanning Mutagenesis

Lin-Fa Wang

Summary

With the advance of whole genome sequencing for an increasing number of organisms, it becomes clear that many proteins exist in multiple forms whose overall structures are similar despite subtle sequence and functional differences. Although the biological significance may not be known for some homologs in a gene family, they nevertheless provide a useful tool for mapping of functional domains or sites involved in inter-molecular interactions. This is also true for epitope mapping. Determination of antibody-binding sites using serial chimeric proteins of different homologs, a technique termed *homolog-scanning mutagenesis* (HSM), has proven to be especially useful in mapping conformational epitopes.

Key words: Homologous proteins, Chimeric protein, PCR, Mutagenesis, Conformational epitope.

1. Introduction

The high specificity of monoclonal antibodies (MAbs) enables them to discriminate subtle sequence and structural differences among homologous proteins. MAbs have thus become an important tool for serotyping of viruses and bacteria *(1–3)*, and for studying protein homologs from multigene families, such as subtypes of different human interferon molecules *(4)* and different immunological markers *(5)*.

With the advances in whole genome sequencing, more homologous proteins are being discovered which share extensive structural and functional similarities *(6)*. Although the biological implications for these homologs may not be easy to define, it is often possible to detect their subtle structural differences using MAbs. These observations have led to the development of the

Ulrich Reineke and Mike Schutkowski (eds.), *Methods in Molecular Biology, Epitope Mapping Protocols, vol. 524*

DOI: 10.1007/978-1-59745-450-6_21

homolog-scanning strategy *(7, 8)* which is useful in identifying sequences that cause functional variation among homologous proteins, such as enzyme activity and epitope antigenicity.

Epitope mapping by homolog-scanning mutagenesis (HSM) is based on systematic or random replacement of sequence segments in a MAb-binding homolog by cognate sequence segments from homologs known to be a nonbinder or vice versa. By testing the binding ability of the recombinant hybrid mutant protein and comparing its sequence with those of the parent molecules, it is possible to identify the sequence segment or amino acid residues essential for MAb binding. Although HSM can, in theory, be used in mapping linear epitopes, it is designed mainly for mapping conformational epitopes which are difficult to map by other means (*see* Chapter "What Is a B-Cell Epitope?"). Basically, epitope mapping by HSM involves three steps: (1) construction of hybrid genes; (2) expression of recombinant proteins either in vivo or in vitro; and (3) analysis of these recombinant proteins for MAb binding.

There are several approaches reported for constructing hybrid genes either by in vitro gene manipulation *(9–11)* or by in vivo homologous recombination *(12, 13)*. The method to be described in this chapter is a PCR-based approach, termed *template-coupled PCR (8)*, which is convenient to operate and very efficient for creating hybrid genes with either defined or random cross-over points. As shown in **Fig. 1**, this approach uses two homologous genes cloned in two different vectors: one coding for a protein reactive with the MAb of interest and the other nonreactive. Of the two vectors, one is an expression vector suitable for production of recombinant hybrid proteins to be used in MAb-binding studies. The pET vector system *(14)* is ideal for this purpose, because the T7 RNA polymerase-directed expression makes it possible to produce proteins both in vivo and in vitro. The other vector is usually a pUC-based plasmid *(15)* or any general cloning vector frequently used in the laboratory. Four "universal primers" (**Ef, Er, Cf, Cr**) common to the two vectors are synthesized, which anneal to the regions flanking the cloning sites in each vector. After cutting one template (I) with restriction enzyme X (or cutting template II with Y), the other template is added together with either pair of the primers (**Ef+Cr** or **Cf+Er**). The whole mixture is then subjected to a standard PCR amplification to obtain the hybrid gene. In the first cycle of PCR, the coupling template (template I digested with restriction enzyme X) actually functions as an "elongated primer" to form the first hybrid molecule, which is further amplified in successive cycles by the flanking PCR primers. The final configuration of the hybrid gene is determined by the pair of primers used as well as by the nature of the coupling template(s).

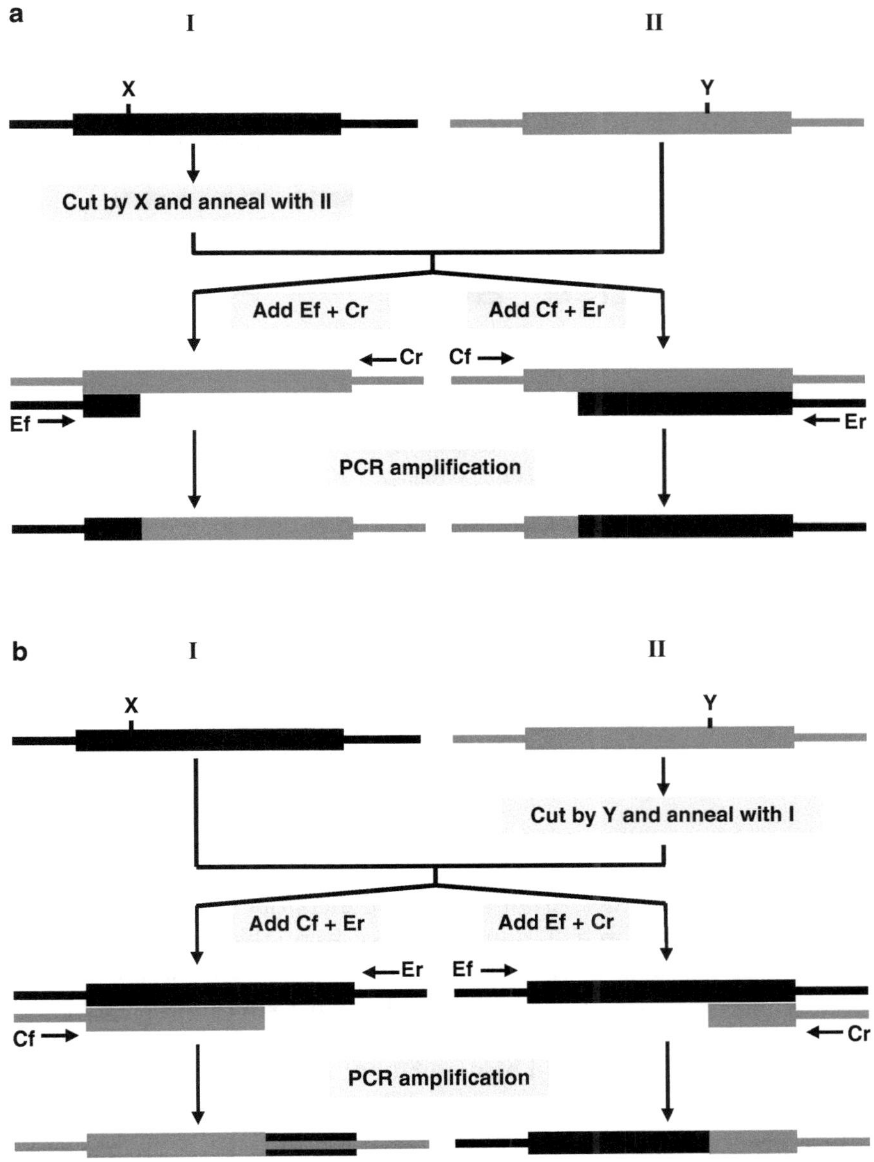

Fig. 1. Generation of homolog hybrid genes by a single step template-coupled PCR amplification. The coupling templates shown here are produced by digestion with restriction enzymes. The black and grey bars represent two templates I and II (i.e., two homologous genes cloned in two different vectors). Vector specific primers are labelled as Ef, Er, Cf, and Cr (E, expression vector; C, cloning vector; f, forward primer; r, reverse primer), respectively. X and Y are restriction enzyme cleavage sites located within the gene coding regions. (**a**) Generation of hybrid genes using template I (cut with X) as coupling templates. (**b**) Generation of hybrid genes using template II (cut with Y) as coupling templates.

Fig. 2 illustrates an example of mapping a conformational epitope using HSM. Monoclonal antibody I-4-A reacts with human interferon-α4a (IFN-α4a), but not with IFN-α14 *(8, 16)*. Although I-4-A reacted with the intact IFN-α4a molecule in Western blotting, it failed to react with truncated recombinant

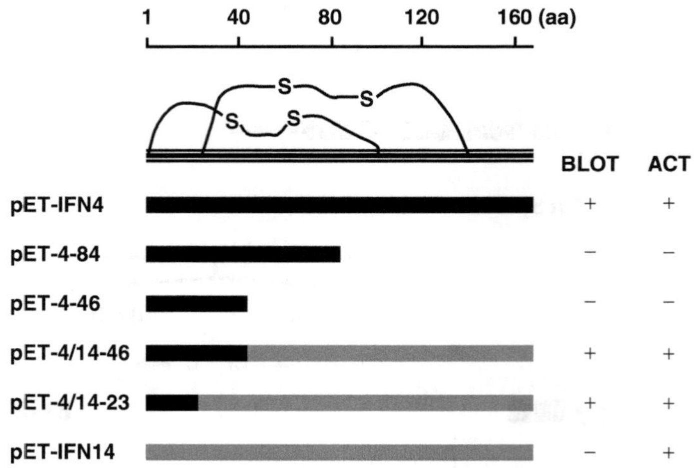

Fig. 2. Antigenic and biological properties of recombinant INF proteins expressed from the pET system. The numbers at the top are amino acid (aa) residue numbers of the IFN-α protein with the diagram underneath representing the two disulfide bonds between residues 1 and 99, and 29 and 139, respectively. At the right is a summary of the results for Western blotting (BLOT) and antiviral activity (ACT): "+," positive result; "–," negative result. The black and grey bars represent the coding sequences for IFN-α4a and IFN-α14, respectively (*see* ref. 8 for more details).

polypeptides expressed from the cloned IFN-α4a gene. This prevented us from mapping the epitope using the serial truncation approach. Using HSM, we were able to show that the critical MAb-binding site is located within the N-terminal 23 aa residues *(8)*. Mapping of conformational epitopes using HSM has also been reported by other groups *(17, 18)*.

It should be pointed out that the strategy outlined in **Fig. 1** is not the only method for generating the coupling template. It is also possible to use PCR and an internal primer to produce a coupling template using any of the internal amino acid residues as a cross-over point *(8)*. One can also generate coupling templates with random cross-over points along the molecule by using partial digestion with enzymes such as DNase I, exonuclease III or Bal 31, thus creating a library of random hybrid molecules. Furthermore, if the hybrid genes can be expressed and displayed on filamentous phage surface as described *(19, 20)* (*see* Chapter "Epitope Mapping Using Phage-Display Random Fragment Libraries"), it will be possible to construct a phage display random library of hybrid molecules. The phage display expression will make the screening process much more efficient and will also make it possible to examine a large number of hybrid molecules simultaneously.

2. Materials

All reagents should be of AR grade. All solutions and buffers should be autoclaved or filter-sterilized where appropriate. Sterile tubes and filter tips should be used. Unless otherwise stated, all molecular biology reagents are obtained from Promega (Madison, WI) and all chemicals from Sigma (St. Louis, MO).

2.1. General

1. Antibodies: MAb(s) of interest, and alkaline phosphatase (AP)- and horse radish peroxidase (HRP)-conjugated anti-mouse antibodies.
2. Recombinant genes: At least two cloned homolog genes are required for this mapping approach. It is preferable to engineer the two homolog genes to have the same cloning sites at each end so that they can be conveniently moved from one vector to another.
3. Bacterial strains and plasmids: *E.coli* BL21[DE3] *(14)*, vectors pET-3a *(14)*, and pUC18 *(15)*.
4. Oligonucleotide primers: (a) flanking primers for pUC18, USP (5′ GTA AAA CGA CGG CCA GT 3′) and RSP (5′AAC AGC TAT GAC CAT G 3′); (b) flanking primers for pET-3a, ET5 (5′ CCT CTA GAA ATA ATT TTG TTT 3′) and ET3 (5′ CAG CCA ACT AAG CTT CCT TTC 3′).
5. Equipment: Power supply, horizontal agarose gel electrophoresis tank, Mini-PROTEIN II SDS-PAGE system, Mini Trans-Blot Module, gel dryer, and Gene Pulser for eletroporation were all purchased from Bio-Rad (Hercules, CA).
6. Image capture device or dark room facilities for photography and X-ray film development.
7. Rocker and orbital shaker.

2.2. Generation of Hybrid Genes

1. Appropriate restriction enzymes.
2. PCR reagents, use as recommended by supplier. It is essential to use high fidelity DNA polymerases to reduce PCR-introduced error.
3. PCR machine: Perkin Elmer GeneAmp 2400 Thermal Cycler or any thermal cycler.
4. QIAquick PCR Purification Kit and QIAquick Gel Extraction Kit from Qiagen (Hilden, Germany).
5. T4 DNA ligase and ligation buffer, use as recommended.
6. Competent cells: Electro-competent cells are prepared according to the method provided with the Bio-Rad Gene Pulser. Aliquots of 40 μL are quickly frozen in liquid nitrogen and kept at –80°C until use.

7. SOC medium: 0.5% yeast extract, 2% bactotryptone, 10 mM NaCl, 2.5 mM KCl, 10 mM $MgCl_2$, 20 mM $MgSO_4$, 20 mM glucose (glucose is prepared separately as a 2 M stock and added just before use).
8. LB medium: 1% bactotryptone, 0.5% yeast extract, 1% NaCl.
9. LB/Amp plates: LB medium containing 1.5% agar, autoclave to sterile. Cool to 50°C before adding ampicillin, from a 50 mg/mL stock, to a final concentration of 50 μg/mL. Pour approximately 20 mL per 90 mm plate.
10. QIAprep Spin Minipre Kit from Qiagen (Hilden, Germany).

2.3. Production of Recombinant Protein Molecules

2.3.1. Expression in *E. coli*

1. Isopropyl-β-D-thiogalactoside (IPTG): 100 mM stock solution in water, kept at –20°C for up to 6 months.
2. MTPBS buffer: 150 mM NaCl, 16 mM Na_2HPO_4, 4 mM NaH_2PO_4, pH 7.3.
3. Sonicator: Vibra Cell™ – High Intensity Ultrasonic Processor, 50-Watt Model, from Sonics & Materials, Inc (Danbury, CT).

2.3.2. *In Vitro* Translation

1. ^{35}S-Met, 1 Ci/mmol, ICN (Costa Mesa, CA).
2. TnT® T7 Quick Coupled Transcription/Translation System (Promega).
3. SDS-PAGE reagents: *see* vol. 10, Chapter 24 of this series for detailed recipes.
4. X-ray film: for example, Kodak X-Omat AR5.
5. Autoradiography cassettes and film development solutions.

2.4. Antibody-Binding Assays

2.4.1. ELISA

1. Multi channel pipette, ELISA plates, and microplate shaker: all from Titertek® Flow Laboratories (McLean, VA).
2. Microplate reader: Multiskan® MS (Labsystems, Helsinki, Finland).
3. Coating buffer: 50 mM Tris-HCl, 150 mM NaCl, pH 9.3.
4. PBST: Dilute 10× PBS (per litre: 10.7 g Na_2PO_4, 3.9 g NaH_2PO_4, 80 g NaCl, pH 7.2) to 1× with distilled water and add Tween-20 to a final concentration of 0.05% (v/v). Store at room temperature for up to 6 months.
5. Blocking solution: PBST containing 2% skimmed milk powder, prepare fresh.
6. Citrate acetate buffer: Make up 100 mL of 1 M sodium acetate and 10 mL of 1 M citric acid. Adjust the sodium acetate solution to pH 5.9 with approximately 1.5 mL of the citric acid.

7. TMB substrate: Dissolve 100 mg of 3,3,5,5,-tetramethylbenzidine (Sigma, St. Louis, MO) in 10 mL dimethyl sulfoxide (DMSO) to make a 42 mM solution. Store at 4°C in small aliquots (0.5 mL) for up to 12 months. Pre-warm at 37°C for 10 min before use.
8. Substrate solution: Prepare freshly by mixing 18 mL of distilled water with 2 mL of citrate acetate buffer and 0.2 mL of the TMB substrate. Add 2.5 μL 30% H_2O_2 just before use.
9. Stopping solution: 1 M H_2SO_4.

2.4.2. Western Blotting

1. Nitrocellulose membrane: 0.45 μm, Schleicher & Schuell (Dassel, Germany).
2. Whatman 3MM filter paper (Whatman, Maidstone, UK).
3. Plastic bag and heat sealer.
4. Container with flat bottom (e.g., square petri dishes).
5. Tris-glycine transfer buffer: Prepare freshly by mixing 100 mL 10× transfer buffer (250 mM Tris/1.92 M glycine, pH 8.3) with 700 mL distilled water, and then with 200 mL methanol.
6. TBST: dilute 10× TBS (per liter: 90 g sodium chloride, 60 g Tris base, adjust pH to 7.9 with HCl) to 1× with distilled water, and add Tween-20 to a final concentration of 0.05% (v/v). Store at room temperature for up to 6 months.
7. Blotto solution: TBST containing 5% skimmed milk powder, prepare fresh.
8. AP substrate buffer: 100 mM Tris-HCl, pH 9.5, 100 mM NaCl, 5 mM $MgCl_2$.
9. AP substrate solutions: 5-bromo-1-chloro-3-indolyl phosphate (BCIP) and nitro blue tetrazolium (NBT).

2.4.3. Magnetic Immuno Capture

1. Streptavidin magnetic beads (SMB): 1 mg/mL suspension (Promega).
2. Magnetic separation stand (Promega).
3. Biotinylated sheep anti-mouse antibodies (Pierce, Rockford, IL).

3. Methods

3.1. Generation of Hybrid Genes

For convenience of discussion, we assume that homolog I (*see* **Fig. 1**) is reactive with the MAb while homolog II is not, and that homolog I is cloned in pET vector with flanking primers

Ef and **Er** whereas homolog II is cloned in pUC with flanking primers **Cf** and **Cr**. Both homolog genes can be excised from the vectors as a BamHI–EcoRI gene cassette. The procedures below are for template I as the coupling template (*see* **Note 1** and **2**).

1. Digest 100 ng pET plasmid containing homolog gene I with 5 units of restriction enzyme X (*see* **Note 3**) in 20 μL reaction mixture. Incubate at 37°C for 60 min, and then at 65°C for 15 min (*see* **Note 4**).
2. Take 2 μL of the digested template I, mix with 5 ng of undigested template II plasmid DNA, and adjust the volume to 20 μL with water. Boil the mixture for 2 min followed by rapid cooling on ice.
3. Set up two PCR reaction mixtures with the following components:

   PCR-1: 50 pmol each of primers **Ef** and **Cr**, 1 μL of the denatured template mixture prepared as described above in **step 2**, 10 μL 10× PCR buffer, 10 μL 25 mM $MgCl_2$, 16 μL 1.25 mM dNTPs, 2.5 units of polymerase. Adjust the volume to 100 μL with water.

   PCR-2: same as above except that primers **Cf** and **Er** are used instead.

   PCR reactions are carried out for 25 cycles at 94°C/1 min, 50°C/2 min, and 72°C/2 min. (The PCR product from PCR-1 is named I/II-X while that from PCR-2 is named II/I-X to distinguish the two different hybrid molecules obtained as shown in **Fig. 1**. For the remaining steps, the two will be treated exactly the same.)
4. Total PCR products are separated on a 1% agarose–TAE gel, and the corresponding PCR band is excised and purified using the QIAquick Gel Extraction Kit following the procedures provided by the supplier. The purified PCR fragment is eluted in 20 μL water.
5. Digest 10 μL of the purified PCR product in a total volume of 20 μL containing 2 μL 10× reaction buffer and 5 units each of BamHI and EcoRI. The reaction mixture is incubated at 37°C for 60 min, followed by heating at 65°C for 15 min.
6. The reaction mixture is purified using the QIAquick PCR Purification Kit, and eluted in 10 μL of water.
7. Ligate 5 μL of the purified product with 50 ng pET-3a vector DNA, which has been digested with BamHI/EcoRI and treated with calf intestine alkaline phosphatase. The ligation is carried out overnight at 16°C in a total volume of 10 μL containing 1 μL 10× ligation buffer and 1 unit of T4 DNA ligase. The ligation reaction is terminated by heating at 65°C for 15 min.

8. Take 1–2 μL of the above ligation mixture to transform *E. coli* strain BL21[DE3] by electroporation using the Bio-Rad Gene Pulser following the supplied instructions. The transformed cell suspension is immediately transferred to a 10 mL culture tube containing 0.5 mL SOC medium, and incubated at 37°C for 60 min with gentle shaking. Aliquots of 50, 100, and 200 μL of cells are plated onto LB/Amp agar plates, followed by overnight incubation at 37°C.
9. Two to four single colonies from each transformation mixture are picked up for plasmid purification using the QIA Spin Miniprep Kit, followed by restriction enzyme digestion to confirm the presence of the expected insert (*see* **Note 5**).

3.2. Production of Recombinant Protein Molecules

Depending on the solubility of the recombinant protein and the quantity of the protein required, one can choose to express the hybrid gene either in *E. coli* or by in vitro translation in the presence of ^{35}S-Met. It is sometimes necessary to try both methods to achieve the optimal results.

3.2.1. Expression in E. coli

1. Pick up a single colony from a fresh plate to inoculate 1 mL LB/Amp medium (i.e., LB containing 50 μg/mL ampicillin), and incubate the culture at 37°C overnight with shaking.
2. Next morning, transfer the 1 mL overnight culture to a 100 mL flask containing 9 mL pre-warmed LB/Amp medium and shake at 37°C for 60 min.
3. Add 100 μL of 100 mM IPTG (final concentration of 1 mM) to induce the expression, and continue the incubation for additional 3 h.
4. Harvest the cells by centrifugation at 5,000 × *g* for 5 min, and resuspend the cell pellet in 0.5 mL of MTPBS buffer. Transfer the cell suspension to a 1.5 mL Eppendorf tube.
5. Lyse the cells by sonication (5 × 30 s) using the output control setting at 50. Hold the tube on ice in a 100-mL beaker during sonication and leave the tube on ice for 1 min between each sonication. Save 100 μL as the "total lysate fraction" (*see* **Note 6**).
6. Spin the remaining lysate for 5 min. Transfer the supernatant to a clean tube and save as the "soluble fraction".
7. Resuspend the pellet in 0.4 mL MBPBS buffer, and save as the "insoluble fraction".
8. Examine the size, solubility and level of expression for the recombinant hybrid proteins by SDS-PAGE following the standard protocols.

3.2.2. In Vitro Translation

1. Isolate plasmid DNA using the QIAprep Spin Miniprep Kit as described above. If necessary, make DNA from 2–4 minipreps and combine the DNA.
2. Use 2 μg of the purified DNA for each in vitro translation reaction using the Promega TnT® T7 Quick Coupled Transcription/Translation System applying the supplied instructions. Determine the total reaction volume by the amount of the DNA used, and calculate using the ratios given in the supplied protocol.
3. Take 5 μL or 5% of the total reaction mixture to examine the translation efficiency by SDS-PAGE, followed by autoradiography (*see* **Note 7**).

3.3. Antibody-Binding Assays

Depending on the nature of the epitope(s) to be studied and the solubility of the expressed recombinant proteins, one may have to try different assays to optimize the detection of antibody binding. Below are three of the most frequently used assays for monitoring antibody binding. The magnetic immuno capture assay is designed for use with labelled proteins produced by in vitro translation. ELISA is more suitable for detection of soluble proteins produced in *E. coli* while Western blotting can be used with any form of the expressed proteins described above.

3.3.1. ELISA

All incubations, except for substrate development, are carried out at 37°C with gentle shaking on a microplate shaker.

1. Use the soluble fraction produced in *E. coli* (*see* **Subheading 3.2.1**) to make a twofold serial dilution in coating buffer from 1:20 to 1:2,560 (*see* **Note 8**).
2. Use 50 μL each of the diluted solutions to coat an ELISA plate in triplicates. Incubate the plate for 60 min with gentle shaking.
3. Wash the plate three times (5 min each) with PBST.
4. Add 100 μL blocking solution to each of the wells and incubate for 30 min. Discard the solution after incubation.
5. Add 50 μL MAb solution, diluted in blocking solution at 1:10 for tissue culture or 1:100–1:1,000 for ascitic fluid (*see* **Note 9**), followed by incubation for 60 min.
6. Wash as in **step 3**.
7. Add 50 μL HRP-conjugated sheep anti-mouse IgG diluted in blocking solution at 1:2,000, followed by incubation for 60 min.
8. Wash as in **step 3**.
9. Add 50 μL substrate solution and incubate at room temperature for 10 min.

10. Stop the reaction by adding 50 μL stopping solution.
11. Read the absorbance at 450 nm.

3.3.2. Western Blotting

All incubations are carried out at room temperature.

1. Prepare denatured protein samples for loading by mixing equal volumes of 2× SDS-PAGE sample loading buffer and protein samples a concentration of approximately 3–4 mg/mL (*see* **Note 10**).
2. Separate the protein samples by SDS-PAGE using an appropriate gel concentration (*see* vol. 32 of this series for detailed instructions).
3. Carry out electrophoresis at constant voltage of 200 V until the front dye reaches the bottom of the gel. This usually takes about 45 min when the Bio-Rad Mini PPROTEIN II apparatus is used.
4. Remove the transfer cassette assembly, separate the glass plates, and cut off the stacking gel. Assemble the filter paper/gel/membrane/filter paper "sandwich" for electroblotting (*see* vol. 32 of this series).
5. Carry out transfer for 60 min at a constant current of 250 mA.
6. Remove the assembly from the Mini Trans-Blot apparatus, and cut away excess nitrocellulose membrane from around the gel. Carefully peel away the gel from the membrane and place the membrane in a square petri dish containing 30–50 mL blotto solution. Incubate for 15 min with gentle rocking (*see* **Note 11**).
7. Seal the membrane inside a plastic bag with one side open, add 1–2 mL of MAb solution diluted in blotto at 1:5 for hybridoma supernatant or 1:100 for ascitic fluid, and finally seal the remaining side. Incubate the bag for 30 min with gentle rocking (*see* **Note 12**).
8. Wash three times (3–5 min each) with approximately 50 mL TBST.
9. Continue as in **step** 7 except that an AP-conjugated sheep anti-mouse antibody is used at 1:1,000 dilution.
10. Wash as in **step 8**.
11. For color development, incubate the membrane with 10 mL of AP substrate buffer containing 33 μL BCIP and 66 μL NBT, mixed just before use. Dark purple signals should appear within 10–30 min (*see* **Note 13**). Stop color development by washing the membrane in water.

3.3.3. Magnetic Immuno Capture

All incubations are carried out at room temperature.

1. Take 50 μL suspension (1 mg/mL) of streptavidin magnetic beads (SMB), and wash twice with 0.5 mL of the same blotto solution as used in **Subheading 3.3.2**.
2. Resuspend the washed SMB in 100 μL blotto containing 5 μL biotinylated anti-mouse antibodies, and incubate the mixture for 30 min with gentle rocking (*see* **Note 14**).
3. Wash the SMB three times (5 min each) with 0.5 mL TBST.
4. Continue as in **step 2** except that 5 μL MAb solution is used (this can be either 5 μL hybridoma supernatant or 5 μL ascitic fluid at 1:100 dilution).
5. Wash as in **step 3**.
6. Continue as in **step 2** except that 10 μL ^{35}S-labelled protein mixture (produced as described in **Subheading 3.2.2**) is used.
7. Wash as in **step 3**.
8. Resuspend the SMB in 20 μL 1× sample loading buffer and boil for 2 min before taking 5 and 10 μL aliquots for SDS-PAGE analysis.
9. Separate the protein samples using standard SDS-PAGE (e.g., as described in vol. 32 of this series) and transfer the gel onto a pre-cut 3 MM filter paper for drying.
10. Dry the gel under vacuum for 60 min at 60°C.
11. Autoradiography using standard X-ray film and developing methods to reveal the signals on the gel.

4. Notes

1. For simplicity, the protocols presented in this chapter only describe the procedures for the generation of two hybrid genes using the X-digested homolog I as the coupling template. As indicated in **Fig. 1**, the procedures used for the production of the other two hybrid genes using the Y-digested homolog II as coupling template are identical to those presented in this chapter.
2. The template-coupled PCR method presented in this chapter is certainly one of the most efficient strategies available for construction of homolog hybrid genes. However, when there are common restriction enzyme sites present in both of the homolog genes, gene splicing using restriction enzymes may still be a preferred method because the downstream

characterization of the recombinant hybrid genes will be much simpler.

3. Restriction enzyme cleavage site X (or Y for homolog II) need not to be a unique site in the plasmid as long as there is no internal cut between the flanking primer and the desired cross-over point.

4. Most restriction enzymes can be inactivated by heating. However, for certain heat resistant restriction enzymes, it might be necessary to inactivate the enzyme activity by other means.

5. The method of choice for characterizing the recombinant hybrid genes will vary depending upon the sequences of the parental genes and on the nature of the antibody assay method. If there are gene-specific restriction enzyme sites available in the parental molecules, they can be conveniently used in analyzing the hybrid genes (e.g., *see* **ref.** *7*). If such restriction sites are also present in the vector DNA, one may wish to simplify the digestion pattern by carrying out the diagnostic restriction digestion on the insert DNA only. This can be easily achieved by PCR amplification of the insert DNA using two flanking primers, followed by direct digestion of the PCR product using appropriate enzyme(s). We found this approach very efficient because high quality insert DNA can be produced by direct colony PCR (e.g., *see* **ref.** *18*), eliminating the need for plasmid minipreps and generating results within 4–5 h. If there are gene-specific internal primers available from other studies (e.g., primers made during the initial sequencing analysis of the genes), they can be directly used for hybrid gene analysis by PCR amplification using one vector-specific primer (i.e., one of the two flanking vector primers) and one gene-specific primer. Finally, since the hybrid genes are generated by PCR, a full characterization of these genes can only be carried out by DNA sequencing.

6. For some "partially soluble proteins," it might be useful to add Triton in the MTPBNS buffer to a final concentration of 1% (v/v) to increase the solubility during sonication and subsequence centrifugation.

7. If overnight exposure gives an easily visible or strong signal from 5% of totally translated product, the in vitro translation reaction is considered to be successful, and 10 μL of the translation product should be enough for immuno capture assays detailed in **Subheading 3.3.3**. If the signal is invisible or very weak after overnight exposure, it might be worth to repeat the in vitro translation reaction before going to the next step.

8. As the ELISA is carried out using unpurified protein samples, it is essential to include proper controls in this type of assay. It is recommended to use the following three controls in all assays: (a) protein sample from *E. coli* containing vector alone (e.g., pET-3a) as a negative antigen control; (b) protein sample from *E. coli* containing the expression plasmid for the MAb-reactive homolog (e.g., pET-I) as a positive antigen control; and (c) an unrelated MAb as a negative antibody control.
9. The antibody dilution given is only to be used as a general guidance. It is found that certain MAbs give better results when diluted in PBST in the absence of skimmed milk proteins. If the supply of antibody is not a major limiting factor, it is recommended to carry out a titration for the antibody as well as the recombinant antigen to determine the optimal assay conditions.
10. As most conformational epitopes are sensitive to treatment by heat, SDS and/or reducing agents such as β-mercaptothanol and dithiothreitol (DTT), one may wish to try different conditions for sample treatment to increase the chance of epitope detection by Western blotting. One starting point is to take out the reducing agent from the conventional SDS-PAGE sample buffer and not to boil the sample before loading.
11. Membrane left in the blotto solution can be kept at 4°C for up to 48 h without significant impact on the overall performance.
12. We found that a convenient way of keeping the membrane flat is to put the bag in the middle of a thick heavy book (e.g., a telephone directory readily available in every laboratory), which is in turn placed on top of a rocker. The 30 min incubation time is the minimum time required, which can be extended or changed to incubation at 4°C overnight to fit in with other on going experiments. This can also be applied for the incubation with conjugated antibody in **step 10**.
13. It is sometimes necessary to carry out an overnight incubation in a light-protected area for very weak signals to appear. For extremely high sensitivity, one may use a HRP-conjugated secondary antibody and a chemiluminescence substrate (e.g., using the ECL Western Blotting Substrate from Pierce) to reveal weak signals.
14. It should be pointed out that the use of a biotinylated anti-mouse antibody is optional in this assay. It is possible to biotinylate the MAb of interest so that it can be used directly without the bridging antibody.

References

1. Robert-Hebmann, V., Emiliani, S., Resnicoff, M., Jean, F. and Devaux, C. (1992) Subtyping of human immunodeficiency virus isolates with a panel of monoclonal antibodies: Identification of conserved and divergent epitopes on p17 and p25 core proteins. *Mol. Immunol.* 29, 1175–1183.
2. Coulson, B. S. (1996) VP4 and VP7 typing using monoclonal antibodies. *Arch. Virol. Suppl.* 12, 113–118.
3. Gabelish, C., Harbour, C., Beard-Pegler, M. A., Stubbs, E., Steffe, R., Large, M., Vickery, A., and Benn, R. (1991) Serological typing of coagulase-negative *staphylococci* using monoclonal antibodies. *Epidemiol. Infect.* 106, 231–237.
4. Kwok, A. Y. C., Zu, X., Yang, C., Alfa, M. J., and Jay, F. T. (1993) Human interferon-gamma has three domains associated with its antiviral function: A neutralizing epitope typing scheme for human interferon-gamma. *Immunology* 79, 131–137.
5. Tsuchiya, N., Kyogoku, C., Miyashita, R., and Kuroki, K. (2007) Diversity of human immune system multigene families and its implication in the genetic background of rheumatic diseases. *Curr. Med. Chem.* 14, 431–439.
6. Wolf, K. H. and Li, W. H. (2003) Molecular evolution meets the genomics revolution. *Nat. Genet.* 33(Suppl.), 255–265.
7. Cunningham, B. C., Jhurani, P., Ng, P., and Wells, J. A. (1989) Receptor and antibody epitopes in human growth hormone identified by homolog scanning mutagenesis. *Science* 243, 1330–1336.
8. Wang, L., Hertzog, P. J., Galanis, M., Overall, M. L., Waine, G. J., and Linnane, A. W. (1994) Structure-function analysis of human IFN-α. Mapping of a conformational epitope by homolog scanning. *J. Immunol.* 152, 705–715.
9. Wang, L.-F., Scanlon, D. B., Kattenbelt, J. A., Mecham, J. O., and Eaton, B. T. (1994) Fine mapping of a surface-accessible, immunodominant site on the bluetongue virus major core protein VP7. *Virology* 204, 811–814.
10. Horton, R. M., Hunt, H. D., Jo, S. N., Pullen, J. K., and Pease, L. R. (1989) Engineering hybrid genes without the use of restriction enzymes: Gene splicing by overlap extension. *Gene* 77, 61–68.
11. Shigaki, T. and Hirschi, K. D. (2002) Chimeric gene construction without reference to restriction sites. *Biotechniques* 32, 736–740.
12. Caramori, T., Albertini, A. M., and Galizzi, A. (1991) In vivo generation of hybrids between two *Bacillus thuringienesis* insect-toxin-encoding genes. *Gene* 98, 37–44.
13. Gritz, L., Destree, A., Cormier, N., Day, E., Stallard, V., Caiazzo, T., Massara, G., and Panicali, D. (1990) Generation of hybrid genes and proteins by vaccinia virus-mediated recombination: Application to human immunodeficiency virus type 1 env. *J. Virol.* 64, 5948–5957.
14. Studier, F. W., Rosenberg, A. H., Dunn, J., and Dubendorff, J. W. (1990) Use of T7 RNA polymerase to direct expression of cloned genes. *Methods Enzymol.* 185, 60–89.
15. Yanisch-Perron, C., Vieira, J., and Messing, J. (1985) Improved M13 phage cloning vectors and host strains: Nucleotide sequences of the M13mp18 and pUC19 vectors. *Gene* 33, 103–119.
16. Overall, M. L. and Hertzog, P. J. (1991) Functional analysis of interferon-α subtypes using monoclonal antibodies to interferon-α4a: Subtypes reactivity, neutralisation of biological activities and epitope analysis. *Mol. Immunol.* 29, 391–399.
17. Bijnens, A. P., Ngo, T. H., Gils, A., Dewaele, J., Knockaert, I., Stassen, J. M., and Declerck, P. J. (2001) Elucidation of the binding regions of PAI-1 neutralizing antibodies using chimeric variants of human and rat PAI-1. *Thromb. Haemostasis* 85, 866–874.
18. Sogabe, S., Stuart, F., Henke, C., Bridges, A., Williams, G., Birch, A., Winkler, F. K., and Robinson, J. A. (1997) Neutralizing epitopes on the extracellular interferon gamma receptor (IFNgammaR) alpha-chain characterized by homolog scanning mutagenesis and X-ray crystal structure of the A6 fab-IFNgammaR1-108 complex. *J. Mol. Biol.* 273, 882–897.
19. Wang, L.-F., Du Plessis, D. H., White, J. R., Hyatt, A. D., and Eaton, B. T. (1995) Use of a gene-targeted phage display random epitope library to map an antigenic determinant on the bluetongue virus outer capsid protein VP5. *J. Immunol. Methods* 178, 1–12.
20. Wang, L.-F. and Yu, M. (2004) Epitope identification and discovery using phage display libraries: Applications in vaccine development and diagnostics. *Curr. Drug Targets* 5, 1–15.

Chapter 22

Epitope Mapping by Region-Specified PCR-Mutagenesis

Tsutomu Mikawa, Masayuki Ikeda, and Takehiko Shibata

Summary

We will describe a procedure to map epitopes on a protein against monoclonal IgGs. In this procedure, we amplified and mutagenized the entire or a part of the protein. Then, a DNA region encoding the protein was cut out by a restriction enzyme and ligated into a lambda-gt11 expression vector to construct a library. Thus, the protein is expressed as a fusion protein with β-galactosidase. Protein in plaques obtained by phages derived from the library were tested for cross-reactivities by means of immunoblotting experiments.

Key words: Random mutagenesis, Targeted mutagenesis, Point mutation, Base substitution, Multiplex PCR, RecA protein, Primer design.

1. Introduction

This technique was developed for mapping of epitopes of monoclonal antibodies at the amino acid sequence level, i.e., the identification of amino acids of an antigenic protein involved in specific interactions with each antibody. The epitope mapping includes:

1. Construction of a random base substitution DNA Library for the gene encoding the antigenic protein in the lambda-gt11 phage expression vector;
2. Tests for the cross-reactivities of proteins expressed in plaques formed by phages in the library; and
3. An analysis of DNA sequences of the mutant genes encoding proteins with altered cross-reactivities.

Ulrich Reineke and Mike Schutkowski (eds.), *Methods Molecular in Biology, Epitope Mapping Protocols, vol. 524*

DOI: 10.1007/978-1-59745-450-6_22

An efficient tool to generate a mutagenized DNA library is the polymerase chain reaction (PCR) *(1)* under conditions that cause highly increased errors in DNA synthesis due to base substitutions, resulting in amino acid substitution mutations (*see* **Note 1**). We added deoxyinosine 5′-triphosphate (dITP) to the reaction mixture of the PCR to increase misincorporation of bases, i.e., base substitutions (*see* **Notes 2** and **3**). In addition, PCR is an excellent tool for specific mutagenesis within a defined DNA region, i.e., a DNA region to be mutagenized is specified by a pair of PCR primers flanking the region. One can amplify an entire gene for the antigenic protein or a subregion of the gene, if one has information about the approximate region of the epitope(s) (**Fig. 1**). The addition of *Thermus thermophilus*

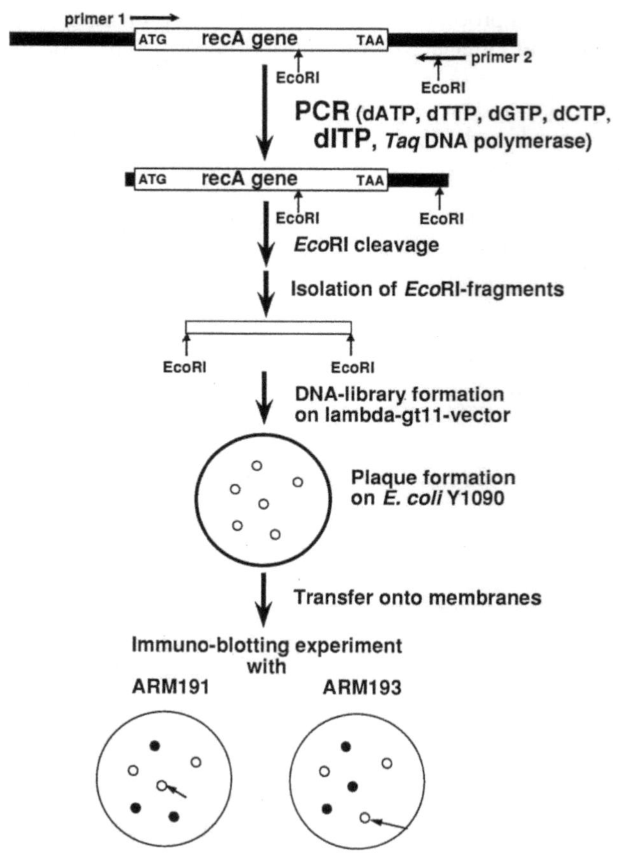

Fig. 1. Outline of the mapping by region-specified PCR-mutagenesis. This figure shows the outline of the procedure used for epitope mapping on RecA protein against anti-RecA protein monoclonal IgGs, ARM191 and ARM193 *(2)*. *Open* and *closed circles* in *two big circles* represent plaques expressing protein showing cross-reaction and those showing no cross-reaction with the indicated IgG, respectively. Modified from ***ref.*** (2).

RecA (TthRecA) to the PCR system (RecA-PCR) extensively suppresses background (or nonspecific) DNA amplification *(3)*, and thus, essentially one can choose any pair of sequences flanking the sequence to be amplified as the primers for PCR (*see* **Note 4**).

The application of this technique requires:

1. A cloned gene encoding the antigenic protein;
2. Sequence data of the gene and the flanking regions; and
3. At least two monoclonal antibodies that cross-react with the protein at different epitopes, or a control serum against the same protein.

One can use other techniques for random mutagenesis of a specified gene, such as in vitro mutagenesis by use of a mixture of oligonucleotides including random replacements of bases or oligonucleotide cassettes *(4, 5)*. The current PCR-mutagenesis has merits over other methods as follows:

1. A region to which mutations are introduced is easily specified by use of a pair of DNA primers;
2. Restriction sites and a linker sequence required for the insertion of the amplified gene into an expression vector are easily introduced by designing the primer sequence;
3. The mutation rate is easily controlled by simple modifications of conditions for PCR and high enough to obtain mutations to locate epitopes;
4. As analyzed so far, all mutations obtained by PCR are base substitutions; and
5. Sufficient amounts of DNA for the construction of a library are obtained through the procedure.

We will describe a procedure applied to map epitopes on *Escherichia coli* RecA protein against anti-RecA protein monoclonal IgGs *(2)*. In this procedure, we amplified and mutagenized the entire *recA* gene. Then, a DNA region encoding a C-terminal 94 amino acid region was cut out by *Eco*RI restriction enzyme and ligated into an *Eco*RI site of lambda-gt11 expression vector to construct a library (**Fig. 1**). Thus, the C-terminal region of the RecA protein is expressed as a fusion protein with β-galactosidase. Protein in plaques obtained by phages derived from the library were transferred onto membranes, and the cross-reactivities were tested by means of immunoblotting experiments against a couple of monoclonal IgGs, ARM191 and ARM193 (**Fig. 2**).

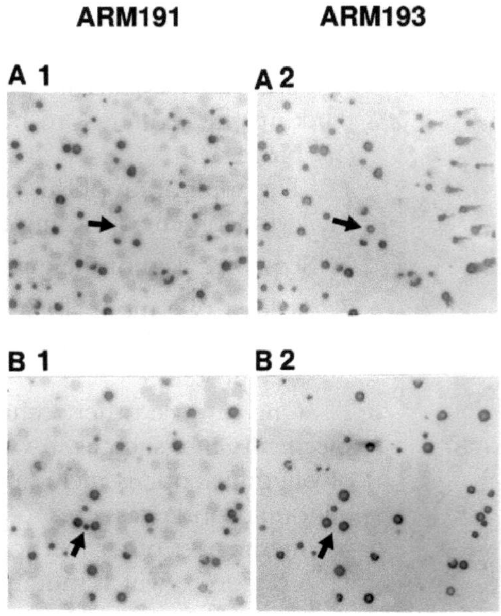

Fig. 2 Examples of immunoblotting experiments to detect plaques expressing mutant protein for cross-reaction. A plaque indicated by an arrow head in each panel for A1 and A2 (Library 1) showed cross-reaction with ARM193 but not with ARM191. That for B1 and B2 (Library 2) showed cross-reaction with ARM191, but not with ARM193. Reproduced from ***ref.*** (2).

2. Materials

1. 10× PCR buffer: 15 mM $MgCl_2$, 500 mM KCl, 100 mM Tris-HCl buffer (pH 8.3 at 25°C after diluted to 10 mM).
2. TBS buffer: 20 mM Tris-HCl, pH 7.5, 150 mM NaCl.
3. TY-plate: 1% (w/v) Difco tryptone, 0.5% Difco yeast extract, 1% NaCl, 1.5% Difco agar.
4. Plastic petri dishes: Square-shaped dishes (10 × 14 cm^2) for screening of plaques of DNA Library.
5. Phage dilution buffer: 10 mM Tris-HCl, pH 7.5, 10 mM $MgSO_4$, 0.01% gelatin.
6. Nitrocellulose membranes for immunoblotting experiments: BA85 type (pore size 0.45 μm, Schleicher & Schuell Bio-Science Inc., Sanford, ME) that are autoclaved at 121°C for 20 min, soaked in 50 mM isopropyl-β-D-thiogalactoside (IPTG), and dried before use.
7. 1% Bovine serum albumin: Bovine serum albumin (Fraction 5, Sigma Aldorich Co., St. Louis, MO) dissolved in TBS buffer.
8. Anti-mouse IgG antibody labelled with horseradish peroxidase: An affinity-purified preparation (Kirkegaard & Perry Laboratories Inc., Gaithersburg, MD).

9. 4-Chloro-1-naphthol-H_2O_2 solution: Prepared just before the assay as follows: Dissolve 4-chloro-1-naphthol in methanol at 3 mg/mL and dilute sixfold in TBS buffer. Add 30% H_2O_2 to the solution to give the final concentration of 0.5 μL/mL.

3. Methods

General methods for restriction enzyme treatment, gel electrophoresis, recovery of DNA from the gel, DNA ligation, phage experiments, cloning of DNA fragments into a sequencing vector, and DNA sequence analysis are described in detail in published laboratory manuals *(6, 7)*.

3.1. Design of Primers for PCR

A pair of primers should be designed to have a cutting site for a suitable restriction enzyme and a linker sequence to connect the amplified DNA in frame to an N-terminal portion of the *LacZ* gene at the unique *Eco*RI site on the lambda-gt11 vector. However, a primer often anneals multiple DNA regions, which results in nonspecific DNA amplifications. Since RecA promotes precise priming in PCR, RecA-PCR which contains heat-stable RecA, TthRecA, effectively eliminates nonspecific products and enables us to choose any sequences to design primers *(3)*. In the mapping of epitopes of RecA protein, we amplified the entire *recA* gene and ligated the C-terminal fragment generated by *Eco*RI digestion into the *Eco*RI site of the lambda-gt11 vector (**Fig.** 1). Thus, only one primer (primer 2) complementary to a region outside the 3′-terminus of the gene was designed to have an *Eco*RI site. The primers used in this mapping were primer 1 (5′-ATGGCTATCGACGAAAACAA-3′) and primer 2 (5′-GAATTCTGTCATGGCATATCCTT-3′).

3.2. PCR

1. Prepare a reaction mixture (50 μL) containing 1 μM each of primers flanking the sequence to be amplified, *ca.* 3 ng of the template DNA (linearized), 200 μM each of dATP, dTTP, dGTP, and dCTP, 200 μM deoxyinosine 5′-triphosphate (dITP), and 0.025 units of *Taq* DNA-polymerase per μL in 1× PCR buffer (tenfold dilution of 10× PCR buffer). One can use a DNA amplification kit with *Taq* DNA-polymerase by adding 200 μM dITP to the reaction. If you employ the RecA-PCR, you should further add 0.44 μM TthRecA and 400 μM ATP to the reaction and reduce the primer concentration stepwise (up to 0.01 μM) until the best result is obtained. In the mapping of epitopes of RecA protein, we used pBEU14 DNA *(8)* linearized by the treatment with *Bam*HI as a template for PCR.
2. For each cycle of PCR, anneal primers onto the template DNA by incubation at 55°C for 30 s, synthesize DNA at 68°C

for 90 s, and denature the synthesized DNA by incubation at 94°C for 10 s (an example, use your condition).

3. After 35 cycles of PCR, purify the amplified DNA by ethanol precipitation or a commercial kit (Wizard SV Gel and PCR Clean-Up System, Promega, Madison, WI).
4. Treat the amplified DNA with *Eco*RI restriction enzyme and separate a fragment encoding the mutagenized by gel electrophoresis.
5. Recover the DNA fragment from the gel by a kit (Wizard SV Gel and PCR Clean-Up System, Promega).

3.3. Construction of a Library of Mutagenized DNA in Lambda-gt11 Vector

1. Mix the isolated DNA fragments with *Eco*RI fragments of lambda-gt11 vector DNA (*ca.* 1 μg) at a 1:1 molar ratio and ligate the fragments into the *Eco*RI site of the *lacZ* gene by incubation at 4°C overnight in a 5 μL reaction mixture. We used the lambda-gt11 cloning kit (Lambda gt11/*Eco*RI/CIAP-Treated Vector Kit, Stratagene, La Jolla, CA) in which lambda-gt11 DNA had been cut by *Eco*RI restriction enzyme and the cut sites had been treated with phosphatase to prevent self-ligation.
2. Package the ligated DNA into a lambda phage particle. We used a lambda in vitro packaging kit (Amersham, Buckinghamshire, UK) for this process. Some other commercial products are available from Epicentre Technologies (Madison, WI) or Stratagene (La Jolla, CA). Since ligated DNA takes either of the two orientations relative to the vector DNA, half of the phage particles contains the amplified DNA in frame with the *lacZ* gene and will express a fusion protein with β-galactosidase, but the rest of the particles will not express the amplified DNA.

3.4. Immunoblotting Experiments

1. Dilute the packaged phage suspension so that it will give *ca.* 10^3 plaques/100 cm^2 of the TY-plate (10 × 14 cm^2).
2. Plate the phages with *E coli* Y1090 strain as host on TY-plates by use of soft-agar (0.7%) overlay technique.
3. Transfer proteins in plaques on the plate onto two or more nitrocellulose membranes by placing the membrane on the plate at 37°C for 2 h.
4. Soak the membranes in 1% bovine serum albumin dissolved in TBS buffer for 2 h at room temperature for blocking.
5. Soak each membrane in a solution of a tested monoclonal antibody (at an appropriate concentration depending on each antibody) dissolved in TBS buffer containing 0.1% bovine serum albumin for about 2 h at room temperature.
6. Wash the membranes with TBS buffer three times, and soak in TBS buffer containing 0.1% bovine serum albumin and an antibody against the tested antibodies labelled with horseradish peroxidase at room temperature for 2 h. In the mapping

of epitopes of RecA protein against anti-RecA protein mouse monoclonal IgG, we used an anti-mouse IgG antibody labelled with horseradish peroxidase.

7. Soak the washed membranes in 4-chloro-1-naphthol-H_2O_2 solution until appropriate expression of coloring reactions, wash the membranes with water, and dry them.
8. Compare the two or more membranes. Plaques that contain wild-type protein (with respect to the cross-reaction with the tested antibodies) give a positive coloring signal on all membranes, and those that do not express the amplified gene give no positive signal on any membrane.
9. Mark plaques that give no positive signal on one (or some) of the membranes and a positive signal on the other(s). **Fig. 2** shows examples in which the wild-type or mutant RecA protein expressed in plaques as fusion proteins reacted with an anti-RecA protein monoclonal lgG, ARM191 or ARM193, followed by a coloring reaction to detect the bound IgGs to the fusion proteins.

3.5. DNA Sequence Analysis

1. Pick up the phages in the marked plaques and suspend them in a least volume of phage dilution buffer.
2. Purify these mutant phages by a series of single plaque isolations, and amplify the isolated phages.
3. Prepare DNA samples from the phage particles.
4. Cut out DNA fragments encoding a mutated gene by use of appropriate restriction enzymes.
5. Reclone the DNA fragments into a sequencing vector (such as pUC119) for DNA sequence analysis.
6. Analyze the DNA sequences of both strands by the dideoxyribonucleotide chain termination technique *(9)* to locate base substitutions. **Fig. 3** shows amino acid substitutions of RecA protein that prevent cross-reaction with each anti-RecA protein IgG *(2)*.

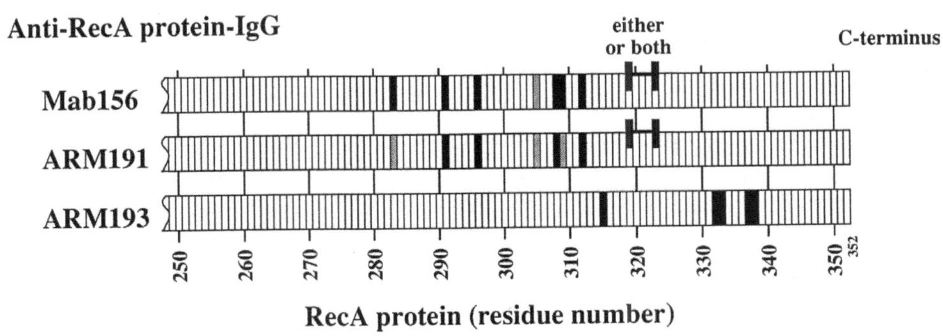

Fig. 3. A map of amino acid residues included in epitopes against anti-RecA protein IgGs *(2)*. Each *box* represents an amino acid residue. *Filled boxes* indicate amino acids of which replacement resulted in a loss of cross-reactivity against the indicated IgG. *Shadowed boxes* indicate amino acids of which replacement resulted in a partial loss of cross-reactivity.

4. Notes

1. *Thermus aquaticus* (*Taq*) DNA polymerase is known to show higher rate of errors in DNA synthesis, because of the absence of a proofreading exonuclease activity and the incubation at higher temperature *(10)*. The higher frequencies of errors during DNA synthesis are a serious problem for gene amplification and DNA sequence analysis, but *Taq* DNA polymerase provides a simple technique for in vitro mutagenesis.
2. We added dITP to increase errors in the incorporation of nucleotides during PCR. We tested the concentration of dITP from 0.2 μM to 200 μM and found that 200 μM dITP gave a tenfold increase in the yield of mutant RecA protein that showed altered cross-reactivities *(2)*. By PCR in the presence of 200 μM dITP, we picked up 21 candidate plaques among 2,000 plaques expressing the C-terminal 94 amino acid region (*ca.* 280 nucleotides) of RecA protein, and finally obtained ten kinds of mutant *recA* genes for the cross-reaction against anti-RecA protein IgGs. We detected 18 kinds among 25 mutations in 19 mutant *recA* genes obtained from several experiments, and found that all of them were base substitutions. Fifteen of the mutant *recA* genes had single base substitutions and the other four had two or three base substitutions *(2)*. Under normal conditions for DNA synthesis (in the absence of dITP), errors in synthesis by *Taq* DNA polymerase were reported to be caused by single base substitution mutations and less frequently (about a quarter of the rate of base substitutions) by frameshift mutations *(10)*. We have not tested the concentrations of dITP higher than 200 μM, which might be worth to be tested.
3. Other conditions that increase errors in DNA synthesis by *Taq* DNA polymerase are available. These include an increase in the concentration of $MgCl_2$ relative to the four dNTPs (dATP, dGTP, dTTP, and dCTP), a decrease in the dNTP concentrations, and higher pH *(11)*. When the $MgCl_2$ concentration was increased from 4 mM to 20 mM in the presence of 250 μM each of dNTPs, the mutation rate was shown to increase by 71-fold, and an increase in pH from 5.1 to 8.2 increased the mutation rate by 56-fold *(11)*. Variations in the relative concentrations among dNTPs and replacement of $MgCl_2$ by $MnCl_2$ were shown to reduce fidelity of DNA synthesis in PCR by *Taq* DNA polymerase *(12)*. These conditions would be useful to increase the mutation frequency during PCR.
4. Though PCR is a potent tool to amplify specific DNA sequences, several technical problems still remain to be solved. The most serious technical problem in this method

was nonspecific amplified products, which were derived from false priming. One may select appropriate priming sites whose complementary sequences are least likely to produce wrong priming. However, particular priming sites are often required to complete subsequent manipulations. RecA, which catalyzes pairing between homologous DNA molecules with great fidelity extensively, reduces false priming in PCR. Therefore, the addition of heat-stable RecA, TthRecA, to PCR not only eliminates nonspecific PCR products but also enables us to choose any sequences to design primers *(3)*.

References

1. Saiki, R. K., Scharf, S., Faloona, F., Mullis, K. B., Horn, G. T., Erlich, H. A., and Arnheim, N. (1985) Enzymatic amplification of beta-globin genomic sequences and restriction site analysis for diagnosis of sickle cell anemia. *Science* 230, 1350–1354.
2. Ikeda, M., Hamano, K., and Shibata, T. (1992) Epitope mapping of anti-recA protein IgGs by region specified polymerase chain reaction mutagenesis. *J. Biol. Chem.* 267, 6291–6296.
3. Shigemori, Y., Mikawa, T., Shibata, T., and Oishi, M. (2005) Multiplex PCR: use of heat-stable *Thermus thermophilus* RecA protein to minimize non-specific PCR products. *Nucleic Acids Res.* 33, e126.
4. Kramer, K., and Fritz, H.-J. (1987) Oligonucleotide-directed construction of mutations via gapped duplex DNA. *Methods Enzymol.* 154, 350–367.
5. Reidhaar-Olson, J. F., and Bowie, J. U. (1991) Random mutagenesis of protein sequences using oligonucleotide cassettes. *Methods Enzymol.* 208, 564–587.
6. Berger, S. L., and Kimmel, A. R. (1987) Guide to molecular cloning techniques. *Methods Enzymol.* 152, 1–812.
8. Uhlin, B. E., and Clark, A. J. (1981) Overproduction of the *Escherichia coli* recA protein without stimulation of its proteolytic activity. *J. Bacteriol.* 148, 386–390.
9. Sanger, F., Nicklen, S., and Coulson, A. R. (1977) DNA sequencing with chain-terminating inhibitors. *Proc. Natl. Acad. Sci. USA* 74, 5463–5467.
10. Tindall, K. R., and Kunkel, T. A. (1988) Fidelity of DNA synthesis by the *Thermus aquaticus* DNA polymerase. *Biochemistry* 27, 6008–6013.
11. Eckert, K., and Kunkel, T. A. (1990) High fidelity DNA synthesis by the *Thermus aquaticus* DNA polymerase. *Nucleic Acids Res.* 18, 3739–3744.
12. Leung, D. W., Chen, E., and Goeddel, D. V. (1989) A method for random mutagenesis of a defined DNA segment using a modified polymerase chain reaction. *Technique* 1, 11–15.

Chapter 23

Epitope Mapping Using Phage-Display Random Fragment Libraries

Lin-Fa Wang and Meng Yu

Summary

Phage-display has become a method of choice for epitope mapping and has been successfully used in numerous published studies. Although the inaugural studies were all done with random peptide libraries (*see* Chapter "Epitope Mapping Using Phage Display Peptide Libraries"), gene- or genome-targeted random fragment libraries have proven to be a more effective epitope mapping approach for some antibodies. In this chapter, we describe the mapping of linear and conformational epitopes of the major African swine fever virus capsid protein using monoclonal as well as polyclonal antibodies.

Key words: Gene-targeted, Phage library, Discontinuous epitope, Linear epitope.

1. Introduction

Phage-display random peptide libraries *(1–3)* are powerful tools for identification and characterization of peptide mimics that bind to specific selector molecules, such as antibodies *(4–6)*. The technology depends on random peptide sequences, displayed on the surface of filamentous bacteriophages, being allowed to interact with antibodies or other ligates. Ligates are usually immobilized on a solid support, such as a petri dish, microplate, or microbeads, and binding phages are specifically enriched by several cycles of affinity selection *(7)*. The displayed peptide(s) responsible for binding to the antibody can be identified by directly sequencing the encoding insert in the genome of the recombinant phage. This random peptide library approach has the potential advantage of being able to identify critical residues within an epitope *(8)*

Ulrich Reineke and Mike Schutkowski (eds.), *Methods in Molecular Biology, Epitope Mapping Protocols, vol. 524*

DOI: 10.1007/978-1-59745-450-6_23

and of providing mimotopes *(9)*, which can mimic discontinuous epitope structures *(10)* (*see* Chapters "Epitope Mapping Using Phage Display Peptide Libraries," "Antibody Epitope Mapping Using De Novo Generated Synthetic Peptide Libraries," and "Epitope Mapping Using Randomly Generated Peptide Libraries" of this volume for more details).

In this chapter, we will describe a different random expression strategy for epitope mapping using phage-display technology. Rather than expressing totally random synthetic peptide sequences, this approach relies on the construction of a random fragment expression library using small DNA fragments generated by partial digestion of target gene fragment(s) using DNase I. Although such a gene-targeted fragment library does not have the diversity exhibited by random synthetic peptide libraries, its limited complexity and presentation of the authentic peptide sequence, rather than a mimotope sequence, makes it an effective method for mapping epitopes *(11)*. Depending on the size of fragments selected, it is possible to construct recombinant phages displaying relatively large peptide fragments, which may be useful in other applications, such as development of diagnostic reagents and phage-based recombinant vaccines.

Although there have been reports that describe the mapping of epitopes using polyclonal antibodies and phage-display random peptide libraries, it is a technically difficult task owing to a high level of nonspecific binding. In contrast, we find that gene-targeted fragment libraries are able to provide much more conclusive results when polyclonal antibodies are used. One example is given in **Fig. 1**. The target antigen under investigation was the 646-amino acid (aa) residue major capsid protein p72 of African swine fever virus (ASFV) *(12)*. From this particular mapping experiment using DNase I fragments in the range of 150–300 bp (coding for peptides of 50–100 aa), several interesting observations were made:

1. It seems that the N-terminal region of p72 is more immunogenic than the remaining part of the molecule.
2. Pig polyclonal antibodies reacted with the epitope fragment displayed in phage clone A-9 in an ELISA, but not in Western blotting, indicating that this epitope is conformation-dependent.
3. The antigenic region defined by phage clones A-10 and A-11 also overlaps with a 7-aa epitope defined by a mouse monoclonal antibody (MAb) raised against p72 (*see* **Fig. 1c**). This indicates that the region is immunogenic not only in the target animals (pigs), but also in mice.
4. Clones A-11 and A-12 contained a hybrid peptide insert that was derived from two different parts of the p72 protein, randomly joined together in-frame before the recombined fragment was inserted, in-frame again, into the vector (*see* **Fig. 1b** for more details). It is not clear whether both of the original peptide

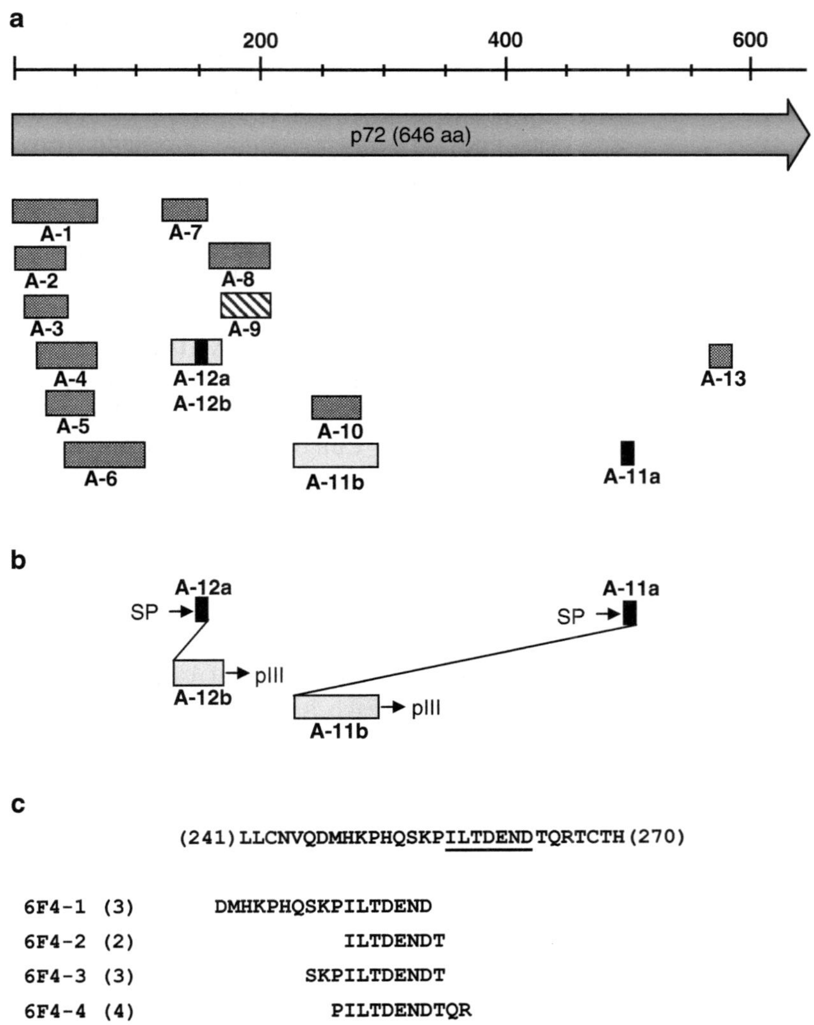

Fig. 1. Summary of epitope mapping results for the major capsid protein p72 of ASFV. (**a**) The numbers given on top of the figure represent the amino acid residue numbers of p72. The bars (*boxes*) underneath the p72 protein are epitope fragments selected using several different ASFV-infected pig sera. The slanted region of clone A-9 indicates it contained a conformational epitope that reacted with the pig antibodies in ELISA, but not in Western blotting. Clones A-11 and A-12 are composed of two fragments (shown as *light gray* and *black bars*). (**b**) Schematic representation of the two hybrid epitope fragments obtained for clones A-11 and A-12. The arrows indicate the direction of gene fusion from N- to C-terminus. In each case, the signal peptide (SP) region (at the N-terminus of gene III) is first fused to peptide-a, then to peptide-b and finally to the coding region for the mature pIII protein. Clone A-11 contains two peptides from different regions of p72, whereas clone A-12 contains a tandem repeat structure with peptide 12a being part of the larger peptide12b. (**c**) Sequence alignment of four different epitope fragments isolated by affinity selection using MAb 6F4. The numbers given in parentheses on the left indicate the total number of independent clones isolated within each class of insert. Shown on top is the amino acid sequence from the region (aa 241–270) covering the 7-aa consensus sequence (*underlined*) shared by all of the clones.

fragments from the hybrid insert are antigenic, but it is conceivable that, if the library size is large enough, it is possible to isolate hybrid epitopes that may contain different determinants of a discontinuous epitope.

If high-resolution mapping of antibody-binding sites is desired, one may use smaller DNase I fragments for library construction. **Fig. 1c** illustrates one of such mapping experiments that we have carried out to map a MAb-defined epitope for ASFV p72. In this case, smaller DNase I fragments in the range of 50–100 bp were used for library construction. Affinity selection using MAb 6F4, a MAb known to react with p72 in Western blotting, led to the isolation of four different classes of positive phage clones containing overlapping peptides as shown in **Fig. 1c**. From this, a 7-aa antibody-binding site was determined. This same antigenic region was also detected by pig antisera as described in phage clones A-10 and A-11.

It should be noted that, although the given examples were based on relatively simple gene-targeted random fragment libraries, the same principle can be applied to much more complex genome-targeted random fragment libraries, since the phage-display system has a capacity to generate a library of 10^7–10^9 independent clones *(4, 7)*. This has been demonstrated by Jacobsson and Frykberg *(13)* in isolation of IgG- and fibronectin-binding domains using a genome-targeted phage-display library constructed from the total genomic DNA of *Staphylococcus aureus*.

For the two different phage-display libraries, random peptide vs. random fragment, each has their own advantages and disadvantages *(14)*. If necessary, it is possible to increase the chance of success by combined phage-display of random gene fragments and random peptides *(15)*.

2. Materials

All reagents should be of AR grade. All solutions and buffers should be autoclaved or filter-sterilized where appropriate. Sterile tubes and filter tips should be used. Unless otherwise stated, all molecular biology reagents are obtained from Promega (Madison, WI) and all chemicals from Sigma (St. Louis, MO).

2.1. General

1. Antibodies: MAb(s) or polyclonal antibodies of interest. Mouse anti-M13-HRP conjugate, biotinylated sheep anti-mouse antibodies and HRP-conjugated secondary antibodies were purchased from Amersham Biosciences (Sydney, Australia).
2. Bacterial strains and plasmids: *Escherichia coli* MC1061 and K91Kan, phage expression vector fUSE1, all obtained from G. Smith *(7)*.
3. Equipment from Bio-Rad (Hercules, CA): power supply, horizontal agarose gel electrophoresis tank, Bio-Dot Microfiltration unit, gel dryer, and Gene Pulser electroporator.

4. Camera and imaging system for capturing DNA gel and Western blot images.
5. Darkroom facilities for X-ray film development, X-ray film (e.g., Kodak X-Omat AR5) and light box for viewing X-ray films.
6. Rocker and orbital shaker.

2.2. Library Construction

1. Recombinant gene: Usually in a plasmid clone containing the gene of interest.
2. Enzymes: RNase-free DNase I, T4 DNA polymerase, restriction enzyme PvuII, calf intestinal alkaline phosphatase (CIP), T4 DNA ligase.
3. DNase I buffer: 50 mM Tris-HCL pH 7.6, 10 mM $MnCl_2$, kept frozen at –20°C. Stable for at least 1 year.
4. EDTA: 0.5 M at pH 8.0.
5. TE buffer: 10 mM Tris-HCl, pH 7.6, 1 mM EDTA.
6. Phenol:chloroform: 1:1 mixture of TE-saturated phenol and chloroform.
7. Sodium acetate: 3 M at pH 5.2.
8. Ethanol: 100% and 70%, kept at –20°C.
9. 50× TAE buffer (per liter): 242 g Tris-base, 57 mL glacial acetic acid, 100 mL 0.5 M EDTA, pH 8.0. Make 1× solution every month.
10. End repairing buffer: 40 mM Tris-HCl, pH 8.5, 10 mM $(NH_4)_2SO_4$, 5 mM $MgCl_2$, 5 mM DTT, 0.5 mM EDTA, 150 μg/mL BSA, and 100 μM dNTPs. Make each of the components separately as a 10× stock and store at –20°C. Before use, make the 1× solution by diluting in water and use on the same day.
11. QIAquick kits for DNA purification and clean up (QIAgen, Hilden, Germany).
12. Gene Pulser cuvet (0.2 cm) from Bio-Rad.
13. Antibiotic stock solutions: Tetracycline (Tet) at 20 mg/mL in absolute ethanol and kanamycin (Kan) at 50 mg/mL in water are kept at –20°C. Unless otherwise stated, the working concentration of antibiotics is at a 1:1,000 dilution, i.e., 20 μg/mL for Tet and 50 μg/mL for Kan (e.g., LB/Tet/Kan medium represents LB containing Tet at 20 μg/mL and Kan at 50 μg/mL).
14. SOC medium (per liter): bacto-tryptone 20 g, yeast extract 5 g, NaCl 0.5 g, 1 M KCl 2.5 mL. Adjust pH to 7.0 with 10N NaOH, autoclave to sterilize, add 20 mL of sterile 1 M glucose immediately before use.
15. LB medium: 1% bacto-tryptone, 0.5% yeast extract, 1% NaCl.

16. Terrific broth (TB) medium: Dissolve 12 g bacto-tryptone, 24 g yeast extract, and 4 mL (5.04 g) glycerol in 900 mL water. Autoclave 90 mL portions in 125 mL bottles. When cooled, add 10 mL of separately autoclaved potassium phosphate solutions (0.17 M KH_2PO_4, 0.72 M K_2HPO_4) to each bottle.
17. PEG solution: 20% (w/v) polyethylene glycol-8000, 2.5 M NaCl (maybe necessary to heat at 65°C to dissolve), autoclave to sterilize, and keep at 4°C.
18. TBS buffer: Dilute l0× TBS (per liter: 90 g NaCl, 60 g Tris-base, adjust pH to 7.9 with HCl) to 1× with distilled water. Store at room temperature for up to 6 months.
19. NaN_3 solution: Make a 20% (w/v) stock solution in water and keep at 4°C. Use at 1:1,000 dilution to a final concentration of 0.02%. *Caution:* this is a toxic chemical. Handle with gloves and label the tube with an appropriate warning sign.
20. Dimethyl sulfoxide (DMSO) from Sigma.
21. TBS/gelatine: Dissolve 0.1 g gelatine in 100 mL TBS by autoclaving, and store at room temperature for up to 6 months.

2.3. Library Screening

1. TBST buffer: TBS containing 0.5% (v/v) Tween-20.
2. Blocker solution: TBST containing 5% (w/v) skimmed milk powder and 1% of M13 phage solution (approximately 10^{13} phage particles/mL). Make fresh before use.
3. Streptavidin magnetic beads (SMB): 1 mg/mL suspension (Promega).
4. Magnetic separation stand (two-hole) (Promega).
5. Elution buffer: 0.1N HCl (pH adjusted to 2.2 with glycine), 1 mg/mL BSA. Store at 4°C up to 6 months. (Optional: Add 0.1 mg/mL phenol red to monitor pH of the solution).
6. 1 M Tris-HCl, pH 9.5.
7. Nitrocellulose membrane: 0.45 μm from Schleicher & Schuell (Dassel, Germany).
8. Blotto solution (for Western blot): TBST containing 5% (w/v) skimmed milk powder. Make fresh before use.
9. Plastic bag and heat sealer.
10. Container with flat-bottom (e.g., square petri dishes).
11. Supersignal West Pico Trail Kit (Pierce, Rockford, IL), for enhanced chemiluminescence (ECL) detection.

2.4. Characterization of Epitope-Displaying Phage Clones

1. Multichannel pipet, ELISA plates, and microplate shaker: all form Titerteck Flow Laboratories (McLean, VA).
2. Microplate reader: Multiscan MS, from Labsystems (Helsinki, Finland).

3. Coating buffer: 50 mM Tris-HCl, 150 mM NaCl, pH 9.3.
4. PBST: Dilute 10× PBS (10.7 g/L Na_2HPO_4, 3.9 g/L NaH_2PO_4, 80 g/L NaCl, pH 7.2) to 1× with distilled water and add Tween-20 to a final concentration of 0.05% (v/v). Store at room temperature for up to 6 months.
5. Blocking solution: PBST containing 2% (w/v) skimmed milk powder, make fresh before use.
6. Citrate acetate buffer: Make 100 mL of 1 M sodium acetate and 10 mL of 1 M citric acid. Adjust the sodium acetate solution to pH 5.9 with approximately 1.5 mL of the citric acid.
7. TMB substrate: Dissolve 100 mg of 3,3,5,5-tetramethylbenzidine (Sigma) in 10 mL DMSO to make a 42 mM solution. Store at 4°C in small aliquots (0.5 mL) for up to 12 months. Pre-warm at 37°C for 10 min before use.
8. Substrate solution: Make fresh by mixing 18 mL of distilled water with 2 mL of citrate acetate buffer and 0.2 mL of the TMB substrate. Add 2.5 μL 30% H_2O_2 just before use.
9. Stopping solution: 1 M H_2SO_4.
10. *Taq* polymerase and PCR reagents, use as recommended by supplier.
11. Oligo primers: gIII-5 (5′-GGT TGG TGC CTT CGT AGT-3′), gIII-3 (5′-CCA TGT ACC GTA ACA CTG-3′), and ^{35}S (5′-CCC TCA TAG TTA GCG TAA CG-3′).
12. PCR machine: any thermal cycler capable of carrying out PCR in 96-well format.
13. BigDye Terminator v1.1 Cycle Sequencing kit from Applied Biosystems (Foster City, CA).

3. Methods

3.1. Library Construction

3.1.1. Generation of Random Fragments by DNase I Partial Digestion

The procedure given below can be used with plasmid DNA containing the target gene insert, PCR-amplified gene fragment(s), or chromosomal DNA. If more DNA is required, we recommend setting up multiple tubes rather than increasing the volume of each reaction (*see* **Note 1**).

1. Resuspend the DNA sample in water at a concentration of 200 μg/mL, divide into four 50 μL aliquots, and keep on ice.
2. Dilute DNase I in ice-cold DNase I buffer at final concentrations of 4, 2, 1 and 0.5 U/mL, respectively.
3. Start the DNase I digestion by transferring 17.5 μL of diluted DNase I solution from each of the above four concentrations

into a tube containing 50 µL of the DNA sample prepared in **step 1**.

4. Incubate at 15°C for 10 min.
5. Stop the digestion by adding in 2.5 µL of 0.5 M EDTA solution, and transfer the tubes onto ice.
6. Take 2 µL of the digested mixture from each of the four tubes and analyze the digestion patterns on a 2% agarose-TAE gel.
7. Combine the two that give optimal digestion patterns (i.e., most DNA fragments are distributed in the range of 50–300 bp) and bring the total volume to 500 µL with TE buffer. Extract once with an equal volume of phenol:chloroform, followed by precipitation with two volumes of absolute ethanol in the presence of 0.3 M sodium acetate. Leave the tube at –20°C for 30–60 min.
8. Pellet the DNA by centrifugation in an Eppendorf centrifuge at 4°C for 10 min, wash the pellet twice with 70% cold ethanol and dry under vacuum for 15 min.
9. Resuspend the pellet in 50 µL of end-repairing buffer, add 10 U of T4 DNA polymerase, and incubate at 15°C for 60 min.
10. Separate DNA fragments by electrophoresis in a 2% agarose-TAE preparative gel and cut out the gel slice containing DNA fragments in the range of 100–300 bp.
11. Purify DNA fragments from the gel slice using the QIAquick Gel Extraction Kit following the supplied instructions. Elute the DNA in a final volume of 20 µL water (*see* **Note 2**).

3.1.2. Vector Preparation

1. Digest 2 µg of fUSE1 vector DNA with 10 U of PvuII enzyme in a total volume 20 µL. Incubate at 37°C for 60 min.
2. Dilute the digestion mixture by adding 24 µL of water and 5 µL of 10× CIP buffer. Add 1 µL of CIP enzyme (1 U/µL) and incubate at 37°C for 30 min. Add another 1 µL of CIP enzyme, followed by 30 min of further incubation.
3. Bring the total volume to 100 µL by adding 50 µL water, extract this diluted mixture with an equal volume of phenol:chloroform, followed by ethanol precipitation in the presence of 0.3 M sodium acetate.
4. Resuspend the DNA in 5 µL water.

3.1.3. Ligation

1. Mix 5 µL vector DNA with 20 µL end-repaired DNase I fragments, followed by addition of 3 µL 10× ligase buffer containing ATP.
2. Start the ligation reaction by adding 1 µL of T4 DNA ligase (3 U/µL), followed by incubation at 16°C for 2–4 h.

3. Add an additional 1 μL of ligase, followed by overnight incubation at 16°C.
4. Inactivate ligase by heating at 65°C for 15 min.
5. Bring the total volume to 100 μL with water, followed by phenol:chloroform extraction and ethanol precipitation.
6. Resuspend the DNA in 5 μL water and keep on ice until use.

3.1.4. Electroporation

1. Prepare electro competent cells of *E. coli* strain MC1061 using protocols provided with the Gene Pulser electroporator. Quickly freeze 120 μL aliquots in liquid nitrogen and keep at −80°C until use.
2. Thaw three tubes of competent cells on ice, combine the cell suspension, and transfer to the tube containing the 5 μL of ligated DNA mixture prepared in **Subheading 3.1.3**.
3. Conduct six separate electroporations, each with approximately 60 μL of cell–DNA mixture in a 0.2-cm cuvet, using settings at 2.4 kV, 25 μF and 200 Ω (*see* **Note 3**).
4. After each electroporation, immediately transfer the mixture into a 100-mL flask containing 10 mL pre-warmed SOC medium with tetracycline at 0.2 μg/mL. After the completion of the last electroporation, incubate the flask at 37°C for 60 min with gentle shaking (at 150 rpm).
5. Plate 25, 50, and 100 μL aliquots onto LB/Tet plates for colony counting and incubate the plates at 37°C overnight.
6. Transfer the rest of the culture to a 1-L flask containing 190 mL pre-warmed LB/Tet medium and incubate for 12–16 h at 37°C with vigorous shaking (at 300 rpm).
7. Transfer the 200 mL culture to a centrifuge bottle and spin for 15 min at 10,000 × *g*. Transfer the supernatant to a clean centrifuge bottle and repeat the spin.
8. Collect the supernatant from the second spin in a clean bottle and add 0.15 volumes of PEG solution (i.e., 30 mL for 200 mL supernatant). Invert the bottle several times and incubate on ice for at least 2 h (*see* **Note 4**).
9. Centrifuge at 12,000 × *g* for 30 min at 4°C. Completely remove the supernatant.
10. Resuspend the phage pellet in 10 mL of TBS buffer by pipeting, followed by incubation at room temperature for approx 30 min to completely resuspend the pellet (*see* **Note 5**).
11. Transfer the phage solution to a 50-mL centrifuge tube and spin for 10 min at 10,000 × *g* to remove insoluble materials.
12. Repeat the PEG precipitation by adding in 1.5 mL PEG solution, followed by incubation on ice and centrifugation as in **steps 8** and **9**.

13. Completely resuspend the phage pellet in 1.6 mL of TBS as in **step 10**, and transfer the phage solution to a 2-mL Eppendorf tube.
14. Spin for 5 min to remove insoluble materials and transfer the supernatant to a clean 2-mL tube.
15. Repeat PEG precipitation as above by adding in 240 μL of PEG solution, followed by incubation on ice and centrifugation.
16. Finally resuspend the phage pellet in 1 mL of TBS containing 0.02% NaN_3.
17. Spin again to remove insoluble materials, collect the phage solution (the supernatant), add DMSO to a final concentration of 7% (v/v), and store 0.1-mL aliquots at –80°C until use (*see* **Note 6**).

3.1.5. Titering Phage Transducing Units (TU)

1. Inoculate 1 mL LB/Kan with *E. coli* strain K91Kan and shake overnight at 37°C.
2. Use 100 μL of the overnight culture to inoculate 10 mL TB/Kan medium in a 100 mL flask. Shake vigorously at 37°C until mid- to late-log phase (*see* **Note 7**).
3. Slow the shaking down to around 100 rpm to allow sheared F-pili to regenerate. Use the cells within approx 60 min.
4. During the slow shaking of the bacterial culture, make a serial dilution of phage solution in TBS/gelatin covering the range $1{:}10^7$, $1{:}10^8$, and $1{:}10^9$.
5. Mix 10 μL of K91kan cells prepared in **step 3** in a 1.5 mL Eppendorf tube with 10 μL each of the diluted phage solutions and incubate at room temperature for 10 min for phage infection.
6. Add 1 mL of LB medium containing 0.2 μg/mL tetracycline and incubate at 37°C for 30 min (with gentle shaking if convenient).
7. Plate 50 and 100 μL aliquots onto LB/Tet/Kan plates, followed by overnight incubation at 37°C. Count the colony numbers to determine phage titer.

3.2. Library Screening

3.2.1. Affinity Selection Using Streptavidin Magnetic Beads (SMB)

The protocol given below is to be used for mouse monoclonal antibodies (MAb). The same protocol can be used for antibodies from other species with appropriate biotinylated anti-species antibodies in **step 4**.

1. Prepare K91Kan cells as in **Subheading 3.1.5, steps 1–3**.
2. Set up a phage–antibody incubation solution in a 2-mL flat-bottom Eppendorf tube by mixing 180 μL blocker, 10 μL antibody, and 10 μL phage. Incubate at room temperature for 45–60 min (*see* **Note 8**).

3. Meanwhile, place 100 μL of SMB suspension (1 mg/mL) in a 2-mL flat-bottom Eppendorf tube and wash the beads three times with 0.5 mL blocker solution.
4. Resuspend the washed SMB in 190 μL blocker solution, followed by the addition of 10 μL biotinylated anti-mouse IgG antibodies. Incubate the solution at room temperature for 30 min (*see* **Note 9**).
5. Wash the beads three times (2 min each) with 1 mL TBST.
6. Transfer the 200 μL phage–antibody incubation mixture from **step 2** to the tube containing the washed beads from **step 4**, followed by a further incubation at room temperature for 30 min.
7. Wash the beads five times as in **step 5**.
8. Resuspend the beads in 200 μL TBS and transfer to a 0.5-mL cone-shaped Eppendorf tube. Position the tube bottom against the magnetic separation stand, so that the magnetic beads will "swim" toward the bottom of the tube rather than toward the side of the tube, as in a normal operation. Completely remove the TBS buffer using a thin pipet tip, so that a small compact pellet is formed at the bottom of the tube.
9. Elute bound phages by incubating the beads with 40 μL elution buffer at room temperature for 10 min.
10. Place the tube bottom against the magnetic separation stand as in **step 8**, remove the 40 μL supernatant using a thin pipet tip, and immediately transfer to a 2-mL tube containing 16 μL 1 M Tris-HCl, pH 9.5, for neutralization of the eluted phage solution.
11. Add 100 μL K91 Kan cells prepared as in **Subheading 3.1.5**, and incubate the tube at room temperature for 10 min for phage infection.
12. Add 1 mL LB containing 0.2 μg/mL tetracycline and incubate at 37°C for 30 min with shaking.
13. Take 100 μL for making dilutions of 1:10, 1:100, and 1:1,000 in pre-warmed LB medium.
14. Plate the cells onto LB/Tet/Kan plates, plate in duplicate 200 μL aliquots of the undiluted culture as well as the three diluted cultures made above in **step 13**. Incubate the plates (eight in total) at 37°C overnight.
15. Transfer the remaining culture from **step 12** into a 250 mL flask containing 30 mL pre-warmed LB/Tet/Kan medium, and shake vigorously at 37°C overnight.
16. Purify phages from the supernatant of this 30 mL culture by three times PEG precipitation as described in **Subheading 3.1.4**. This can be used as an enriched library for a second cycle of affinity selection, if required (*see* **Note 10**).

3.2.2. Colony Lift Immunoblotting

All incubation steps are carried out at room temperature.

1. Select two to four plates, from **step 14** of **Subheading 3.2.1**, which have a colony density in the range of 100–500 colonies/plate (*see* **Note 11**).
2. Place a piece of precut circular nitrocellulose membrane onto the surface of the plate. Make sure that the membrane makes an even contact with the plate so that it will be completely wet by the moisture from the plate within 1–2 min. Mark the orientation of the membrane in relation to the plate so that it can later be correctly superimposed (*see* **Note 12**).
3. Lift the membrane and immediately transfer to a container (e.g., a square petri dish) containing 30–50 mL of blotto solution. Gently rock for 15 min.
4. Change the blotto solution and rock for an additional 15 min.
5. Seal the membrane inside a plastic bag with one side open, add 2 mL of MAb solution diluted in blotto solution at 1:5 for MAb tissue-culture supernatant or 1:100 for ascitic fluid, and seal the remaining side. Incubate the bag for 30 min with gentle rocking (*see* **Note 13**).
6. Wash three times (5 min each) with approx 50 mL TBST.
7. Proceed as in **step 5**, except that an HRP-conjugated sheep anti-mouse antibody is used at 1:1,000 dilution.
8. Wash as in **step 6**.
9. Develop the blot using ECL and X-ray films following the instructions given by the supplier (*see* **Note 14**).
10. After drying the film, place it on a light box so that the plate containing the corresponding bacterial colonies can be aligned with the signals (black dots) on the film. Pick up the positive clones using individual toothpicks and patch onto a fresh LB/Tet/Kan plate, followed by overnight incubation at 37°C. Keep the plate at 4°C as the master plate for positive phage clones.

3.3. Characterization of Epitope-Displaying Phage Clones

3.3.1. Phage Minipreparation

1. Pick up a single colony from the above master plate and innoculate 2 mL TB/Tet medium in a 10 mL culture tube. Vigorously shake the tubes overnight at 37°C (*see* **Note 15**).
2. Remove cells by centrifugation in a 2 mL tube for 5 min at room temperature.
3. Transfer the supernatant to a clean tube and repeat the centrifugation as above.
4. Carefully take 1.6 mL supernatant from the final spin and transfer to a 2-mL tube containing 240 μL PEG solution. Invert the tube several times and either incubate the tube on ice for at least 2 h or leave the tubes at 4°C overnight.
5. Collect phage precipitate by centrifugation at 4°C for 15 min.

6. Resuspend the phage pellet in 30 μL TBS by pipeting, followed by 30 min incubation at 4°C. Spin for 5 min to remove insoluble materials. The phage solution is ready to be used in ELISA, dot-blotting, or Western blotting analysis.

3.3.2. ELISA

All incubations, except for substrate development, are carried out at 37°C with gentle shaking on a microplate shaker.

1. Make a 1:5,000 dilution of rabbit anti-M13 antiserum in coating buffer and use 50 μL/well to coat an ELISA plate (*see* **Note 16**). Incubate the plate for 60 min.
2. Wash the plate three times (5 min each) with PBST.
3. Add 100 μL blocking solution to each of the wells and incubate for 30 min. Discard after incubation.
4. Add 50 μL phage solution serially diluted in blocking solution starting from 1:100 (*see* **Note 17**), followed by incubation for 60 min.
5. Wash as in **step 2**.
6. Add 50 μL MAb, diluted in blocking solution at 1:10 for tissue-culture supernatant or 1:100–1:1,000 for ascitic fluid, followed by incubation for 60 min.
7. Wash as in **step 2**.
8. Add 50 μL HRP-conjugated sheep anti-mouse IgG diluted in blocking solution at 1:2,000, followed by incubation for 60 min.
9. Wash as in **step 2**.
10. Add 50 μL TMB substrate solution and incubate at room temperature for 10 min.
11. Stop the reaction by adding 50 μL stopping solution.
12. Determine the absorbance at 450 nm.

3.3.3. Dot Blotting

All incubations are carried out at room temperature.

1. Wet a precut nitrocellulose membrane (10 × 14 cm^2) in TBS buffer and assemble the membrane into the Bio-Dot Microfiltration unit following the given instructions.
2. Serially dilute phage solutions in PBS starting at 1:100 (*see* **Note 17**). The dilution can be conveniently carried out in an ELISA plate, since the Bio-Dot unit has the same 96-dot/well format.
3. Under vacuum, apply each 20 μL of diluted phage solution into a corresponding well. Wait for 1–2 min after all samples are applied. Slowly release the vacuum and disassemble the unit.
4. Remove the membrane from the unit and immediately transfer to a container with 50 mL blotto solution. Gently rock for 30 min.

5–10. Follow steps 5–10 described in **Subheading 3.2.2** (*see* **Note 18**).

3.3.4. Colony PCR and DNA Sequencing

We find that it is most convenient to carry out PCR and sequencing reactions in microplates. However, the following procedures are equally applicable for using tubes as long as minor adjustments are made accordingly for PCR conditions.

1. Place 14 μL water into each well of a microplate that is suitable for plate PCR application.
2. Use a toothpick with a sharp tip to transfer cells from a colony to a corresponding well by gently touching, rather than digging, the colony, and then mixing in water.
3. Heat the plate at 100°C for 2 min, followed by immediate cooling on ice. Keep on ice until the next step.
4. Set up a PCR reaction cocktail as follows (the volumes given are for one reaction):
   - 2.5 μL 10× PCR buffer
   - 2.5 μL 25 mM $MgCl_2$
   - 4 μL dNTPs, 1.25 mM
   - 1 μL forward primer gIII-5, 10 pmol/μL
   - 1 μL reverse primer gIII-3, 10 pmol/μL
   - 0.25 U of *Taq* polymerase

   Mix by pipetting and add 11 μL of this cocktail mixture into each well.
5. Carry out a PCR amplification for 25 cycles at 94°C for 1 min, 50°C for 2 min, and 72°C for 2 min (*see* **Note 19**).
6. After amplification, run the PCR product on a 1% agarose-TAE gel to check for insert size and the quality of PCR products.
7. For those PCR products with inserts and having satisfactory quality, purify the PCR fragment using the QIAquick Gel Extraction Kit, and elute in 30 μL elution buffer.
8. Take 2–3 μL for sequencing using the BDT v1.1 Sequencing kit and the internal primer ^{35}S, following procedures provided with the kit.

4. Notes

1. It should be noted that there are other ways of generating random DNA fragments other than DNase I partial digestion. The other commonly used method is by sonication (e.g., in **ref.** *13*). A detailed protocol for generation of random

fragments by sonication can be found in Chapter "Epitope Mapping by Proteolysis of Antigen—Antibody Complexes" of vol. 23 of this series. The remaining procedures, from **step 9** of **Subheading 3.1.1**, are equally applicable to sonicated DNA fragments. Recently, Kawamura et al. *(16)* described a new strategy for construction of gene fragment libraries by reverse transcription and random priming. To increase the number of productive in-frame fused gene fragments in the phage-display library, one can also apply a "filtering" strategy by constructing a library containing random fragments fused in-frame with an antibiotic selective gene first, before moving the random "productive" fragments into a phage-display vector *(17)*.

2. To achieve the best yield from purification using the QIAquick gel extraction kit, an effort should be made to reduce the sample volume and, hence, reduce the size of the agarose gel slice after electrophoresis.
3. We recommend the use of new cuvets for this. The same cuvet can be used for multiple electroporations of the same sample. For optimal performance, cool the cuvet on ice for 1 min between each usage. Under these conditions, we normally get a pulse of 4.0–4.6 ms.
4. Although a 2-h incubation on ice is usually enough for PEG precipitation of phages, overnight incubation at 4°C can give a slightly better yield. It is also a convenient break point in the protocol. This applies for all other PEG precipitation steps described in this protocol.
5. Owing to the filamentous shape, phage particles in PEG precipitates are hard to resuspend. The 30-min incubation is essential for phage particles to diffuse completely. If convenient, the phage solution may also be kept at 4°C overnight in the presence 0.02% NaN_3.
6. The phage solution purified by 3x PEG precipitation is usually good enough for the application described here. However, further purification by $CsCl_2$ gradient centrifugation (*see* **ref.** *7* for details) is recommended if the phage library is to be kept for a long period and for multiple applications. Although the phage particles purified by 3 × PEG precipitation are stable at 4°C, we find that the antigenicity of certain recombinant phages displaying foreign epitopes decreases with time, probably because of degradation by trace amounts of contaminating proteases.
7. The growth of the bacterial culture is best monitored by a spectrophotometer at 600 nm. However, we find that 5 h vigorous shaking at 300 rpm, followed by 30 min incubation at 100 rpm, usually gives satisfactory results.

8. The optimal ratio of antibody vs. phage is hard to determine because both the phage titer and antibody titer vary from one experiment to another. Our common practice is to use the phage solution at approx 10^{12} TU/mL, and the antibody solution at approx 0.1 μg/mL for purified antibodies or 10–100 μg/mL for crude antibodies.
9. The use of a biotinylated anti-species antibody as a bridge is not absolutely necessary. It is possible to biotinylate the antibody of interest and bind it directly to the SMB. However, we recommend the use of a biotinylated bridging antibody for two reasons:
   (a) It is convenient to operate since there is no need to biotinylate individual antibody molecules.
   (b) This may also help the later elution process, since the binding between the bridging antibody and the primary antibody provides an additional break point during elution, and is more homogeneous in binding affinity than the interaction between the primary antibody and individual phage-displayed epitope fragments.
10. Although multiple panning can usually enrich the population of binding phages, there is an associated danger of losing relatively weak binders or slow-growth phage clones. When polyclonal antibodies are used for affinity selection, we especially recommend carrying out a single panning, followed by colony lift immunoblotting.
11. If possible, the plates should be used within 1–2 days for colony lift. However, old plates, kept at 4°C up to a week, have also been used in our laboratory with satisfactory results.
12. There are several methods one can use to mark the membrane for correct superimposition. The method we use is as follows:
   (a) Cut three small triangles off the edge of a circular membrane at positions approx corresponding to 12, 1, and 4 o'clock.
   (b) After placing the membrane onto the plate surface, mark the corresponding positions on the plate using a permanent marking pen.
   (c) Overexpose one film in **step 9** of **Subheading 3.2.2** to reveal the three triangles on the edge of the membrane.
   (d) Superimpose the overexposed film to the film with the best exposure and mark the triangles. Then superimpose this second film onto the plate to identify the positive colonies.
13. A convenient way of keeping the membrane flat is to put the bag in the middle of a thick heavy book (e.g., a telephone directory readily available in every laboratory),

which is in turn placed on top of a rocker. The 30-min incubation time is the minimum time required, but can be extended to incubation at 4°C overnight to fit in with other ongoing experiments. This is also true for the incubation with conjugated antibody in **step 7** of **Subheading 3.2.2**.

14. The exposure time required may vary from one experiment to another. We normally carry out three exposures at 10, 30, and 60 s, and then a 5-min exposure while the first three films are being developed. Less-sensitive methods can also be used if the expected signals are strong (e.g., using 4-chloro-l-naphthol as substrate for direct development of signal on the membranes). However, it is better to avoid the use of alkaline phosphatase (AP)-conjugated antibodies, since the endogenous AP activity from *E. coli* can cause a high background.
15. The phage production yield can be increased by extending the incubation time to 24–36 h.
16. The use of anti-M13 antibodies to capture the phage particles is optional. Phage particles can also be directly coated in an ELISA plate. However, we found the capture ELISA gives more consistent results than direct coating.
17. It is essential to include a control phage in the binding assays. Since the cloning vector fUSE I is a nonproductive phage, which contains a nonfunctional gene III, we normally use a closely related fUSE2 vector (*see* **ref.** *7*) for production of control phages. For initial screening studies, a single well at each dilution is acceptable. Duplicate or triplicate wells should be used in more detailed binding analysis later on.
18. Owing to space limitation, we presented only two of the most convenient binding assays here. There are other assays that can be used in confirming antibody-binding. One frequently used method is Western blotting (e.g., *see* **refs.** *11* and *13*). Dyson et al. *(18)* have also reported a method for direct measurement of phage-ligate binding via phage titering.
19. If primers other than those described here are used, the PCR conditions may have to be optimized accordingly.

Acknowledgments

We thank G. P. Smith of University of Missouri, Columbia for providing the fUSE expression system and a set of comprehensive protocols.

References

1. Scott, J. K. and Smith, G. P. (1990) Searching for peptide ligands with an epitope library. *Science* 249, 386–390.
2. Cwirla, S. E., Peters, E. A., Barrett, R. W., and Dower, W. J. (1990) Peptides on phage: a vast library of peptides for identifying ligands. *Proc. Natl. Acad. Sci. USA* 87, 6378–6382.
3. Devlin, J. J., Panganiban, L. C., and Devlin, P. E. (1990) Random peptide libraries: a source of specific protein binding molecules. *Science* 249, 404–406.
4. Cortese, R., Monaci, P., Luzzago, A., Santini, C., Bartoli, F., Cortese, I., Fortugno, P., Galfrè, G., Nicosia, A., and Felici, F. (1996) Selection of biologically active peptides by phage-display of random peptide libraries. *Curr. Opin. Biotech.* 7, 616–621.
5. Scott, J. K. and Craig, L. (1994) Random peptide libraries. *Curr. Opin. Biotech.* 5, 40–48.
6. Wang, L.-F. and Yu, M. (2004) Epitope identification and discovery using phage-display libraries: applications in vaccine development and diagnostics. *Curr. Drug Targets* 5, 1–15.
7. Smith, G. P. and Scott, J. K. (1993) Libraries of peptides and proteins displayed on filamentous phage. *Method. Enzymol.* 217, 228–257.
8. Du Plessis, D. H., Wang, L.-E., Jordaan, E. A., and Eaton, B. T. (1994) Fine mapping of a continuous epitope on VP7 of Bluetongue Virus using overlapping synthetic peptides and a random epitope library. *Virology* 198, 346–349.
9. Geysen, H. M., Rodda, S. J., and Mason, T. J. (1986) *A priori* delineation of a peptide which mimics a discontinuous antigenic determinant. *Mol. Immunol.* 23, 709–715.
10. Balass, M., Heldman, Y., Cabilly, S., Givol, D., Katchalski-Katzir, E., and Fuchs, S. (1993) Identification of a hexapeptide that mimics a conformation-dependent binding site of acetylcholine receptor by use of a phage-epitope library. *Proc. Natl. Acad. Sci. USA* 90, 10638–10642.
11. Wang, L.-F., Du Plessis, D. H., White, J. R., Hyatt, A. R., and Eaton, B. T. (1995) Use of a gene-targeted phage-display random epitope library to map an antigenic determinant on the bluetongue virus outer capsid protein VP5. *J. Immunol. Methods* 178, 1–12.
12. Lopez-Otin, C., Freije, J. M., Parra, F., Mendez, E., and Vinuela, E. (1990) Mapping and sequence of the gene coding for protein p72, the major capsid protein of African swine fever virus. *Virology* 175, 477–484.
13. Jacobsson, K. and Frykberg, L. (1995) Cloning of ligand-binding domains of bacterial receptors by phage-display. *Biotechniques* 18, 878–885.
14. Coley, A. M., Campanale, N. V., Casey, J. L., Hodder, A. N., Crewther, P. E., Anders, R. F., Tilley, L. M., and Foley, M. (2001) Rapid and precise epitope mapping of monoclonal antibodies against *Plasmodium falciparum* AMA1 by combined phage-display of fragments and random peptides. *Protein Eng.* 14, 691–698.
15. Fack, F., Hügle-Dörr, B., Song, D., Queitsch, I., Petersen, G., and Bautz, E. K. (1997) Epitope mapping by phage-display: random versus gene-fragment libraries. *J. Immunol. Methods* 206, 43–52.
16. Kawamura, M., Shibata, H., Kamada, H., Okamoto, T., Mukai, Y., Sugita, T., Abe, Y., Imai, S., Nomura, T., Nagano, K., Mayumi, T., Nakagawa, S., Tsutsumi, Y., and Tsunoda, S. I. (2006) A novel method for construction of gene fragment library to searching epitopes. *Biochem. Biophys. Res. Commun.* 346, 198–204.
17. Di Niro, R., Ferrara, F., Not, T., Bradbury, A. R., Chirdo, F., Marzari, R., and Sblattero, D. (2005) Characterizing monoclonal antibody epitopes by filtered gene fragment phage-display. *Biochem. J.* 388, 889–894.
18. Dyson, M. R., Germaschewski, V., and Murray, K. (1995) Direct measurement via phage titre of the dissociation constants in solution of fusion phage-substrate complexes. *Nucleic Acids Res.* 23, 1531–1535.

Subsection E

B-Cell Epitope Prediction

Chapter 24

Prediction of Linear B-cell Epitopes

Ulf Reimer

Summary

The prediction of B-cell epitopes is desirable for designing peptide-based vaccines, or generating antibodies especially if the purified protein is difficult to obtain and immunization has to be performed with protein-derived synthetic peptides. A number of freely available tools predict epitopes from protein sequence or structural information. The handling of these tools is described and the predictive power is assessed using test data based on the proteome of HIV, where comprehensive epitope mapping data are available.

Key words: B-cell, Antibody, Epitopes, Sequence, 3D structure, Prediction.

1. Introduction

The humoral immune response is based on the amazing ability of antibodies to recognize and bind to antigens presented by intruding organisms, such as bacteria or viruses. Antibodies bind specifically to either linear stretches of amino acids on the surface of the intruders, or to surface patches on the protein formed by different amino acids not necessarily in consecutive sequence. This recognition event enables the immune system to clear the pathogens from the invaded organism.

The specific interaction between antibodies and their antigens is also exploited extensively in biochemical research tools. Specific antibodies are indispensable for a multitude of experimental techniques. In many cases it is difficult to obtain a pure preparation of the protein of interest for immunization purposes. However, to raise antibodies it is not necessary to present the complete protein but only the immunogenic fractions. Specific antibod-

Ulrich Reineke and Mike Schutkowski (eds.), *Methods in Molecular Biology, Epitope Mapping Protocols, vol. 524*

DOI: 10.1007/978-1-59745-450-6_24

ies can be generated by immunization of animals with a peptide/template protein fusion if the peptide is an antibody binding site from the protein of interest. If the peptide is well chosen and presents an effective epitope of the protein, the resulting antibodies often cross-react with the entire protein they are derived from.

Attempts to predict B-cell epitopes started about 30 years ago. The first approaches were based on evaluating different properties of amino acids along a protein's sequence. However, an exhaustive benchmark procedure *(1)* confirmed that single-scale amino acid propensity profiles cannot be used to reliably predict an epitope's location.

More advanced methods take into account available information about epitopes and nonepitopes, leading to better predictions *(2)*. A major breakthrough in the prediction of antibody epitopes came from structural biology (*see* **Note 1**). An antibody usually recognizes and binds amino acids on the surface of a natively folded protein. Therefore, when the three-dimensional structure of a protein is known there is a much better chance of correctly predicting the epitopes. Even if no experimental structural information of a particular protein of interest is available, it is possible to gather useful information from structures of homologous proteins to automatically build structural models of the chosen protein. More than 40,000 protein structures are available today, representing 1,054 different folds as defined by the SCOP classification system (http://scop.mrc-lmb.cam.ac.uk/scop/). Easy to use services such as SWISS-MODEL (http://swissmodel.expasy.org/) make protein sequence comparison and automatic model-building very convenient.

Different approaches based on sequence and structure information are now freely available as web services. The use of these services is described here, illustrated with results for a set of examples taken from the well-investigated proteome of HIV (*see* **Note 2**).

2. Materials

2.1. Input Data

Crucial for choosing the best method for predicting B-cell epitopes is assessing the available information on the protein of interest. If structural information on the protein, or a related protein with high sequence homology, similar function or a similar fold, is available one should use this information. This can be rapidly checked on the Swiss Model Server at swissmodel.expasy.org. Using the "First Approach Mode" will quickly deliver information on available structural data for this or related proteins upon providing solely the sequence of the protein of interest (*see* **Subheading 3.2.1**).

2.2. Input Data for the Presented Example

The comparative prediction example uses sequence and epitope mapping information from the HIV epitope map (at http://www.hiv.lanl.gov/content/immunology/maps/maps.html; April 11, 2007). Epitope information is available for 11 of the HIV proteins (p17, p24, p2p7p1p6, protease, reverse transcriptase (RT), integrase, Vif, Tat, Rev, gp160, and Nef). The performance of prediction algorithms was evaluated for these proteins.

Structural data were used when available (protein name: pdb code; p17:2H3V; p24:1E6J; protease:1B6K; RT:1TVR; integrase:1K6Y; gp160:2NXY; Nef:2NEF). Models for the proteins p2p7p1p6 and Tat were built using the SWISS-MODEL server *(3)*.

2.3. Statistical Measures Used for Comparison

Methods using the example dataset described in **Subheading 2.2** were evaluated by calculating two measures for each prediction: sensitivity and specificity. Sensitivity is the fraction of correctly predicted epitopes calculated as the number of true-positive hits divided by the sum of true-positives and true-negatives. Specificity is the fraction of epitopes correctly predicted as nonepitopes and results from dividing the number of true-negatives by the sum of true-negatives and false positives. Both values are multiplied by 100 to give percent values. For an ideal prediction both values would reach 100%. If the sensitivity is higher than 100 minus specificity, the prediction is better than a random selection (*see* **Note 2**).

3. Methods

3.1. Methods Based on Sequence Information

3.1.1. ABCpred

ABCpred *(4)* is located at http://www.imtech.res.in/raghava/abcpred. It uses recurrent neural networks (RNN) and was trained with a dataset of 700 experimentally detected B-cell epitopes from the Bcipep database *(5)* and 700 random peptides from the Swiss-Prot database for which no antibody binding is reported as a negative dataset.

At the submission page the sequence of the protein of interest has to be entered in one-letter code. The parameters thres hold and window length can be defined. A higher threshold (from 0.1 to <1) leads to fewer predicted epitopes with a higher probability of correct prediction. The available window length (or epitope length) ranges from 10 to 20 residues. To understand this parameter the length of the epitopes used to train the RNN has to be considered. Epitopes of varying length were collected in the training dataset, and if shorter than 20 amino acids, extended to this size by adding amino acids from the original sequences of the respective epitope at the N- and C-terminus until reaching the length of 20 amino acids. Thus, using a small window length will cut off information from the peripheral sequence regions of epitopes longer than the window size in the training dataset.

The prediction of epitopes in the example dataset gave evidence for an overprediction of epitopes at all scores. At the score of 0.51 almost all residues were predicted as epitopes. However, at higher scores above 0.7 the rate of correctly predicted non-epitopes increased, but at the same time the number of correctly predicted residues dropped. The best results were achieved using a window length of 16 residues. Together with a threshold of 0.6 the prediction was better than a random guess for 6 out of 11 proteins of the HIV dataset (**Fig. 1**, **Table 1**).

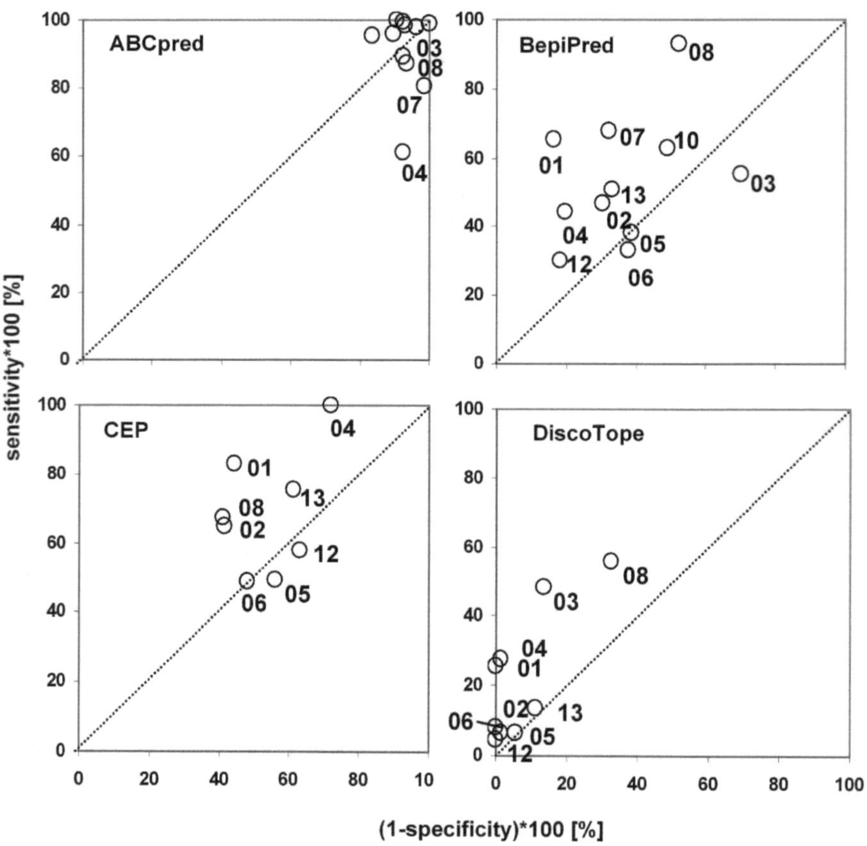

Fig. 1. Plots of prediction results from different methods for the HIV data set in Receiver Operating Characteristic (ROC) space. The ideal prediction would be at a sensitivity of 1 (100% on the *Y*-axis, all true-positives predicted) and a specificity of 1 (0% on the *X*-axis, no false-positives predicted) in the *upper left* area of the graph (*see* Note 2). The *dotted line* represents random guesses. Predictions in the *upper left triangle* from this line are better than a random guess, predictions in the *lower right triangle* are worse than a random guess. The *points* in the diagrams represent single proteins from the HIV data set (01 = p17; 02 = p24; 03 = p2p7p1p6; 04 = protease; 05 = RT; 06 = integrase; 07 = Vif; 08 = tat; 10 = Rev; 12 = gp160; 13 = Nef).

Table 1
True-positive (TP), true-negative (TN), false-positive (FP) and false-negative (FN) numbers for the prediction of B-cell epitopes using several freely available B-cell epitope prediction tools

	Epi	ABCpred				BepiPred				CEP				Discotope −3.1				DiscoTope −7.7			
	Percent[a]	TP	TN	FP	FN	TP	TN	FP	FN	TP	TN	FP	FN	TP	TN	FP	FN	TP	TN	FP	FN
p17	67	85	7	36	4	58	36	7	31	74	24	19	15	23	43	0	66	51	36	7	38
p24	84	193	0	36	2	91	25	11	104	113	21	15	61	14	36	0	160	75	32	4	99
p2p7p1p6[b]	34	42	7	83	5	26	27	63	21	c	c	c	c	12	26	4	13	21	13	17	4
Protease	18	11	6	75	7	8	65	16	10	18	22	57	0	5	78	1	13	18	56	23	0
RT	16	87	18	453	2	34	288	183	55	44	207	262	45	6	442	27	83	20	279	190	69
Integrase	41	115	12	159	2	39	106	65	78	22	76	71	23	3	145	2	42	16	121	26	29
Vif	16	25	2	159	6	21	109	52	10	d	d	d	d	d	d	d	d	d	d	d	d
tat[b]	54	48	3	43	7	51	22	24	4	35	20	14	17	29	23	11	23	37	22	12	15
Rev	63	73	4	39	0	46	22	21	27	d	d	d	d	d	d	d	d	d	d	d	d
gp160	62	411	15	127	17	128	116	26	300	132	28	48	96	10	76	0	218	50	60	16	178
Nef	87	178	2	25	1	91	18	9	88	89	7	11	29	16	16	2	102	58	8	10	60

[a]Percentage of amino acids that are known to be part of an epitope in this protein
[b]Structural models generated using SWISS-MODEL were used as input for CEP and Discotope
[c]The structural model was not accepted as an input for CEP
[d]No structural information available

3.1.2. Bepipred

Hidden Markov models (HMM) and the propensity scale methods of Parker et al. *(6)* and Levitt *(7)* were combined to develop the BepiPred method *(8)*. A dataset of 127 protein sequences with annotated epitopes taken from the AntiJen database *(9)* was used for building the HMM. The web server is located at http://www.cbs.dtu.dk/services/BepiPred. The input sequence has to be pasted into the sequence window in one-letter amino acid code. Multiple sequences can be pasted using FASTA format or submitted directly in this format from a local computer. A single threshold parameter can be adjusted. The default value is 0.35. An increase in this threshold increases specificity at the cost of decreased sensitivity. Decreasing the threshold leads to the opposite result. After submitting the task the output is shown as a table with scores for every amino acid of the submitted protein. Residues with assigned scores above the threshold are annotated as epitopes. If the scores are generally high in the prediction, and thus a high number of residues are predicted to be part of an epitope, the threshold can be increased for a prediction of higher specificity and lower sensitivity.

For the prediction of epitopes in the HIV dataset, a threshold of 0.35 was used and resulting predictions were better than a random guess for 8 out of 11 proteins. For two of the less accurately predicted proteins an increase in the threshold to 0.7 (reverse transcriptase) or a decrease to 0 (integrase) led to predictions better than a random guess (**Fig. 1**, **Table 1**).

3.2. Methods Based on Structural Information

3.2.1. Searching Structural Information or Building Homology Models

Structural data of proteins are collected in the Research Collaboratory for Structural Bioinformatics Protein Data Bank (PDB at http://www.pdb.org). The search page offers an option for searching for the sequence of the protein of interest using BLAST or FASTA. If an experimental structure is available it can be downloaded and used as an input for the epitope prediction. Even though more than 40,000 experimental structures are currently deposited in the protein database, there is a high probability that no structural data are available for a protein of interest. However, the wealth of experimental information on protein structure and the available computing power make structural modeling attractive. This led to databases of structural models, such as MODBASE *(10)* (at http://modbase.compbio.ucsf.edu/modbase), FAMSBASE (at http://www.pharm.kitasato-u.ac.jp/fams), or the SWISS-MODEL Repository (at http://swissmodel.expasy.org/repository/). It is inherent in modeling that the details in the structures predicted show considerable deviations from experimental structures, but in most cases the general architecture of the fold is correct. So these models can contribute valuable information to a prediction of residues on the surface of a protein and those buried in the inside and thus not accessible for antibody binding.

If no model is available, easy to use interfaces for modeling servers allow the straightforward generation of structural models.

If structural information on a related sequence can be used there is a good chance of obtaining a useful model. Such services are offered at SWISS-MODEL (http://swissmodel.expasy.org/), ModWeb (http://alto.compbio.ucsf.edu/modweb-cgi/main.cgi), and FAMS (http://www.pharm.kitasato-u.ac.jp/fams) (*see* **Note 3**).

3.2.2. CEP

The CEP server *(11)* (at http://bioinfo.ernet.in/cep.htm) predicts conformational epitopes and linear epitopes from a 3D protein structure. About 1,800 structures with a resolution better than 1.5 Å are available as a precalculated dataset. The PDB code is sufficient as input in these cases, otherwise a coordinate file in PDB format needs to be uploaded to the server. If more than one chain is stored in the file information, the ID of the chain of interest also has to be entered.

The heart of the prediction algorithm is calculating the accessibility of residues to the solvent. When the accessibility of a residue is ≥25% the residue is considered to be accessible for antibody binding. Stretches of at least three accessible amino acids in a row are considered antigenic determinants (ADs). When other accessible residues are separated from the AD by a single inaccessible residue these are added to the AD. Conformational epitopes (CE) are predicted by combining ADs in a spatial proximity of 6 Å. ADs and CEs are listed in separate tables with the sequence number of the first and last amino acid in AD as well as its sequence. Amino acids within the predicted ADs that do not fulfill the accessibility criterion are shown in lower case. Reference ADs for the CE are shown in green and accessible residues within the spatial range of the CE are also given. The predicted epitopes can be visualized on the structure of the protein using an implemented JAVA viewer.

With the HIV dataset, the CEP server performed considerably better than a random guess for 5 out of 8 proteins. However, the predictions for reverse transcriptase, integrase and gp160 were in the range of a random guess (**Fig. 1**, **Table 1**).

3.2.3. DiscoTope

The most advanced service for the prediction of epitopes is DiscoTope *(12)* (at http://www.cbs.dtu.dk/services/DiscoTope/). As the CEP server, DiscoTope evaluates the surface localization of residues. Here, residue contact numbers are used as a measure for surface localization and structural protrusion. Additionally, amino acid statistics from a training set of 76 X-ray structures of antibody/antigen pairs are used to improve the predictions. Although DiscoTope was designed especially for the prediction of discontinuous or conformational epitopes, its predictions are also valid for potential linear epitopes.

There are three ways of submitting one or multiple protein queries to the DiscoTope server: (1) by entering a single PDB code and chain ID, (2) by uploading a file containing a list of PDB codes and chain IDs, or (3) by uploading a coordinate file in PDB format and specifying the chain ID. If not already included in the coordinate file, a chain ID can be added as described in **Note 3**.

The only input parameter to be set is the threshold for epitope identification. By increasing the default value of –7.7 toward –3.1 the sensitivity of the prediction decreases, but the specificity increases. Thus, the number of residues predicted to be part of an epitope will drop, but the accuracy of prediction will improve.

The output table contains the following parameters for each amino acid of the input: chain ID, residue number, name of the amino acid in the three letter code, contact number (a value for the surface accessibility), propensity score (a knowledge-based statistical value), DiscoTope score (a combined value from the contact number and propensity score) and, if the residue is predicted as a B-cell epitope, the symbol < = B. This annotation is automatically set when the DiscoTope score is higher than the threshold for epitope identification on the input screen.

For the 7 structures and 2 models used from the HIV dataset all predictions were of high specificity but low sensitivity. At a threshold of –3.1 a prediction better than a random guess was achieved for all structures. Except for reverse transcriptase, the number of residues correctly predicted as part of epitopes (true-positives, TP) exceeded the number of false positives (FP). All predictions were better than a random guess at this threshold for the HIV dataset. A lower score down to –7.7 led to a higher number of correctly predicted epitope residues. However, in parallel the number of correctly predicted nonepitope residues decreased (**Fig. 1**, **Table 1**).

3.3. Summary of Results

In summary, there is currently no reliable way to exactly predict B-cell epitopes based solely on the sequence or the three-dimensional structure of a protein. Sequence-based methods hardly give better predictions than guessing randomly. Adding structural information improves the quality considerably. In particular, DiscoTope seems to be a valuable tool to predict peptides that are B-cell epitopes with reasonable confidence, as well as improve the chances of generating valuable peptide-based antibodies for detecting a given antigenic protein.

4. Notes

1. Any structural information will considerably increase the reliability of epitope prediction for B-cells. If there is no experimental data, structure prediction as described in **Subheading 3.2.1** can be valuable.

2. All prediction tools described here have adjustable parameters for the prediction, such as scores or window length. Trying different settings will provide a better understanding of the results.

3. For some prediction tools that use structural information a chain identifier must be present in the coordinate file (pdb-file). A number of coordinate files of single chain proteins in the PDB do not have a chain ID assigned to the chain. In such cases it is necessary to add this, which can be easily achieved using the Deep View Swiss-PDB Viewer (freeware at http://expasy.org/spdbv). After opening the coordinate file in this viewer the chain must be selected by double clicking to the right of the three letter amino acid code in the control window. Then under "Edit/Rename Current Layer" the chain of selected groups can be renamed by simply entering a letter. Finally, the coordinate file is saved and can be used as an input file for DiscoTope (*see* **Subheading 3.2.3**).

Acknowledgments

The author thanks Dr. Tatiana Reimer and Prof. Maik Kschischo for their advice.

References

1. Blythe, M. J. and Flower, D. R. (2005) Benchmarking B cell epitope prediction: underperformance of existing methods. *Protein Sci.* 14, 246–248.
2. Greenbaum, J. A., Andersen, P. H., Blythe, M., Bui, H. H., Cachau, R. E., Crowe, J., Davies, M., Kolaskar, A. S., Lund, O., Morrison, S., Mumey, B., Ofran, Y., Pellequer, J. L., Pinilla, C., Ponomarenko, J. V., Raghava, G. P., van Regenmortel, M. H., Roggen, E. L., Sette, A., Schlessinger, A., Sollner, J., Zand, M., and Peters, B. (2007) Towards a consensus on datasets and evaluation metrics for developing B-cell epitope prediction tools. *J. Mol. Recognit.* 20, 75–82.
3. Schwede, T., Kopp, J., Guex, N., and Peitsch, M. C. (2003) SWISS-MODEL: An automated protein homology-modeling server. *Nucleic Acids Res.* 31, 3381–3385.
4. Saha, S. and Raghava, G. P. (2006) Prediction of continuous B-cell epitopes in an antigen using recurrent neural network. *Proteins* 65, 40–48.
5. Saha, S., Bhasin, M., and Raghava, G. P. S. (2005) Bcipep: A database of B-cell epitopes. *BMC Genomics* 6, 79.
6. Parker, J., Guo, D., and Hodges, R. (1986) New hydrophilicity scale derived from high-performance liquid chromatography peptide retention data: correlation of predicted surface residues with antigenicity and X-ray-derived accessible sites. *Biochemistry* 25, 5425–5432.
7. Levitt, M. (1978) Conformational preferences of amino acids in globular proteins. *Biochemistry* 17, 4277–4285.
8. Pontoppidan Larsen, J. E., Lund, O., and Nielsen, M. (2006) Improved method for predicting linear B-cell epitopes. *Immunome Res.* 2, 2.
9. McSparron, H., Blythe, M., Zygouri, C., Doytchinova, I., and Flower, D. (2003) Jen-Pep: a novel computational information resource for immunobiology and vaccinology. *J. Chem. Inf. Comp. Sci.* 43, 1276–1287.

10. Pieper, U., Eswar, N., Davis, F. P., Braberg, H., Madhusudhan, M. S., Rossi, A., Marti-Renom, M., Karchin, R., Webb, B. M., Eramian, D., Shen, M. Y., Kelly, L., Melo, F., and Sali, A. (2006) MODBASE: a database of annotated comparative protein structure models and associated resources. *Nucleic Acids Res.* 34, D291–D295.

11. Kulkarni-Kale, U., Bhosle, S., and Kolaskar, A. S. (2005) CEP: a conformational epitope prediction server. *Nucleic Acids Res.* 33, W168–W171.

12. Andersen, P. H., Nielsen, M., and Lund, O. (2006) Prediction of residues in discontinuous B cell epitopes using protein 3D structures. *Protein Sci.* 15, 2558–2567.

Section II

T-Cell Epitope Mapping

Chapter 25

Molecular recognition of diverse ligands by T-Cell receptors

Eric J. Sundberg

Summary

T-cell receptors (TCRs) are structurally related to antibodies, and also interact with a diverse set of ligands. TCRs recognize foreign peptide antigens displayed by major histocompatibility complex (MHC) molecules and foreign lipid-based antigens presented by CD1. These interactions initiate an immune response through T-cell activation. These critical surveillance and response initiation functions of the adaptive immune system are not perfect, though, as TCR interactions with self antigens can lead to autoimmune disease. Mutated peptides can also be recognized specifically by TCRs, and may be important in tumor immunity. TCRs are also bound specifically by a family of bacterial toxins called superantigens, which over-stimulate the immune system to cause numerous human diseases.

Key words: T-cell receptor, Peptide antigen, Lipid antigen, Superantigen, X-ray crystallography.

1. Introduction

T-cell receptors (TCRs) are structural cousins to antibodies, and also interact with a diverse set of ligands. Their primary function is to recognize antigens displayed by specialized molecules on the cell surface, binding to foreign peptide antigens in the context of major histocompatibility complex (MHC) and to foreign lipid-based antigens presented by CD1, thereby initiating an immune response through T-cell activation. In this way, TCRs serve a critical surveillance and response initiation function in the adaptive immune system. These recognition events are not perfect, however, as TCR interactions with self antigens lead to autoimmune disease. Mutated peptides or altered-self epitopes

Ulrich Reineke and Mike Schutkowski (eds.), *Methods in Molecular Biology, Epitope Mapping Protocols, vol. 524*

DOI: 10.1007/978-1-59745-450-6_25

can also be recognized specifically by TCRs, and may serve an important immunosurveillance role in tumor immunity. TCRs are also bound specifically by a family of bacterial toxins called superantigens, which over-stimulate the immune system to cause numerous human diseases.

2. Anatomy of the T-Cell Receptor

TCRs are composed of multiple immunoglobulin (Ig) domains and are structurally equivalent to a single Fab fragment of an antibody (**Fig. 1**). Just like the variable domains of antibodies (V_H and V_L), those of TCRs (Vα and Vβ) contain complementarity-determining regions (CDRs) that connect the framework of β-strands in the Ig domain. These CDR loops lie in close spatial proximity on the surface of the molecule and together form a contiguous hypervariable

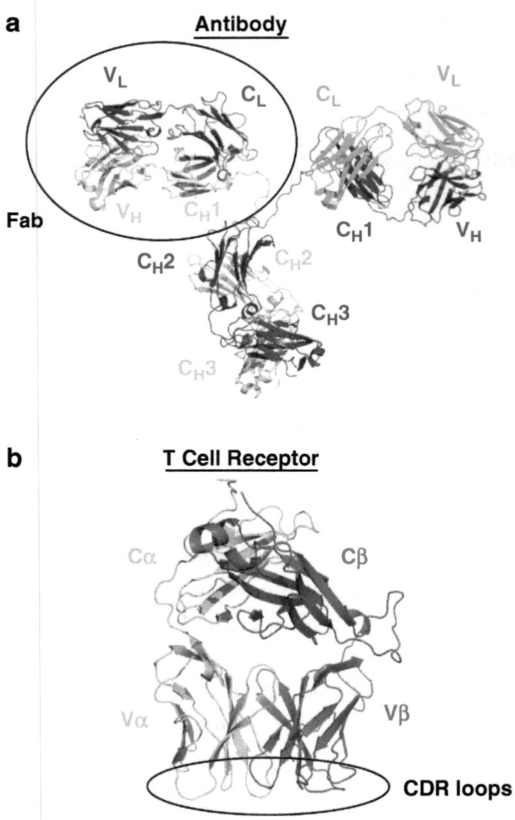

Fig. 1. Structural similarities between antibodies and T-cell receptors. (**a**) An intact antibody molecule in which the Fab fragment is highlighted by the *oval*. (**b**) A T-cell receptor. The CDR loops are highlighted by the *oval* (*see Color Plates*).

surface that is able to recognize specifically a nearly limitless array of antigenic epitopes. One major difference between antibodies and TCRs is that the former undergo an affinity maturation process via somatic hypermutation after the initial encounter with the antigen, while the latter are genetically static once they have undergone recombination and selection in the developing thymus. This reflects their distinct antigen-binding affinity requirements, approximately nm for antibodies, but only in the μm range for TCRs.

3. TCR Recognition of Peptide–MHC Complexes

T-cell receptors (TCRs) are an integral part of the adaptive immune system that has evolved to distinguish nonself pathogens from self tissues. Whereas, T-cell recognition of foreign peptides is essential for immune defense against invading microorganisms, recognition of self-peptides is thought to cause autoimmune disease, and T-cell epitopes involving altered-self peptides resulting from mutations accumulated during aging or disease are often associated with immunity to cancer *(1, 2)*. In terms of T-cell recognition, the boundaries separating foreign, self, and altered-self epitopes are somewhat blurred. Here, the structural characteristics of TCR recognition of these three broad classes of peptide–major histocompatibility complexes (pMHCs) are discussed.

3.1. Recognition of Foreign Peptide Antigens by TCRs

The earliest structural studies of TCR/pMHC complexes targeted TCRs specific for microbial and other foreign epitopes, or displaying alloreactivity *(3)*. These studies demonstrated remarkable similarities in the overall topology of TCR binding to pMHC, irrespective of MHC class I or class II restriction. In general, the TCR is positioned diagonally across the compound surface created by the peptide and the MHC α-helices that flank the peptide-binding groove, although some class I-restricted TCRs adopt a more orthogonal binding mode *(4)*. The orientation angle, defined as the angle between the line formed by the peptide direction and a line between the centers of mass of the Vα and Vβ domains, for all reported foreign pMHC class I- or class II-restricted TCRs is 45–80° *(3)*. The diagonal orientation is exemplified by the structure of human TCR HA1.7 bound to an influenza virus hemagglutinin (HA) peptide and HLA-DR1 (**Fig. 2a**) *(5)*. The most structurally diverse CDR loops, CDR3α and CDR3β, are generally located over the central peptide residue at position P5, and form a pocket that accommodates the P5 side chain. This docking mode maximizes interactions between the CDR3 loops and the MHC-bound peptide. This concentration of CDR3 loop contacts with foreign antigen is reminiscent of

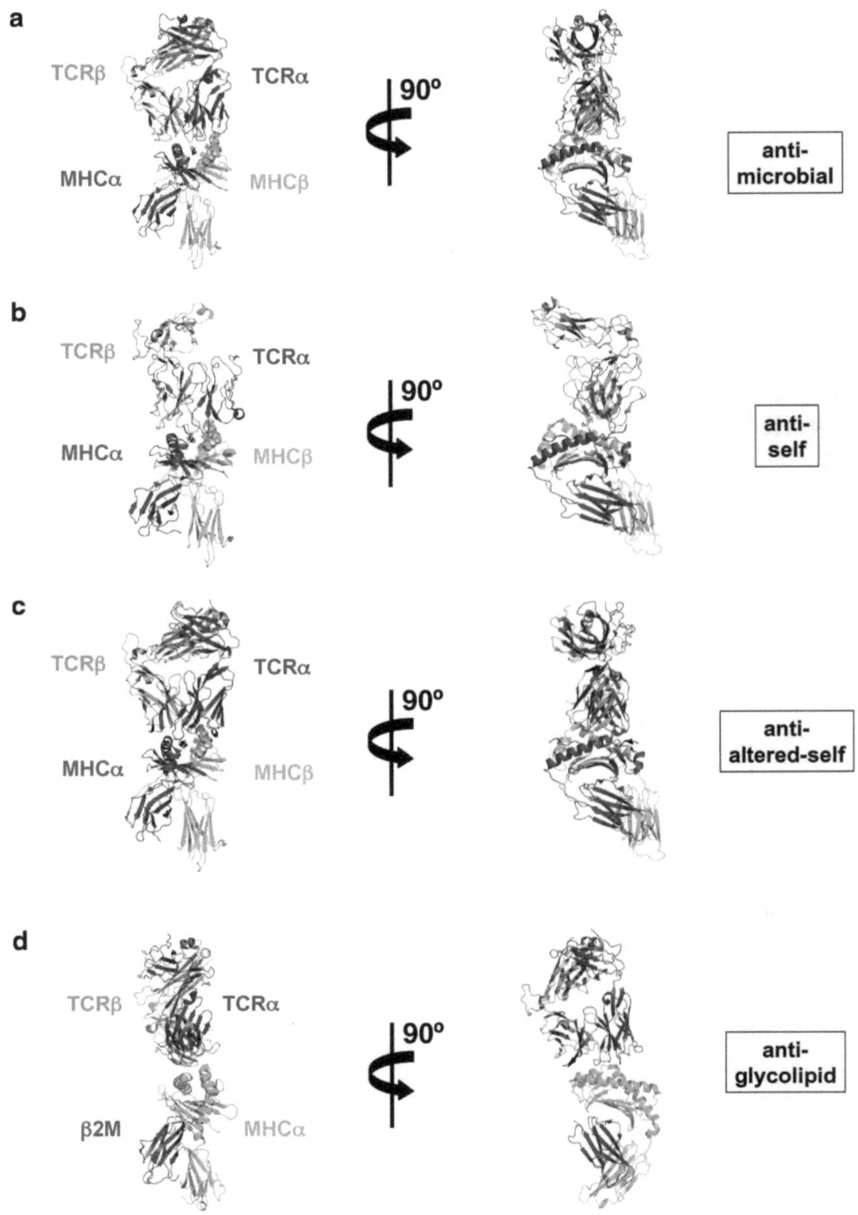

Fig. 2. TCR recognition of peptidic and lipid-based antigens displayed by MHC and CD1 molecules. (**a**) A representative anti-microbial complex, the structure of TCR HA1.7 bound to an influenza hemagglutinin peptide displayed by the MHC class II molecule HLA-DR1. (**b**) A representative anti-self complex, the structure of TCR Ob.1A12 bound to the MBP_{85-99} peptide displayed by the MHC class II molecule HLA-DR2b. (**c**) A representative anti–altered-self complex, the structure of TCR E8 bound to the mutant TPI peptide displayed by the MHC class II molecule HLA-DR1. (**d**) A representative anti-glycolipid complex, the structure of NKT TCR bound to the α-GalCer glycolipid displayed by CD1d (*see Color Plates*).

antibody–antigen interactions. The central diagonal orientation commonly observed for TCRs recognizing microbial epitopes represents an optimal binding mode for maximizing interactions between TCR and the MHC-bound peptide, resulting in relatively high affinity for pMHC (K_D ~ 1–100 μM) *(6)*.

3.2. Recognition of Self Peptide Antigens by TCRs

The overall similarities among the initial structures of TCRs bound to MHC class I and II that display foreign peptide antigens created the expectation that all TCRs bind pMHC complexes similarly, and that pMHC recognition by autoreactive TCRs would be qualitatively indistinguishable from these. Several recent structures of autoimmune TCR/pMHC complexes, though, have revealed that autoimmune TCRs engage pMHC with distinct unconventional binding topologies compared with TCRs specific for foreign antigens.

In one such autoimmune complex between the TCR Ob.1A12 and a peptide composed of residues 85–99 of myelin basic protein (MBP) presented by the MHC class II molecule HLA-DR2b *(7)*, the TCR is not centered over pMHC and only contacts the N-terminal portion of the self peptide (**Fig. 2b**). Furthermore, this TCR exhibits a counter-clockwise rotation relative to pMHC compared with anti-foreign TCRs, resulting in a highly asymmetrical interaction with MHC and an orientation angle of 110°. In another autoimmune complex between human TCR 3A6 and MBP 89–101 presented by HLA-DR2a *(8)*, the orientation angle of TCR to peptide/MHC is within the range of anti-foreign TCR/pMHC complexes, although the CDR footprint of this TCR on pMHC is shifted markedly towards the N-terminus of the bound peptide, and towards the MHC β1 α-helix, compared with the CDR footprint on representative foreign pMHC complexes. A third autoimmune TCR/pMHC complex is unusual in that the N-terminal one-third of the binding groove is empty *(9)*, and as a consequence, this TCR recognizes only six peptide residues and uses only two CDRs to engage the peptide.

It is possible that autoimmune TCRs are intrinsically more cross-reactive than anti-foreign TCRs, which would increase the probability of self pMHC recognition, and the pathogenic potential of T cells expressing such TCRs would be enhanced, thus resulting in autoimmunity. Besides their sub-optimal binding topologies, these autoimmune TCR/pMHC complexes are commonly characterized by a scarcity of intermolecular hydrogen binding interactions. Additionally, these interactions exhibit much weaker affinities than do anti-foreign TCR/pMHC complexes and/or exceedingly short half-lives.

3.3. Recognition of Altered-Self Peptide Antigens by TCR

Recently, the crystal structure of a human tumor-specific TCR bound to a melanoma peptide epitope derived from the enzyme triosephosphate isomerase (TPI), in which a single-site mutation has occurred, and an MHC class II molecule has been determined *(10)*. This complex reveals a number of features intermediate between those of anti-foreign and autoimmune TCR-pMHC class II complexes that may reflect the hybrid nature of altered-self (**Fig. 2c**). These include a shift of the TCR toward the N-terminus of the bound peptide relative to anti-foreign TCRs, though not as extreme as for autoimmune TCRs, while maintaining the generally diagonal binding orientation of anti-foreign TCRs. As a consequence of this shift, the CDR3 loops of the TCR are positioned directly over the mutated residue of the altered-self peptide epitope. This focus on the N-terminal half of self-peptides, which may be prevalent among both anti-self and anti-altered-self TCRs that have escaped negative selection during thymic development, implies that the N-terminal site is intrinsically less favorable for TCR binding than the central site typically utilized by TCRs recognizing foreign epitopes *(3, 5, 11)*. As with autoimmune TCRs, the TCR/altered-self pMHC complex exhibits very low affinity, although the altered-self mutation at the TCR-contacting position of the peptide epitope results in a modest increase in binding strength. The TCR in this complex is tilted toward the DR β-chain, with which it makes many more contacts (~80% of total contacts) than it does with the MHC α-chain *(10)*, a feature that also generally distinguishes autoimmune from anti-microbial TRC/pMHC complexes.

4. TCR Recognition of Glycolipid Antigens Displayed by CD1 Molecules

CD1 molecules comprise a large cluster of nonpolymorphic MHC class I-like molecules that present lipid-based antigens, such as glycolipids, to TCRs. The antigen-binding clefts of CD1s are characterized by vastly larger and more hydrophobic pockets than those formed in MHC molecules that accommodate the lipid moieties of these antigens. This allows the hydrophilic portions of the antigen, such as sugars, to protrude from the top surface of the CD1 molecule where they can be easily encountered by TCRs. The lipids, sugars, and peptides synthesized by microbes are significantly different from those made by vertebrates, thereby providing a basis for antigenicity.

4.1. Structural Characteristics of TCR/Glycolipid–CD1 Complexes

Certain T-cells, known as invariant natural killer T (NKT) cells, in part because of their expression of a semi-invariant TCR (NKT TCR), in which the TCR α chain is invariant while the β chain is restricted, are known to recognize the glycolipid α-galactosylceramide (α-GalCer) displayed by CD1d. A recent structure of the NKT TCR/α-GalCer/CD1d complex *(12)* provides the first view of how a TCR recognizes nonpeptidic antigens (**Fig. 2d**). In this structure, NKT TCR is positioned over the end of the CD1d antigen-binding cleft, at the F' pocket, analogous to the C-terminus of antigenic peptides displayed by MHC molecules. Furthermore, NKT TCR binds essentially parallel to the antigen-binding cleft, in stark contrast to the vast array of diagonal to orthogonal binding modes of

TCR/pMHC complexes. These two distinctive features of the NKT TCR/α-GalCer/CD1d complex results in the sugar moiety of the antigen being contacted by only the TCR Vα domain and not at all by the Vβ domain. These TCR-antigen contacts are roughly split between the CDR1α and CDR3α loops, the latter of which form contacts as well with both α-helices of the CD1d molecule. The TCR Vβ domain, conversely, accounts for less than 30% of all intermolecular contacts in the complex and only forms contacts with the CD1d molecule confined to the extreme C-terminal end of the α1 α-helix. By comparing the structure of the complex with that of the unbound NKT TCR *(13, 14)*, it is evident that the NKT TCR/α-GalCer/CD1d complex exhibits the hallmarks of a lock-and-key interaction, with no significant conformational changes induced upon complex formation. Accordingly, binding of this complex has been shown to be insensitive to changes in temperature *(15)*. This is quite unlike TCR/pMHC complexes, for which protein plasticity is a common feature.

TCR recognition of glycolipid antigen presented by CD1 seems to blur the line between innate and adaptive immunity. Like mechanisms of innate immunity, CD1 molecules exhibit a high degree of cross-species recognition. Also, the exclusive invariant domain usage for antigen recognition, at least in the case of the NKT TCR/α-GalCer/CD1d complex is, by definition, innate-like. However, the recognition of distinct CD1d/antigen complexes by various NKT cells, provides an adaptive component to immunity to lipid-based antigens.

5. TCR Recognition of Superantigens

Bacterial superantigens (SAGs) comprise a large family of disease-associated proteins that are produced predominantly by *Staphylococcus aureus* and *Streptococcus pyogenes (16)*, as well as by a number of other bacteria and viruses. SAGs function by simultaneously interacting with class II MHC and TCR molecules on antigen presenting cells and T lymphocytes, respectively *(17)*. Contrary to the peptidic and lipid-based antigens discussed above, SAGs bind to MHC molecules outside of their peptide-binding grooves and interact predominantly with only the Vβ domains of TCRs, resulting in the stimulation of up to 20% of the entire T-cell population. In this way, SAGs initiate a systemic release of inflammatory cytokines that results in various immune-mediated diseases including a condition known as toxic shock syndrome (TSS) that can ultimately lead to multi-organ failure and death. SAGs have also been implicated in the pathogeneses of arthritis, asthma and inflammatory bowel syndrome, and are classified as bioterror reagents.

5.1. Superantigen-TCR Specificity and Cross-Reactivity

With some 30 SAGs from *S. aureus* and *S. pyogenes* and more than 50 TCR Vβ domains encoded by the human genome, SAG–TCR interactions constitute a complex molecular recognition problem, in which some SAGs are strictly specific for a single Vβ domain, while others bind much more promiscuously to a multitude of Vβ domain targets. The recently expanded database of SAG–TCR Vβ domain crystal structures allows the construction of a paradigm for how SAGs confer specificity and cross-reactivity in TCR recognition.

The least specific SAGs (including staphylococcal enterotoxins B (SEB) and C (SEC3)) depend primarily on a common conformation adopted by the CDR2 loop and the fourth hypervariable (HV4) loop in many Vβ domains *(18, 19)*. In these complexes (**Fig. 3a**), hydrogen bonds are made only to Vβ main chain atoms, such that numerous combinations of amino acid

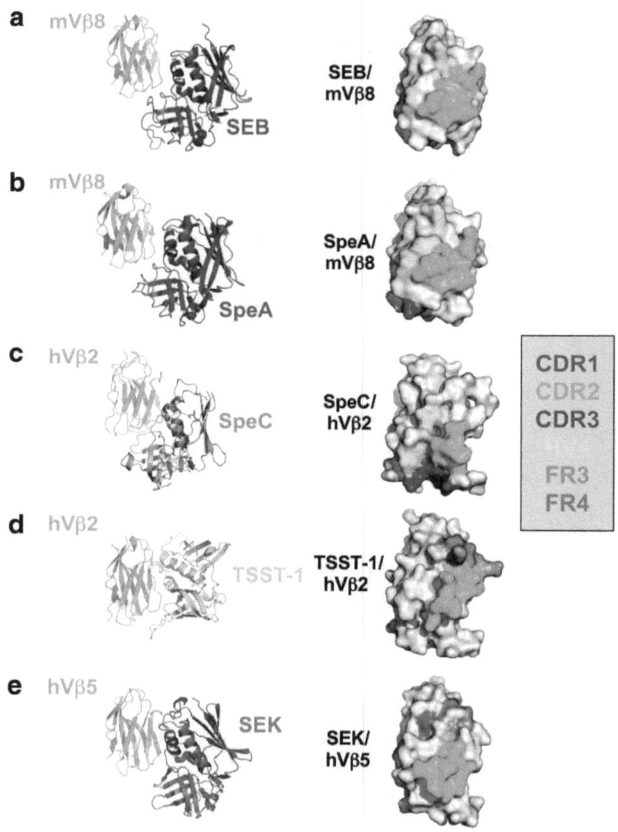

Fig. 3. Superantigen engagement of the T-cell receptor Vβ domain. *Left*, structures of the (**a**) SEB/mVβ8, (**b**) SpeA/mVβ8, (**c**) SpeC/hVβ2, (**d**) TSST-1/hVβ2, and (**e**) SEK/hVβ5 complexes. The Vβ domains are aligned to one another to highlight the distinct orientations by which these SAGs engage their TCR ligands. *Right*, TCR Vβ domain molecular surface buried by various SAGs. Hypervariable and framework region surface residues buried in the interface formed by TSST-1, SEB, SpeC, and SEK are color-coded as follows: CDR1 (*red*); CDR2 (*green*); CDR3 (*blue*); HV4 (*yellow*); FR3 (*orange*); and FR4 (*magenta*). The Vβ domains in the *right* are rotated counter-clockwise approximately 90? about the vertical axis of the page relative to their positions in the *left* (*see Color Plates*).

sequences in CDR2 and HV4 can satisfy the binding requirements for these SAGs, as long as they do not change the lengths of these hypervariable loops nor disrupt the common structural conformation adopted.

As TCR specificity increases (e.g., for streptococcal pyrogenic exotoxin A (SpeA)), the number of hypervariable loops with which the SAG interacts increases beyond CDR2 and HV4 to include CDR1 (**Fig. 3b**). Additionally, the interface becomes increasingly populated by hydrogen bonds formed directly between side chain atoms from both SAG and TCR *(20)*.

As TCR Vβ domain binding partners become restricted even further (e.g., for SpeC), the engagement of the entire repertoire of TCR hypervariable elements is observed (**Fig. 3c**). The CDR loops with which the SAG interacts also have incorporated non-canonical residue insertions that alter both their length and conformation to provide highly unique binding sites *(20)*.

SAG–TCR specificity is thus accomplished with increased side chain-to-side chain hydrogen bond interactions, an expanded set of hypervariable elements engaged, and an accumulation of non-canonical CDR loop structures, which is effectively exhausted at this point. In order to exhibit even greater specificity than SpeC, toxic shock syndrome toxin-1 (TSST-1) appears to target a structural element, a loop in the framework region (FR3), that adopts a common conformation in all but a few Vβ domains, at the expense of interacting with each of the hypervariable structures (**Fig. 3d**). The fine specificity of TSST-1 for TCR Vβ domains is enhanced by requiring a particular residue at a particular position in FR3 in order to bind and efficiently activate T cells.

This targeting of rarely variable regions, at the expense of canonical hypervariable regions, in Vβ domains as a means for TCR specificity may constitute a general mechanism for enhancing SAG-TCR specificity, as the structural analysis of SEK in complex with one of its Vβ ligands, hVβ5.1, shows similar characteristics *(21)* (**Fig. 3e**). SEK appears to derive its specificity, at least in part, through interactions with relatively uncommon residues in both FR3 and FR4, with which a single residue in SEK forms side chain-to-side chain hydrogen bonds *(21)*.

The distinct orientations with which each of these SAGs engage the TCR Vβ domain result in unique patterns of hypervariable and framework region surfaces that are buried (**Fig. 3e**, right panel). Binding to the TCR Vβ CDR2 loop is a requirement for all bacterial SAGs, and the proportion of the SAG-TCR interface that is contributed by the CDR2 loop is invariably the greatest in any SAG-TCR complex, relative to any other single hypervariable or framework region. Involvement of Vβ domain regions beyond the CDR2 loop, however, plays a significant role in the TCR Vβ domain specificity and cross-reactivity of a SAG *(22–24)*. SEK and TSST-1 engage one or more framework region apical loops, at the expense of contacting the hypervariable elements. SEK buries significant molecular surface belonging to both the FR3 and FR4, while TSST-1 contacts only residues from FR3. The lower relative positions of SEB and SpeC on the Vβ domain result in their engagement of hypervariable elements at the expense of binding the apical loops of the framework regions. SEB buries molecular surface belonging to HV4, while SpeC contacts residues from CDR1, CDR3, and HV4.

5.2. Superantigen-Mediated T-cell Signaling Complexes

There exist three known binding modes for SAGs to interact with pMHC complexes. These binding modes are exemplified by the following SAGs: TSST-1, which binds predominantly to the MHC α subunit at a site that overlaps with that of SEB but also extends over the surface of the peptide to make contacts with the β subunit *(25)*; SEB, which binds MHC exclusively to its α subunit with no contacts made with the antigenic peptide *(26)*; and SpeC, which binds the MHC β subunit through coordination of a zinc ion and makes numerous contacts with the displayed peptide *(21, 27, 28)*. Crystal structures of TSST-1 *(29)*, SEB *(18)*, and SpeC *(20)* in complex with their TCR β chain ligands have allowed the construction of models of those MHC/SAG/TCR ternary complexes that are necessary for efficient T-cell activation that are distinct from pMHC-TCR complexes.

TSST-1 bridges the pMHC and TCR molecules such that two protein–protein interfaces, SAG/MHC and SAG/TCR, are formed (**Fig. 4a**). No direct MHC–TCR contacts are made. The relative orientation of the TCR and pMHC is such that a plane that passes through both the TCR α and β chains and one that is aligned with the MHC-displayed peptide are approximately perpendicular to one another.

In the SEB-dependent T-cell signaling complex (**Fig. 4b**), SEB acts as a wedge between the pMHC and TCR molecules, effectively rotating the TCR about a contact point between the MHC β subunit and the TCR α chain. This removes the antigenic peptide from any possible contacts with the TCR. The relative orientation of pMHC and TCR is otherwise akin to that observed in the TSST-1–mediated T-cell signaling complex model. In this supramolecular complex there exist three protein–protein interfaces: SEB/MHC, SEB/TCR and MHC/TCR. The presence of the direct MHC/TCR interaction (as indicated by the arrow in **Fig. 4b**) has been verified biochemically *(30)*.

SpeC, in contrast to SEB but similar to TSST-1, bridges the MHC and TCR molecules (**Fig. 4c**). There exists no direct interaction between MHC and TCR, and thus only two distinct protein–protein interfaces (i.e., SAG/MHC and SAG/TCR) comprise this complex. However, the TCR and pMHC are oriented such that planes passing through the TCR α and β chains and the antigenic peptide are approximately parallel to one another.

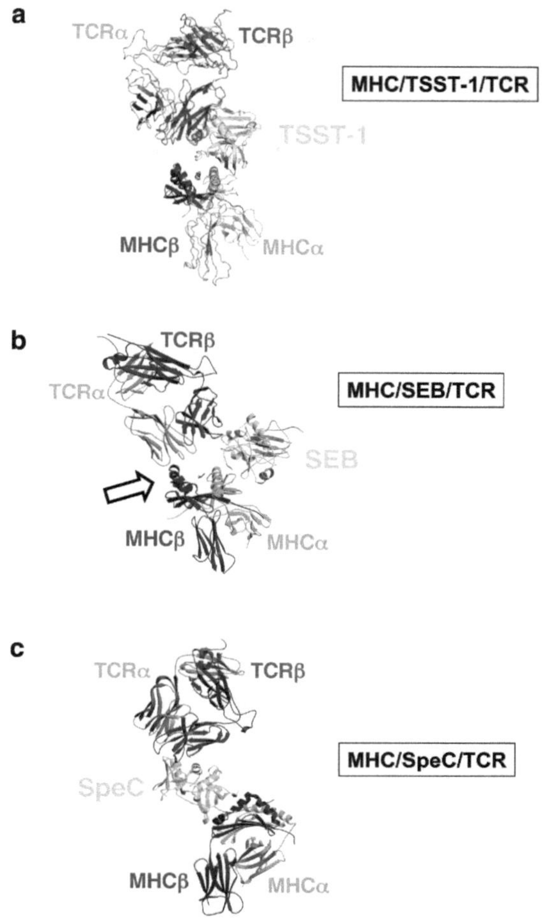

Fig. 4. MHC/SAG/TCR ternary signaling complexes mediated by (**a**) TSST-1, (**b**) SEB, and (**c**) SpeC. Colors are as follows: MHC α subunit, *green*; MHC β subunit, *blue*; antigenic peptide, *gray*; TCR α chain, *orange*; TCR β chain, *red*; SAGs, *yellow*. For clarity, the MHC/SAG/TCR complexes mediated by SpeC (**c**) is rotated approximately 90? clockwise about the vertical axis of the page relative to those mediated by TSST-1 (**a**) and SEB (**b**) (*see Color Plates*).

References

1. Houghton, A. N. and Guevara-Patino, J. A. (2004) Immune recognition of self in immunity against cancer. *J. Clin. Invest.* 114, 468–471.
2. Rosenberg, S. A. (2001) Progress in human tumour immunology and immunotherapy. *Nature* 411, 380–384.
3. Rudolph, M. G., Stanfield, R. L., and Wilson, I. A. (2006) How TCRs bind MHCs, peptides, and coreceptors. *Ann. Rev. Immunol.* 24, 419–466.
4. Clements, C. S., Dunstone, M. A., MacDonald, W. A., McCluskey, J., and Rossjohn, J. (2006) Specificity on a knife-edge: the alphabeta T cell receptor. *Curr. Opin. Struct. Biol.* 16, 787–795.
5. Hennecke, J., Carfi, A., and Wiley, D. C. (2000) Structure of a covalently stabilized complex of a human alphabeta T-cell receptor, influenza HA peptide and MHC class II molecule, HLA-DR1. *EMBO J.* 19, 5611–5624.
6. van der Merwe, P. A. and Davis, S. J. (2003) Molecular interactions mediating T cell antigen recognition. *Annu. Rev. Immunol.* 21, 659–84.
7. Hahn, M., Nicholson, M. J., Pyrdol, J., and Wucherpfennig, K. W. (2005) Unconventional topology of self peptide-major histocompatibility

complex binding by a human autoimmune T cell receptor. *Nat. Immunol.* 6, 490–496.

8. Li, Y., Huang, Y., Lue, J., Quandt, J. A., Martin, R., and Mariuzza, R. A. (2005) Structure of a human autoimmune TCR bound to a myelin basic protein self-peptide and a multiple sclerosis-associated MHC class II molecule. *EMBO J.* 24, 2968–2979.
9. He, X. L., Radu, C., Sidney, J., Sette, A., Ward, E. S., and Garcia, K. C. (2002) Structural snapshot of aberrant antigen presentation linked to autoimmunity: the immunodominant epitope of MBP complexed with I-Au. *Immunity* 17, 83–94.
10. Deng, L., Langley, R. J., Brown, P. H., Xu, G., Teng, L., Wang, Q., Gonzales, M. I., Callender, G. G., Nishimura, M. I., Topalian, S. L., and Mariuzza, R. A. (2007) Structural basis for the recognition of mutant self by a tumor-specific, MHC class II-restricted T cell receptor. *Nat. Immunol.* 8, 398–408.
11. Reinherz, E. L., Tan, K., Tang, L., Kern, P., Liu, J., Xiong, Y., Hussey, R. E., Smolyar, A., Hare, B., Zhang, R., Joachimiak, A., Chang, H. C., Wagner, G., and Wang, J. (1999) The crystal structure of a T cell receptor in complex with peptide and MHC class II. *Science* 286, 1913–1921.
12. Borg, N. A., Wun, K. S., Kjer-Nielsen, L., Wilce, M. C., Pellicci, D. G., Koh, R., Besra, G. S., Bharadwaj, M., Godfrey, D. I., McCluskey, J., and Rossjohn, J. (2007) CD1d-lipid-antigen recognition by the semi-invariant NKT T-cell receptor. *Nature* 448, 44–49.
13. Gadola, S. D., Koch, M., Marles-Wright, J., Lissin, N. M., Sheperd, D., Matulis, G., Harlos, K., Villiger, P. M., Stuart, D. I., Jakobsen, B. K., Cerundolo, V., and Jones, E. Y. (2006) Structure and binding kinetics of three different human CD1d-alpha-galactosylceramide-specific T cell receptors. *J. Exp. Med.* 203, 699–710.
14. Kjer-Nielsen, L., Borg, N. A., Pellicci, D. G., Beddoe, T., Kostenko, L., Clements, C. S., Williamson, N. A., Smyth, M. J., Besra, G. S., Reid, H. H., Bharadwaj, M., Godfrey, D. I., Rossjohn, J., and McCluskey, J. (2006) A structural basis for selection and cross-species reactivity of the semi-invariant NKT cell receptor in CD1d/glycolipid recognition. *J. Exp. Med.* 203, 661–673.
15. Cantu, C. 3rd, Benlagha, K., Savage, P. B., Bendelac, A., and Teyton, L. (2003) The paradox of immune molecular recognition of alpha-galactosylceramide: low affinity, low specificity for CD1d, high affinity for alpha beta TCRs. *J. Immunol.* 170, 4673–4682.
16. McCormick, J. K., Yarwood, J. M., and Schlievert, P. M. (2001) Toxic shock syndrome and bacterial superantigens: an update. *Annu. Rev. Microbiol.* 55, 77–104.
17. Sundberg, E. J., Li, Y., and Mariuzza, R. A. (2002) So many ways of getting in the way: diversity in the molecular architecture of superantigen-dependent T-cell signaling complexes. *Curr. Opin. Immunol.* 14, 36–44.
18. Li, H., Llera A., Tsuchiya, D., Leder, L., Ysern, X., Schlievert, P. M., Karjalainen, K., and Mariuzza, R. A. (1998) Three-dimensional structure of the complex between a T cell receptor beta chain and the superantigen staphylococcal enterotoxin B. *Immunity* 9, 807–816.
19. Fields, B. A., Malchiodi, E. L., Li, H., Ysern, X., Stauffacher, C. V., Schlievert, P. M., Karjalainen, K., and Mariuzza, R. A. (1996) Crystal structure of a T-cell receptor beta-chain complexed with a superantigen. *Nature* 384, 188–192.
20. Sundberg, E. J., Li, H., Llera, A. S., McCormick, J. K., Tormo, J., Schlievert, P. M., Karjalainen, K., and Mariuzza, R. A. (2002) Structures of two streptococcal superantigens bound to TCR beta chains reveal diversity in the architecture of T cell signaling complexes. *Structure* 10, 687–699.
21. Günther, S., Varma, A. K., Moza, B., Kasper, K. J., Wyatt, A. W., Zhu, P., Rahman, A. K., Li, Y., Mariuzza, R. A., McCromick, J. K., and Sundberg, E. J. (2007) A novel loop domain in superantigens extends their T cell receptor recognition site. *J. Mol. Biol.* 371, 210–221.
22. Moza, B., Buonpane, R. A., Zhu, P., Herfst, C. A., Rahman, A. K., McCormick, J. K., Kranz, D. M., and Sundberg, E. J. (2006) Long-range cooperative binding effects in a T cell receptor variable domain. *Proc. Natl. Acad. Sci. USA* 103, 9867–9872.
23. Buonpane, R. A., Moza, B., Sundberg, E. J., and Kranz, D. M. (2005) Characterization of T cell receptors engineered for high affinity against toxic shock syndrome toxin-1. *J. Mol. Biol.* 353, 308–321.
24. Rahman, A. K., Herfst, C. A., Moza, B., Shames, S. R., Chau, L. A., Bueno, C., Madrenas, J., Sundberg, E. J., and McCormick, J. K. (2006) Molecular basis of TCR selectivity, cross-reactivity, and allelic discrimination by a bacterial superantigen: integrative functional and energetic mapping of the SpeC-Vbeta2.1 molecular interface. *J. Immunol.* 177, 8595–8603.
25. Kim, J., Urban, R. G., Strominger, J. L., and Wiley, D. C. (1994) Toxic shock syndrome

toxin-1 complexed with a class II major histocompatibility molecule HLA-DR1. *Science* 266, 1870–1874.

26. Jardetzky, T. S., Brown, J. H., Gorga, J. C., Stern, L. J., Urban, R. G., Chi, Y. I., Stauffacher, C., Strominger, J. L., and Wiley, D. C. (1994) Three-dimensional structure of a human class II histocompatibility molecule complexed with superantigen. *Nature* 368, 711–718.
27. Li, Y., Li, H., Dimasi, N., McCormick, J. K., Martin, R., Schuck, P., Schlievert, P. M., and Mariuzza, R. A. (2001) Crystal structure of a superantigen bound to the high-affinity, zinc-dependent site on MHC class II. *Immunity* 14, 93–104.
28. Fernandez, M. M., Guan, R., Swaminathan, C. P., Malchiodi, E. L., and Mariuzza, R. A. (2006) Crystal structure of staphylococcal enterotoxin I (SEI) in complex with a human major histocompatibility complex class II molecule. *J. Biol. Chem.* 281, 25356–25364.
29. Moza, B., Varma, A. K., Buonpane, R. A., Zhu, P., Herfst, C. A., Nicholson, M. J., Wilbuer, A. K., Seth, N. P., Wucherpfennig, K. W., McCormick, J. K., Kranz, D. M., and Sundberg, E. J. (2007) Structural basis of T cell receptor specificity and activation by the bacterial superantigen TSST-1. *EMBO J.* 26, 1187–1197.
30. Andersen, P. S., Lavoie, P. M., Sékaly, R. P., Churchill, H., Kranz, D. M., Schlievert, P. M., Karjalainen, K. and Mariuzza, R. A. (1999) Role of the T cell receptor alpha chain in stabilizing TCR-superantigen- MHC class II complexes. *Immunity* 10, 473–483.

Chapter 26

Identification of Human MHC Class I Binding Peptides using the iTOPIA™– Epitope Discovery System

Markus Wulf, Petra Hoehn, and Peter Trinder

Summary

CD8+ T cells recognize antigenic peptides presented in the context of MHC class I. They play a key role in cellular immunity and are crucial for longterm protective immunity to many infectious diseases. The quest for new and enhanced vaccines requires improved means for identification of relevant antigens and the epitopes present within these. While there are several algorithms available for epitope prediction (all of which work to differing degrees of success), the definition of actual MHC class I-binding epitopes is very reliant on time-consuming and difficult to perform functional assays using often very limited biological material. The iTOPIA assay is quick and easy to perform and determines real-binding to MHC class I molecules. It provides an excellent platform for screening and elimination of potential epitopes and identification of novel epitopes prior to validation with a relevant functional assay.

Key words: MHC class I, T-cell epitopes, MHC-binding, CD8+ T-cells, Epitope mapping, Vaccine design.

1. Introduction

The iTopia™– epitope discovery assay from Beckman Coulter provides a powerful tool for the identification and mapping of epitopes. It helps to quickly eliminate the peptides that are not likely to be viable epitopes due to their inability to bind to, or lack of affinity to maintain a stable, MHC class I complex. The iTOPIA system assay provides real experimental data for peptide binding by MHC class I allele specificity. The system is available for use with human class I molecules (HLA), and allows rapid screening of large numbers of peptides for their binding characteristics and affinities for HLA class I molecules and relative rates of

Ulrich Reineke and Mike Schutkowski (eds.), *Methods in Molecular Biology, Epitope Mapping Protocols, vol. 524*

DOI: 10.1007/978-1-59745-450-6_26

dissociation from the respective HLA class I molecule. The assay can also be utilized to assess the enhanced or altered binding characteristics of modified epitopes. Currently kits covering eight of the most frequent HLA class I alleles in the Caucasian and Asian populations are available (HLAs A*0101, A*0201, A*0301, A*1101, A*2402, B*0702, B*0801, B*1501) and development of kits for some of the most frequent African HLA class I alleles is in progress. Examples of use for epitope mapping (iTOPIA only) and for epitope discovery and validation (iTOPIA followed by custom tetramer synthesis and validation by tetramer analysis of patient material) can be found in references *(1)* and *(2)*. Although designated to work optimally with 9-mer peptides, we have successfully used some peptides consisting of 8–10 amino acids in this assay.

1.1. Terms and Definitions

Affinity Assay: characterizes the relative binding affinity of each peptide for the HLA class I molecule, expressed in terms of an ED50 value.

- β_2M: Beta-2 Microglobulin.
- *t1/2 value:* value obtained for each peptide during the off-rate assay. Represents the time point at which 50% of peptide remains bound to the HLA class I molecule. Expressed as a percentage of *T0* (time zero).
- *ED 50 value*: value obtained for each peptide during affinity assay. Represents concentration at which peptide is bound to the HLA class I molecule at 50% of maximum binding efficiency. It is the mid-point of a sigmoidal dose response curve.
- *iScore:* multi-parametric calculation for *t1/2* and ED50 values. Permits integration of off-rate and affinity parameters into an index or iScore. The iScore reflects ability of peptide to fold and remain in stable peptide–HLA class I complex. This ultimately defines its overall level of binding.
- *Off-rate assay:* characterizes relative rate of dissociation of peptide from HLA class I molecule. Expressed in terms of half life (*t1/2*) value.

1.2. Overview

The iTopia Epitope Discovery System consists of three assays. Each of these makes use of a fluorescently labeled anti-HLA class I-FITC antibody, which indicates successful binding of the test peptide to the HLA class I molecule. Detection of bound peptide is via a microplate reader with a fluorescence detector such as the DTX 880 multi-mode detector from Beckman Coulter.

The first of these assays is the peptide binding assay. It provides a rapid screening of all test peptides in up to eight alleles. The assay utilizes optimal, standardized binding conditions in order to perform this initial screening. Once peptides that have successfully bound to the HLA class I molecule have been identified, these peptides or "binders" can be further analyzed using the off-rate and affinity assay. Nonbinders are eliminated from further evaluation.

Off-rate and affinity assays are performed using the binder peptides identified in the peptide binding assay. Their affinity for, and rate of dissociation from the HLA class I molecule are each evaluated independently and ED50 and *t1/2* values are determined for each peptide. These are incorporated into a multi-parametric calculation to yield an iScore for the individual peptide. The test peptides can then be ranked in descending order permitting the investigator to easily eliminate peptides deemed unsuitable for validation in functional studies.

1.3. Peptide-Binding Assay

iTOPIA kits containing plates and reagents are available for purchase from Beckman Coulter. The microtiter plates have HLA class I molecules, containing "placeholder" peptides, bound in a set orientation to the surface of each well. A proprietary low pH buffer is introduced into each well in accordance with the kit instructions, and under these "stripping" conditions, the HLA class I complex dissociates. The placeholder peptide and β_2M are washed away leaving the denatured HLA class I heavy chain bound to the reaction well. Individual test peptides from the target protein library along with fresh ß$_2$M and the fluorescently labeled anti-HLA antibody are added to each well and incubated in the renaturation buffer provided. This promotes peptide binding and refolding of the complex. The anti-HLA antibody binds only to a properly folded peptide–HLA class I complex. This is the first step of all of three assays.

The fluorescently labeled, refolded HLA class I molecule, with the newly bound test peptide is identified by fluorimetric analysis with a microplate reader. If the peptide binds with low affinity or does not bind the HLA class I molecule, the result is a weak or negative fluorimetric read out, relative to a positive control peptide that is a known strong binder. Peptides that do not bind to the HLA class I molecule are clearly identified and can be eliminated from further investigation. The amount of fluorescence generated is therefore related to the amount of test peptide bound. As a guideline, 30% binding of the test peptide, relative to the positive control peptide, is defined as the limit to discriminate peptide "binders" from peptide "nonbinders". The iTopia System software automatically establishes a threshold of ≤30% of the fluorescence generated by the positive control peptide (provided with the relevant iTOPIA kit).

1.4. Off-Rate Assay

Binder peptides previously identified in the peptide binding assay are incubated, washed and the bound peptide detected using the fluorimeter as in the binding assay. The peptide is then incubated in a time dependent manner on each allele specific plate, for up to 8 h depending on the individual peptide. The amount of peptide remaining bound is determined by washing the plates at each timepoint and taking fluorometric readings with the plate reader. Data are fitted to a one-phase exponential decay curve and the data are normalized relative to the nominal value of the control peptide. The results are expressed as the amount of time needed to achieve 50% dissociation of the peptide from the HLA class I complex. This is *t1/2* value and is expressed in hours.

1.5. Affinity Assay

In this assay, each "binder" previously identified in the peptide binding assay, is incubated in increasing concentrations with each allele. The bound peptide is detected using the plate reader (fluorimeter) and the degree of binding at each concentration is determined by level of fluorescence. The data are fitted to a sigmoidal dose response curve and the results are expressed as the concentration of the peptide needed to achieve 50% binding (compared with the positive control peptide), or the ED50 value.

1.6. iScore Calculation

The off-rate and affinity values (*t1/2* and ED50, respectively) for each test peptide are evaluated using a multi-parametric calculation to generate the iScore. Using the iScore to rank the peptides permits determination of which peptides should be considered for further functional analysis. An iScore of 0 is automatically assigned to the peptides that had binding characteristics of <30% of the positive control peptide, and were therefore not analyzed by off-rate and affinity. The peptides that are termed "binders" with high affinity and slow rates of dissociation can be easily discriminated from the residual peptides, eliminating a large number from the need for further analysis (validation). Based on the iScore, epitopes can then be validated by one or more of several of methods (such as tetramer analysis, ELISpot or ICS), some of which are presented in other chapters within this volume.

2. Materials

1. Peptides: Minimum 80% pure, HPLC purified (e.g. JPT Peptide Technologies, Berlin, Germany) (*see* **Note 1**) (e.g. SigmaAldrich, Munich, Germany).
2. DMSO (e.g. SigmaAldrich, Munich, Germany).
3. Uncoated 96-well microtiter plates for dilution of peptides.
4. iTOPIA Kits for relevant alleles (Beckman Coulter, Krefeld, Germany).
5. Each kit contains:
   - Five plates (12 × 8-well strips) coated with single HLA class I monomer
   - Positive control peptide (HLA matched)

- Renaturation buffer (10×)
- Anti-HLA-ABC-FITC conjugated mAb (50×)
- β_2Microglobulin (50×)
- Wash solution (20×)
- Stripping buffer (10×)
- Dliution buffer (10×).

6. iTOPIA Software (Beckman Coulter, Krefeld, Germany).
7. Aluminium foil microtiter plate sealers (e.g., Greiner, Heidelberg, Germany).
8. Temperature controlled incubator 21°C (e.g., Aqualytic, Dortmund, Germany).
9. Temperature controlled incubator 37°C (e.g., Heraeus Function Line, Waltham, MA).
10. 2× Rockers (e.g., IKA MTS 2/4, Staufen, Germany).
11. Microplate washer (e.g., BIO-TEK EL-50, Winooski, VT).
12. Microplate reader with fluorescence detector (e.g., DTX 880 Multimode Detector from Beckman Coulter, Krefeld, Germany), capable of reading from top of plates.

3. Methods

3.1. Binding Assay

1. Dissolve peptides in DMSO: 10 mM stock (*see* **Note 1**).
2. 1:90 peptide dilution in renaturation buffer (uncoated microtiter plates) (*see* **Notes 2** and **3**).
3. Strip allele-specific iTOPIA plates: 5 min 1× strip buffer.
4. Wash plates three times with 300 μL 1× wash buffer.
5. Add 180 μL/well renaturation mixture + 20 μL 1:90 peptide dilution (duplicates).
6. Cover plates with aluminum foil.
7. Incubate 18 h at 21°C (on rocker).
8. Flick plates to remove well contents.
9. Wash plates three times with 300 μL 1× wash buffer.
10. 200 μL of 1× dilution buffer/well.
11. Read plates in DTX 880 reader (excitation wavelength 495 nm, emission wavelength 525 nm) (*see* **Note 4**).
12. Transfer of data to PC and analysis using iTOPIA Software (*see* **Note 5**).

3.2. Off-Rate

Seven timepoints (*see* **Note 6**).

1. Dissolve peptides in DMSO: 10 mM stock (i.e., use stock dilutions mentioned in **Subheading 3.1, step 1** above).

2. 1:90 peptide dilution in renaturation buffer (uncoated microtiter plates).
3. Strip allele-specific iTOPIA plates: 5 min 1× strip buffer.
4. Wash plates three times with 300 µL 1× wash buffer.
5. Add 180 µL/well renaturation mixture + 20 µL 1:90 peptide dilution (duplicates).
6. Cover plates with aluminum foil.
7. Incubate 18 h at 21°C (on rocker).
8. Flick plates to remove well contents.
9. Wash plates three times with 300 µL 1× wash buffer.
10. 200 µL of 1× dilution buffer/well.
11. Read all plates in DTX 880 reader, this is $T = 0$ h.
12. Incubate plates at 37°C in dilution buffer.
13. At following time points, wash plates three times with 300 µL 1× wash buffer, add 200 µL dilution buffer and read plates in DTX 880 reader: 30 min/60 min/90 min/2 h/4 h/6 h/8 h (*see* **Note 6**).
14. Transfer of data to PC and analysis using iTOPIA Software.

3.3. Affinity Assay

Peptide dilutions in 96-well plates (uncoated microtiter plates) from stock solutions as in **Subheading 3.1**, **step 1** above):

- Well A: 10 µL (10 mM) + 90 µL renaturation buffer
- Well B: 40 µL from A + 80 µL renaturation buffer
- Well C: 40 µL from B + 80 µL renaturation buffer
- Well D: 40 µL from C + 120 µL renaturation buffer
- Well E: 40 µL from D + 120 µL renaturation buffer
- Well F: 40 µL from E + 120 µL renaturation buffer
- Well G: 40 µL from F + 120 µL renaturation buffer
- Well H: 40 µL from G + 120 µL renaturation buffer
- Positive control peptide: Dilute out to well G. Well H is without peptide = > negative control.

1. Strip allele-specific iTOPIA plates: 5 min 1× strip buffer.
2. Wash plates three times with 300 µL 1× wash buffer.
3. Add 180 µL/well renaturation mixture + 20 µL 1:90 peptide dilution (duplicates).
4. Cover plates with aluminum foil.
5. Incubate 18 h at 21°C (on rocker).
6. Flick plates to remove well contents.
7. Wash plates three times with 300 µL 1× wash buffer.
8. 200 µL of 1× Dilution buffer/well.

9. Read plates in DTX 880 reader.
10. Transfer of data to PC and analysis using iTOPIA software.

4. Notes

1. It is recommended for ease of use to purchase peptides supplied in Micronic– tubes and racks. These have a 96-well microplate format and greatly simplify solubilization, transfer, and dilution procedures.
2. Uncoated (blank) 96-well plates are ideal for preparing dilutions and permit transfer to iTOPIA plates with a multichannel pipette.
3. The use of electronic multichannel pipettes with built-in reservoirs for all dilution and wash steps saves time and frustration and also your thumb! (e.g., Eppendorf "Research Pro").
4. Microtiter plate reader must read from the top as plates have opaque bottoms. The fluoresceine is FITC, therefore excitation wavelength should be 495 nm and the emission wavelength should be 525 nm.
5. Data should be in a tab-delimited text file format for easy transfer to the iTOPIA Software. The iTOPIA– software from Beckman Coulter is essential for simple and optimal analysis of data, calculation of binding efficiencies, off-rates, affinity, and iScores.
6. Careful planning is required to avoid performance of off-rate analysis at socially unacceptable hours.

References

1. Bachinsky, M. M., Guillen, D. E., Patel, S. R., Singleton, J., Chen, C., Soltis, D. A., and Tussey L. G. (2005) Mapping and binding analysis of peptides derived from the tumor-associated antigen survivin for eight HLA alleles. *Cancer Immunol.* 22, 6–11.
2. Weichold, F. F., Mueller, S., Kortsik, C., Hitzler, W. E., Wulf, M., Hone, D. M., Sadoff, J. C., and Maeurer, M. J. (2007) Impact of MHC class I alleles on the *M. tuberculosis* antigen-specific CD8+ T-cell response in patients with pulmonary tuberculosis. *Genes Immunol.* 8, 334–343.

Chapter 27

T-Cell Epitope Mapping in *Mycobacterium tuberculosis* Using PepMixes Created by Micro-Scale SPOT™– Synthesis

Marisa Frieder and David M. Lewinsohn

Summary

Mycobacterium tuberculosis (Mtb) remains a major threat to human health worldwide. Although treatment of infection is an important part of tuberculosis control, an improved vaccine is essential for the elimination of this disease. Control of infection with Mtb is dependent on the cellular immune system, which in turn requires an understanding of those antigens that are capable of stimulating CD4+ and CD8+ T-cell responses. Peptide libraries provide a high-throughput system for identifying novel T-cell epitopes. They can also be used to assess the hierarchy of immunodominance of these novel antigens and epitopes that are associated with infection with Mtb. This T-cell-driven means of antigen discovery is well adapted to vaccine development as well as developing the tools necessary to understand the natural history of this important human pathogen.

Key words: Tuberculosis, CD8+ T cells, Immunodominance, Antigen/epitope discovery, Elispot, Peptide library.

1. Introduction

Mtb-specific CD8+ T-lymphocytes play an important role in the host response to infection *(1–9)*; yet the repertoire and dominance pattern of human CD8+ Mtb antigens remain poorly characterized (*see* **Table 1**). Until recently, attempts to identify CD8+ T-cell epitopes have focused on peptides with strong predicted HLA binding. A major limitation to these peptide-driven studies is that they fail to place these epitopes in the broader context of an Mtb infection, and are often focused on a limited number of HLA alleles. To characterize the immune response to Mtb and to define candidate antigens for future vaccines, it is necessary to determine whether an epitope or antigen is immunodominant

Ulrich Reineke and Mike Schutkowski (eds.), *Methods in Molecular Biology, Epitope Mapping Protocols, vol. 524*

DOI: 10.1007/978-1-59745-450-6_27

Table 1
Known CD8+ T-cell antigens in *Mycobacterium tuberculosis*

Gene	Common name	Amino acid position	Frequency	Method (references)
Rv0129c	Ag85C	204–212	1/13,700–1/22,200 PBMCs	Synthesis of predicted HLA-B35 binding peptides *(10)*
Rv0287	Mtb9.8	3–11 53–61	<1/25,000 CD8+ 1/2,840 CD8+	T-cell clones from infected donor *(11)*
Rv0341		33–42 33–44 33–45	ND[a]	Infection of HLA-A2 cell line, mass spectrometric analysis of bound peptides *(12)*
Rv0440	Hsp65	369–377	ND	Synthesis of predicted HLA-A2 binding peptides DNA immunization of HLA-A2 transgenic mice *(13)*
Rv0512	AlaDH	160–169	ND	Synthesis of predicted HLA-A2 binding peptides *(14)*
Rv0934	38 kDa antigen	8–17	1/331 CD8+	Screening of overlapping peptides spanning the entire protein *(15)*
Rv1174c	Mtb8.4	5–15 32–40	1/10,416 CD8+ 1/1,190 CD8+	T-cell clones from infected donor *(11)*
Rv1196	Mtb39	144–153 346–355	ND	T-cell clones from infected donor *(16)*
Rv1694	Hemolysin	73–82	1/353 CD8+	Screening of overlapping peptides spanning the entire protein *(15)*
Rv1886c	Ag85B	143–152 199–207 264–272	ND Undetectable (tetramer) ND	Synthesis of predicted HLA-A2 binding peptides Immunization of HLA-A2 transgenic mice with protein or peptides *(17)* Synthesis of predicted HLA-B35 binding peptides *(10)*

Rv2031c	16 kDa protein	21–29	1/1,000 CD8+	Synthesis of predicted HLA-A2 binding peptides *(18)*
		120–128	1/800 CD8+	
Rv2220	GlnA1	308–316	ND	Synthesis of predicted HLA-A2 binding peptides *(14)*
Rv2903c		201–209	Undetectable	Synthesis of predicted HLA-B35 binding peptides *(19)*
Rv3763	19 kDa lipoprotein	88–97	ND	Synthesis of predicted HLA-A2 binding peptides *(20)*
Rv3804c	Ag85A	48–56	1/20,693 PBMCs (Elispot) 1/3,300 PBMCs (tetramer)	Screening of overlapping peptides spanning the entire protein *(21)*
		242–250	1/23,779 PBMCs (Elispot) 1/3,750 PBMCs (tetramer)	
Rv3846	SodA	160–168	ND	Synthesis of predicted HLA-A2 binding peptides *(14)*
Rv3874	CFP-10	2–9	1/101 CD8+	T-cell clones from infected donor *(11)*
		2–12	1/125 CD8+	
		3–11	1/645 CD8+	
		75–83	1/145 CD8+	
		2–11	1/700 CD8+	T-cell clones from infected donor *(22)*
		49–58	1/7,000 CD8+	
		85–94	1/2,100 CD8+	
		71–79	1/438–1/1,602 PBMCs	Screening of overlapping peptides spanning the entire protein *(23)*
		76–85	1/437–1/2,427 PBMCs	
Rv3875	ESAT-6	69–76	1/700 CD8+	Synthesis of predicted HLA-A2, -B7, -B8, -B35, and -B53 binding peptides *(24)*
		82–90	1/2,500 PBMCs	
		21–29	1/2,100 CD8+	Screening of overlapping peptides spanning the entire protein *(25)*

ND no data

within an individual and whether that epitope is commonly recognized by Mtb-infected individuals.

These questions can be addressed with an alternative epitope-discovery strategy, using T-cell clones obtained from Mtb-infected donors or through direct ex-vivo assessment. A key advantage to this T-cell-driven approach is that antigens and epitopes identified in this way often reflect immunodominant responses in the immune response to Mtb. To this end, our lab generated a peptide library comprising 10% of the Mtb proteome, which we hypothesized would most likely be enriched for CD8 antigens *(11)*. Mtb proteins were assigned evidence-based weights based on the three lines of evidence available: proteomics, genomics, and known absence from Bacille Calmette-Guérin (BCG) strains. Our weighting scheme favored proteins that have been identified in secreted or cell wall preparations, or those that were expressed differentially in one or more growth condition (i.e., hypoxic stress) *(26)*. Additionally, weights were assigned based on the functional categories of the TubercuList *(27)*, a complete annotated dataset of DNA and protein sequences derived from Mtb H37Rv. Categories either known to contain or likely to contain Mtb antigens (i.e., "PPE/PE," "cell wall and cell processes," "virulence, detoxification, adaptation") were given a greater weight than categories unlikely to contain Mtb antigens (i.e., "information pathways"). A finalized list contained 389 proteins, consisting of 39,499 15meric peptides overlapping by 11 amino acids for each gene product (*see* **Table 2**).

Table 2
Functional category of *Mycobacterium tuberculosis* genes represented in a peptide library

Functional category	Number of genes
PE/PPE	168
Cell wall and cell processes	134
Conserved hypotheticals	30
Virulence, detoxification, adaptation	29
Intermediary metabolism and respiration	10
Lipid metabolism	9
Regulatory proteins	5
Conserved hypotheticals with an orthologue in *M. bovis*	4
Total	389

CD8+ T cells are screened for responses to the peptide library using an IFN-γ Elispot assay, which allows for the quantification of effector cells. The library is used in two sets of experiments. First, to identify the cognate antigens recognized by classically-restricted T-cell clones. In this case, the library is used to identify the peptide pool containing the antigen of interest. Additional peptides are then synthesized and tested to define the minimal epitope (defined as the peptide eliciting a response at the lowest concentration), and the restricting allele. Using this information, the immunodominance of each epitope can be ascertained by determining the ex-vivo frequency of the CD8+ T-cell response in peripheral blood mononuclear cells (PBMCs) from the original donor. Second, the library is used for direct ex-vivo screens of CD8+ T-cell responses from Mtb-infected donors, to assess the diversity of the immune response within and between individuals. Data obtained from multiple donors can be compared to identify those antigens that are both commonly and strongly recognized in those with latent and active infection with Mtb, and thus are possible vaccine candidates.

2. Materials

2.1. Donors and Cells

2.1.1. Donor Selection

PBMCs from those with Mtb infection, both tuberculin skin test (TST)-positive donors as well as those with active TB, are prescreened prior to use in screens of the peptide library. To be a candidate for the peptide library screen, the donor must have documented infection with Mtb, or a demonstrably positive CD4+ T-cell response to either CFP-10 (Rv3874) or ESAT-6 (Rv3875). These antigens are absent from BCG and thus may be used to distinguish true Mtb infection from exposure to BCG or atypical mycobacteria. Additionally, CD8+ T cells must have minimal IFN-γ production in the presence of autologous DCs and culture medium, as this may predict unacceptable background in screens of the peptide library.

2.1.2. Cell Culture and Preparation

1. Human PBMCs are obtained by leukaphoresis, according to IRB-approved protocols.
2. Culture media: RPMI containing a total of 884 mg/L L-glutamine, supplemented with 10% fetal bovine serum (FBS), 10% human serum (HuS) or 2% HuS. Store all at 4°C.
3. Cytokine stocks: 3 μg/mL IL-2 (Chiron, Emeryville, CA). Store at –80°C. 3 mg/mL GM-CSF (Amgen, Thousand Oaks, CA) and 3 mg/mL IL-4 (Amgen, Thousand Oaks, CA). Store at –20°C.

4. 3 mg/mL DNase. Store at –20°C. Once thawed, store at 4°C and discard after 2 weeks.
5. Sodium phosphate-buffered saline (PBS), pH 7.4 (Invitrogen Corp., Carlsbad, CA).
6. PBS–EDTA: PBS with 2 mM EDTA.
7. Histopaque 1077 (Sigma-Aldrich, St. Louis, MO).
8. Freezing medium: RPMI supplemented with 50% FBS and 10% dimethyl sulfoxide.

2.1.3. Dendritic Cells

1. Cell dissociation solution, nonenzymatic, in PBS without calcium or magnesium (Sigma-Aldrich, St. Louis, MO). Store at 4°C.
2. T-150 flasks.

2.1.4. CD8+ T Cells

1. MACS buffer: PBS–EDTA with 5 g/L BSA. Store at 4°C.
2. CD8 microbeads (Miltenyi Biotec Inc., Auburn, CA). Store at 4°C.
3. Midi MACS columns (Miltenyi Biotec Inc., Auburn, CA).
4. MACS separator (Miltenyi Biotec Inc., Auburn, CA).
5. Lymphoblastoid cell lines (LCL): B cell lines transformed with Epstein-Barr virus using supernatants from the cell line 9B5–8 (American Type Culture Collection, Manassas, VA). Cell lines are maintained in RPMI with 10% FBS.

2.2. Peptide Library

1. Peptides. 39,499 peptides were synthesized by JPT Peptide Technologies GmbH (Berlin, Germany). The peptides are 15mers overlapping by 11 amino acids for each gene product. Each peptide (50 nmol) was synthesized individually and then pooled into 777 pools (50 peptides per pool) in a 96-well format (nine plates). Five blank wells and one well of an irrelevant peptide pool representing SIV gag were included on each of the nine plates. The pools were synthesized, divided into five replicates to limit the possibility of contamination during later use, and lyophilized before shipment.
2. 96-well sealable plates, sealing film, plate sealer (Eppendorf Inc., Westbury, NY).

2.3. Elispot

1. Filter plates: 45-µm filter plates with mixed cellulose esters, either Multiscreen or Multiscreen HTS (Millipore, Billerica, MA).
2. Antibodies: Anti-IFN-γ monoclonal antibody (1-D1K, Mabtech AB, Cincinnati, OH), biotinylated anti-IFN- γ monoclonal antibody (7-B6-1, Mabtech AB, Cincinnati, OH).
3. Carbonate buffer: To make carbonate buffer, 4.3 g $NaHCO_3$ is added to 515 mL distilled water. The mixture is brought to pH 9.6 with approximately 185 mL 0.1 M Na_2CO_3.

4. Elispot buffer: PBS with 5 mg/mL BSA and 0.05% Tween 20.
5. Horseradish peroxidase: Vectastain Elite ABC kit (Vector Laboratories, Burlingame, CA) contains biotinylated horseradish peroxidase and avidin-DH complex. One drop reagent A and one drop reagent B are added per 10 mL PBS–Tween.
6. Horseradish peroxidase substrate: Vectastain AEC kit (Vector Laboratories, Burlingame, CA) contains horseradish peroxidase substrate 3-amino-9-ethylcarbazole. Four drops buffer solution, six drops AEC, and four drops hydrogen peroxide are added per 10 mL dH_2O.
7. PBS–Tween: PBS supplemented with 0.05% Tween 20. Prepare large quantities for washing steps, store at room temperature.
8. Controls: Phytohemagglutinin (PHA) stock solution (5 mg/mL), store at –20°C. Optional: *Mycobacterium tuberculosis* strain H37Rv in Middlebrook 7H9 with 50% glycerol. Store at –80°C (*see* **Note 1**).
9. Elispot plate reader: AID Elispot high-resolution reader system (CellTechnology Inc., Colombia, MD).

3. Methods

3.1. Cells

3.1.1. PBMCs

1. Following collection by leukaphoresis, cells are washed twice with PBS–EDTA in order to reduce platelet contamination.
2. Cells are resuspended in PBS and layered onto Histopaque 1077. Cells are then centrifuged for 20 min at 863 × *g* in a table-top Sorvall centrifuge at room temperature.
3. The interface is collected and washed three times in PBS to further remove contaminating platelets.
4. Following a final wash in RPMI with 10% FBS, cells are suspended in freezing medium, frozen to –80°C at a controlled rate of 1°C/min (i.e., controlled rate freezer or isopropanol chamber), and transferred to liquid nitrogen for long-term storage.

3.1.2. Preparation of Dendritic Cells

1. PBMCs are thawed at 37°C and transferred to conical tubes containing RPMI with 10% FBS. Cells are centrifuged for 10 min at 311 × *g* in a tabletop Sorvall centrifuge at 4°C (*see* **Notes 2–4**).
2. Supernatants are aspirated and each pellet resuspended in 60 mL RPMI with 2% HuS and 30 μg/mL DNase. Cells are transferred to T150 flasks and allowed to incubate, lying flat, at 37°C for 60 min (*see* **Note 5**).

3. Following incubation, cells are washed twice, very gently, with 30 mL PBS before the addition of 60 mL RPMI with 10% HuS and 30 μg/mL each GM-CSF and IL-4. Cells are γ-irradiated for 30 min (15 Gy) and allowed to mature for 5–7 days (*see* **Notes 6 and 7**).

3.1.3. Harvesting of Dendritic Cells

1. DCs are harvested on the day that the Elispot will be performed. Culture supernatant is first collected in a conical tube. Cells are washed with PBS, and the wash is pooled with the culture supernatant. 20 mL cell dissociation solution is placed on the cells and allowed to incubate at 37°C for 20–30 min (*see* **Note 8**).
2. Flasks are agitated violently to detach cells. The cell solution is collected and added to the solution collected in **step 1**. An additional PBS wash is performed to remove residual cells from the flask, and the wash is added to the collected cell suspension.
3. The resulting cell suspension is centrifuged for 15 min at 699 × *g* in a tabletop Sorvall centrifuge at 4°C.
4. Following centrifugation, the supernatant is aspirated, and the pellet is resuspended in RPMI with 10% HuS (*see* **Note 8**).

3.1.4. CD8+ T Cell Clones

Mtb-specific CD8+ T-cell clones are generated from PBMCs by mini-limiting dilution analysis, as described earlier *(28)*. 10–14 days prior to use, clones are expanded with anti-CD3 and heterologous feeder stimulation as described earlier *(29)*. Briefly, T-cell clones (5×10^5) are co-cultured with irradiated heterologous lymphoblastoid cell lines (5×10^6) and heterologous irradiated PBMCs (2.5×10^7). Cells are grown in RPMI with 10% HuS, in the presence of 30 ng/mL anti-CD3 monoclonal antibody. IL-2 (2 ng/mL) is added the day after initiation of culture. Three days later, cells are washed to remove excess anti-CD3 monoclonal antibody, and IL-2 (2 ng/mL) is added to the culture medium. IL-2 (2 ng/mL) is added again every other day.

3.1.5. Magnetic Separation of CD8+ T Cells from PBMCs

1. PBMCs are thawed and added to conical tubes containing RPMI with 10% HuS and 80 μg/mL DNase. Cells are centrifuged for 10 min at 365 × *g* in a tabletop Sorvall centrifuge (*see* **Notes 2–4 and 9**).
2. Each pellet is resuspended in RPMI with 10% HuS and 40 μg/mL DNase. All PBMCs are then combined and counted.
3. Following counting, cells are divided into 50 mL tubes and centrifuged 10 min at 365 × *g*.
4. Each pellet is resuspended with 80 μL cold MACS buffer per 10^7 cells. 20 μL CD8 microbeads are then added per 10^7 cells. This mixture is allowed to incubate at 4°C for 15 min (*see* **Note 10**).

5. An excess of cold MACS buffer is added to each tube, which is then centrifuged for 10 min at 4°C at 365 × *g*.
6. Following centrifugation, supernatants are removed and the pellets are each thoroughly resuspended in 50 μL cold MACS buffer per 10^7 cells (*see* **Note 11**).
7. Cells are then slowly applied to Midi MACS columns, which have been pre-rinsed with MACS buffer and placed inside the magnetic separator (*see* **Note 11**).
8. Flow-through, containing CD4+ and other cells, is collected and stored at 4°C.
9. Columns are then rinsed three times with cold MACS buffer, and the flow-through is collected for storage at 4°C.
10. To elute CD8+ cells, columns are removed from the magnet, and MACS buffer is added. A plunger is then used to force cells out of the column into a clean tube.
11. The eluted CD8+ cells are centrifuged and the purification process is repeated from **step 6** using fresh columns.
12. Following the second purification, CD8+ cells are centrifuged at 311 × *g*, 4°C for 10 min. Pellets are each resuspended and combined in RPMI with 10% HuS.
13. Cells are counted and a sample is taken to be stained using anti-CD4 and anti-CD8 antibodies, and analyzed by flow cytometry to assess the purity of the cell solution (*see* **Note 12**).

3.2. Peptide Library

Prior to use, lyophilized peptides are resuspended to 5 mg/mL in DMSO and then diluted to a working concentration in PBS. Stocks are stored at –80°C (*see* **Notes 13 and 14**).

3.3. Elispot

3.3.1. Plating

1. On the day before the Elispot will be performed, filter plates are coated with 50 μL/well of monoclonal anti-IFN-γ antibody (10 μg/mL in sodium carbonate buffer). Plates are incubated overnight at 4°C (*see* **Notes 15 and 16**).
2. On the day that the Elispot will be performed, the plates are washed three times for 15 min each with PBS at room temperature. Plates are then blocked for at least 1 h in RPMI with 10% HuS.
3. DCs (2 × 10^4) are added to each well of the plates, followed by 1 μg peptides in an appropriate (5–25 μL) volume of PBS. Appropriate controls are added to designated wells. Plates are incubated for 1 h (*see* **Notes 1, 13, 17, and 18**).
4. CD8+ T cells or T-cell clones are added and the plates are incubated at 37°C for 18 h (*see* **Notes 13 and 19**).

3.3.2. Development

1. After 18 h incubation, culture medium is removed. Plates are rinsed six times in PBS–Tween (*see* **Note 20**).
2. 100 μL/well biotinylated monoclonal anti-IFN-γ antibody (1 μg/mL in Elispot buffer) is added and the plates are allowed to incubate for 2 h.
3. The horseradish peroxidase (Vectastain ABC) solution is prepared and allowed to incubate at room temperature for 30 min (*see* **Note 21**).
4. Following the 2 h incubation, plates are washed six times in PBS–Tween, allowed to incubate 15 min, and washed six additional times in PBS–Tween.
5. 100 μL/well of the horseradish peroxidase (Vectastain ABC) solution from **step 3** is added to the plate, which is then allowed to incubate at room temperature for 60 min (*see* **Note 15**).
6. Plates are washed six times in PBS–Tween, and allowed to incubate 15 min in PBS–Tween. They are then washed three times in PBS.
7. 100 μL/well of substrate (Vectastain AEC) solution is added to the plates, and allowed to incubate up to 10 min (*see* **Note 14**).
8. Plates are then washed copiously with dH_2O and allowed to dry completely before reading and analysis (*see* **Notes 21–24**).

4. Notes

1. *Mycobacterium tuberculosis* is a biosafety level 3 pathogen. It must be handled by trained personnel and Elispots using Mtb as a control are performed in a biosafety level 3 facility. The use of Mtb as a positive control is optional. When available, other controls for T-cell specificity should be used in the place of Mtb.
2. Cell yields differ between donors and leukaphoresis products. The number of frozen PBMCs aliquots required for a given step must be determined experimentally.
3. DNase is used to prevent clumping of PBMCs.
4. Thawed cells should not be allowed to sit for any length in freezing medium containing DMSO. Cells can be thawed and prepared in batches for rapid handling. When diluting thawed cells, care must be taken to ensure that the concentration of DMSO is no more than 1%.
5. PBMCs must not be allowed to incubate for longer than 60 min in the flask. After this time, dendritic cell precursors begin to detach from the flask wall.

6. Post-adhesion washing must be done carefully to avoid detaching precursor cells. To aspirate, the flask is gently turned upright, and the medium is aspirated from a corner of the flask opposite to the adhered cells. PBS is then pipetted onto the opposite side of the flask. The flask is then laid flat and gently rocked.
7. Washing and γ-irradiation eliminate contaminating T cells from DC preparations, thereby reducing background IFN-γ release in the Elispot. However, these steps can reduce DC yields. DCs can be washed 1–3 times, and γ-irradiation can be performed at any time up to the day of use.
8. Harvested DCs must be kept at 4°C as much as possible to prevent adherence to plasticware. The harvesting should be performed quickly, centrifugation must be at 4°C, and cells should be kept refrigerated when not in use.
9. The percentage of CD8+ T cells in PBMCs varies between donors. To estimate the number of PBMCs required for recovery of sufficient CD8+ T cells, we assume an average yield of 15%.
10. MACS buffer and cells should be kept on ice throughout the separation.
11. Care must be taken to avoid producing clumps or bubbles, which will impede binding and elution.
12. Following two rounds of magnetic bead separation, purity of CD8+ cells should be at least 96%.
13. Elispots are performed with a final volume of 200–225 μL/well. Volumes of individual components can be adjusted as desired, however DMSO concentrations in the assay should not be more than 0.5%. T cells should be prepared in medium containing sufficient IL-2 such that the final concentration in the Elispot well is 0.5 ng/mL.
14. Peptides are most easily stored in a 96-well format. If a large library is to be generated, it would be highly beneficial to use a plate sealing system. Care must be taken during the removal of sealing film to prevent cross-contamination.
15. During plate coating and development, the antibodies should be added in relatively large (50–100 μL) volumes to ensure full coverage of the membrane. Antibody and staining solutions should be prepared just prior to use.
16. Plates can also be coated for 3 h at 37°C.
17. Experiments are performed with two technical replicates per screen. Important positive controls for these experiments include 0.5 μg/mL phytohemagglutinin (PHA) and an appropriate specificity control for a given T cell (such as Mtb for the experiments described here), each added to wells

containing DC and T cells. Negative controls include wells containing DC and T cells without antigen, DC alone, and T cells alone.

18. To prevent bacterial clumping and ensure a consistent multiplicity of infection, Mtb must be resuspended by repeated passage through a 27-gauge needle. Extreme caution must be taken with this step to avoid accidental needle sticks and infection with Mtb.
19. 2.5×10^5 purified CD8+ T cells or 5,000 CD8+ T-cell clones.
20. Washes with PBS–Tween are performed by immersion, to ensure complete washing of the membrane.
21. Vectastain ABC reagent must incubate 30 min prior to use to allow complexes to form between avidin and the biotinylated horseradish peroxidase. These complexes are stable for several hours. We find it most efficient to prepare this reagent during the final 15 min of the 2 h antibody incubation step.
22. Count settings are set to minimize background in negative control wells while maintaining detection of spots in positive control wells. These settings must be determined by the user.
23. For each plate, the mean background response is calculated from five blank wells. A well is considered positive if it has more than ten spots above mean background response, and is also at least twice the mean background response.
24. Because these screens are used for antigen discovery, a well may be considered positive if at least one of the technical replicates meets the above criteria. However, in selection of those responses most suitable for further testing, weight is given to consistent technical replicates, those antigens in the top 5% of the donor response, and those shared between different donors.

References

1. Muller, I., Cobbold, S. P., Waldmann, H., and Kaufmann, S. H. (1987) Impaired resistance to *Mycobacterium tuberculosis* infection after selective in vivo depletion of L3T4+ and Lyt-2+ T cells. *Infect. Immun.* 55, 2037–2041.
2. Flynn, J. L., Goldstein, M. M., Triebold, K. J., Koller, B., and Bloom, B. R. (1992) Major histocompatibility complex class I-restricted T cells are required for resistance to *Mycobacterium tuberculosis* infection. *Proc. Natl. Acad. Sci. USA* 89, 12013–12017.
3. Silva, C. L., Silva, M. F., Pietro, R. C., and Lowrie, D. B. (1994) Protection against tuberculosis by passive transfer with T-cell clones recognizing mycobacterial heat-shock protein 65. *Immunology* 83, 341–346.
4. Stenger, S., Mazzaccaro, R. J., Uyemura, K., Cho, S., Barnes, P. F., Rosat, J. P., Sette, A., Brenner, M. B., Porcelli, S. A., Bloom, B. R., and Modlin, R. L. (1997) Differential effects of cytolytic T cell subsets on intracellular infection. *Science* 276, 1684–1687.
5. Tan, J. S., Canaday, D. H., Boom, W. H., Balaji, K. N., Schwander, S. K., and Rich, E. A. (1997) Human alveolar T lymphocyte responses to *Mycobacterium tuberculosis* antigens: role for CD4+ and CD8+ cytotoxic T cells and relative resistance of alveolar macrophages to lysis. *J. Immunol.* 159, 290–297.

6. Ladel, C. H., Daugelat, S., and Kaufmann, S. H. (1995) Immune response to *Mycobacterium bovis* Bacille Calmette Guerin infection in major histocompatibility complex class I- and II-deficient knock-out mice: contribution of CD4 and CD8 T cells to acquired resistance. *Eur. J. Immunol.* 25, 377–384.
7. Behar, S. M., Dascher, C. C., Grusby, M. J., Wang, C. R., and Brenner, M. B. (1999) Susceptibility of mice deficient in CD1D or TAP1 to infection with *Mycobacterium tuberculosis. J. Exp. Med.* 189, 1973–1980.
8. Sousa, A. O., Mazzaccaro, R. J., Russell, R. G., Lee, F. K., Turner, O. C., Hong, S., Van Kaer, L., and Bloom, B. R. (2000) Relative contributions of distinct MHC class I-dependent cell populations in protection to tuberculosis infection in mice. *Proc. Natl. Acad. Sci. USA* 97, 4204–4208.
9. Rolph, M. S., Raupach, B., Kobernick, H. H., Collins, H. L., Perarnau, B., Lemonnier, F. A., and Kaufmann, S. H. (2001) MHC class Ia-restricted T cells partially account for beta2-microglobulin-dependent resistance to *Mycobacterium tuberculosis. Eur. J. Immunol.* 31, 1944–1949.
10. Klein, M. R., Smith, S. M., Hammond, A. S., Ogg, G. S., King, A. S., Vekemans, J., Jaye, A., Lukey, P. T., and McAdam, K. P. (2001) HLA-B*35-restricted T cell epitopes in the antigen 85 complex of *Mycobacterium tuberculosis. J. Infect. Dis.* 183, 928–934.
11. Lewinsohn, D. A., Winata, E., Swarbrick, G., Tanner, K. E., Cook, M. S., Null, M. D., Cansler, M. E., Sette, A., Sidney, J., and Lewinsohn, D. M. (2007) Immunodominant tuberculosis CD8 antigens preferentially restricted by HLA-B. *PLoS Pathog.* 3, 1240–1249.
12. Flyer, D. C., Ramakrishna, V., Miller, C., Myers, H., McDaniel, M., Root, K., Flournoy, C., Engelhard, V. H., Canaday, D. H., Marto, J. A., Ross, M. M., Hunt, D. F., Shabanowitz, J., and White, F. M. (2002) Identification by mass spectrometry of CD8(+)-T-cell *Mycobacterium tuberculosis* epitopes within the Rv0341 gene product. *Infect. Immun.* 70, 2926–2932.
13. Charo, J., Geluk, A., Sundback, M., Mirzai, B., Diehl, A. D., Malmberg, K. J., Achoulr, A., Huriguchi, S., van Meijgaarden, K. E., Drijfout, J. W., Beekman, N., van Veelen, P., Ossendorp, F., Ottenhoff, T. H., and Kiessling, R. (2001) The identification of a common pathogen-specific HLA class I A*0201-restricted cytogoxic T cell epitope encoded within the heat shock protein 65. *Eur. J. Immunol.* 31, 3602–3611.
14. Dong, Y., Demaria, S., Sun, X., Santori, F. R., Jesdale, B. M., De Groot, A. S., Rom, W. N., and Bushkin, Y. (2004) HLA-A2-restricted CD8+-cytotoxic-T-cell responses to novel epitopes in *Mycobacterium tuberculosis* superoxide dismutase, alanine dehydrogense, and glutamine synthetatse. *Infect. Immun.* 72, 2412–2415.
15. Shams, H., Barnes, P. F., Weis, S. E., Klucar, P., and Wizel, B. (2003) Human CD8+ T cells recognize epitopes of the 28-kDa hemolysin and the 38-kDa antigen of *Mycobacterium tuberculosis. J. Leukoc. Biol.* 74, 1008–1014.
16. Lewinsohn, D. A., Lines, R. A., and Lewinsohn, D. M. (2002) Human dendritic cells presenting adenovirally expressed antigen elicit Mtb-specific CD8+ T cells. *Am. J. Respir. Crit. Care* 166, 843–848.
17. Geluk, A., van Meijgaarden, K. E., Franken, K. L., Drijfhout, J. W., D'Souza, S., Necker, A., Huygen, K., and Ottenhoff, T. H. (2000) Identification of major epitopes of *Mycobacterium tuberculosis* AG85B that are recognized by HLA-A*0201-restricted CD8+ T cells in HLA-transgenic mice and humans. *J. Immunol.* 165, 6463–6471.
18. Caccamo, N., Milano, S., Di Sano, C., Cigna, D., Ivanyi, J., Krensky, A. M., Dieli, F., and Salerno, A. (2002) Identification of epitopes of *Mycobacterium tuberculosis* 16-kDA protein recognized by human leukocyte antigin-A*0201 CD8(+) T lymphocytes. *J. Infect. Dis.* 186, 991–998.
19. Klein, M. R., Hammond, A. S., Smith, S. M., Jaye, A., Lukey, P. T., and McAdam, K. P. (2002) HLA-B*35-restricted CD8(+)-T-cell epitope in *Mycobacterium tuberculosis* Rv2903c. *Infect. Immun.* 70, 981–984.
20. Mohagheghpour, N., Gammon, D., Kawamura, L. M., van Vollenhoven, A., Benike, C. J., and Engleman, E. G. (1998) CTL response to *Mycobacterium tuberculosis*: identification of an immunogenic epitope in the 19-kDa lipoprotein. *J. Immunol.* 161, 2400–2406.
21. Smith, S. M., Brookes, R., Klein, M. R., Malin, A. S., Lukey, P. T., King, A. S., Ogg, G. S., Hill, A. V., and Dockrell, H. M. (2000) Human CD8+ CTL specific for the mycobacterial major secreted antigen 85A. *J. Immunol.* 165, 7088–7095.
22. Lewinsohn, D. M., Zhu, L., Madison, V. J., Dillon, D. C., Fling, S. P., Reed, S. G., Grabstein, K. H., and Alderson, M. R. (2001) Classically restricted human CD8+ T lymphocytes derived from *Mycobacterium tuberculosis*-infected cells: definition of antigenic specificity. *J. Immunol.* 166, 439–446.

23. Shams, H., Klucar, P., Weis, S. E., Lalvani, A., Moonan, P. K., Safi, H., Wizel, B., Ewer, K., Nepom, G. T., Lewinsohn, D. M., Andersen, P., and Barnes, P. F. (2004) Characterization of a *Mycobacterium tuberculosis* peptide that is recognized by human CD4+ and CD8+ T cells in the context of multiple HLA alleles. *J. Immunol.* 173, 1966–1977.

24. Lalvani, A., Brookes, R., Wilkinson, R. J., Malin, A. S., Pathan, A. A., Andersen, P., Dockrell, H., Pasvol, G., and Hill, A. V. (1998) Human cytolytic and interferon gamma-secreting CD8+ T lymphocytes specific for *Mycobacterium tuberculosis. Proc. Natl. Acad. Sci. USA* 95, 270–275.

25. Pathan, A. A., Wilkinson, K. A., Wilkinson, R. J.,Latif, M., McShane, H., Pasvol, G., Hill, A. V., and Lalvani, A. (2000) High frequencies of circulating IFN-gamma-secreting CD8 cytotoxic T cells specific for a novel MHC class I-restricted *Mycobacterium tuberculosis* epitope in M. tuberculosis-infected subjects without disease. *Eur. J. Immunol.* 30, 2713–2721.

26. Schnappinger, D., Ehrt, S., Voskuil, M. I., Liu, Y., Mangan, J. A., Monahan, I. M., Dolganov, G., Efron, B., Butcher, P. D., Nathan, C., and Schoolnik, G. K. (2003) Transcriptional Adaptation of *Mycobacterium tuberculosis* within macrophages: Insights into the phagosomal environment. *J. Exp. Med.* 198, 693–704.

27. http://genolist.pasteur/tuberculist/.

28. Lewinsohn, D. M., Briden, A. L., Reed, S. G., Grabstein, K. H., and Alderson, M. R. (2000) *Mycobacterium tuberculosis*-reactive CD8+ T lymphocytes: the relative contribution of classical versus nonclassical HLA restriction. *J. Immunol.* 165, 925–930.

29. Heinzel, A. S., Grotzke, J. E., Lines, R. A., Lewinsohn, D. A., McNabb, A. L., Streblow, D. N., Braud, V. M., Grieser, H. J., Belisle, J. T., and Lewinsohn, D. M. (2002) HLA-E dependent presentation of Mtb-derived antigen to human CD8+ T cells. *J. Exp. Med.* 196, 1473–1481.

Chapter 28

High-Throughput T-Cell Epitope Discovery Through MHC Peptide Exchange

Sine Reker Hadrup, Mireille Toebes, Boris Rodenko, Arnold H. Bakker, David A. Egan, Huib Ovaa, and Ton N.M. Schumacher

Summary

Recombinant major histocompatibility complex (MHC) class I molecules complexed with pathogen-specific or other disease-associated antigens have become essential reagents for the analysis of adaptive T-cell responses. However, conventional techniques for the production of recombinant peptide-MHC (pMHC) complexes are highly involved and thereby limit the use of pMHC complexes in terms of antigen diversity. To make pMHC-based techniques suitable for high-throughput analyses we developed an MHC peptide exchange technology based on the use of conditional MHC ligands. This technology enables the parallel production of thousands of different pMHC complexes within hours, allowing the development of high-throughput MHC-based assay systems to identify MHC ligands and cytotoxic T-cell responses. These high-throughput assays should prove valuable for the screening of entire disease-associated proteomes, including pathogen-encoded proteomes, tumor-associated antigens, and autoimmune antigens.

Key words: Conditional ligands, MHC peptide exchange, High-throughput screening, T-cell epitopes, CD8+ T cells.

1. Introduction

The adaptive immune system has the remarkable potential to generate antigen-specific T-cell responses against a very wide variety of antigens from intruding microorganisms. Unlike antigen recognition by B cells, T-cell recognition of antigen does not involve direct binding to an offending antigen, but rather an interaction of T-cell-resident T-cell receptors with the composite surface of a pathogen-derived peptide epitope and the major histocompatibility

Ulrich Reineke and Mike Schutkowski (eds.), *Methods in Molecular Biology, Epitope Mapping Protocols, vol. 524*

DOI: 10.1007/978-1-59745-450-6_28

complex (MHC) molecule that carries the epitope to the cell surface. The extreme functional diversity of the T-cell-based immune system relies on the presence of clone-specific T-cell receptors on T cells, and it is estimated that the human immune system harbors some 25 million T-cell clones with distinct specificities *(1)*. Unfortunately, the broad recognition potential of the human T-cell repertoire is poorly matched by the currently established methods for immune monitoring and T-cell epitope discovery. The methodology that is frequently used to analyze antigen specific T-cell responses involves the staining of antigen-specific T cells with fluorescently labeled multimeric peptide-MHC (pMHC) complexes. This MHC multimer technology, originally developed by Altman and colleagues *(2)*, has become a mainstay of cellular immunology as evidenced by some 2,000 citations over the past decade. While MHC multimer-based T-cell detection is used extensively for the analysis of T-cell responses against a small set of antigens, it is unfortunately less suited for high-throughput applications, due to the cumbersome production process that needs to be performed for each specific pMHC complex. To overcome this problem and make pMHC complex-based techniques compatible with high-throughput analyses we developed an MHC peptide exchange technology *(3)*. This technology enables the production of large arrays of different pMHC complexes within hours, and thereby allows the development of pMHC based high-throughput screening assays for the identification of new MHC ligands and antigen specific T-cell responses. Such assays might serve valuable not only to screen for new T-cell epitopes, but also to monitor immune responses in a high-complexity manner, thereby better reflecting the complexity of the T-cell-based immune system.

1.1. The Strategy of MHC Class I Conditional Ligands

Conditional ligands (p*) are peptide ligands for MHC molecules that can be used to produce large quantities of recombinant p*MHC complexes and that can be triggered to fall into pieces on command, while bound to the MHC structure. As the resultant peptide fragments have a much reduced affinity for the peptide binding groove of the MHC molecule, these fragments dissociate, thereby generating empty, peptide-receptive MHC complexes under physiological conditions. The conditional ligands that are used in the protocols included here are cleaved when subjected to long-wavelength (366 nm) UV-light. UV-sensitivity of these conditional ligands is imparted by the introduction of one non-natural amino acid that carries a nitrophenyl side chain within the peptide (*see* **Fig. 1a**). When cleavage of the conditional ligand is performed in the absence of another peptide ligand, dissociation of the resulting peptide fragments leads to disintegration of the MHC complex. However, when cleavage is performed in the presence of a new MHC ligand, this ligand can bind to and thereby stabilize the MHC molecule (*see* **Fig. 1b**) *(3)*.

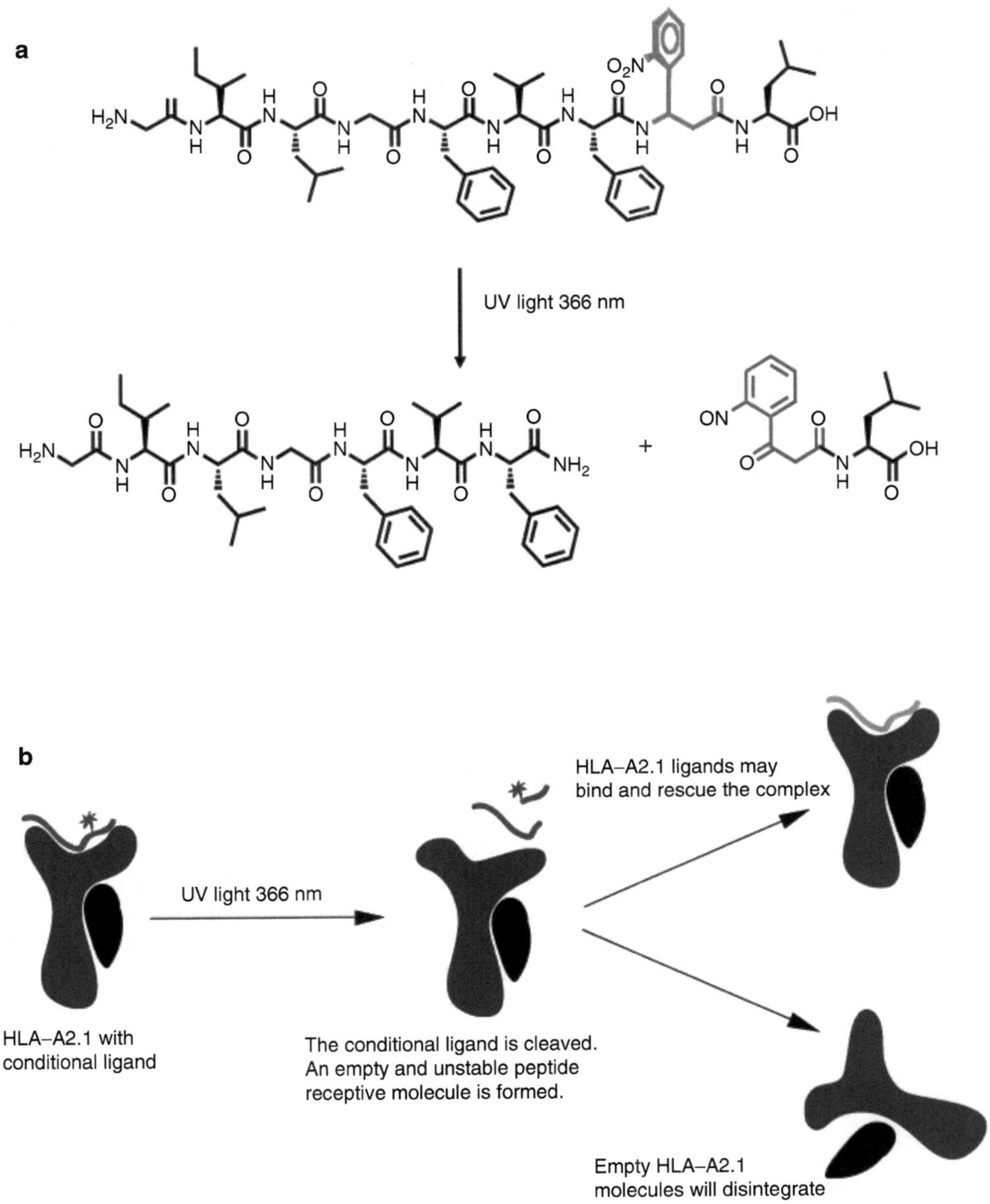

Fig. 1. (**a**) The UV-sensitive HLA-A2.1 restricted conditional ligand KILGFVFJV, is based on the HLA A2.1 restricted influenza A Matrix-1(58–66) epitope GILGFVFTL, where Gly1 is replaced with lysine to increase solubility, and Thr8 is replaced with a 2-nitrophenyl containing amino acid residue termed J. Anchor residue Leu9 is replaced with valine to increase HLA A2.1 affinity. Upon UV irradiation the 2-nitrophenyl moiety rearranges. This results in the formation of two fragments: a carboxamido terminal 7-mer and a 2-nitrosoaceptophenone-containing dimeric peptide. (**b**) A schematic overview of the UV-mediated exchange reaction. The conditional ligand is cleaved by UV-irradiation to form two fragments that dissociate from the HLA A2.1 binding groove, resulting in an empty, unstable peptide receptive molecule. The empty HLA A2.1 molecule will disintegrate if not stabilized by binding of an HLA A2.1 ligand.

1.2. The Use of Conditional MHC Class I Ligands

Peptide-receptive MHC molecules generated upon UV-cleavage of conditional ligands can be used for two main purposes. First, MHC exchange reactions can be used to rapidly identify novel MHC ligands for the allele under investigation. Because MHC exchange reactions can be performed in multiwell formats and under native conditions, high-throughput analyses of putative ligands for MHC complexes are straightforward. Binding of such putative ligands may either be revealed by techniques such as ELISA, that monitor the stability of the MHC structure (*see* **Subheading** 3.4), or by biophysical techniques that monitor peptide binding, such as fluorescence polarization *(4)*. This type of MHC binding assays have proven very useful to scan relevant peptide sets for MHC ligands (Reker Hadrup et al., unpublished). In addition, it may be possible to identify nonpeptide ligands for MHC molecules using such screens.

The second main application of the conditional exchange method lies in its use for the production of pMHC complexes of interest. pMHC complexes produced by MHC exchange reactions can be used to prepare MHC tetramers (or other multimeric MHC reagents) for the detection of antigen-specific T cells by flow cytometry or for T-cell purification. Because the MHC exchange technology allows the production of very large collections of pMHC multimers, this technology is well-suited for the rapid identification of cytotoxic T-cell antigens. The feasibility of such large-scale screening has been demonstrated by the definition of cytotoxic T-cell epitopes within the H5N1 influenza A/Vietnam/1194/04 genome *(3)*. In addition, because of the simplicity of producing different pMHC complexes by MHC exchange, this technology may prove valuable for the production of a set of clinical grade MHC reagents from a batch of clinical grade MHC complexes occupied with a UV-sensitive ligand. The clinical potential of MHC multimer-based T-cell isolation has been shown by Moss and colleagues *(5)* and based on the rapid developments in adoptive T-cell therapies *(6)* we consider it likely that MHC multimer-based adoptive T-cell therapies will become more widespread. The methods described here are also applicable to the production of MHC class II-peptide complexes as well as other peptide-bound complexes. However, as the conditional ligands that have been developed to date are not genetically encoded, this is under the provision that the complex of MHC or other macromolecule with the conditional ligand is produced in vitro rather than in vivo *(7, 8)*.

In the following sections, all steps necessary to produce complexes of soluble MHC class I molecules with conditional ligands are described. In addition, the use of the UV-mediated exchange technology to identify new MHC ligands and to produce pMHC complexes for high-throughput screening purposes is detailed. This encompasses protocols for the synthesis of conditional ligands, refolding of such conditional ligands with MHC class I heavy chain and β2m, purification, biotinylation, UV-mediated peptide exchange, multimerization with fluorescence labeled streptavidin, as well as a high-throughput screening method for identifying new MHC ligands.

These protocols are described for the human MHC allele HLA-A2.1 and its conditional ligand KILGFVFJV (in which J is the photolabile amino acid residue). However, similar procedures can be used to form p*MHC complexes for other human MHC alleles, and with minor modifications in the refolding procedure, also for mouse MHC alleles. A conditional ligand has been described for the mouse MHC alleles H2-D^b *(3)* and K^b *(23)*. In addition, conditional ligands have also been identified and validated for the human MHC alleles HLA-A1, A3, A11, and B7 *(9)*.

A schematic overview of the use of the exchange technology in MHC ligand and T-cell epitope discovery is given in **Fig. 2**.

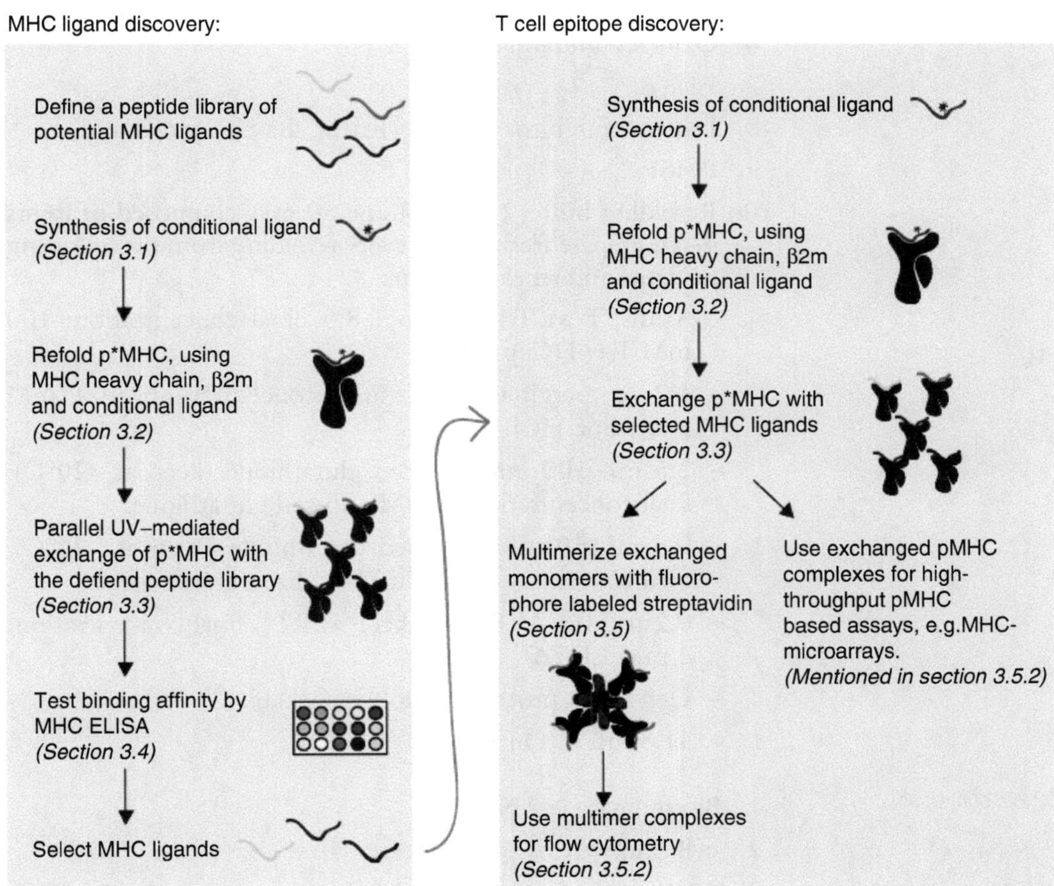

Fig. 2. A schematic overview of the steps and protocols used to perform high-throughput discovery of MHC ligands and T-cell epitopes.

2. Materials

2.1. Synthesis of the Conditional Ligand

1. 3-Amino-3-(2-nitro)phenyl-propionic acid
2. Fluorenylmethylchloroformate (Fmoc-chloride)
3. Dioxane
4. Deoxycholic acid
5. Fmoc-protected amino acids and other reagents and solvents for peptide synthesis

2.2. Refolding, Biotinylation, and Purification of MHC Complexes with Conditional Ligands

2.2.1. Refolding

1. MHC class I heavy chain inclusion bodies isolated from the *E. coli.* strain BL21 (DE3) pLysS *(10)*.
2. β2m in inclusion bodies isolated from E. coli strain XL1 Blue *(10)*.
3. Denaturation buffer: 8 M urea in 100 mM Tris-HCl pH 8.0.
4. L-Arginine-HCl.
5. Reduced glutathione.
6. Oxidized glutathione.
7. EDTA.
8. Protease inhibitor cocktail (Roche, Basel, Switzerland).
9. PMSF.
10. Refolding buffer (total volume 50 mL, composed of **items 4–8**). Prepare fresh before every refolding from the following reagents and stock solutions:
    - 5 mL, 1 M Tris-HCl pH 8.0, final concentration: 100 mM Tris-HCl pH 8.0.
    - 4.2 g L-arginine-HCl, final concentration: 400 mM L-arginine HCl.
    - 2.5 mL 100 mM reduced glutathione (keep at –20°C), final concentration: 5 mM reduced glutathione.
    - 0.5 mL 50 mM oxidized glutathione (keep at –20°C), final concentration: 0.5 mM oxidized glutathione.
    - 0.2 mL 0.5 M EDTA (keep at RT), final concentration: 2 mM EDTA.
    - One tablet protease inhibitor cocktail.
    - 41.8 mL H_2O (milliQ).

2.2.2. Biotinylation

1. Biotin ligase (*see* **Note** 1).
2. D-Biotin.
3. ATP.
4. EDTA-free protease inhibitor cocktail, tablets (Roche, Basel, Switzerland).

5. Ligase buffer (10×): 50 mM $MgCl_2$, 0.2 M Tris-HCl pH 7.5.
6. Biotinylation solution (2×): (prepare fresh before biotinylation in a volume equal to the final volume of pMHC solution after purification):
   - 0.04 vol 5 mM D-biotin dissolved in 100 mM sodium-phosphate buffer, pH 7.5 (keep at –20°C).
   - 0.04 vol 0.5 M ATP dissolved in 1 M Tris-HCl pH 9.5 (prepare fresh) (*see* **Note** 2).
   - 0.01 vol biotin ligase (20 µM) (keep at –20°C).
   - 0.2 vol 10 × ligase buffer.
   - 0.08 vol EDTA-free protease inhibitor cocktail (one tablet in 2 mL H_2O, prepare fresh).
   - 0.63 vol water (MilliQ).

2.2.3. Concentration and Purification

1. Amicon ultrafiltration membranes, MW cut-off 30 kDa (Millipore, Billerica, MA).
2. Amicon Centriplus filters ym30 MW cut-off 30 kDa (Millipore).
3. Amicon Centricon filters MW cut-off 30 kDa (Millipore).
4. HPLC or FPLC system with a gel-filtration column, preferably a Phenomenex Biosep SEC S3000 column, 300 × 21.2 mm (Phenomenex, Torrance). Running buffer is PBS, pH 7.4.

2.3. UV Mediated Peptide Exchange

1. Polypropylene microtubes 1.5 mL with cap (*see* **Note** 3).
2. Polypropylene microplate 96-well V shape plate (*see* **Note** 4).
3. UV-lamp 366 nm: CAMAG UV lamp long-wave UV, 366 nm, 2 × 8 W in a protection cabinet (CAMAG, Muttenz, Switzerland) (*see* **Note** 5).

2.4. Measuring Peptide Mediated MHC Stabilization by MHC ELISA

1. Coating solution: 2 µg/mL streptavidin (Invitrogen, Carlsbad, CA) in PBS.
2. Wash buffer: 0.05% Tween 20 in PBS.
3. Blocking buffer: 2% BSA in PBS.
4. Horseradish peroxidase (HRP)-conjugated polyclonal rabbit anti-human β2m antibody (DakoCytomation).
5. ABTS (2′,2′-azino-bis-(3-ethylbenzthiazoline-6-sulphonic acid) diammonium salt) developing solution (Sanquin Reagents, Amsterdam, The Netherlands) (*see* **Note** 6).
6. Stop buffer (Sanquin Reagents, Amsterdam, The Netherlands).
7. Immuno Maxisorp plate 96-wells (Nunc, Roskilde, Denmark) or High Bind 384-well ELISA plate (Greiner Bio-one, Frickenhausen, Germany).
8. Absorbance plate reader (405 nm).

2.5. Multimerization of MHC Monomers

1. PE-streptavidin solution 1 mg/mL (Invitrogen).
2. or APC-streptavidin solution 1 mg/mL (Invitrogen) (*see* **Note** 7).
3. Glycerol.
4. BSA, >96% purity.
5. FACS buffer: PBS, 0.5% BSA, 0.02% NaN_3.

3. Methods

To make a collection of pMHC complexes by MHC exchange technology three basic things are required: (1) a batch of refolded p*MHC complexes, (2) (potential) MHC ligands of interest and (3) a UV lamp (366 nm). In the following section, we have described the procedure of synthesizing the conditional ligand, the refolding, biotinylation, and purification of the p*MHC complex, the peptide exchange reaction, and the use of the resultant complexes in pMHC based assays. Main focus is on the identification of new MHC ligands by the combined use of MHC exchange technology and an MHC ELISA, and on the generation of collections of pMHC multimer complexes for flow cytometry.

3.1. Synthesis of Conditional Ligands

For production of conditional ligands an Fmoc-protected photolabile building block is required to allow the production of any conditional ligand by standard Fmoc-peptide synthesis methodology. It is noted that Fmoc-(nitrophenyl)propionic acid, the photolabile building block used to date *(3, 9)*, is now also available commercially (Advanced Chemtech).

3.1.1. Production of the Photolabile Building Block (2 Days)

1. Dissolve fluorenylmethylchloroformiate (4.0 g, 15.5 mmol, 1.2 Eq) in dioxane (60 mL) in a 100-mL addition funnel.
2. To a 500-mL round-bottom flask add 3-amino-(2-nitro)phenyl propionic acid (3 g, 14.3 mmol), dioxane (60 mL) and 10% (w/w) aqueous sodium carbonate (60 mL). Place the flask in an ice-bath and stir vigorously.
3. Add the Fmoc-chloride solution dropwise to the cooled suspension over 15 min. Allow the reaction mixture to warm up to room temperature and leave stirring overnight, covering the reaction vessel with aluminum foil.
4. Pour the reaction mixture into 600 mL of water and extract twice with diethyl ether (2 × 200 mL) to remove Fmoc residues.
5. Cool the aqueous layer in an ice bath and adjust the pH to 1 by the addition of 2 M aqueous HCl.

6. Filter off the resulting precipitate and dry under high vacuum overnight, to obtain *N*-fluorenylmethyloxycarbonyl 3-amino-3-(2-nitro)phenyl-propionic acid as a white solid (5.5 g, 12.7 mmol, 86% yield) (*see* **Note** 8). Cover the flask with aluminum foil. *N*-fluorenylmethyloxycarbonyl 3-amino-3-(2-nitro)phenyl-propionic acid can be stored in the dark at 20°C for years.

3.1.2. Peptide Synthesis of Conditional Ligands Using Photolabile Building Block (3 Days)

1. Prepare the HLA A2.1 restricted conditional ligand KILGFVFJV (J is the photolabile amino acid residue) by standard automated Fmoc-peptide synthesis *(11)* using the UV-labile building block Fmoc-3-amino-(2-nitro)phenyl propionic acid (*see* **Notes** 9 and 10).

3.2. Refolding, Biotinylation, and Purification of p*HLA A2.1 Complexes

The production of p*HLA A2.1 complexes is largely based on the refolding procedures established by Garboczi *(12)*. As a prerequisite for performing the refolding reactions as described in this chapter, recombinant MHC class I heavy chain and β2m are needed. These proteins are obtained via bacterial expression systems and isolated as inclusion bodies. A detailed description of this procedure can be found in *(10)*. The refolding of p*MHC following the procedure described below results in approximately 600 μg of biotinylated p*MHC, depending on the refolding efficiency. The scale of this procedure can easily be adjusted. The refolding protocol can be used to produce complexes of the extracellular domains of most human MHC alleles, and the biotinlylation protocol can be used to biotinylate recombinant versions of all human and mouse alleles that carry a biotinylation sequence as a COOH-terminal tag to the MHC class I heavy chain.

3.2.1. Refolding of p*HLA A2.1 Complexes (3 Days)

1. Prepare fresh, cold denaturation buffer.
2. Dissolve approximately 0.2 μmol MHC heavy chain and approximately 0.4 μmol β2m in separate tubes using 1 mL denaturing buffer for each (*see* **Note** 11).
3. Spin down the remaining aggregates for 2 min at 16,000 × *g*.
4. Take off supernatant and place in fresh 1.5-mL polypropylene tubes.
5. Measure the protein concentration in a spectrophotometer at 280 nm, by diluting 1/100 in denaturing buffer: 6 μL protein with 594 μL denaturing buffer (extinction coefficient at 280 nm for HLA A2.1 with biotag is 80,750/M cm and extinction coefficient at 280 nm for β2m is 19,180/M cm). For example: if OD HLA A2.1 = 0.24/cm, then the protein concentration of the sample is $\{(0.24/\text{cm} \times 100)/80{,}750/\text{M cm}\} \times 10^6 = 297.2$ μM. For each refolding an aliquot of MHC heavy chain corresponding to 1 μM final concentration is added on 3 consecutive days (resulting

in an MHC heavy chain concentration of 3 μM on day 3). Thus, in this example, a total of 3 × 1 μM × 50 mL/297.2 μM × 1,000 = 505 μL of the dissolved HLA-A2 heavy chain is required to perform a 50-mL refolding. One third of this, e.g.,168 μL, is added each day on 3 consecutive days. The concentration of β2m is calculated in the same fashion. For each refolding an aliquot of β2m corresponding to 2 μM final concentration is added on 3 consecutive days (resulting in a β2m concentration of 6 μM on day 3). The fractions to be added on day 2 and 3 should be stored at –20°C in denaturation buffer until use.

6. To set up the refolding reaction, add the following components to cold refolding buffer in the critical order indicated below. Leave the tube slowly stirring in a cold room (approximately 10°C) for a total of 3 days. Keep it in dark.
   (a) Conditional ligand (p*) dissolved in 100% DMSO (50 mM stock), final concentration 60 μM.

PMSF, final concentration 1 mM, from 100 mM stock in isopropanol. PMSF is very unstable in aqueous solution, proceed with the subsequent steps immediately.

   (b) β2m, to a final concentration of 2 μM.
   (c) MHC heavy chain, dropwise, to a final concentration of 1 μM.
   (d) Stir overnight.

Repeat addition of β2m and MHC heavy chain and add fresh PMSF to a final concentration of 1 mM.

   (e) Stir overnight.
   (f) Repeat addition of β2m and MHC heavy chain and add fresh PMSF to a final concentration of 1 mM.
   (g) Stir overnight.
   (h) Centrifuge 4000 × *g* for 20 min to removed aggregates.

7. Wash an Amicon ultrafiltration membrane (MW cut-off 30 kDa) with milliQ water.
8. Concentrate the refolded sample to approximately 6 mL by nitrogen flow over the ultrafiltration membrane. Alternatively, centriplus filters can be used. The need for concentration depends on the maximum volume and protein that can be loaded on the HPLC or FPLC system used for purification.
9. Rinse filter twice with 1 mL of PBS and combine with the 6 mL sample to give 8 mL in total (*see* **Note** 12).
10. Purify the 8 mL sample by gel-filtration HPLC or FPLC. We typically use a 5 mL loop for a Biosep 3000 Phenomenex column, 4 mL/min flow rate. After HPLC purification (2 runs of 4 mL on this system) you will have approximately 16 mL of p*MHC sample. Keep the sample on ice.

Following this step, the sample is generally used immediately for biotinylation. The use of MHC class I complexes stored in glycerol is not recommended as the presence of glycerol adversely affects biotinylation efficiency (*see* http://www.avidity.com/tech-protein.html). However, p*MHC complexes may be stored by snap-freezing in liquid nitrogen without the addition of glycerol offering a possible route for biotinylation at a later time-point.

3.2.2. Biotinylation of p*HLA A2.1 Complexes (2 Days)

1. Add 1 vol of fresh biotinylation solution to 1 vol of p*MHC monomer (*see* **Note** 13).
2. Incubate overnight at 25°C. Keep it in dark.
3. Wash centriplus filter (MW cut-off 30 kDa), by placing 10 mL milliQ water on filter and centrifuging at 4°C, 4000 × *g* for 20 min.
4. Discard water from the centriplus filter, add biotinylation solution and concentrate the sample to a volume of 3 mL.
5. Rinse the filter twice with 500 μL PBS.
6. Purify by gel-filtration HPLC or FPLC and collect the biotinylated p*MHC monomer (depending on the efficiency of p*MHC refolding and the chromatography system used, this may require multiple runs). In our hands the biotinylated p*MHC monomer is recovered in a volume of approximately 8 mL.
7. Wash centricon filter (MW cut-off 30 kDa): place 2 mL milliQ water on filter and centrifuge at 4°C, 4000 × *g* for 20 min.
8. Discard H_2O from the centriplus filter and add the purified biotinylated pMHC solution. Concentrate the 8 mL to approximately 400 μL on the centricon filter. A final concentration of approximately 20–40 μM (1–2 mg/mL) is obtained, depending on the refolding efficiency and recovery during purification.
9. Add glycerol to a final concentration of 16% and divide in 50 μL aliquots in 1.5-mL Sarstedt tubes. The biotinylated p*MHC monomer (MW, 47 kDa) with a final concentration of approximately 1 mg/mL can be stored for at least 1 year at –20°C.

3.2.3. Determination of Biotinylation Efficiency

Biotinylation reactions performed according to the protocol above should give complete or near-complete biotinylation of the tagged heavy chain. It is nevertheless recommended to verify the degree of biotinylation for every batch of p*MHC before further use. The degree of biotinylation can be determined by coupling the biotinylated monomers to streptavidin to form multimers and subsequent analysis by gel-filtration chromatography. In case of full biotinylation, all monomers should be bound to streptavidin when an excess of streptavidin is used (more than one streptavidin molecule per four monomers), resulting in a shift in retention time of streptavidin and the complete disappearance of the monomer peak in gel-filtration chromatography (*see* **Fig. 3**).

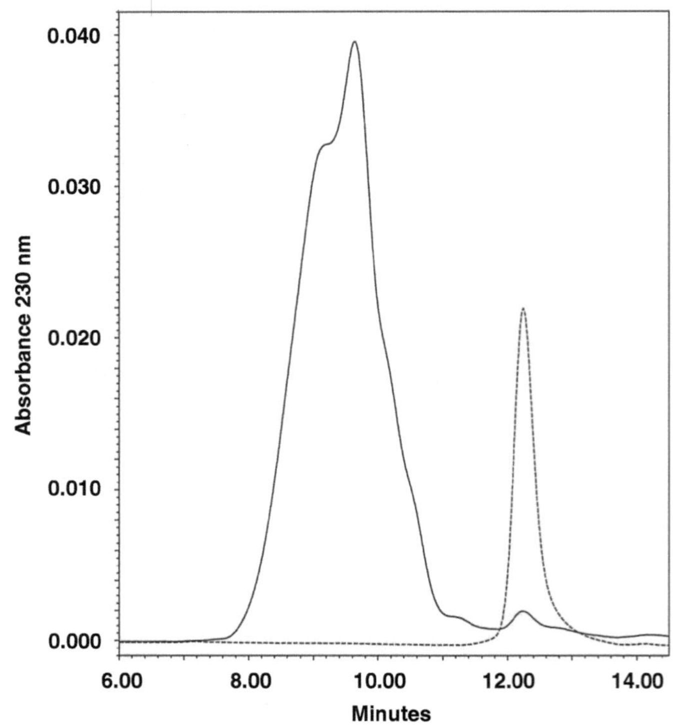

Fig. 3. Gel-filtration chromatography of biotinylated p*MHC (*dashed line*) and biotinylated p*MHC incubated with excess streptavidin-PE (*solid line*). Note the essentially complete disappearance of the p*MHC monomer peak after incubation with excess streptavidin-PE, indicating 100% biotinylation.

1. Prepare two samples: (1). Mix 10 μL p*MHC (250 μg/mL) with 5 μL Streptavidin-PE (1 mg/mL) in a total volume of 100 μL PBS (corresponding to a molar ratio of 3 monomers per Streptavidin-PE) (*see* **Note** 14) (2). Dilute 10 μL p*MHC (250 μg/mL) in a total volume of 100 μL PBS.
2. Incubate 20 min at 4°C and analyze by gel-filtration chromatography.

3.3. UV-Mediated Peptide Exchange

The UV-mediated cleavage of the conditional ligand is time dependent. With the lamp used here, peptide cleavage can be detected after 1 min and is essentially complete after approximately 15 min. A 30–60 min incubation time is routinely used to ensure optimal exchange of the conditional ligand with the peptide of interest. Notably, protein concentration can influence the rate of UV-mediated cleavage, as both the nitrophenyl moiety and the reaction product absorb long wavelength UV light. In addition, it seems possible that the path length would have some effect, although this has not been tested. Empty, peptide receptive MHC molecules that are formed upon UV exposure can be rescued by performing the UV-mediated cleavage in the presence of an MHC ligand of interest. In routine experiments, a 100-fold molar excess of peptide over MHC is used. UV-induced peptide exchange is routinely performed using 25 μg/mL of p*MHC. However, exchange reactions have been performed with p*MHC concentrations up to 200 μg/mL:

1. In a 96-well plate, add the following reagents to each well (*see* **Note** 15):
   - 100 μL PBS.
   - 12.5 μL 500 μM exchange peptide in PBS.
   - 12.5 μL 250 μg/mL UV-sensitive MHC monomer.
2. Place the 96-well plate under a UV lamp (366 nm) for 1 h, with a distance between the UV lamp and sample of approximately 5 cm (*see* **Note** 16).
3. Spin the plate at 3300 × *g* for 5 min, transfer 100 μL of supernatant (keep the plate at an angle to avoid transferring any pellet) to a fresh 96-well plate for subsequent measurements.

Resulting complexes may either be used to determine exchange efficiency by gel-filtration chromatography or by MHC ELISA (*see* **Subheading 3.4**, *see* **Note** 17). Alternatively, the exchanged complexes can be used for multimerization or for other pMHC-based assays (*see* **Note 18**).

3.4. Measuring Peptide Mediated MHC Rescue by MHC ELISA

The MHC ELISA measures the concentration of correctly folded pMHC molecules after the UV-mediated peptide exchange (*see* **Fig. 4a**). This assay can be used to verify the success of routine exchange reactions, i.e., to ensure that the MHC ligand of choice indeed rescued the MHC complex, but can also be used for high-throughput screening of potential MHC ligands. Depending on the number of epitopes that is screened, the assay can be performed both in a 96- or a 384-well format..

The MHC ELISA measures the concentration of MHC class I complexes in the range of 0.5–10 nM (*see* **Fig. 4b**). It is recommended to include a standard MHC titration curve to ensure that the obtained absorbance values are within the linear range of the assay (*see* **Note 19**). UV-irradiation of the HLA A2.1 p*MHC complex (HLA A2.1-KILGFVFJV) leads to a reduction of the absorbance signal of approximately 70–80% and an estimated reduction in concentration of folded MHC of 85% (*see* **Fig. 4c**, **d**). This reduction is comparable to the reduction in MHC concentration as detected by gel-filtration chromatography (*see* **Fig. 5**).

Inclusion of high affinity HLA A2.1 restricted ligands in UV-mediated exchange reactions leads to recovery of 80–100% of the original ELISA signal, shown in **Fig. 4c, d** for a number of viral T-cell epitopes. Lower rescue is observed using peptides with a moderate to low affinity for HLA A2.1, such as the tumor associated antigens MART-1 and Survivin *(13, 14)*. The native MART-1 (EAA) and Survivin-96 (LTL) peptide show very low

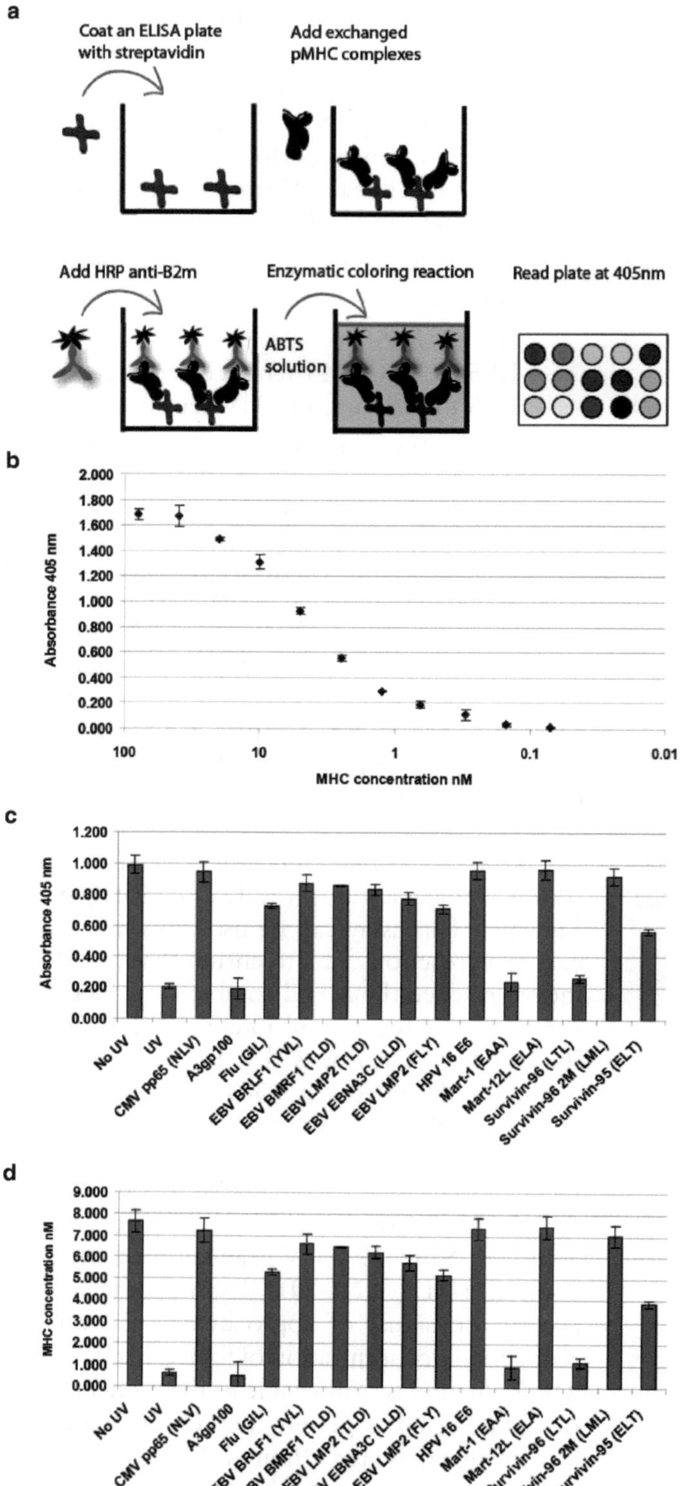

Fig. 4. (**a**) Schematic overview of the MHC ELISA. (**b**) Standard curve of the ELISA assay obtained using HLA A2.1/GILGFVFTL. The linear range of the assay is approximately 0.5–10 nM pMHC. (**c**) MHC rescue measured by ELISA after UV-mediated exchange using a panel of peptides with varying HLA A2.1 affinities. Rescue is determined by measurement of absorbance at 405 nm. (**d**) Concentration of the pMHC complexes obtained after UV-mediated exchange shown in C, as calculated based on the standard curve displayed in B.

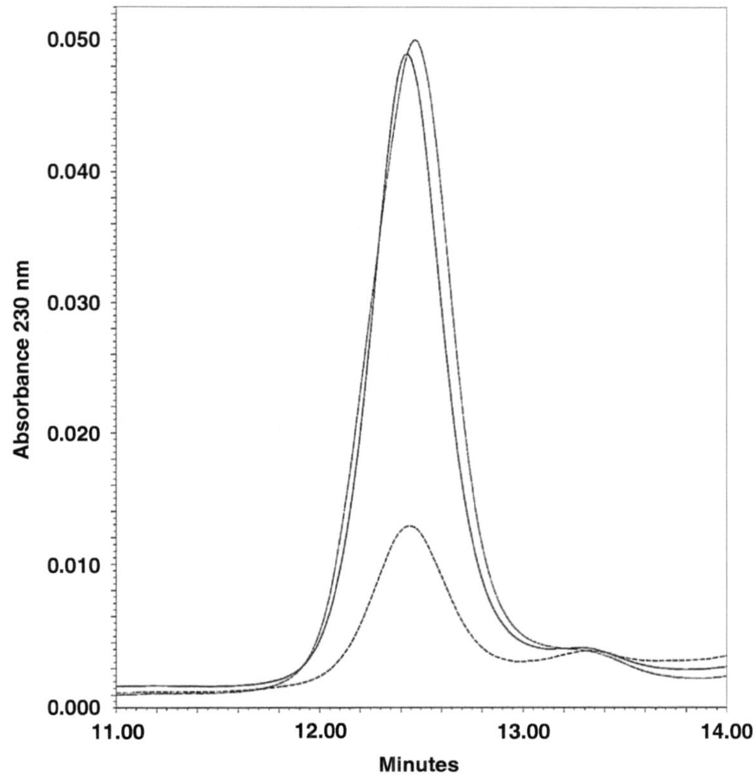

Fig. 5. Gel-filtration chromatography profile of p*MHC without UV exposure (*dotted line*), after 60 min UV-irradiation without addition of a rescue peptide (*dashed line*), and after exchange using the high affinity HLA A2.1 ligand CMV pp65 (NLV) (*solid line*). For all three samples, exchange was performed using 25 μg/mL p*MHC in a 125 μL exchange reaction. After exchange, 100 μL was injected for analysis by gel-filtration chromatography.

HLA A2.1 affinity and correspondingly very low rescue. Changing the P2 anchor residue of these epitopes into residues that are preferred by HLA A2.1 increases HLA affinity, and leads to an increased rescue as detected by the MHC ELISA. As expected, addition of an irrelevant peptide (the HLA-A3 restricted gp100 epitope) does not lead to an appreciable increase in MHC complex recovery (*see* **Fig. 4c, d**).

3.4.1. Protocol for 96-Well MHC ELISA

1. In a 96-well plate add 2 μg/mL streptavidin in PBS, 100 μL per well.
2. Incubate for 2 h at 37°C or overnight at RT. Cover the plate to avoid evaporation and contamination.
3. Wash four times with wash buffer, 300 μL per well, discarding the wash buffer after each wash.
4. Add blocking buffer, 300 μL per well.
5. Incubate for 30 min at RT. Cover the plate with a lid.

6. Dilute 2.5 μL of the exchange reactions (*see* **Subheading** 3.3) 100-fold in blocking buffer, based on 100% rescue this results in a 5 nM concentration in the ELISA. Prepare 250 μL diluted MHC to analyze each complex in duplicate.
7. Tip out blocking buffer. Add 100 μL per well of diluted MHC monomer solutions from the exchange reactions or standards to obtain a standard-titration curve. Standard titration-curve: Add dilutions of a stock of standard MHC monomer in a concentration range of 0.08–40 nM. We have used refolded HLA A2.1 – GILGFVFTL (influenza A Matrix 1(58–66) epitope) as a standard MHC monomer (but any stable HLA complex should suffice) (*see* **Note** 20). Concentration of the standard MHC monomer is determined by absorbance at 280 nm.
8. Incubate the plate for 1 h at 37°C. Cover the plate with a lid.
9. Wash four times with wash buffer, 300 μL per well, discarding the wash buffer after each wash. Carefully dry plates by tipping out excess liquid on a tissue.
10. Add HRP conjugated anti-β2m antibody solution (1 μg/mL in blocking buffer), 100 μL per well.
11. Incubate the plate for 1 h at 37°C. Cover the plate with a lid.
12. Wash four times with wash buffer, 300 μL per well.
13. Prepare 10 mL ABTS coloring solution for each 96-well plate by mixing 8.7 mL H_2O, 1 mL 10× substrate buffer, 200 μL 50× ABTS solution, and 100 μL 100× H_2O_2 solution (3%).
14. Add 100 μL of ABTS coloring solution per well.
15. Incubate for 10–15 min at RT. Monitor color development by eye.
16. Block the reaction by adding 50 μL of stop buffer per well.
17. Measure absorbance at 405 nm in a plate reader.

3.4.2. Protocol for 384-Well MHC ELISA

The general setup of the 384-well MHC ELISA is similar to that of the 96-well MHC ELISA, except for the fact that in our hands the assay is somewhat more sensitive in this format and pMHC complexes should be diluted 300-fold after the exchange reaction, to a final concentration of 1.6 nM in the ELISA plate (possibly a reflection of a different dynamic range of the plate reader). All reaction steps are performed using 25 μL/well, blocking and washing steps using 100 μL/well, and the stop buffer is added in 12.5 μL/well.

3.4.3. Alternative Method: Analysis of Peptide Exchange Reactions by Gel-Filtration Chromatography

1. As an alternative method, analysis of p*MHC exchange reactions may be performed by gel-filtration chromatography. Prepare three samples as described in **Subheading 3.3**: (1) p*MHC not subjected to UV-light, (2) UV-exposed p*MHC

without addition of peptide, (3) UV-exposed p*MHC in the presence of peptide.

2. Analyze 100 μL of each sample by gel-filtration chromatography (monitoring protein at 230 or 280 nm).

Anticipated results (*see* **Fig. 5**): Without addition of peptide a residual MHC signal of approximately 20% is observed. This residual signal observed after UV-exposure is not further reduced by prolonged UV-exposure. This indicates that the residual signal is not derived from MHC complexes containing the intact UV-sensitive ligand. In line with this, reverse-phase HPLC analysis of p*HLA A2.1 after a 1 h UV exposure reveals a complete cleavage of the p* peptide *(9)*. Consequently, the residual signal most likely reflects incomplete dissociation of empty or partially occupied MHC class I heavy chain-β2m complexes. Exchange with a high-affinity MHC ligand leads to close to 100% rescue of the pMHC complex.

3.5. Multimerization of Exchanged MHC Monomers

Exchanged pMHC complexes may be complexed with fluorophore-labeled streptavidin to form MHC tetramers for T-cell analyses. The amount of streptavidin conjugate required to generate MHC tetramers is calculated on the basis of the molecular weight of PE-streptavidin (MW, 300 kDa), APC-streptavidin (MW, 164 kDa) and MHC (MW, 47 kDa), assuming four biotin binding sites per fluorophore-streptavidin conjugate, as detailed below. We recommend adding the fluorophore-streptavidin conjugate in 3 or 4 steps with 10–15 min intervals to ensure that in the initial steps the MHC is present in excess (thereby avoiding the formation of lower order MHC multimers, *see* **Note** 21).

3.5.1. Multimerization of Exchanged Monomers Using Fluorophore-Labeled Streptavidin

Microtiter plates with exchanged MHC monomers prepared according to **Subheading 3.3** contain 25 μg/mL of pMHC in 100 μL per well. This corresponds to 2.5 μg or 0.05 nmol monomer per well.

With four binding sites per streptavidin molecule, 0.0125 nmol fluorophore-streptavidin conjugate is needed to bind all monomers, assuming 100% rescue. For streptavidin-PE (MW, 300 kDa), this corresponds to 3.75 μg. To ensure optimal MHC tetramer formation even for peptides that do not fully stabilize MHC class I, and to correct for loss of MHC class I upon centrifugation/transfer, only 70% of this amount is added to each well: 0.70×3.75 μg = 2.6 μg PE-streptavidin. When the resulting MHC tetramer complexes are analyzed by gel-filtration chromatography a small residual pMHC monomer peak should still be observed indicating that MHC was present in excess (*see* **Fig. 6**). As a control, this residual peak should disappear upon addition of an excess of streptavidin (*see* **Note** 22).

To obtain MHC multimers at a concentration suitable for flow cytometry, dilute the 2.6 μg PE-streptavidin in 100 μL PBS

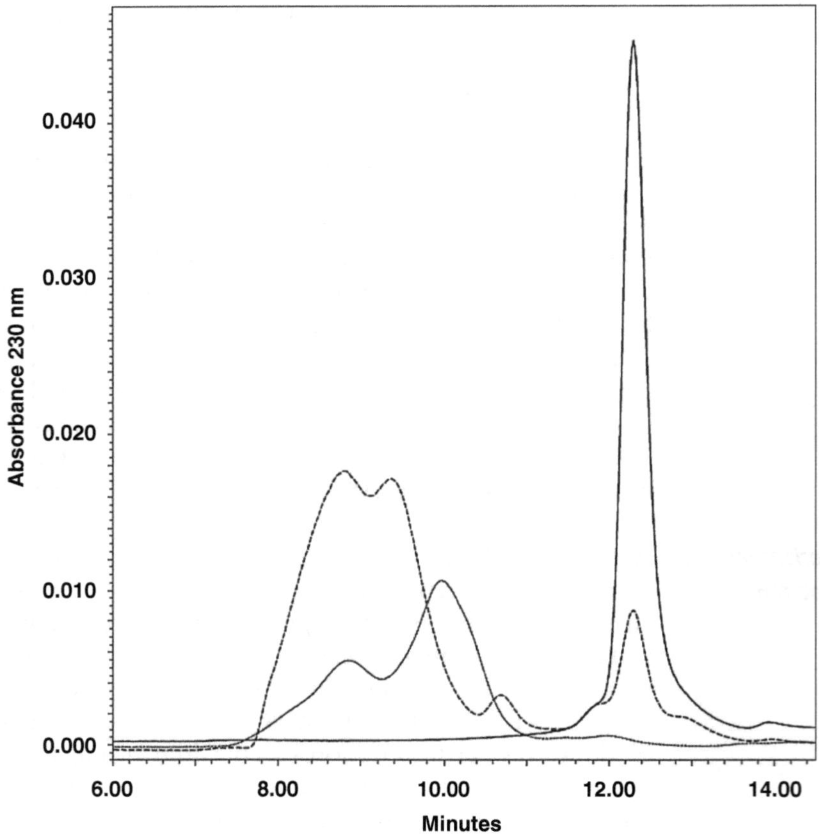

Fig. 6. Gel-filtration chromatography analysis of 2.5 μg pMHC (*solid line*), 2.5 μg pMHC complexes with 2.6 μg streptavidin-PE (*dashed line*), and 2.6 μg streptavidin-PE (*dotted line*). After coupling of monomers to streptavidin-PE a small residual MHC monomer fraction should remain to ensure the formation of MHC tetramers, rather than lower order MHC multimers.

and add it in four steps (4 × 25 μL) with 10 min intervals to each well of the plate containing the exchanged MHC monomers (resulting in 200 μL of MHC tetramer solution per well).

MHC tetramers are directly used for T-cell staining or stored at –20°C. In the latter case, first add BSA and glycerol to a final concentration of 0.5% BSA and 16% glycerol. Typically, 2–4 μL of MHC tetramer is used for the staining of approximately 10^6 cells resuspended in 50 μL FACS buffer.

3.5.2. Using Exchanged Multimer-Complexes for T-Cell Analysis by Flow Cytometry

Multimerized forms of pMHC complexes obtained by UV-mediated peptide exchange can be used to stain antigen-specific T cells with the same selectivity as conventional pMHC multimers (*see* **Fig. 7**) *(3, 9)*. T-cell staining may be performed with pMHC multimers conjugated with the classical fluorophores phycorerythrin and allophycocyanin. However, in view of the high complexity of T-cell responses and the possibility of creating large arrays of pMHC complexes by MHC exchange, it is of interest to create

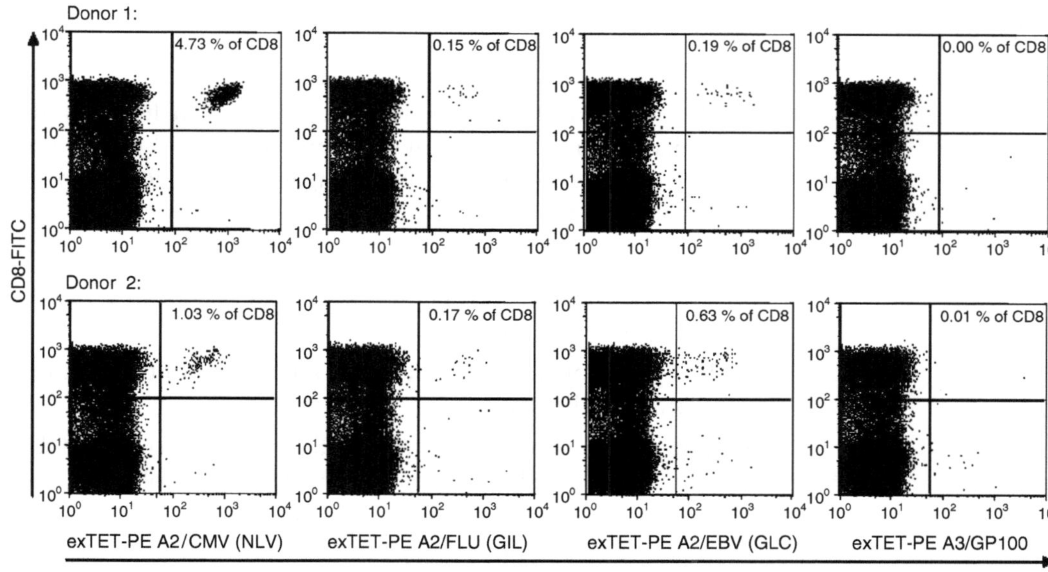

Fig. 7. Flow cytometry based detection of antigen specific T cells using PE-labeled HLA A2.1 tetramers obtained by peptide exchange. Peripheral blood samples of two HLA-A2.1$^+$ donors displaying T-cell responses against CMV, EBV and influenza A were analyzed for the occurrence of T cells specific for the CMV_{NLV}, EBV_{GLC} and FLU_{GIL} epitopes. Background was determined by use of MHC tetramer complexes of HLA A2.1 exchanged with the A3gp100 peptide.

multiplex platforms for T-cell identification. We are currently establishing multicolor flow cytometry protocols using MHC multimers labeled with quantum dots *(15, 16)*. Quantum dots are nanometer-sized cadmium crystals with very favorable (i.e., narrow) emission spectra, and have previously been used for multiparametric measurements by flow cytometry. Quantum dots are available with a streptavidin coating and can therefore directly be used to form multimers of biotinylated pMHC complexes. Ongoing experiments suggest that six quantum dots (QD800, 705, 655, 605, 585, 565) are particularly suited for T-cell analysis, and combined with the classical PE and APC labeled streptavidin conjugates, this allows the detection of eight different T-cell specificities within a single sample (*see* **Note 23**). Conceivably, this complexity may be further increased by "combinatorial coding schemes," in which a combination of multiple fluorescent labels is used as a code for each specific pMHC complex.

Other approaches to obtain multiplex platforms for T-cell identification include the coupling of pMHC molecules to different types of microarray slides to obtain a spatial separation between T cells specific for distinct pMHC complexes *(17–20)*. Evidently, large collections of pMHC complexes obtained by UV-mediated peptide exchange may also be used in this type of technology and it may in the future become possible to create such complexes in situ.

4. Notes

1. We purify our own biotin ligase, but biotin ligase is also available from commercial suppliers. A protocol for the production of recombinant biotin ligase can be found in the supplementary data from *(10)*.
2. The use of Tris-HCl 9.5 is essential, do not dissolve in water.
3. The use of polypropylene tubes minimizes sticking of peptide/protein to the tube.
4. Use polypropylene material to avoid sticking of the peptide/protein to the plate.
5. The use of short wavelength (254 m) or broad band UV lamps is highly detrimental to the pMHC complex.
6. We use an ABTS developing solution from Sanquin Reagents, Amsterdam. Alternatively, the developing solution can be made as follows: Dissolve 0.5 mg/ml ABTS in 0.1 M citrate buffer, pH 4.2 containing 0.03% hydrogen peroxide.
7. The quality of the APC- or PE-streptavidin batch is a critical factor in the success of tetramer formation, and presumably depends on the stoichiometric composition of the fluorochrome-streptavidin conjugates *(21, 22)*.
8. Production of *N*-fluorenylmethyloxycarbonyl3-amino-3-(2-nitro)phenyl-propionic acid can be verified either by mass spectrometry or NMR. For details on NMR analysis *see (10)*.
9. Alternatively, conditional MHC complexes will become commercially available from Sanquin, Amsterdam, The Netherlands (http://www.sanquin.nl/Sanquin-eng/Pelimers.nsf/All/Pelichange-Screen--Measure-Peptide-Mhc-Affinity.html).
10. During handling, the conditional peptide should be kept away from direct long wave UV light (e.g., sunlight). It is stable on the bench exposed to artificial light, but we advise to keep stocks in the dark.
11. MHC heavy chain and β2m are obtained via bacterial expression systems and isolated as inclusion bodies. A detailed description of this procedure can be found in *(10)*. After purification, pellets of inclusion bodies are kept at –80°C. One or more pellets are dissolved to obtain the amounts needed for a 50 mL refolding. If the pellets are difficult to resuspend it is helpful to first resuspend them in a volume of 200 μL, and then add the remaining 800 μL.
12. The volumes given for concentration and washing of the filters are only meant as guidelines. If concentration proceeds further, use a higher volume in the washing steps. The requirement for concentration obviously depends on the

injection loop and protein load maximally allowed on the chromatography system that is available.

13. We reach complete biotinylation by addition of 0.03 μmol biotin ligase per μmol MHC heavy chain used within the original refolding reaction.
14. Use PE- or APC-labeled streptavidin to obtain a maximal change in elution time when analyzing samples by gel-filtration chromatography.
15. Performing the UV exposure in samples with a higher column height (i.e., a large volume in a well with the same dimension) is expected to give less rapid/efficient UV mediated exchange reactions, due to filtering of the incident UV light by the aqueous solution.
16. Some UV lamps produce excessive heat. Ensure that no substantial liquid evaporation occurs during UV exposure.
17. For the analysis of parallel exchange reactions involving high numbers of peptides, the MHC ELISA offers a rapid test. In addition, pMHC complexes can be analyzed by HPLC gel-filtration chromatography (*see* **Subheading 3.4.3** and **Fig. 5**) or mass spectrometry. These approaches are less suited for high-throughput assays and require the use of a substantial amount of pMHC. However, these techniques may be valuable to follow the peptide exchange process by an independent method.
18. Plates containing exchanged MHC monomers may be directly used for tetramer formation and/or ELISA. Alternatively, they can be stored at 4°C. Storage for up to 1 week is possible. Depending on the affinity of the newly introduced ligand for the MHC molecule, it is likely that exchanged MHC monomers can be stored at 4°C for substantially longer periods (i.e., months). However, this has not been tested. It is also possible to quick-freeze the MHC monomers in liquid nitrogen and store them at –80°C until further use.
19. The standard curve is obligatory if absolute concentration values are needed. However, for most purposes relative absorbance values are sufficient. In this case, the standard curve is necessary in the set-up phase, but can be omitted at later stages. It is important to realize that the linear window of this assay is narrow, and that slight changes in input concentrations can therefore push the values out of the linear range.
20. For this purpose, the non-UV irradiated p*MHC complex can also be used.
21. The ratio of fluorophore-streptavidin conjugate to MHC determines the quality of the resulting tetramers. An excess

of fluorophore-streptavidin over MHC results in the formation of MHC multimers with a lower valency. This affects the avidity of the T-cell–pMHC multimer interaction and results in poor binding. In addition, a failure to saturate the available biotin-binding sites may lead to spurious (background) signal in cases where biotinylated antibodies are used for costaining.

22. Analysis of tetramer formation by gel-filtration chromatography is particularly helpful to reveal incomplete MHC biotinylation or proteolytic cleavage of the peptide tag (use excess amount of streptavidin as described in **Subheading 3.2.3** and shown in **Fig. 3**), and to determine the quality of new fluorophore-streptavidin batches (perform analysis as described in **Subheading 3.5.1** and shown in **Fig. 6**).

23. The use of quantum dots in flow cytometry depends on the available flow cytometry instrument. For the analyses of eight different multimers in parallel, we use an LSR-II flow cytometer (BD), containing four different lasers (UV, violet, blue and red).

References

1. Arstila, T. P., Casrouge, A., Baron, V., Even, J., Kanellopoulos, J., and Kourilsky, P. (1999) A direct estimate of the human alphabeta T cell receptor diversity. *Science* 286, 958–961.
2. Altman, J. D., Moss, P. A., Goulder, P. J., Barouch, D. H., McHeyzer-Williams, M. G., Bell, J. I., McMichael, A. J., and Davis, M. M. (1996) Phenotypic analysis of antigen-specific T lymphocytes. *Science* 274, 94–96.
3. Toebes, M., Coccoris, M., Bins, A. D., Rodenko, B., Gomez, R., Nieuwkoop, N. J., van de Kasteele, W., Rimmelzwaan, G., Haanen, J. B., and Schumacher, T. N. (2006) Design and use of conditional MHC class I ligands. *Nat. Med.* 12, 246–251.
4. Checovich, W. J., Bolger, R. E., and Burke, T. (1995) Fluorescence polarization–a new tool for cell and molecular biology. *Nature* 375, 254–256.
5. Cobbold, M., Khan, N., Pourgheysari, B., Tauro, S., McDonald, D., Osman, H., Assenmacher, M., Billingham, L., Steward, C., Crawley, C., Olavarria, E., Goldman, J., Chakraverty, R., Mahendra, P., Craddock, C., and Moss, P. A. (2005) Adoptive transfer of cytomegalovirus-specific CTL to stem cell transplant patients after selection by HLA-peptide tetramers.*J. Exp. Med.* 202, 379–386.
6. Dudley, M. E., Wunderlich, J. R., Yang, J. C., Sherry, R. M., Topalian, S. L., Restifo, N. P., Royal, R. E., Kammula, U., White, D. E., Mavroukakis, S. A., Rogers, L. J., Gracia, G. J., Jones, S. A., Mangiameli, D. P., Pelletier, M. M., Gea-Banacloche, J., Robinson, M. R., Berman, D. M., Filie, A. C., Abati, A., and Rosenberg, S. A. (2005) Adoptive cell transfer therapy following non-myeloablative but lymphodepleting chemotherapy for the treatment of patients with refractory metastatic melanoma. *J. Clin. Oncol.* 23, 2346–2357.
7. Bakker, A. H. and Schumacher, T. N. (2005) MHC multimer technology: current status and future prospects. *Curr. Opin. Immunol.* 17, 428–433.
8. Grotenbreg, G. M., Nicholson, M. J., Fowler, K. D., Wilbuer, K., Octavio, L., Yang, M., Chakraborty, A. K., Ploegh, H. L., and Wucherpfennig, K. W. (2007) Empty class II major histocompatibility complex created by peptide photolysis establishes the role of DM in peptide association. *J. Biol. Chem.* 282, 21425–21436.
9. Bakker, A. H., Hoppes, R., Linnemann, C., Toebes, M., Rodenko, B., Berkers, C. R., Hadrup, S. R., van Esch, W. J., Heemskerk, M. H., Ovaa, H., and Schumacher, T. N. (2008) Conditional MHC class I ligands and peptide exchange technology for the human

MHC gene products HLA-A1, -A3, -A11 and -B7. *Proc. Natl. Acad. Sci. U. S. A.* 105, 3825–3830.

10. Rodenko, B., Toebes, M., Hadrup, S. R., van Esch, W. J., Molenaar, A. M., Schumacher, T. N., and Ovaa, H. (2006) Generation of peptide-MHC class I complexes through UV-mediated ligand exchange. *Nat. Protoc.* 1, 1120–1132.
11. Wellings, D. A. and Atherton, E. (1997) Standard Fmoc protocols. *Meth. Enzymol.* 289, 44–67.
12. Garboczi, D. N., Hung, D. T., and Wiley, D. C. (1992) HLA-A2-peptide complexes: refolding and crystallization of molecules expressed in Escherichia coli and complexed with single antigenic peptides. *Proc. Natl. Acad. Sci. U. S. A.* 89, 3429–3433.
13. Coulie, P. G., Brichard, V., van Pel, A., Wolfel, T., Schneider, J., Traversari, C., Mattei, S., De Plaen, E., Lurquin, C., Szikora, J. P., Renauld, J. C., and Boon, T. (1994) A new gene coding for a differentiation antigen recognized by autologous cytolytic T lymphocytes on HLA-A2 melanomas. *J. Exp. Med.* 180, 35–42.
14. Andersen, M. H., Pedersen, L. O., Becker, J. C., and thor Straten, P. (2001) Identification of a cytotoxic T lymphocyte response to the apoptosis inhibitor protein surviving in cancer patients. *Cancer Res.* 61, 869–872.
15. Bruchez, M. Jr., Moronne, M., Gin, P., Weiss, S., and Alivisatos, A. P. (1998) Semiconductor nanocrystals as fluorescent biological labels. *Science* 281, 2013–2016.
16. Chattopadhyay, P. K., Price, D. A., Harper, T. F., Betts, M. R., Yu, J., Gostick, E., Perfetto, S. P., Goepfert, P., Koup, R. A., De Rosa, S. C., Bruchez, M. P., and Roederer, M. (2006) Quantum dot semiconductor nanocrystals for immunophenotyping by polychromatic flow cytometry. *Nat. Med.* 12, 972–977.
17. Soen, Y., Chen, D. S., Kraft, D. L., Davis, M. M., and Brown, P. O. (2003) Detection and characterization of cellular immune responses using peptide-MHC microarrays. *PLoS Biol.* 1, 429–438.
18. Chen, D. S., Soen, Y., Stuge, T. B., Lee, P. P., Weber, J. S., Brown, P. O., and Davis, M. M. (2005) Marked differences in human melanoma antigen-specific T cell responsiveness after vaccination using a functional microarray. *PLoS Med.* 2, 1018–1030.
19. Stone, J. D., Demkowicz, W. E. Jr., and Stern, L. J. (2005) HLA-restricted epitope identification and detection of functional T cell responses by using MHC-peptide and costimulatory microarrays. *Proc. Natl. Acad. Sci. U. S. A.* 102, 3744–3749.
20. Deviren, G., Gupta, K., Paulaitis, M. E., and Schneck, J. P. (2007) Detection of antigen-specific T cells on p/MHC microarrays. *J. Mol. Recognit.* 20, 32–38.
21. Guillaume, P., Legler, D. F., Boucheron, N., Doucey, M. A., Cerottini, J. C., and Luescher, I. F. (2003) Soluble major histocompatibility complex-peptide octamers with impaired CD8 binding selectively induce Fas-dependent apoptosis. *J. Biol. Chem.* 278, 4500–4509.
22. Ramachandiran, V., Grigoriev, V., Lan, L., Ravkov, E., Mertens, S. A., and Altman, J. D. (2007) A robust method for production of MHC tetramers with small molecule fluorophores. *J. Immunol. Meth.* 319, 13–20.
23. Grotenbreg, G. M., Roan, N. R., Guillen, E., Meijers, R., Wang, J. H., Bell, G. W., Starnbach, M. N., and Ploegh H. L. (2008) Discovery of CD8+ T cell epitopes in Chlamydia trachomatis infection through use of caged class I MHC tetramers. Proc. Natl. Acad. Sci. U. S. A. 105, 3831–3836.

Chapter 29

T-Cell Epitope Processing (The Epitope Flanking Regions Matter)

Alejandra Nacarino Martinez, Stefan Tenzer, and Hansjörg Schild

Summary

Epitopes presented by major histocompatibility complex (MHC) class I molecules for cytotoxic T-lymphocyte (CTL) recognition are derived mainly from cytosolic proteins. Antigen presentation on the cell surface requires correct processing of epitopes by the proteasome, cytosolic and endoplasmic reticulum (ER) aminopeptidases, efficient TAP transport, and sufficient binding to MHC class I molecules. The efficiency of the epitope generation depends not only on the epitope itself but also on its flanking regions. To investigate preferences at the C-terminal epitope extension on processing and presentation, the SIINFEKL (S8L) epitope can be used as a model epitope. By exchanging the amino acids at the C-terminus of S8L, their influence on the presentation of S8L can be analyzed.

Key words: Antigen processing, Antigen presentation, MHC class I epitope, Proteasome, Aminopeptidases.

1. Introduction

Epitopes presented by MHC class I molecules for CTL antigen recognition are derived mainly from cytosolic proteins *(1–3)*. Most of these intracellular proteins are normal self proteins, which may be wrongly folded or no more needed. Besides they can also be foreign antigens derived from viral products or in tumor cells from mutated self genes or oncogenes. The degradation of most cellular proteins occurs by the ubiquitin–proteasome pathway *(4–6)*. The first step in this pathway is the conjugation of an ubiquitin chain to the protein substrate, which serves as a molecular "tag" for rapid degradation by the proteasome. The proteasome is responsible for most of the nonlysosomal protein

Ulrich Reineke and Mike Schutkowski (eds.), *Methods in Molecular Biology, Epitope Mapping Protocols, vol. 524*

DOI: 10.1007/978-1-59745-450-6_29

degradation in both the nucleus and cytosol *(7)*. The 20 S proteasome is a 700-kDa complex composed of 14 different subunits, which are arranged in four stacked rings with the stoichiometry of $\alpha 7\beta 7\beta 7\alpha 7$. The proteolytically active subunits are located in the inner hollow center of the 20 S ß-rings *(8)*. The proteasome creates peptide fragments in a length of 3–20 residues *(9–11)*. It generates the correct C-terminus of epitopes, which can be N-terminally extended *(12–14)*. These proteasomal products are further broken down by cytosolic and ER aminopeptidases into smaller peptides or single amino acids. A small part of these trimmed peptides is translocated into the ER by the TAP transporter. Once here, peptides can bind into the groove of the MHC class I molecules, which are then transported to the cell surface for antigen presentation and CTL recognition *(15)*.

The efficiency of the epitope generation depends not only on the epitope itself but also on its flanking regions. To investigate preferences at the C-terminal epitope extension on processing and presentation, the SIINFEKL (S8L) epitope from chicken ovalbumin (aa 257–264) can be used as a model epitope, which is presented on the murine MHC class I molecule H-$2K^b$. The Flp-In $293K^b$ cell line was transfected with different constructs each enabling the expression of the S8L sequence with different defined C-terminal flanking regions. By consecutively exchanging the amino acids at position P1' and P2' after the S8L epitope, it is possible to analyze their influence on H-$2K^b$/S8L processing/presentation on the cell surface of the Flp-In $293K^b$ cells. The detection of surface expression of this complex was performed by immunostaining and flow cytometry.

2. Materials

If nothing else stated, all chemicals were obtained from Sigma (Taufkirchen, Germany) or Roth (Karlsruhe, Germany).

2.1. Transfection of the S8L-Vector with Fugene

1. Flp-In system (Invitrogen GmbH, Karlsruhe, Germany): Flp-In 293 cell line (transfected for H-$2K^b$), the pOG44, and the pcDNA5/FRT plasmid.
2. 1× PBS from stock 10× PBS, 1.4 M NaCl, 0.1 M NaH_2PO_4, pH 7.2.
3. DMEM supplemented with 10% fetal calf serum (FCS), both purchased from Vitromex GmbH, Geilenkirchen, Germany.
4. 0.25% trypsin /1 mM EDTA (Invitrogen GmbH, Karlsruhe, Germany).
5. Fugene (Roche Diagnostics GmbH, Mannheim, Germany).

6. Selection medium for the transfectants: DMEM (10% FCS) + 100 μg/mL Hygromycin B (Invitrogen GmbH, Karlsruhe, Germany).

2.2. Stripping the Cell Surface of the Transfectants ("Acid Wash")

1. "Acid wash" solution: 0.263 M citric acid and 0.132 M NaH_2PO_4 (mix 1:1, pH 3.0).
2. Stop-solution: 0.15 M Na_2HPO_4, pH 7.5.
3. DMEM (Vitromex GmbH, Geilenkirchen, Germany) supplemented with 10% fetal calf serum (Vitromex GmbH, Geilenkirchen, Germany).
4. 10 mM lactacystin (Biomol GmbH, Hamburg, Germany) diluted in distilled H_2O.

2.3. Staining of the "Acid Wash" Aliquots for H-2K^b/S8L Detection by Flow Cytometry

1. 25-D1.16 (primary antibody): a mouse IgG1 monoclonal antibody specific for SIINFEKL peptide associated with H-2K^b *(16)*. Working solution of 1:100, diluted with FACS-buffer.
2. gam-APC (secondary antibody): APC conjugated, goat anti-mouse IgG(H + L) F(ab)2-fragment (gam-APC), Dianova GmbH, Hamburg, Germany. Working solution of 1:400, diluted in FACS-buffer.
3. FACS-buffer: 1% BSA and 0.05% sodium azide in 1× PBS.
4. PFA-buffer: 1% paraformaldehyde in 1× FACS-buffer.

3. Methods

To analyze the influence of the C-terminal flanking region on MHC class I restricted antigen presentation, the Flp-In System was chosen, because it allows a single integration and expression of the gene of interest in mammalian cells at a specific genomic location. Our working cell line was the Flp-In 293K^b (the Flp-In 293 cell line was purchased from Invitrogen GmbH, Karlsruhe/Germany, and then transfected for the H-2K^b expression), which was transfected for S8L epitope expression. For this purpose, several constructs were generated containing the S8L sequence (*see* **Fig. 1**), but each encoding different C-terminal amino acids at the first (P1') and/or second (P2') position after the leucine of the S8L epitope. After transfecting (*see* **Note 1**) a specific cell line (Flp-In 293K^b) with the construct coding for S8L, the H-2K^b/S8L complex can be detected on the cell surface by staining with a specific monoclonal antibody, the 25-D1.16 *(16)*. By comparing the presentation of this epitope on the cell surface for all the different transfectants, it is possible to analyze the influence of

Fig. 1. Construct for the S8L expression and presentation in the Flp-In 293K^b cells. The construct starts with an Ubiquitin, which is needed for proteasomal targeting. This is followed by a cassette bearing the chicken ovalbumin S8L epitope (aa 257–264). The *X* at the C-terminus of this epitope stands for the exchange of amino acids at its specific positions. The S8L cassette is directly followed by an IRES–EGFP cassette, which allows expression of the S8L and the EGFP from the same mRNA. Hence, the SIINFEKL:EGFP expression rates in the cells are directly connected. Thus, the EGFP levels measured for the corresponding transfectants can be used to normalize the measured S8L presentation.

different amino acids at the P1' and P2' positions on processing and presentation of S8L. To compare the S8L presentation, it is necessary to have the same turnover rates in these cells. For this purpose, the level of the EGFP expression was used to normalize the detected SIINFEKL presentation. The S8L cassette is directly followed by an IRES–EGFP cassette, which guarantees for uninterrupted, consecutive expression of the S8L and the EGFP. Hence, the SIINFEKL:EGFP expression rates in all transfected cells are equal (*see* **Note 12**). Thus, the EGFP levels measured for the corresponding transfectants can be used to normalize the measured S8L presentation for possible variations in expression levels. Furthermore, this system allows investigating the influence of proteasomes and other proteases by adding specific inhibitors to investigate their role in generation of the S8L epitope out of the different constructs.

3.1. Transfection with Fugene of the S8L-Vector into the Cells

1. 24 h before transfection of the S8L plasmid, 8×10^5 adherent Flp-In 293K^b cells/well were seeded into six-well plates (*see* **Note 2**).
2. On the next day, cells were washed with 1× PBS (pH 7.2) to remove the serum and 1 mL DMEM serum free medium was added to each well. Then the cells were incubated at 37°C.
3. Ratio of the DNA/Fugene complexes was 1:3 for one well of a six-well plate (*see* **Note 1**).
4. 12 μL of Fugene were diluted into 88 μL of DMEM serum free medium and incubated for 5 min. Then, 4 μg DNA diluted in 100 μL DMEM serum free medium was added to the Fugene dilution and mixed together.
5. After 20 min incubation at room temperature the DNA/Fugene complex was added drop wise to the cells seeded in one well of the six-well plate.
6. After 4 h, 2 mL of DMEM with 10% FCS was added to the cells.
7. 24 h after transfection, selection media was given to the transfected cells, adding hygromycin B to the DMEM (10% FCS).

8. Approximately 3–4 weeks after transfection, clones of the stable transfectants can be visualized in the microscope.

3.2. Stripping the Cell Surface of the Transfectants ("Acid Wash")

This cell surface stripping technique is based on the dissociation of the peptides from the MHC class I molecule binding groove by incubating the cells with an acid solution of pH 3 for 1 min and 30 s. After this acid treatment, the MHC class I molecules on the cell surface are free of any peptide. Nevertheless, new loaded MHC class I molecules from the ER are transported to the cell surface through the Golgi apparatus for CTL recognition. This new presentation of epitopes coming up is then analyzed by a kinetic to compare the transfectants re-presentation rates.

3.2.1. "Acid Wash" Standard

1. 24 h before the assay the adherent Flp-In 293K^b transfectants (*see* **Note 8**) were seeded onto six-well-plates: 1.5 × 10^6 cells/well in 2 mL DMEM (10% FCS).
2. One well contains sufficient numbers of cells to analyze four different time points, e.g. 0, 120, 240, and 360 min.
3. After one day, the Flp-In 293K^b transfectants were washed with 1× PBS (pH 7.2) and removed from the plate bottom with EDTA, not trypsin (*see* **Note 10**).
4. After centrifugation of the cell suspension (2 min at 600 × *g*, 4°C) the pellet was resuspended in 500 μL (for 1 × 10^6 cells) isotonic "acid wash"-buffer (pH 3.0) and incubated for 1.5 min on ice (*see* **Note 3**).
5. 1 mL of the stop-solution was added to neutralize the pH and the cells were pelleted by centrifugation (2 min at 600 × *g*, 4°C).
6. Then the cells were washed twice with 1 mL DMEM (*see* **Note 4**).
7. The cells were allowed to recover in the presence or absence of inhibitors (*see* **Subheading 3.2.2**) (*see* **Notes 5–7**) adding fresh DMEM (10% FCS) incubating at 37°C for 6 h. Samples were taken for the staining to perform a kinetic analysis of the H-2K^b/S8L presentation using the following time points: 0, 120, 240, and 360 min. Additionally, untreated cells (control) were also taken as a reference value for the re-presentation kinetic of the S8L epitope (*see* **Figs. 2** and **3**).

3.2.2. "Acid Wash" with Inhibitors

By adding specific inhibitors (*see* **Notes 5–7**) to the cell media, the influence of proteases on generation of the SIINFEKL epitope can be analyzed. For example, to analyze the proteasomal dependence on the generation of the S8L epitope upon detection of the H-2K^b/S8L presentation, the proteasomal inhibitor lactacystin (LC) can be used.

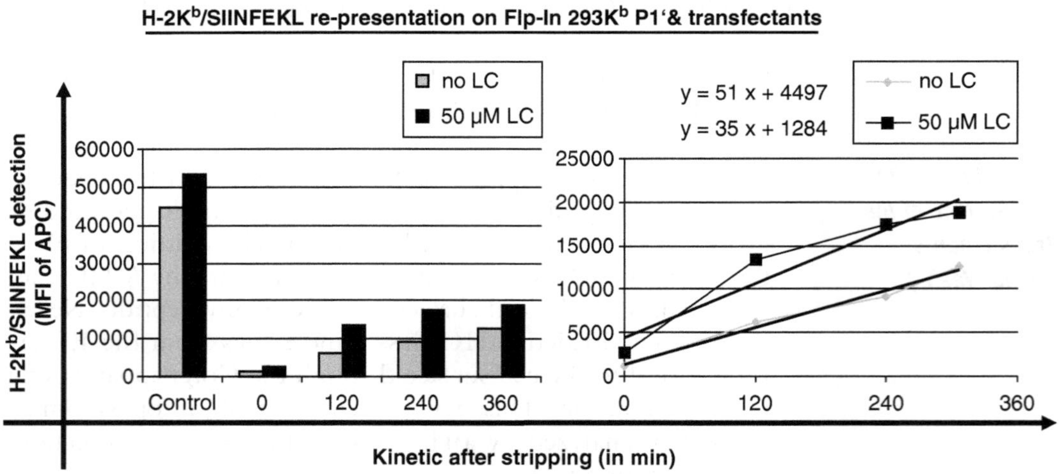

Fig. 2. H-2K^b/SIINFEKL re-presentation kinetic on the cell surface of the Flp-In 293K^b_P1'& transfectant after stripping. The "&" of the Flp-In 293Kb_P1'& transfectant stands for a stop-codon at the first position after the leucine of the S8L epitope. This implicates that the SIINFEKL precursor is not C-terminally extended. Thus, this precursor peptide needs no proteasomal processing. The *y*-axis shows the APC mean fluorescence intensity (MFI) measured by flow cytometry using the γ-APC secondary antibody after having stained with the 25-D1.16 antibody first. The *x*-axis represents the different time points taken for the kinetic. The kinetic was performed adding 50 µM lactacystin (LC) to analyze the proteasomal dependence on generation of the S8L. "Control" stands for the cells not treated with the acid solution. These show the steady state presentation of the H-2K^b/SIINFEKL on the cell surface. On the *left*, a *bar-diagram* is shown. On the *right*, the same H-2K^b/SIINFEKL re-presentation levels are analyzed on a *dot chart*. This diagram is used to calculate the slope of each kinetic.

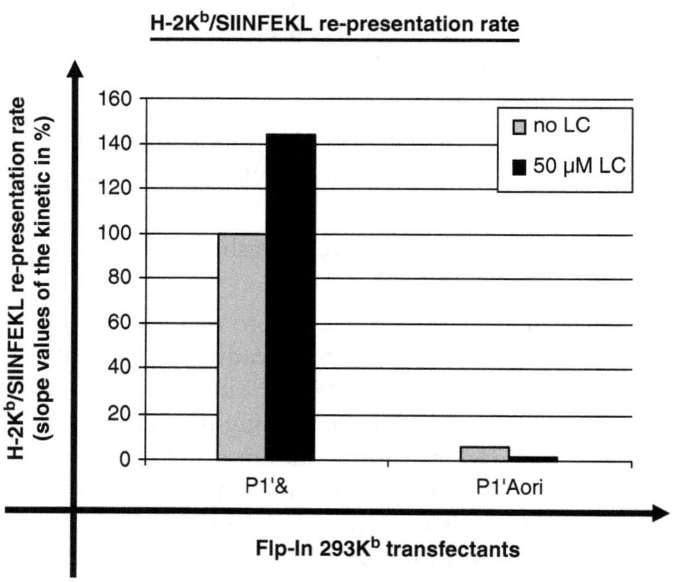

Fig. 3. H-2K^b/SIINFEKL re-presentation rate on the cell surface of two Flp-In 293K^b transfectants. The *y*-axis shows the S8L re-presentation rates calculated out of the slope of each kinetic, with and without adding lactacystin. On the *x*-axis the two different Flp-In 293K^b transfectants, P1'& and P1'A_{ori} are represented. P1'& carries a stop-codon at the first position after the leucine of the S8L epitope and doesn't need proteasomal processing to generates the correct C-terminus. P1'Aori carries an alanine (A) at the first position after the S8L and is followed by the original sequence (ori) of the ovalbumin protein after the exchange. For the proteasome independent S8L precursor peptide (P1'&), an increase of S8L presentation is observed after inhibiting the proteasome. This implicates that the proteasome not only generates the C-terminus of potential epitopes, but can also destroy them.

1. One day after seeding the Flp-In 293K^b transfectants, before starting the stripping assay, the inhibitor lactacystin was added to the cells at a concentration of 50 µM.
2. After 45 min of inhibitor incubation at 37°C, the cells were washed with 1× PBS (pH 7.2) and removed from the plate bottom with EDTA (not trypsin) as described above (*see* **Subheading 3.2.1**).
3. After the stripping of the cell surface and the further washing steps, the cells were resuspended in DMEM (10% FCS) supplemented with 50 µM lactacystin for further inhibition of the proteasome during the 6 h incubation. The kinetic was performed as described in **Subheading 3.2.2**, **step 7**.

3.3. Staining of the "Acid Wash" Aliquots for H-2Kb/S8L Detection by Flow Cytometry

The aliquots taken every 2 h for the re-presentation kinetic of H-2K^b/S8L on the cell surface of the transfectants were stained as follows:

1. 3 × 10^5 cells were incubated in 96-well round bottom plates for 30 min with the primary antibody, 25-D1.16 diluted in FACS-buffer (1/100) at 4°C.
2. Then, the cells were centrifuged (3 min at 600 × *g*, 4°C), washed twice with 200 µL FACS-buffer and centrifuged (3 min at 600 × *g*, 4°C) again.
3. Afterwards, the cells were incubated with a fluorochrome-conjugated secondary antibody (gam-APC) for 30 min at 4°C in the dark.
4. Subsequently, they were centrifuged (3 min at 600 × *g*, 4°C) and washed 2× in 200 µL FACS-buffer.
5. Finally, if the samples are going to be analyzed on the same day by flow cytometry, they are resuspended in 100 µL of FACS-buffer (*see* **Note 11**). If necessary, the cells can be fixed with PFA-buffer, so they can be stored at 4°C in the dark for up to 1 week).
6. Flow cytometry was performed using a FACSCanto flow cytometer (Becton Dickinson, Heidelberg, Germany).

3.4. Analysis of the Flow Cytometry Values

1. In flow cytometry the mean fluorescence intensity (MFI) of the APC (corresponds to the amount of H-2K^b/S8L) and EGFP (to monitor expression levels of the constructs) of the transfectants was analyzed (*see* **Note 11**).
2. After having determined the APC mean fluorescence intensity (MFI) values of the samples from different time points by flow cytometry, the corresponding values were plotted in a dot plot to calculate the slope (which correlates to the respective re-presentation rate) of each kinetic out of the trend line.
3. The steady state expression of the S8L peptide in the transfectants was monitored by the expression level of the EGFP

protein, because the S8L cassette is followed by an IRES–EGFP cassette, which allows the same expression level of S8L and EGFP. Therefore, the EGFP MFI value was used to normalize the measured S8L levels (APC level) of the transfectants (*see* **Note 12**).

4. With the calculated slopes of the different cell lines in each setting, it is possible to compare the S8L re-presentation of each cell line (*see* **Note 9**). Out of these values it is possible to state if the specific C-terminal amino acids in the S8L-precursor of the cell lines are generated with higher frequency then precursors carrying other amino acids. Additionally, using inhibitors in the stripping assay it is also possible to examine, if some S8L-precursor is dependent on the cleavage activity of a certain protease (e.g., the proteasome).

4. Notes

1. Transfection with Fugene efficiency should be examined first with a control plasmid carrying a reporter gene.
2. Transfection should be performed before the cells are >50–70% confluent.
3. All steps of the stripping assay have to be performed on ice.
4. Short and carefull resuspending of the cells during the stripping assay, after acid wash treatment, as the cells are very sensitive.
5. When using inhibitors titrate them first on the cells for the optimal inhibitory concentration, as cells differ from each other. Too high concentrations can be toxic.
6. It should be tested if the inhibitors used are able to cross the cell membrane. Additionally, the stability of the inhibitors should be verified.
7. FCS in the medium may impair the action of the inhibitors in use. FCS might stick to the inhibitor blocking its action. This depends on the inhibitors added.
8. If using other cell lines for the acid wash, check first acid treatment time and viability of the cells for an efficient dissociation of the peptides from the MHC class I molecules without killing them.
9. If there is no re-presentation after 120, 240, and 360 min examine the viability of the cells.
10. Do not use trypsin to remove the cells from the plates, because it will cleave the MHC class I molecules and the 6 h

(360 min) incubation won't be sufficient for re-presentation, as the cell has to build up new MHCs.

11. For FACS analysis it is important to have enough cells in the sample (minimum 2 × 10^5 cells/sample).
12. EGFP levels directly after the acid treatment (time point 0 min) are very low. The cell still expresses EGFP, but the fluorescence capacity of the EGFP molecule is impaired by the acid, which presumably enters through channels into the cytoplasm.

References

1. Pamer, E. and Cresswell, P. (1998) Mechanisms of MHC class I-restricted antigen processing. *Annu. Rev. Immunol.* 16, 323–358.
2. Princiotta, M. F., Finzi, D., Qian, S. B., Gibbs, J., Schuchmann, S., Buttgereit, F., Bennink, J. R., and Yewdell, J. W. (2003) Quantitating protein synthesis, degradation, and endogenous antigen processing. *Immunity* 18, 343–354.
3. Yewdell, J. W. (2005) The seven dirty little secrets of major histocompatibility complex class I antigen processing. *Immunol. Rev.* 207, 8–18.
4. Goldberg, A. L., Cascio, P., Saric, T., and Rock, K. L. (2002) The importance of the proteasome and subsequent proteolytic steps in the generation of antigenic peptides. *Mol. Immunol.* 39, 147–164.
5. Goldberg, A. L. (2003) Protein degradation and protection against misfolded or damaged proteins. *Nature* 426, 895–899.
6. Rock, K. L., York, I. A., Saric, T., and Goldberg, A. L. (2002) Protein degradation and the generation of MHC class I-presented peptides. *Adv. Immunol.* 80, 1–70.
7. Groll, M., Heinemeyer, W., Jager, S., Ullrich, T., Bochtler, M., Wolf, D. H., and Huber, R. (1999) The catalytic sites of 20S proteasomes and their role in subunit maturation: a mutational and crystallographic study. *Proc. Natl. Acad. Sci. USA* 96, 10976–10983.
8. Groll, M., Ditzel, L., Lowe, J., Stock, D., Bochtler, M., Bartunik, H. D., and Huber, R. (1997) Structure of 20S proteasome from yeast at 2.4 A resolution. *Nature* 386, 463–471.
9. Kisselev, A. F., Akopian, T. N., Woo, K. M., and Goldberg, A. L. (1999) The sizes of peptides generated from protein by mammalian 26 and 20 S proteasomes. Implications for understan-ding the degradative mechanism and antigen presentation. *J. Biol. Chem.* 274, 3363–3371.
10. Reits, E., Neijssen, J., Herberts, C., Benckhuijsen, W., Janssen, L., Drijfhout, J. W., and Neefjes, J. (2004) A major role for TPPII in trimming proteasomal degradation products for MHC class I antigen presentation. *Immunity* 20, 495–506.
11. York, I. A., Chang, S. C., Saric, T., Keys, J. A., Favreau, J. M., Goldberg, A. L., and Rock, K. L. (2002) The ER aminopeptidase ERAP1 enhances or limits antigen presentation by trimming epitopes to 8–9 residues. *Nat. Immunol.* 3, 1177–1184.
12. Craiu, A., Akopian, T., Goldberg, A., and Rock, K. L. (1997) Two distinct proteolytic processes in the generation of a major histocompatibility complex class I-presented peptide. *Proc. Natl. Acad. Sci. USA* 94, 10850–10855.
13. Mo, X. Y., Cascio, P., Lemerise, K., Goldberg, A. L., and Rock, K. (1999) Distinct proteolytic processes generate the C and N termini of MHC class I-binding peptides. *J. Immunol.* 163, 5851–5859.
14. Stoltze, L., Dick, T. P., Deeg, M., Pommerl, B., Rammensee, H. G., and Schild, H. (1998) Generation of the vesicular stomatitis virus nucleoprotein cytotoxic T lymphocyte epitope requires proteasome-dependent and -independent proteolytic activities. *Eur. J. Immunol.* 28, 4029–4036.
15. Rock, K. L. and Goldberg, A. L. (1999) Degradation of cell proteins and the generation of MHC class I-presented peptides. *Annu. Rev. Immunol.* 17, 739–779.
16. Porgador,A.,Yewdell,J.W.,Deng,Y.,Bennink,J.R., and Germain, R. N. (1997) Localization, quantitation, and in situ detection of specific peptide-MHC class I complexes using a monoclonal antibody. *Immunity* 6, 715–726.

Chapter 30

Identification of MHC Class II Binding Peptides: Microarray and Soluble MHC Class II Molecules

Simani Gaseitsiwe and Markus J. Maeurer

Summary

CD4+ T-helper cells recognize antigenic peptides presented by MHC class II molecules. The binding of the nominal peptide to the MHC class II allele is dependent on the amino acid sequence of the peptide as well as on amino acid (aa) residues in the peptide binding groove of the MHC class II allele. MHC class II alleles can either be associated with protection or susceptibility to disease (coined as "MHC class II-associated diseases"). A detailed knowledge about the nature, composition, and biochemical interaction of peptides with MHC class II molecules aids to link individual peptide species with MHC class II presentation and ultimately with CD4+ T-cell recognition. Several methods have been described to identify potential MHC class II candidate binding peptides. We present here a high content screening for MHC class II (HLA-DR) binding to a peptide library in a chip-format. Binding of soluble MHC class II molecules to individual peptides can be visualized using an anti-DR directed monoclonal antibody (mAb). Positive events (MHC class II/peptide complexes) are normalized and available for pattern analysis.

Key words: MHC class II molecules, T-cell epitopes, MHC binding, CD4+ T-cells.

1. Introduction

CD4+ T-helper cells play a central role in the immune system *(1)*, they provide help for the induction of CD8+ cytotoxic T-lymphocytes (CTLs), produce survival and maturation factors for B-cells and act as effector T-cells either through direct cell contact or cytokine production *(2)*. CD4+ T-helper cells recognize antigenic peptides presented by MHC class II molecules which are endogenously expressed by B-cells, monocytes, and dendritic cells. Pro-inflammatory cytokines, i.e., IFNγ or TNFα are able to induce MHC class II protein expression on nonantigen presenting cells.

Ulrich Reineke and Mike Schutkowski (eds.), *Methods in Molecular Biology, Epitope Mapping Protocols, vol. 524*

DOI: 10.1007/978-1-59745-450-6_30

The binding of the nominal peptide to the MHC class II allele is dependent on the amino acid sequence of the peptide as well as on specific amino acid (aa) residues in the peptide-binding groove of the MHC class II allele *(3)*. The MHC class II peptide-binding site, in contrast to the "closed end" peptide-binding site of MHC class I molecules, allows for the binding of longer peptides: In some instances even entire protein molecules *(4, 5)* provided there exists a stretch of 9–15 amino acid residues, which can settle into the MHC class II peptide-binding pockets *(6)*. The crystal structure of the MHC class II allele molecule suggests an "open" peptide-binding groove, a direct structural evidence for the observed ability of MHC class II molecules to bind longer peptides.

It is well known that some MHC class II binding peptides are "promiscuous": the same peptide is capable of interacting with several MHC class II alleles *(7–10)*. The identification of both conserved and allele specific anchors in MHC class II binding peptides *(11)* together with the fact that not all the anchors need to be used by individual ligands *(12)* provides the molecular basis for "promiscuity" and allele specificity of peptide binding to MHC class II alleles.

CD4+ T-cells play a central role in humoral and cellular immune responses, it is therefore of interest to identify their nominal ligands, i.e., peptide antigens which bind to MHC class II (HLA-DR, -DP, and –DQ) molecules. The identification of MHC class II ligands is pivotal for a mechanistic understanding of diseases. For instance, the MHC class II molecule DQ0602 confers *susceptibility* to narcolepsy, a sleep disorder of unknown origin. Ninety to hundred percent of patients with narcolepsy carry DQ0602 which, in turn, confers dominant *protection* against type 1 diabetes *(13)*. A detailed structural analysis revealed that the presentation of a "broader," diverse peptide repertoire is critical for protection against type 1 diabetes. Of note, the closely related DQ0601 allele *protects* against the development of narcolepsy, the peptide-binding characteristics of these closely related MHC class II alleles (DQ0601/0602) account for these differences. In general, MHC class II-linkage has been reported for *susceptibility* or *protection* in the context of "MHC-associated disorders" which encompasses infectious diseases as well as autoimmune disorders (e.g. multiple sclerosis, sarcoidosis).

Thus, it is desirable to (1) identify potential peptide ligands to individual MHC class II alleles which are either associated with protection or enhanced risk to develop disease and (2) to test the impact of allelelic MHC class II variants to candidate peptides species which serve as targets for CD4+ T-cells associated with specific diseases.

A number of assays are currently implemented to identify peptide binding to different MHC class II alleles, these include:

1. Mass spectrometric sequencing of peptides eluted from purified MHC alleles *(14, 15)*. The MHC class II allele is purified from cells and peptides are eluted from the binding site of the MHC allele by acid treatment. This is a costly process since the entire spectrum of peptides binding to the HLA allele will be eluted and sequenced, the focus on the peptide species of interest is challenging.
2. The test peptide can be "competed out" with a labeled "control-peptide" of biochemically defined binding affinity to the MHC allele *(16–18)*. This is a popular assay but it appears to be limited to a few test peptides and MHC alleles, it is time-consuming and costly due to peptide synthesis.
3. Defined MHC class II alleles loaded with candidate peptides can be screened for the induction of cytokine production in T-cells using ICS or ELISPOT. If a peptide binds, it may be presented to MHC class II-restricted CD4+ T-cells *(19)*. The main disadvantage with this assay is that a high number of responder T-cells is needed to establish the associations of the HLA alleles and the test peptide(s). In addition, this assay reflects more the nature of the responding T-cell population and not the biochemical MHC class II-peptide interaction.
4. Computer-based approaches to predict peptide binding to HLA alleles: a number of computer-based algorithms have been developed to predict the binding of peptides to different MHC alleles *(20–23)*. These algorithms are convenient since they do not require biological samples and they save time. They may not be very reliable for predicting peptide binding to certain MHC class II alleles *(2)*.

We provide below the protocols to identify peptide binding to different soluble HLA class II alleles (e.g., DR1*0101 and DR1*1501). Candidate peptides are produced using SPOT synthesis, followed by printing on epoxy functionalized glass slides. This work was performed by JPT Peptide Technologies GmbH (Berlin, Germany). Peptide microarrays are incubated with soluble MHC class II molecules, followed by several washing steps. MHC class II-peptide binding interactions are visualized using an antibody directed against MHC class II DR-molecules conjugated to a fluorochrome. The fluorescence signals are analyzed to identify the nature of the spotted peptides (i.e., the formation of an MHC class II/peptide complexes) that bind to MHC class II alleles. Soluble MHC class II alleles were obtained from Beckman Coulter and produced as described in detail *(24, 25)*.

2. Materials

2.1. Equipment

1. Vortex.
2. Shaker.
3. Slide centrifuge (Euro Tech, UK).
4. Microarray scanner, i.e., Genepix4000B.
5. High-density peptide microarray, manufacturer: JPT Peptide Technologies GmbH, Berlin, Germany, stored at 4°C and in a dry environment.
6. Coverslip, (ABgene, Surrey, UK).
7. Liquid blocker pen, (Dako Cytomation, Glostrup, Denmark).
8. Dark box with slide holder, (VWR, Stockholm, Sweden).
9. Polypropylene 15-mL screw cap tubes.

2.2. Reagents

1. PBS stored at room temperature.
2. Tween 80 (VWR, Stockholm, Sweden), stored at room temperature.
3. Washing solution (PBS + 0.05% Tween 80), stored at 4°C, 70% ethanol.
4. Soluble MHC class II monomers (Beckman Coulter):
   (a) HLA DRB1*0101, stock concentration 1.35 mg/mL
   (b) HLA DRB1*1501, stock concentration 1.40 mg/mL to be used at working concentration of 1 μg/mL, stored at –70°C
5. Cy5-labeled L243 monoclonal antibody (Beckman Coulter) stock concentration 2.38 mg/mL, working concentration 5 μg/mL, stored at 4°C in the dark.
6. MHC class II binding buffer: (Sodium phosphate 36 mM, citrate 14.4 mM, BSA 0,15%, Octyl β-D-glucopyranoside (OG), sigma #O8001, 0.25%, NaN3 0.02%, pH 5.5), stored at +4°C in the dark.

3. Methods

3.1. Slide Preparation and Incubation with MHC Class II Monomeric Molecules

1. Dilute each monomer to 1 μg/mL using the MHC class II binding buffer.
2. Ensure that the surface of the slide is dry and clean.
3. Use a liquid blocker pen to circumscribe and define the area of incubation.

4. Care should be taken not to touch the peptide-surface of the slide.
5. Mark the slide with the blocker pen as close as possible to the edge of the slide.
6. Pipette 300 μL of the diluted monomer evenly across the slide (*see* **Note 1**).
7. Cover the slide by placing a cover slip over the slide.
8. Place the slide in a humid incubation chamber, incubate at 37°C for 48 h (*see* **Note 2**).
9. After 48 h, remove the cover slip.
10. Start the 3× washing procedure by placing the slide in a box containing a slide holder. Use sufficient volume of washing solution to cover the entire slide.
11. Wash the slide in the washing solution on a shaker two times (5-min each).
12. Wash the slide in PBS on a shaker one time (5 min).
13. Tap the slide on a dry tissue to remove any washing solution droplets.
14. Pipette 300 μL of the diluted Cy5-labeled L243 antibody evenly onto the slide, still using the area marked by the blocker pen to keep the antibody solution within the peptide array area (*see* **Note 3**).
15. Fluorescent dyes are light sensitive and therefore all work with the fluorochrome-labelled antibody should be performed in the dark. The slide should also be protected from light after this step.
16. Cover the slide by placing a cover slip over the slide.
17. Incubate the slide 1 h at room temperature in a humid chamber in the dark.
18. After the incubation is complete, repeat the 3× washing procedure.
19. Tap the slide on a dry tissue to remove any washing solution.
20. Dry the slide by spinning for 10 s in the slide centrifuge (*see* **Note 3**).
21. Place slides in a light-proof box (cleaned with 70% alcohol) and keep at 4°C until ready for scanning.

3.2. Slide Scanning

Slides are scanned in the GenePix Pro scanner according to the instructions given in the manual *(26)*. The scanning is performed at two wavelengths, i.e., 635 and 532 nm (*see* **Note 4**). The anti-MHC class II directed mAb (detecting peptide-bound MHC class II molecules) is conjugated to Cy5, which emits fluorescence at 635 nm (*see* **Fig. 1a**). Cy3, which emits fluorescence

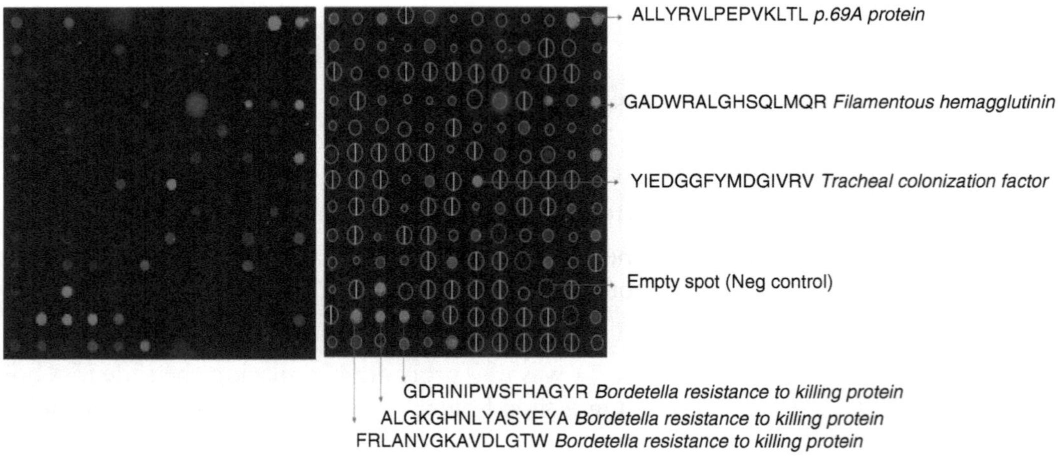

Fig. 1. Example of a peptide library from *B. pertussis* proteins probed for binding to DRB*0101. (**a**) Reactive (*red*) and nonreactive (*dark*) spots signify individual peptide species binding to MHC class II molecules visualized by the anti-DR mAb. Nonreactive spots (*dark*) indicate the absence of a peptide/MHC class II molecule complex. (**b**) Overlay with the GAL file enables the identification of binding peptides to DR*0101 molecules (*see Color Plates*).

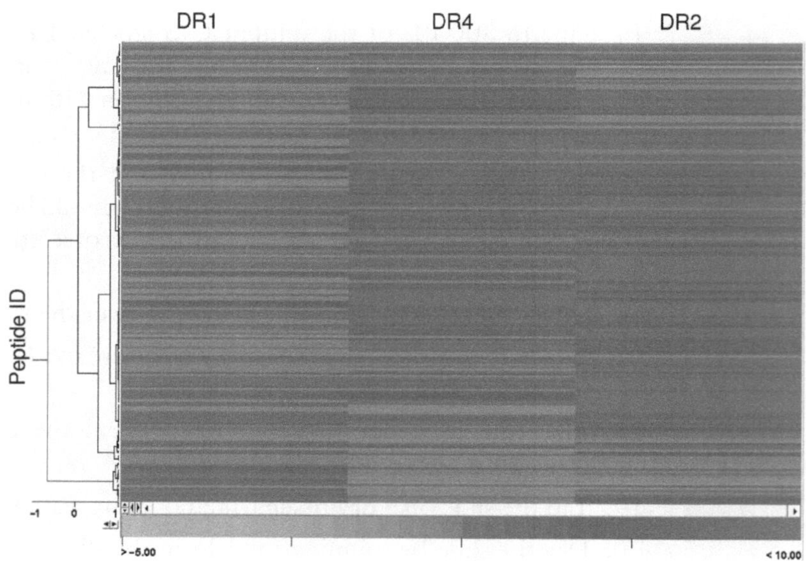

Fig. 2. Example of hierarchical cluster analysis of HIV-1 peptides binding to different DR molecules. Note the different peptide binding profile to individual MHC class II molecules (*see Color Plates*).

at 532 nm, is printed at specific slide positions to assist alignment of the GenePix Array List (GAL) file. The scanned images of the slides are saved in tiff-format.

3.3. Analysis of Tiff Files in GenePix Pro

1. Perform image analysis on the saved tiff files using GenePix Pro 5.1 software (Axon Instruments) and GAL files supplied by the manufacturer of the slides used in the experiment.

2. Use the GenePix Pro manual to identify and test the most suitable criteria. We provide here the analysis protocol used in our laboratory.
3. Prior to analysis, the following "conditions" are set:
   - Find circular feature.
   - Resize features during alignment; minimum 25%, maximum 400%.
   - Limit feature movement during alignment; maximum translation: 60 μm.
   - Flag features that fail background threshold criteria; NOT FOUND.
   - Composite Pixel Intensity (CPI) threshold to include a pixel in a feature during alignment → 10.
   - Maximum translation pixel →10
4. We defined the criteria that will flag spots with nonuniform background and foreground.

   ([F635 Mean] > (1.5*[F635 Median])) AND

   ([F635 Median] > 40) OR

   ([B635 Mean] > (1.5*[B635 Median])) AND

   ([B635 Median] > 40)
5. The entire analysis is performed using the above conditions and criteria.
6. The saved tiff files are opened individually in GenePix, the corresponding GAL file to the slide batch is opened as well.
7. The Cy3 controls which are printed at specific positions on the slide are used to fit the GAL file on the image (*see* **Fig.1b**). When properly placed, the green spots should match the Cy3 controls as indicated by the GAL file.
8. When the GAL file has been correctly placed on the slide image, the alignment of the spots can be initiated by pressing the F5 key. The GenePix software will align the spots according to the GAL file, and also flag the spots accordingly (*see* **Fig. 1b**) (*see* **Note 5**).
9. This should be followed by visual inspection of the slide to ensure that the spots have been aligned correctly by the software. In case if spots have not been correctly aligned, the GAL file has to be correctly placed.
10. If visual inspection confirms that all spots have been found to be well aligned, the next step represents spot analysis. This is initiated by pressing the Alt+A key. The software will calculate the (pre-set) features associated with the spots and the corresponding features will be shown in the results window.
11. Before saving results, click on the Flag Features icon which appears on the results windows. This will open a window in

which the criteria we described earlier (*see* above, **Subheading 3.4**) will show up. Click on the "Evaluate icon" so that the software flags all spots that do not meet the (pre-set) criteria.

12. To evaluate whether the correct spots have been flagged, access the image screen and inspect the flagged spots. Unflag spots that have been incorrectly flagged and also flag those spots that are not flagged but do not meet the criteria using visual inspection (i.e., no circular features).
13. After all spots have been visually inspected, press the <Alt+A> key to analyze individual spots, results are saved as GenePix "Results" (GPR) files.

3.4. Identifying Peptides Binding to Soluble HLA Molecules /Normalization

1. The GPR files are imported into the analysis program "R" for data preprocessing and normalization is carried out as described in detail earlier *(26)*. Systematic effects of slide, sub-array and block are removed using a linear model, which also investigates and removes any interaction between slide and sub-array. The residuals from this model provide the "normalized" peptide responses (with random measurement error) that can be subjected to further analysis in the program "Acuity" (Molecular Devices, Sunnyvale, CA).
2. The index response for each peptide (log2 F635/B635) is imported into the Acuity program.
3. Peptides which show a high index response (two SD above the mean of the index response of the "empty" (no peptide printed) spots obtained from the slide incubated with buffer and the anti-MHC class II-detection mAb (i.e., L243) are removed from the analysis and considered as "false-positives."
4. The remaining peptides which show an index response above the cut off (i.e., two SD of the empty spots, see above) are considered to form peptide/MHC class II binding complexes. In contrast, peptide spots with a lower index response (below two SD of the empty spots) are considered to be "not binding" to soluble MHC class II molecules.
5. Hierarchical clustering analysis of the peptides can be performed to identify differential binding patterns of peptides to the different HLA alleles (*see* **Fig. 2**), this analysis is carried out in the program "Acuity".

4. Notes

1. It is crucial to ensure that slides are placed on a flat surface during the incubation, otherwise the soluble MHC class II monomer or the mAb directed against DR-molecules may

be unevenly distributed resulting in discordant results in subarrays of the same slide (peptides in different subarrays will not have similar exposure to the reagents).

2. It is important to ensure that enough water is present in the incubation chamber during the incubation to avoid "drying" of slides and artefacts on the slide surface associated with it.
3. Automated incubation will overcome operator-associated differences, e.g., the Tecan HS Hybridization Station has been proven to reduce the amount of background. Other instruments may also be applicable.
4. Scanning the slides: optimize the "PMT gain" to the point where there is no saturation of the signal. This is crucial since the fluorescence intensity from saturated spots can not be reliably used in the analysis. Saturation indicates that the fluorescence intensity is above the threshold for the scanner.
5. Ensure that during the analysis of the slides in the program "genepix," data from each sub-array is saved as an individual file and *not* together with other subarrays from the same slide. If files are not saved in separate files, the analysis program "Acuity" will create an average of the peptide fluorescence intensity from *all* the subarrays present on the same slide.

Acknowledgements

We thank Marie Reilly, Davide Valentini and Yen Ngo, MEB, Karolinska Institutet for statistical analysis, Emmanuel Gautherot and Felix Montero at Beckman Coulter Marseille, for HLA-DR molecules and JPT Peptide Technologies for peptide microarrays.

References

1. Topalian, S. L. (1994) MHC class II restricted tumor antigens and the role of CD4+ T-cells in cancer immunotherapy. *Curr. Opin. Immunol.* 6, 741–745.
2. Southwood, S., Sidney, J., Kondo, A., del Guercio, M. F., Appella, E., Hoffman, S., Kubo, R. T., Chesnut, R. W., Grey, H. M., and Sette, A. (1998) Several common HLA-DR types share largely overlapping peptide binding repertoires. *J. Immunol.* 160, 3363–3373.
3. Sinigaglia, F. and Hammer, J. (1994) Defining rules for the peptide-MHC class II interaction. *Curr. Opin. Immunol.* 6, 52–56.
4. Sette, A., Adorini, L., Colon, S. M., Buus, S., and Grey, H. M. (1989) Capacity of intact proteins to bind to MHC class II molecules. *J. Immunol.* 143, 1265–1267.
5. Stern, L. J., Brown, J. H., Jardetzky, T. S., Gorga, J. C., Urban, R. G., Strominger, J. L., and Wiley, D. C. (1994) Crystal structure of the human class II MHC protein HLA-DR1 complexed with an influenza virus peptide. *Nature* 368, 215–221.
6. Sercarz, E. E. and Maverakis, E. (2003) Mhc-guided processing: binding of large antigen fragments. *Nat. Rev. Immunol.* 3, 621–629.
7. Gaudebout, P., Zeliszewski, D., Golvano, J. J., Pignal, C., Le Gac, S., Borras-Cuesta, F., and Sterkers, G. (1997) Binding analysis of 95 HIV gp120 peptides to HLA-DR1101

and -DR0401 evidenced many HLA-class II binding regions on gp120 and suggested several promiscuous regions. *J. Acquir. Immune Defic. Syndr. Hum. Retrovirol.* 14, 91–101.

8. Kobayashi, H., Wood, M., Song, Y., Appella, E., and Celis, E. (2000) Defining promiscuous MHC class II helper T-cell epitopes for the HER2/neu tumor antigen. *Cancer Res.* 60, 5228–5236.
9. Panina-Bordignon, P., Tan, A., Termijtelen, A., Demotz, S., Corradin, G., and Lanzavecchia, A. (1989) Universally immunogenic T-cell epitopes: promiscuous binding to human MHC class II and promiscuous recognition by T-cells. *Eur. J. Immunol.* 19, 2237–2242.
10. Sinigaglia, F., Guttinger, M., Kilgus, J., Doran, D. M., Matile, H., Etlinger, H., Trzeciak, A., Gillessen, D., and Pink, J. R. (1988) A malaria T-cell epitope recognized in association with most mouse and human MHC class II molecules. *Nature* 336, 778–780.
11. Hammer, J., Valsasnini, P., Tolba, K., Bolin, D., Higelin, J., Takacs, B., and Sinigaglia, F. (1993) Promiscuous and allele-specific anchors in HLA-DR-binding peptides. *Cell* 74, 197–203.
12. Hammer, J., Takacs, B., and Sinigaglia, F. (1992) Identification of a motif for HLA-DR1 binding peptides using M13 display libraries. *J. Exp. Med.* 176, 1007–1013.
13. Siebold, C., Hansen, B. E., Wyer, J. R., Harlos, K., Esnouf, R. E., Svejgaard, A., Bell, J. I., Strominger, J. L., Jones, E. Y., and Fugger, L. (2004) Crystal structure of HLA-DQ0602 that protects against type 1 diabetes and confers strong susceptibility to narcolepsy. *Proc. Natl. Acad. Sci. U. S. A.* 101, 1999–2004.
14. Dengjel, J., Rammensee, H. G., and Stevanovic, S. (2005) Glycan side chains on naturally presented MHC class II ligands. *J. Mass Spectrom.* 40, 100–104.
15. Wahlstrom, J., Dengjel, J., Persson, B., Duyar, H., Rammensee, H. G., Stevanovic, S., Eklund, A., Weissert, R., and Grunewald, J. (2007) Identification of HLA-DR-bound peptides presented by human bronchoalveolar lavage cells in sarcoidosis. *J. Clin. Invest.* 117, 3576–3582.
16. Newman, M. J., Livingston, B., McKinney, D. M., Chesnut, R. W., and Sette, A. (2002) T-lymphocyte epitope identification and their use in vaccine development for HIV-1. *Front. Biosci.* 7, d1503–d1515.
17. Schaeffer, E. B., Sette, A., Johnson, D. L., Bekoff, M. C., Smith, J. A., Grey, H. M., and Buus, S. (1989) Relative contribution of "determinant selection" and "holes in the T-cell repertoire" to T-cell responses. *Proc. Natl. Acad. Sci. U. S. A.* 86, 4649–4653.
18. Sette, A. and Grey, H. M. (1992) Chemistry of peptide interactions with MHC proteins. *Curr. Opin. Immunol.* 4, 79–86.
19. Altfeld, M. A., Livingston, B., Reshamwala, N., Nguyen, P. T., Addo, M. M., Shea, A., Newman, M., Fikes, J., Sidney, J., Wentworth, P., Chesnut, R., Eldridge, R. L., Rosenberg, E. S., Robbins, G. K., Brander, C., Sax, P. E., Boswell, S., Flynn, T., Buchbinder, S., Goulder, P. J., Walker, B. D., Sette, A., and Kalams, S. A. (2001) Identification of novel HLA-A2-restricted human immunodeficiency virus type 1-specific cytotoxic T-lymphocyte epitopes predicted by the HLA-A2 supertype peptide-binding motif. *J. Virol.* 75, 1301–1311.
20. Tong, J. C., Tan, T. W., and Ranganathan, S. (2007) Methods and protocols for prediction of immunogenic epitopes. *Brief. Bioinform.* 8, 96–108.
21. Zhu, S., Udaka, K., Sidney, J., Sette, A., Aoki-Kinoshita, K. F., and Mamitsuka, H. (2006) Improving MHC binding peptide prediction by incorporating binding data of auxiliary MHC molecules. *Bioinformatics* 22, 1648–1655.
22. DeLuca, D. S., Khattab, B., and Blasczyk, R. (2007) A modular concept of HLA for comprehensive peptide binding prediction. *Immunogenetics* 59, 25–35.
23. Mallios, R. R. (2003) A consensus strategy for combining HLA-DR binding algorithms. *Hum. Immunol.* 64, 852–856.
24. Novak, E. J., Liu, A. W., Nepom, G. T., and Kwok, W. W. (1999) MHC class II tetramers identify peptide-specific human CD4(+) T-cells proliferating in response to influenza A antigen. *J. Clin. Invest.* 104, R63–R67.
25. Ye, M., Kasey, S., Khurana, S., Nguyen, N. T., Schubert, S., Nugent, C. T., Kuus-Reichel, K., and Hampl, J. (2004) MHC class II tetramers containing influenza hemagglutinin and EBV EBNA1 epitopes detect reliably specific CD4(+) T-cells in healthy volunteers. *Hum. Immunol.* 65, 507–513.
26. Nahtman, T., Jernberg, A., Mahdavifar, S., Zerweck, J., Schutkowski, M., Maeurer, M., and Reilly, M. (2007) Validation of peptide epitope microarray experiments and extraction of quality data. *J. Immunol. Methods* 328, 1–13.

Chapter 31

T-Cell Epitope Mapping

Raija K.S. Ahmed and Markus J. Maeurer

Summary

Identification of epitopes defined by T-cell responses aids to (1) monitor antigen-specific cellular immune responses (2) guide rational vaccine design, and (3) understand the nature of protective or harmful T-cell responses in diseases with defined target antigens. The 6-h intracellular cytokine staining (ICS) assay preferentially identifies effector T cells that are readily detectable in the peripheral circulation. In contrast, the whole blood assay (WBA) allows to gauge expansion of antigen-specific T cells over time (7 days), i.e., T cells with lower frequencies (e.g., memory T cells) defined by proliferation and cytokine production. Any cellular immune profile can be measured in the WBA (using the 7 days cell culture supernatants) or directly in responder T cells after antigenic stimulation (in the ICS) with appropriate cytokine-specific detection systems. The choice of the cytokine test panel depends on the nature of the expected immune response. A broad panel of candidate peptides can be tested for T-cell recognition in the WBA due to its simplicity and the low input of (unprocessed, heparinized) blood.

Key words: Epitope mapping, Whole blood assay, Intracellular staining, T-cells, Flow-cytometry.

1. Introduction

Cellular immune responses play an important role in infections, autoimmune diseases, malignancy, and transplantation *(1–3)*. T cells aid, with other components of the immune system, to eliminate foreign antigens and to develop protective immunity. Cellular memory immune responses may be induced (in the case of infectious diseases) by natural infection or by antigen-specific vaccination. Activation of T cells is triggered upon recognition of specific peptide epitopes from disease-related proteins presented by major histocompatibility complex (MHC) class I or II molecules on the surface of antigen presenting cells. In general, proteins contain several T-cell epitopes, but only a few peptide species have been shown to serve

Ulrich Reineke and Mike Schutkowski (eds.), *Methods in Molecular Biology, Epitope Mapping Protocols, vol. 524*

DOI: 10.1007/978-1-59745-450-6_31

as target antigens for CD8+ or CD4+ T-lymphocytes . Several factors may account for the situation that just a few epitopes represent T-cell targets: (1) alternate intracellular processing and presentation of protein antigens, (2) the inflammatory microenvironment which impacts on antigen processing as well as, (3) the molecular composition of the T-cell repertoire that is available in individuals to actively respond with measurable expansion and cytokine production to peptide targets. A detailed understanding which specific peptide epitopes activate and regulate T cells is important for any epitope-based treatment (e.g., in context of autoimmune disease) and for the rational design of vaccines.

Over the last few years, experimental efforts were intensified to identify and locate regions of target antigens for T-cell recognition and to specifically map pathogen specific epitopes in clinical trials. For instance, a recent HIV/SIV vaccine study, conducted in nonhuman primates, revealed that T cells recognize *different* virus specific epitopes *after* immunization as compared with the epitope pattern associated with the HIV (natural) infection *(4)*. The mapping of T-cell epitopes enables to obtain a detailed profile concerning T-cell "epitope shifting" associated with different vaccination schedules, it may escort our efforts to identify markers of immune protection in the context of pathogens (*see* **Fig. 1**).

Several methods have been used to map T-cell epitopes. The IFN-γ ELISPOT assay is widely used to detect antigen-specific immune responses to target antigens, it has been implemented for the identification of T-cell epitopes in human papillomavirus type 16 (HPV 16) antigens *(5)* and the mycobacterial major secreted antigen 85A *(6)*. The disadvantages of this technique are the subjectivity of reading plates manually (which could be overcome by automation) and that the assay does not allow the discrimination between antigen-specific responses derived from CD4+ or CD8+ T-lymphocytes without cell separation procedures. It also requires a high number of immune cells in order to screen for a comprehensive number of candidate peptide species.

A different popular method for identifying T-cell epitopes represents the tetramer-guided epitope mapping *(7–10)*. This assay uses flow cytometry to measure CD8+ T-cells (or CD4+ T-cells) that recognize a specific epitope restricted by a single MHC molecule. Tetramer-based epitope mapping has been used for the identification of peptide epitopes to a broad panel of MHC class I presented epitopes provided by the mycobacterial major secreted antigen 85B *(9)* and for the detection of tetanus-toxin specific CD4+ T-cells *(8)*. MHC-guided epitope mapping allows to identify antigenic epitopes presented by multiple MHC alleles simultaneously *(10, 11)*. The primary advantage with tetramer-guided epitope mapping followed by flow cytometry: it does not require in vitro culture or stimulation of immune cells,

a Infection: before | after

	1	2	3	4	5	6	7	8
A	STD 320 pg/ml	STD 320 pg/ml	1	9	17	1	9	17
B	STD 160 pg/ml	STD 160 pg/ml	2	10	18	2	10	18
C	STD 80 pg/ml	STD 80 pg/ml	3	11	19	3	11	19
D	STD 40 pg/ml	STD 40 pg/ml	4	12	20	4	12	20
E	STD 20 pg/ml	STD 20 pg/ml	5	13	21	5	13	21
F	STD 10 pg/ml	STD 10 pg/ml	6	14	NEG	6	14	NEG
G	STD 5 pg/ml	STD 5 pg/ml	7	15	POS	7	15	POS
H	BLANK	BLANK	8	16		8	16	

b

	1	2	3	4	5	6	7	8	
A	0,986	0,965	−0,006	−0,001	0,008	0,029	0,031	0,037	
B	0,801	0,781	−0,006	−0,004	−0,006	0,029	0,026	0,033	
C	0,599	0,630	−0,005	−0,004	−0,006	0,027	0,026	0,046	
D	0,391	0,385	−0,005	−0,001	−0,006	0,029	0,029	0,253	*
E	0,229	0,218	−0,005	−0,004	−0,007	0,026	0,028	0,070	
F	0,130	0,121	−0,005	→ 0,697	−0,006	0,029	→ 0,947	−0,002	
G	0,070	0,070	−0,004	−0,004	0,571	0,030	0,027	0,357	
H	0,002	−0,002	−0,002	−0,001	−0,009	0,032	0,039	−0,007	

c

	1	2	3	4	5	6	7	8	
A	>max	310,64	<min	<min	<min	<min	<min	<min	
B	168,45	151,71	<min	<min	<min	<min	<min	<min	
C	76,34	83,91	<min	<min	<min	<min	<min	<min	
D	40,44	39,56	<min	<min	<min	<min	<min	23,26	*
E	20,57	19,43	<min	<min	<min	<min	<min	<min	
F	10,42	9,58	<min	→ 209,47	<min	<min	→ 588,63	<min	
G	5,00	5,00	<min	<min	2809,57	<min	<min	717.14	
H	<min	<min	<min	<min	<min	<min	<min	<min	

Fig. 1. T-cell epitope profiling using the WBA. Typical experimental setup to screen for T-cell epitope mapping and a shift in recognition pattern induced by infection. 21 individual test peptides from an MTB protein were evaluated for T-cell recognition prior to and after MTB infection in an animal model. Supernatants were collected after 7 days of peptide stimulation. (**a**) Example for an ELISA plate layout. 1 + 2: standard. 3–5: test peptides, negative and positive control. 6–8: identical layout as compared with 3–5 but supernatants from blood drawn after infection. (**b**) "Raw" OD readings. (**c**) Corresponding IFN-γ values in pg/mL. Note that peptide number 14 *(arrow)* was recognized prior to and after infection. Noteworthy, peptide mapping revealed that peptide number 14 was still dominantly recognized after infection, but also a novel peptide epitope (number 20) was recognized (marked with a *star*). The positive and negative controls are *boxed*.

it is able to directly visualize antigen-specific T cells in peripheral blood and to link enumeration of antigen-specific T cells with clinically and biologically relevant T-cell markers (e.g., markers for memory or effector T cells).

We describe here in detail the protocols for peptide epitope mapping using the Whole Blood Assay (WBA) and the 6-h "intracellular cytokine staining" (ICS) assay. WBA: diluted whole (heparinized) blood which is stimulated with candidate peptides (usually 15 or 20mers) for 7 days. On day 7, supernatants are collected for INF-γ detection. Other cytokines, dependent on the nature of the immune response (i.e., Th1/Th2 or Tc1/Tc2) could also be tested either by ELISA or using other, more comprehensive, cytokine detection panels (e.g., Luminex). During the 7-day incubation period, antigen-activated cells proliferate and form lymphoblasts (*see* **Fig.** 2). Cells are immunophenotyped with monoclonal antibodies against CD3, CD4 and CD8 and the number of CD3+ CD4+/CD3+ CD8+ lymphoblasts generated in the culture in response to peptide stimulation are identified and compared with a medium control and SEA/SEB (as a positive control) *(12, 13)* (*see* **Fig.** 1). This method is simple, requires only a small blood volume without the necessity of cell separation: Blood must be collected using a heparinized container, it should be used within 12 h after the blood draw (stored at room temperature). This assay does not allow to discriminate which cells (e.g., CD4 or CD8+ T-cells) represent the source of INF-γ production. ICS can be independently used to further characterize peptide antigen-specific, polyfunctional (cytokine producing) T cells.

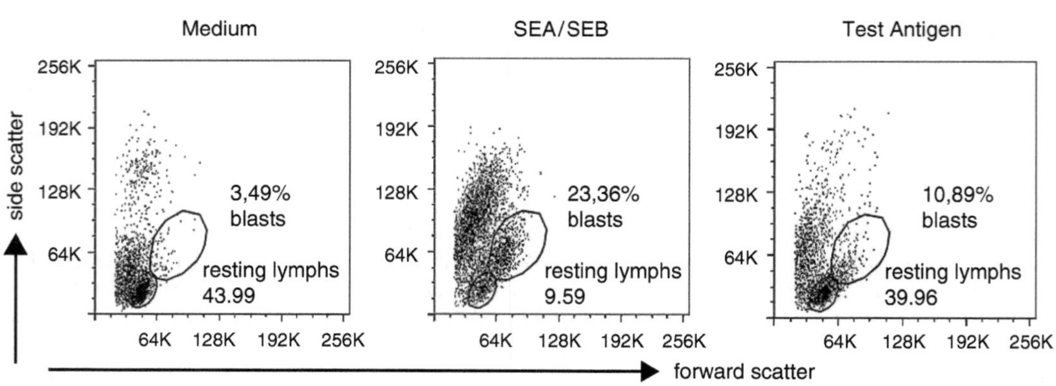

Fig. 2. WBA: lymphoproliferation in responding T-cells. Blood was incubated either with medium (negative control, *left*), positive control (*middle*) or the test peptide antigen (*right*) and T-cell blast formation was evaluated after 7 days of culture based on side and forward scatter analysis. Cells can further be analyzed for CD4+ and CD8+ T-cell markers to define the nature of the proliferative T-cell response.

ICS is a flow cytometry-based method and permits not only the detection of cytokine-positive cells, it also allows the identification of the responding cells: Cytokine-producing CD4+ or CD8+ T-cells. Flow-cytometry allows to analyze cytokine-producing cells at a single cell level and to determine "polyfunctional" CD8+ or CD4+ T-cells producing INF-γ, TNF-α, and IL-2 simultaneously. Peripheral mononuclear cells (PBMC) are stimulated with the specific antigen for 6 h in the presence of Brefeldin A. Cells are stained with monoclonal antibodies to CD3, CD4, and CD8 and after fixation and permeabilization for intracellular cytokines. Peptide-specific T cells producing INF-γ and/or IL-2 and/or TNF-α are detected by flow cytometry (*see* **Fig.** 3). Note that the ICS most likely detects effector T cells, which are readily detectable in the peripheral circulation (there is only a 6-h incubation!). In contrast, the WBA allows to gauge expansion of antigen-specific T cells over time (7 days), which enables to visualize T cells with lower frequencies (e.g., memory T cells) defined by proliferation and cytokine production. Any cellular immune response can be measured either in the WBA (using the supernatants) or directly in responder T cells after antigenic stimulation (in the ICS) with appropriate cytokine-specific detection systems, depending on the nature of the expected immune response. Identification of epitopes defined by T-cell responses will aid to define the nature of protective or harmful T-cell responses in diseases with defined target antigens, it will also escort the rational design of peptide–epitope-based treatment and immune-monitoring strategies.

2. Materials

1. RPMI-1640 W/Glutamax medium (500 mL) (Gibco, Invitrogen) supplemented with 2.5 mL Penicillin/Streptomycin and 5 mL of HEPES. Store at 2–8°C.
2. RPMI-medium (see above) with 10% Foetal Bovine Serum (FBS).
3. Penicillin–Streptomycin (Gibco, Invitrogen) Stock:10,000 IU/mL–10,000 μg/mL. Store at −20 ± 2°C.
4. HEPES 1 M (Gibco, Invitrogen): aliquot into 15-mL centrifuge tubes (5 mL/tube). Store at 2–8°C.
5. BD PharmLyse 10× (Becton Dickinson). Dilute 1:10 in sterile water.
6. Paraformaldehyde (PFA). Prepare a 1% solution in PBS fresh for each experiment.

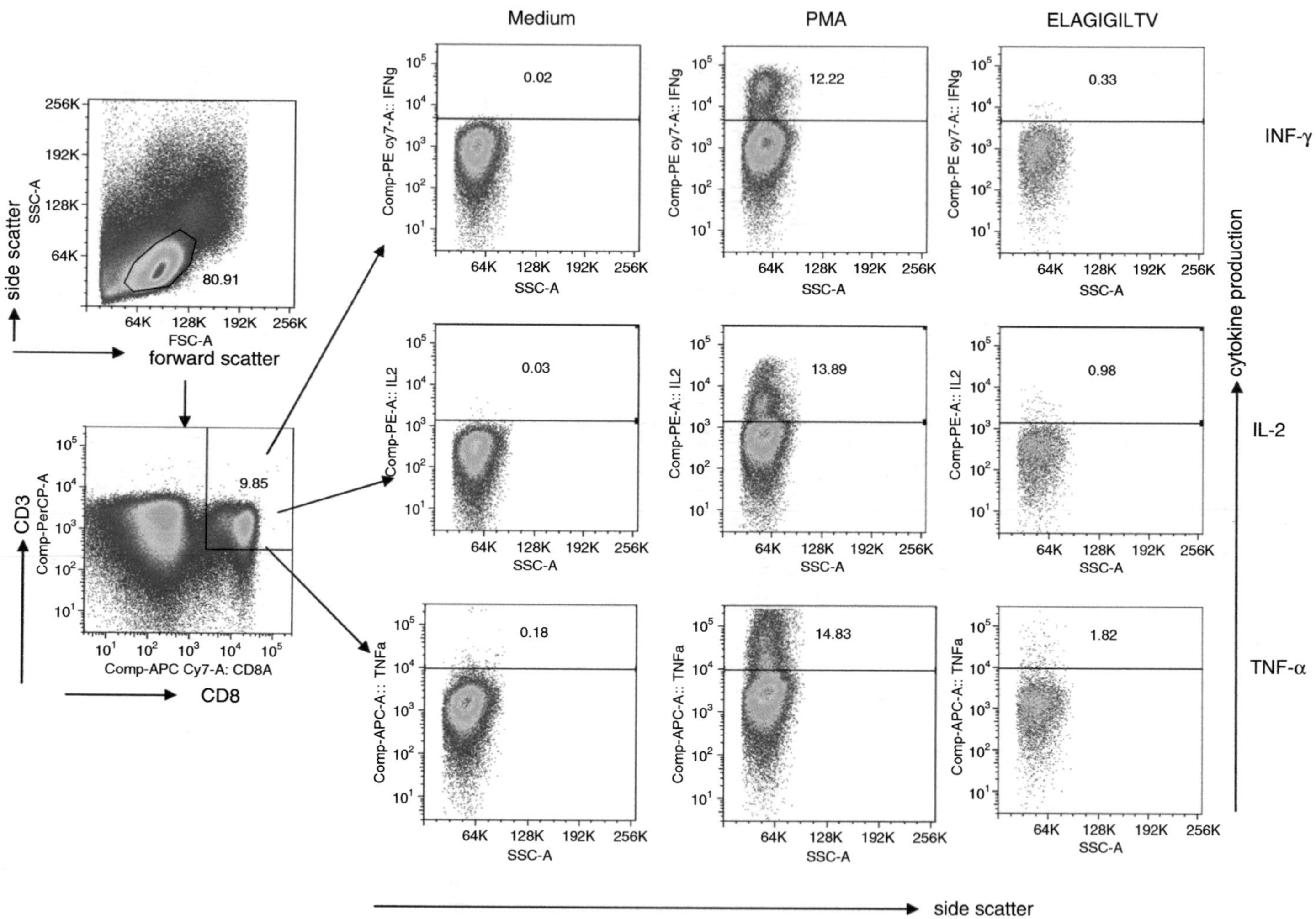

Fig. 3. ICS: PBMCs from a patient with melanoma were stimulated either with medium (*left*), the positive control (PMA) or the test peptide (ELAGIGILTV from the tumor antigen Melan-A/MART-1) for 6 h. Cells were gated based on forward and side-scatter, followed by gating on CD3+ and CD8+ T-cells and single cytokine-producing cells were identified using intracellular staining for IL-2, TNF-α or IFN-γ using appropriate mAbs. Note that different gating strategies also allow (1) detection of CD4+ T-cells (in the presence of an appropriate target antigen) as well as (2) detection of T cells capable of producing simultaneously any of the tested cytokines (i.e., "multi- or polyfunctional T-cells"). (*see Color Plates*)

7. Staphylococcal enterotoxin A (SEA) (Sigma). Dissolve 0.5 mg of SEA in 2.5 mL PBS (resulting in 200 µg/mL). Store at −20°C. (*see* **Note 1**).
8. Staphylococcal enterotoxin B (SEB) (Sigma). Dissolve 1 mg of SEB in 10 mL of PBS (resulting in 100 µg/mL). Store at −20°C (*see* **Note 1**).
9. Monoclonal antibodies for WBA: anti-CD3-APC (UCHT1), anti-CD4 PerCP (SK3) and anti-CD8α-PE (SK1) (all from BD Pharmingen) (*see* **Note 2**). Note that other mAbs or fluorochromes can also be applied after titration.
10. Foetal Bovine Serum (FBS) (Sigma). Heat-inactivate at 56°C for 30 min. Aliquot in 50 mL tubes (50 mL/tube). Store at −20°C (*see* **Note 3**).
11. Sodium phosphate buffered saline (PBS), Ca^{2+} and Mg^{2+} free.
12. Phorbol 12-myristate 13-acetate (PMA) (Sigma). Dissolve 1 mg PMA in 10 mL of DMSO (100 µg/mL) and store in aliquots at −20°C. Avoid repeated freezing and thawing. Prepare a working solution (5 µg/mL) by diluting the stock 1:20 in RPMI-medium.
13. Ionomycin (Sigma). Dissolve 1 mg in 2 mL of 100% ethanol (0.5 mg/mL) and store in aliquots at 2–8°C. Prepare a working solution (0.05 mg/mL) by diluting the stock 1:10 in RPMI-medium.
14. Brefeldin A (BFA, Sigma) is a protein transport inhibitor that is used to retain and accumulate the produced cytokines inside cells as a result of antigenic stimulation. Dissolve 5 mg BFA in 1 mL of DMSO (5 mg/mL) and store in aliquots at −20°C. Avoid repeated freezing and thawing. Prepare a working solution (0.5 mg/mL) by diluting the stock 1:10 in RPMI-medium.
15. IntraPrep Kit (from Beckman Coulter or from any other distributor) is used for fixation to maintain the structure of the cells and for permeabilization to allow access of cytokine-antibodies into the cells. Store at room temperature.
16. Monoclonal antibodies for cell surface staining (ICS): anti-CD3-PerCP (clone SP-34–2) PerCP or anti-TCRab-PerCP (clone WT31), anti-CD4-PerCP Cy5.5 (clone L200), and anti-CD8α- APC-Cy7 (clone SK1) (all from BD Pharmingen). Note that other mAbs (coupled with alternate fluorochromes) can also be applied, after appropriate titration.
17. Monoclonal antibodies for intracellular staining (ICS): anti-IFNγ- PE-Cy7 (clone B27), anti-TNFα-APC (clone Mab11), and anti-IL2-PE (clone MQ1–17H12) (all from BD Pharmingen) (*see* **Note 4**).

3. Methods

3.1. Whole Blood Assay

3.1.1. Assay Setup and Collection of Cell Culture Supernatants

1. Dissolve each peptide of interest in 100% DMSO. Peptides are used at a final concentration of 1 μg/mL (for each peptide-specie). For each individual test prepare the plate accordingly (*see* **Note 5**). Peptides are plated individually (100 μL/well), each in its own well in progressive order. On each plate include positive (SEA/SEB) and negative control (medium only) wells (100 μL/well) (*see* **Fig. 1**).
2. Label plates with the sample ID and date according to the plate layout.
3. Obtain heparinized blood from the test individual. Blood should be used within 12 h after the blood draw and kept at room temperature. Never store it at lower temperature which would result in decrease of reactivity (and "artificial decrease" of CD4+ T-cells due to shedding of CD4 molecules at lower temperatures).
4. Invert the heparinized blood tubes 10 times and dilute the blood 1:2.5 with prewarmed RPMI-medium (without FBS).
5. Invert the diluted blood tubes 10 times and add 100 μL blood to each well (final volume per well: 200 μL) according to plate layout. (Dilution in the wells will be 1:5) (*see* **Note 6**).
6. Incubate the plates at 37°C in a 5% CO_2 incubator for 7 days.
7. Remove the plates from the incubator, collect the supernatants from each well and measure INF-γ production (*see* **Fig. 1**) by using a commercial ELISA-kit according to the manufacturer's instructions. Other cytokines, dependent on the nature of the expected response can also be tested. The same is true if a "multiplex" screening assay (for multiple cytokines) is desired.
8. Cells in the wells can now be stained and processed for detection of lymphoproliferation.

3.1.2. Staining of Cell Surface Antigens

1. Transfer the remaining blood (cells) from the wells into prelabeled round bottom polystyrene tubes.
2. Wash the wells with 150 μL of PBS and transfer it into tubes labeled with the sample ID.
3. Centrifuge for 5 min at 160 × *g*.
4. Aspirate supernatant by using the vacuum aspirator and add 100 μL of PBS + 0.1% FCS containing anti-CD4-PerCP, anti-CD3-APC, anti-CD8α-PE, and anti-CD8β-FITC (*see* **Note 3**).
5. Vortex and incubate for 15 min at 2–8°C.
6. Add 1 mL of 1× Pharm-lyzing buffer in all tubes, vortex and incubate 10 min at room temperature.
7. Centrifuge for 5 min at 160 × *g*
8. Aspirate the supernatant by using the vacuum aspirator.

9. Add 1.0 mL 1× Pharm-lyzing buffer in all tubes, vortex and incubate 10 min at room temperature.
10. Centrifuge the tubes for 5 min at 160 × *g*.
11. Aspirate the supernatant by using the vacuum aspirator.
12. Add 1 mL of PBS to all tubes, vortex and centrifuge the tubes for 5 min at 160 × g.
13. Aspirate the supernatant by using the vacuum aspirator and resuspend the cell pellet in 400 μL of PBS containing 1% paraformaldehyde.
14. Store the tubes at 2–8°C until acquisition (*see* **Fig.** 2). Avoid storage exceeding 24 h. This time frame is dependent on individual mAbs and fluorochromes.

3.2. Intracellular Cytokine Staining

3.2.1. Preparation and Stimulation of Immune Cells

1. PBMCs are isolated from heparinized blood using a standard Ficoll-procedure. After isolation, cells can be directly used in ICS or frozen in liquid nitrogen for later use.
2. Count the cells and adjust the cell concentration to 4 × 10^6 cells/ mL in RPMI-medium with 10% FBS. Add 250 μL of the cell suspension (1 × 10^6 cells) in tubes reserved for each peptide, the negative control (medium only) tube, and for the positive control (PMA/Ionomycin) tube. (We run samples in duplicates). This assay detects low frequency events responding to single peptide antigens. Note that a high number of total events have to be acquired (usually 1 × 10^6 cells) in order to obtain robust results!
3. Add 250 μL of each peptide dilution to the respective tubes (final volume in the tube is then 500 μL). Peptides are used at a final concentration of 1 μg/mL (diluted in RPMI-medium with 10% FBS) for each peptide (*see* **Note** 7).
4. Add 500 μL of RPMI-medium containing 10%FBS to the negative control tube(s) and the positive control (PMA/ Ionomycin) tube(s).
5. Add 2.5 μL of PMA working solution and 10 μL of Ionomycin working solution in the positive control tube(s). (final concentration of PMA will be 25 ng/mL and final concentration of Ionomycin 1 μg/mL).
6. Add 10 μL BFA working solution in each tube (final concentration 10 μg/mL).
7. Incubate tubes with the cells for 6 h in an incubator, 37°C with 5% CO_2.

3.2.2. Staining of Cell Surface Antigens

1. After incubation, centrifuge the tubes for 5 min at 160 × *g* at room temperature.
2. Discard the supernatant and resuspend the cell pellet in 300 μL of PBS + 0.1% FBS.
3. Centrifuge for 5 min at 160 × *g* at room temperature.

4. Discard the supernatant and add 100 μL of PBS + 0.1% FBS containing anti-CD4-PerCP Cy5.5, anti-CD3-PerCP (or anti-TCRαβ-PerCP), and anti-CD8α-APC-Cy7 in each tube (*see* **Note 8**) and vortex resulting solution.
5. Incubate for 15 min at 2–8°C.
6. Add 300 μL of PBS + 0.1% FBS to each tube and centrifuge for 5 min at 160 × *g* at room temperature.
7. Discard the supernatant and vortex the cell pellet.

In order to detect intracellular antigens, cells must now be fixed and permeabilized prior to staining with monoclonal antibodies.

3.2.3. Cell Fixation and Permeabilization for Intracellular Staining

1. Add 70 μL of reagent 1 (IntraPrep Kit) in each tube. Vortex well.
2. Incubate for 10 min at 2–8°C.
3. Add 300 μL of PBS + 0.1% FBS in each tube.
4. Centrifuge for 5 min at 160 × *g* at room temperature.
5. Discard the supernatant and resuspend the cell pellet with 50 μL or reagent 2 (IntraPrep Kit), vortex well and incubate for 15 min at 2–8°C.

3.2.4. Staining of Intracellular Cytokines

1. Directly after incubation with reagent 2, add anti-IFNγ-PE-Cy7, anti-TNFα-APC, and anti-IL2-PE antibodies in each tube and incubate for 30 min at 2–8°C (*see* **Notes 4** and **8**).
2. Add 300 μL of PBS + 0.1% FBS in all tubes and centrifuge for 5 min at 160 × *g* at room temperature.
3. Discard supernatants and resuspend cell pellets in 100 μL of PBS + 1% PFA.
4. Store tubes at +4°C in the dark until acquisition which should be performed within 24 h. The cells were analyzed using a FACSAria flow-cytometer and by FlowJo software (*see* **Fig. 3**). Any other flow-cytometer with appropriate filters and detection systems can be used to detect antigen-specific T cells using the ICS and the WBA. Pilot experiments have to be performed to gauge variability in positive and negative controls and to determine the relevant "cut-off" for positive signals in peptide-specific T-cell responses. Note that the positive events are reported as percentage of the parental population and as the absolute number of epitope-specific responder cells. The cut-off for true-positive events is also associated with the flow-cytometer *(14)*.

4. Notes

1. Mixture of SEA and SEB is used as a positive control. Prepare a working solution of SEA/SEB (20 ng/mL) by adding 1 μL of SEA stock solution and 2 μL of SEB stock solution in 10

mL RPMI-medium. Add 100 μL of working solution per well according to the plate layout.

2. Each individual monoclonal antibody and the entire monoclonal antibody panel must be titrated and calibrated for use with human whole blood. Other monoclonal antibodies may have to be used for nonhuman primates and mice.
3. Heat-inactivation of serum by raising the temperature to 56°C for 30 min will destroy complement and ensure that the cells will not be lysed by antibody binding leading to complement activation.
4. Each monoclonal antibody has to be titrated and evaluated for staining of intracellular antigens. It is important to identify monoclonal antibodies that are compatible with a fixation and permeabilization procedure.
5. Plates with the peptides and other antigens for WB-assay can be prepared in advance and stored at –80°C until the day of the assay.
6. We found 1:5 be the optimal dilution for detection of INF-γ and lymphoproliferation using the WB-assay.
7. Tubes can be prepared in advance and stored at –80°C until the day of the assay.
8. Each individual monoclonal antibody and the entire monoclonal antibody panel must be titrated and calibrated for use with human ICS-assay. These monoclonal antibodies can also be used for nonhuman primates.

Acknowledgements

We thank Isabelle Magalhaes (KI, Sweden) for provision of the data from Melan-A/MART-1 peptide (ELAGIGILTV) stimulated T cells.

References

1. Rouse, B. T. (2007) Regulatory T cells in health and disease. *J. Intern. Med.* 262, 78–95.
2. Romero, P., Certtini, J. C., and Speicer, D. E. (2006) The human T cell response to melanoma antigens. *Adv. Immunol.* 92, 187–224.
3. Boom, W. H., Canaday, D. H., Fulton, S. A., Gehring, A. J., Rojas, R. E., and Torres, M. (2003) Human immunity to *M. tuberculosis*: T cell subsets and antigen processing. *Tuberculosis (Edinb.)* 83, 98–106.
4. Vogel, , T. U., Horton, H., Fuller, D. H., Carter, D. K., Vielhuber, K., O'Connor, D. H., Shipley, T., Fuller, J., Sutter, G., Erfle, V., Wilson, N., Picker, L. J., and Watkins, D. I. (2002) Differences between T cell epitopes recognized after immunization and after infection. *J. Immunol.* 169, 4511–4521.
5. Toberty, T. W., Wang, S., Wang, X. M., Neeper, M. P., Jansen, K. U., McClements, W. L., and Caulfield, M. J. (2001) A simple and efficient method for the monitoring of antigen-specific T cell responses using peptide pool arrays in a modified ELISpot assay. J. Immunol. Methods 254, 59–66.

6. Smith, , S. M., Brookes, R., Klein, M. R., Malin, A. S., Lukey, P. T., King, A. S., Ogg, G. S., Hill, A. V., and Dockrell, H. M. (2000) Human CD8+ CTL specific for the Mycobacterial major secreted antigen 85A. *J. Immunol.* 165, 7088–7095.
7. Reijonen, H. and Kwok, W. W. (2003) Use of HLA class II tetramers in tracking antigen-specific T cells and mapping T-cell epitopes. *Methods* 29, 282–288.
8. James, , E. A., Bui, J., Berger, D., Huston, L., Roti, M., and Kwok, W. W. (2007) Tetramer-guided epitope mapping reveals broad, individualized repertoires of tetanus toxin-specific CD4+ T cells and suggests HLA-based differences in epitope recognition. *Int. Immunol.* 19, 1291–1301.
9. Yang, J., James, E. A., Huston, L., Danke, N. A., Liu, A. W., and Kwok, W. W. (2006) Multiplex mapping of CD4 T cell epitoppes using class II tetramers. *Clin. Immunol.* 120, 21–32.
10. Weichold, F. F., Mueller, S., Kortsik, C., Hitzler, W. E., Wulf, M. J., Hone, D. M., Sadoff, J. C., and Maeurer, M. J. (2007) Impact of MHC class I alleles on the *M. tuberculosis* antigen-specific CD8+ T-cell response in patients with pulmonary tuberculosis. *Genes Immun.* 8, 334–343.
11. Novak, E. J., Liu, A. W., Gebe, J. A., Falk, B. A., Nepom, G. T., Koelle, D. M., and Kwok, W. W. (2001) Tetramer-guided epitope mapping: Rapid identification and characterization of immunodominant CD4+ T-cell Epitopes from complex antigens. *J. Immunol.* 166, 6665–6670.
12. Lagrelius, M., Jones, P., Franck, K., and Gaines, M. (2006) Cytokine detection by multiplex technology useful for assessing antigen specific cytokine profiles and kinetics in whole blood cultured up to seven days. *Cytokine* 33, 156–165.
13. Gody-Ramirez, K., Franck, K., Mahdavifar, S., Andersson, L., and Gaines, H. (2004) Optimum culture conditions for specific and non-specific activiation of whole blood and PBMC for intracellular cytokine assessment by flow cytometry. *J. Immunol. Methods* 292, 1–15.
14. Gauduin, M. C., Kaur, A., Ahmad, S., Yilma, T., Lifson, J. D., and Johnson, P. (2004) Optimization of interacellular cytokine staining for the quantitation of antigen-specific CD4+ T cell responses in rhesus macaques. *J. Immunol. Methods* 288, 61–79.

Chapter 32

Identification and Validation of T-Cell Epitopes Using the IFN-γ ELISPOT Assay

Markus Wulf, Petra Hoehn, and Peter Trinder

Summary

The quest for new and enhanced vaccines requires improved means for identification of relevant antigens and the epitopes present within these. While CD8+ T cells recognize antigenic peptides presented in the context of MHC Class I, and play a key role in cellular immunity, CD4+ helper T-cells recognize antigens in the context of MHC Class II and assist other immune cells in orchestration of the defined immune response. Being a functional assay, ELISPOT can be used to define both MHC Class I and II epitopes as well as to validate them. Providing the quality of the cells used in the assay is sufficiently good, ELISPOT provides a solid platform for both screening and validating T-cell epitopes. While some ELISPOT assays rely upon loading of dendritic cells with antigen followed by incubation of HLA matched or autologous PBMC or enriched T cells, the assay presented here uses PBMC from diseased or vaccinated individuals to simultaneously identify and validate T-cell epitopes.

Key words: T-cell epitopes, MHC-binding, CD8+ T cells, CD4+ T cells, Epitope mapping, Epitope validation, ELISPOT.

1. Introduction

The enzyme-linked immunospot, ELISPOT Assay is a laboratory technique in which the secreted products of a cell are specifically and locally captured on the surface of a filter plate and then detected by an enzyme-conjugated detection reagent, usually an antibody, and soluble substrates that give colored, insoluble products. The ELISPOT methodology was originally developed to detect specific-antibody secreting cells, the ELISPOT has been adapted to detect secretion of cytokines through the use of antibody-pairs that are used in a sandwich format, similar to conventional cytokine ELISA assays.

Ulrich Reineke and Mike Schutkowski (eds.), *Methods in Molecular Biology, Epitope Mapping Protocols, vol. 524*

DOI: 10.1007/978-1-59745-450-6_32

1.1. Requirements of Responder T Cells

Shipment and storage conditions are critical for the successful performance of the ELISPOT assay. The assay measures the frequency of cytokine-secreting cells, T cells have to be both functional and viable, i.e., able to respond to stimulation during the assay. Because it is often preferable to batch serially collected specimens for assay in order to eliminate interassay variability, the use of cryopreserved and thawed mononuclear cells for ELISPOT assays has become common and is strongly advised. Comparisons between fresh and cryopreserved cells responding to stimulation by IFN-γ secretion have shown that ELISPOT assays can be reliably used for testing of banked lymphocytes. However, the methods of cryopreservation and thawing must be carefully controlled to ensure that viability is at least 80% and cell recovery is above 70%. It is crucial that cell viability is as high as possible, since a high rate of dead or dying cells may produce false-negative results, particularly if low frequency events in antigen-reactive T cells are evaluated and since cell debris from dead cells may negatively interfere with the interaction of antigen-specific and MHC-restricted T-lymphocytes (*1, 2*).

1.2. Principle

The enzyme-linked immunospot (ELISPOT) assay, initially used to detect antibody secreting cells has been adapted to enumerate both CD4+ and CD8+ T cells secreting specific cytokine(s) in response to an antigenic or mitogenic stimulus. Although the ELISPOT assay can be formatted to detect a variety of different cytokines, currently the widest application is the interferon gamma (IFN-γ) based assay. For this reason, the procedures comprising this section will refer to reagents specific for the IFN-γ ELISPOT assay, although the more general aspects of the technology can be applied more extensively to encompass detection of other cytokines. By utilizing a short (i.e., <24 h) stimulation time, the assay is structured toward enumeration of antigen-primed T cells, in contrast to naïve T cells that require 3–5 days of antigenic stimulation prior to cytokine secretion (and rely upon presentation via dendritic cells).

The general principle of the ELISPOT assay involves plating responder cell populations into wells of a 96-well nitrocellulose or PVDF plate that has been precoated with an anti-cytokine antibody. Specific MHC Class I and/or II antigens, stimulator cells, or nonspecific mitogens are added to tubes with responder cells and incubated for defined periods, following which the stimulated cells are transferred to the ELISPOT plates (in quadruplicate) and are incubated for a further 16–20 h. Any cytokine produced in response to the stimulation will be captured in the local environment of the cytokine-secreting cell by the membrane-bound antibody. All cells are subsequently removed, and the captured cytokine is detected using a biotinylated, cytokine-specific detection antibody followed by addition of an avidin-conjugated enzyme. Visualization of the resulting complex is achieved through addition of an appropriate enzyme substrate, producing a colored spot representing an individual cell that secreted the specific cytokine. The number of cytokine-secreting cells present within the population of plated responders can then be calculated. The ELISPOT assay has been reported to be capable of detecting one IFN-γ secreting cell per 100,000 cells plated (ten cytokine-secreting cells/million cells) (*3, 4*).

The ELISPOT assay described below is a highly optimized procedure that has been applied to the measurement of antiviral and antitumor immunogenicity. The inclusion of a panel of positive and negative controls with each assay is essential to the maintenance of high performance standards. The most commonly encountered problem in the ELISPOT assay is sporadic occurrence of high background values. The potential causes of high background can involve various aspects of sample preparation as well as changes in specific lots of reagents used in the assay. A carefully planned strategy for identifying the sources of assay variability is essential to the successful application of this highly sensitive assay technology. In our laboratory, only screened batches of heat inactivated human serum which yields low background staining may be utilized in the ELISPOT assay.

2. Materials

1. Laminar flow hood.
2. Incubator (37°C and 5% CO_2).
3. Centrifuge (standard low-speed (up to 500 × *g*), benchtop model).
4. 96-well plate washer (ELX50 Plate Washer, BioTek Instruments, Winooski, VT).
5. AID ELISPOT Reader ELR02 and Analysis Software AID, Strassberg, Germany.
6. 96-well plates, sterile plates with high protein binding capacity. PVDF membrane (Millipore S2EM004M99, Molsheim, France).
7. Repeating pipettors.
8. Sterile tips for repeating pipettors (sterile, 2.5 mL and 5.0 mL).
9. Pipettors.
10. Sterile micropipette tips.

11. Conical, polypropylene centrifuge tubes (sterile, 15 mL and 50 mL (Falcon or Greiner).
12. Pipettes (sterile, 1 mL, 2 mL, 5 mL, 10 mL, and 25 mL).

3. Methods

3.1. Specimen and Sample Acceptability

The specimens analyzed using the ELISPOT technology can vary widely. The responder cells may be whole PBMC, CD4+, and/or CD8+ enriched PBMC fractions; cultured PBMC activated by in vitro stimulation (IVS); or cultured T-cell lines. Specific antigenic stimulation can generally be accomplished through the use of synthetic peptides, peptide pools, or peptide-pulsed antigen presenting cells. Other specialized presentation platforms using transduced cells are also possible. The numbers of responder and presenting cells plated per well will vary, depending on their origin. In general, lower numbers of cultured cells are plated compared with fresh responders. All cells must be free of mycoplasma contamination, and their number and viability must be documented. Generally, cell populations with less than 80% viable cells are unacceptable for ELISPOT analysis. Due to the functional nature of this assay, if possible, we recommend using fresh cells within 8 h of blood draw. Frozen cells are ideally from blood draws processed and frozen within an 8-h window (*see* **Note 1**).

Each sample is potentially infectious; pay attention to the safety precautions promoted by your laboratory and carefully follow the appropriate SOPs, worksheets, and safety instructions.

3.2. Preparation of Cells

Specific antigen stimulation is accomplished through the addition of peptides. For peptide stimulation, individual or pooled synthetic 9-mer (8–10mers) peptides of >80% purity are used at a final (in well) concentration of 10 μg/mL for MHC Class I responses (100 μL is used/well). 15–20mers (20 μg/mL) are used to detect MHC Class I and II-restricted antigen responses. Appropriate controls for peptide-stimulated assays include cells alone, peptide(s) alone, and assay medium alone. Note that the antigen-specific response begins to lose linearity at cell concentrations below 50,000 cells per well due to the limited numbers of APCs. The addition of exogenous stimulator cells (antigen-pulsed APC) can usually overcome the divergence from linearity at low responder cell concentrations but is associated with other events (stimulation of naïve T cells, mixed leukocyte reaction if not completely HLA matched or autologous).

When cryopreserved cells such as PBMC are used in the ELISPOT assay, thawed cells are cultured overnight at 37°C (5% CO_2) in a polypropylene 50-mL centrifuge tube containing 2 ×

10^6 cells/mL in assay medium. The purpose of this step is to permit those cells stressed during the cryopreservation and thawing process to undergo apoptosis. The viability assessment after this brief culture period will be more accurate than one performed immediately post-thaw, and will improve the overall sensitivity and precision of the ELISPOT assay (*see* **Note 2**).

3.3. Day 1: Stimulation of Cells

(Freshly isolated PBMC used immediately or PBMC thawed and incubated overnight prior to use.)

1. Label tubes with sample ID and also with antigen ID.
2. Place 20 µL of peptide (10/20 µg/mL end concentration (*see* **Note 3**) in a 15-mL Falcon (1 peptide/tube). Add 1 × 10^6 cells in 1 ml of complete HL-1 medium (with 5% human AB serum) to each peptide tube (1:50 dilution of peptide).
3. Incubate at 37°C, 5% CO_2 for 48 h.

3.4. Day 2: Coat ElISPOT Plates (Aseptic Conditions Under Laminar Flow)

1. Dilute primary (unlabelled) anti-IFNγ antibody in PBS (final concentration 4 µg/mL) (*see* **Note 4**).
2. Briefly pre-wet PVDF ELISPOT plates with ethanol and flick dry.
3. Add 100 µL diluted anti-IFNγ antibody/well into ELISPOT plates.
4. Incubate overnight at 4°C.

3.5. Day 3: Block ElISPOT Plates and Incubate with Cells (Aseptic Conditions)

1. Wash plates 6× with sterile PBS, 200 µL/well.
2. Block empty sites on PVDF with 1% BSA in PBS for 1 h at 37°C.
3. Add 3 mL of warm (37°C) HL-1 medium (serum-free) to each tube from **Subheading 3.1**.
4. Centrifuge 300g rpm/10 min.
5. Carefully discard medium from each tube, gritch the cells (disrupt pellet), and resuspend cells in 400 µL of warm complete (with AB serum) HL-1 medium.
6. Discard the blocking media from the plates.
7. Add 100 µL of resuspended cells to each well of the ELISPOT plate, carefully labelling the plate for peptide/cell positioning. Quadruplicate wells (*see* **Notes 5 and 6**).
8. Incubate plates for 16 h at 37°C in a CO_2 incubator (*see* **Note 7**).

3.6. Day 4: Removal of Cells and Addition of Secondary Antibody (Detection)

1. Wash plates with sterile ice-cold water (300 µL/well).
2. Add 300 µL/well fresh ice-cold water and incubate on ice for 10 min.
3. Wash ELISPOT plates six times with 200 µL/well PBS/Tween 20.

4. Blot plates on absorbent paper to dry.
5. Dilute secondary antibody (biotin-conjugated anti-IFNγ) to 2 μg/mL in PBS with 1% BSA.
6. Add 100 μL per well.
7. Incubate 2 h at 37°C or overnight at 4°C.
8. Dilute Extravidin-AP 1:1,000 in PBS with 1% BSA (10 mL/plate).
9. Wash plates six times with 200 μL/well PBS/Tween 20.
10. Add 100 μL Extravidin-AP/well and incubate at 37°C for 1.5 h.
11. Prepare developer (BCIP/NBT): one tablet BCIP/NBT in 10 mL deionized water.
12. Wash plates four times with 200 μL/well PBS/Tween/1% BSA.
13. Wash plates four times with 200 μL/well PBS.
14. Blot plates on absorbent paper to dry.
15. Add 100 μL BCIP/NBT solution/well.
16. Incubate until dark blue spots are visible (ca. 1 min) (*see* **Note 8**).
17. Discard substrate solution.
18. Wash the plates under running water to stop the reaction.
19. Blot plates on absorbent paper to dry.
20. Allow the plates to dry overnight.

Count spots using the AID ELISPOT Reader (Model ELR02; see user manual, www.elispot.com). The reader should be preset for use, to ensure reproducibility. Alternatively, a CTL Immunospot Reader (www.immunospot.com) can be use. The information (*see* **Subheading 3.5**) on count settings is for the AID ELR02 Reader and may differ for other readers.

3.7. Count Settings

There are a wide variety of different ELISPOT assays, and even performing the same kind of assay two labs will only rarely get exactly the same characteristics of spots in their respective plates. It is therefore necessary to predefine settings and to use them while interpreting ELISPOT assays (*see* **Notes 9 and 10**). This is possible under the Count settings dialog box, where settings can be edited, deleted, and stored under selected names. To create new settings press the "New" button. Settings can be selected from existing ones by marking the source settings from the list before accessing the "New" button. Step one: Designate a name for the new setting. The new settings will appear in the list box. (This may require closing and opening of the dialog box). If no particular settings from the list are selected, the default settings will be used. The "Edit" button permits modifications in the new settings at this stage. This is possible by entering the dialog box and pressing the "Save" button.

Adjustments can be made on the emphasis function and on the size, intensity, and gradient of the spots. Select Minimum and Maximum values for spots. Enable "Min" to count all spots with the same or higher intensity, size, or gradient than the selected value. "Min" is selected as a default. Enable "Max" to count all spots that have a lower value than selected. Use "Min" and "Max" to define a range of objects to be counted as spots.

3.8. Defining Positive Responses

As with all immunoassays, a positive ELISPOT response is defined in terms of the relationship between the numbers of spot-forming cells in experimental wells vs. those appearing in corresponding background wells (*see* **Note 9**).

4. Notes

The ELISPOT assay provides a sensitive means of detection and enumeration of antigen-specific T cells in freshly harvested or cryopreserved PBMC. However, the following limitations may limit the assay utility:

1. A loss of precursor T cells during processing or washing steps will negatively affect the assay results (i.e., "false-negative" results). We recommend processing fresh cells within 8 h of blood draw (eg Ficoll® enrichment of PBMC). Frozen cells are ideally from blood draws processed and frozen within an 8-h window.
2. When using cryopreserved and thawed PBMC, the viability should be 80% and the recovery should not be less than 70%.
3. Insufficient antigen/peptide concentration will underestimate the response. An excess of antigen could induce unresponsiveness ("Pro-zone phenomenon").
4. This is an antibody-based assay, and selection of the antibodies is the most crucial step.
5. "Crowding" of cells will compromise spot detection and enumeration.
6. The numbers of total cells/well should be identical in triplicate or quadruple wells.
7. While a 16-h assay is usually optimal for IFN-γ, the IL-5 ELISPOT requires 48 h. Optimal performance has to be predetermined for every cytokine/cell product measured.

8. It is necessary to optimize color development. Overdeveloped color will interfere with the spot counts.
9. High number of background spots makes the assay noninterpretable. It is essential to prescreen human AB serum batches prior to use to minimize nonspecific background effects.
10. When the frequency of antigen-responsive T cells is lower than 1/50,000, ELISPOT results may not be reliable. Amplification of the response may be necessary, using in vitro sensitization (IVS).
11. Typical images of ELISPOT wells and examples of the artifacts that may occur can be found on the websites of the main ELISPOT Reader manufacturers:
    - AID: www.elispot.com
    - CTL: www.immunospot.com

References

1. Bull, M., Lee, D., Stucky, J., Chiu, Y.-L., Rubin, A., Horton, H., and McElrath, M. J. (2007) Defining blood processing parameters for optimal detection of cryopreserved antigen-specific responses for HIV vaccine trials. *J. Immunol. Methods* 322, 57–69.
2. Maecker, H. T., Moon, J., Bhatia, S., Ghanekar, S. M., Maino, V. C., Payne, J. K., Kuus-Reichel, K., Chang, J. C., Summers, A., Clay, T. M., Morse, M. A., Lyerly, H. K., De La Rosa, C., Ankerst, D. P., and Disis, M. L. (2005) Impact of cryopreservation on tetramer, cytokine flow cytometry, and ELISPOT. *BMC Immunol.* 6, 17–30.
3. Janetzki, S., Panageas, K. S., Ben-Porat, L., Boyer, J., Britten, C. M., Clay, T. M., Kalos, M., Maecker, H. T., Romero, P., Yuan, J., Kast, W. M., and Hoos, A. (2008) Results and harmonization of guidelines from two large-scale international Elispot proficiency panels conducted by the Cancer Vaccine Consortium (CVC/SVI). *Cancer Immunol. Immunother.* 57, 303–315.
4. Janetzki, S., Cox, J. H., Oden, N., and Ferrari, G. (2006) Standardization and validation issues of the ELISPOT assay. *Methods Mol. Biol.* 302, 51–86.

Index

A

ABCpred, in B-cell epitopes 337–339
α-BTX. *See* α-bungarotoxin
α-bungarotoxin 41
Active peptide mixtures, deconvolution
process of 206–207
Adaptive threshold method, advantage of 271
ADs. *See* Antigenic determinants
African swine fever virus 316
capsid protein p72 of 317
Alexa detection system, in IgE binding detection .... 267. *See also* Allergy diagnosis, microarrayed allergen molecules in
Allergy diagnosis, microarrayed allergen
molecules in 259–261
fluorescent labeled secondary antibody,
preparation of 262–263
immunolabelling with patient serum 265–267
materials in 261–262
microarray scanning and slide alignment 267–268
peptide microarrays, printing of 263–265
Amino acid residues, chemical modification of 123
Amino-oxy-acetylated peptides, role of 249
Anti-acetyllysine antibody, profiling of 173–174
Antibodies
affinity maturation 30–33
antigen cross-reactivity of 27–29
screening and characterization of 68–69
structure of 23–26
Antibody-antigen complexes
analysis of 6
epitope mapping by NMR spectroscopy 37–38
dynamic filtering approach 38–41
^{1}H-^{15} N HSQC peaks intensities, epitope
mapping 42–43
^{1}H-^{15}N NOE measurements, epitope
mapping by 44–45
materials in 47–48
methodology in 48–54
transverse relaxation time measurement,
epitope mapping by 43–44
$T_{1\rho}$ measurements, in epitope mapping 45–46
epitope mapping by proteolysis of 88–92
materials for 93
methodologies for 93–98
Antibody-antigen interactions
conformational flexibilty of 29–30
structural basis of 23–26
Antibody-binding assays .... 298–300. *See also* Epitope mapping, homolog-scanning mutagenesis in
Antibody-binding phage, affinity selection of
materials of 184–185
methodologies for 188–192
Antibody-coated plates preparation. *See also* Solid-phase mutual competition assay, for MAbs epitope specificity determination
materials for 62
methodologies for 63
Antibody epitope mapping, *de novo*
approaches in 203–205
combinatorial peptide libraries 204–208
random generated peptide libraries 208–210
Antibody epitope mapping, SPOT™ peptide
arrays in materials for 149–153
methodologies for
antibody binding detection 162–163
library design 154–157
N-terminus acetylation 159–160
peptide array, reutilization of 163–165
peptide arrays membrane and array
formatting 153–154
peptide arrays screening 161–162
peptide synthesis 157–159
side-chain deprotection and automated
SPOT™ synthesis 160–161
Antibody Fv expression and purification. *See also* Antibody-antigen complexes
materials used in 47
methodologies in 48–50
Antibody signatures, identification of 274
Antibody specificity 12
profiling, functional protein microarrays in 213–215
materials for 215–217
methodologies for 217–221
Anti-CAMKII antibody, protein microarray
specificity profiling for 220
Anti-cyclophilin18.1 antiserum and human
cyclophilin 33, 229
Antigen. *See also* Epitope mapping
biotinylation (*see also* Solid-phase mutual competition assay, for MAbs epitope specificity determination)

Antigen. (*Continued*)
materials for ... 62
methodologies for ... 63
cross-reactivity of ... 27–29
proteolysis of
materials for ... 124–125
methodologies for ... 127
Antigenic cross-reactivity ... 10
Antigenic determinants ... 341
residues identification, chemical modification in ... 104–105
Antigenicity, of proteins ... 13
Antigen, purification of ... 122
Antigen recognition, binding properties of ... 26–27
Antigens and immunogens, difference of ... 13–15
Anti-idiotopic antibodies ... 28
Anti-phosphotyrosine antibody, profiling of ... 170–172
Antiprotein antisera, antibodies in ... 5
ASFV. *See* African swine fever virus
Autoimmune disease, epitope signatures and peptide microarrays in ... 247–248
data management and visualization ... 254–255
image aquisition ... 252
materials used in ... 248–250
peptide microarray, handling of ... 250–251
peptide microarray, staining of ... 251–252
perspectives of ... 255–256
primary data analysis ... 252–253

B

Bacille Calmette-Guérin ... 372
Bacillus subtilis ... 226
BamHI–EcoRI gene ... 296
B-cell epitope mapping ... 137
B-cell epitopes ... 3–4, 14
B cell response monitoring, microarrays technology in ... 273–275
BCG. *See* Bacille Calmette-Guérin
BCIP. *See* 5-bromo-1-chloro-3-indolyl phosphate
Biacore™ instrument, application of ... 67
Biacore systems ... 67–68
Biological specificity, definition of ... 11
Biotinylated peptides, synthesis of. *See also* Linear B-Cell epitope mapping, enzyme-linked immunosorbent assay in
materials for ... 139
methodologies for ... 140–142
Bovine serum albumin ... 230
Brefeldin A, role of ... 433
5-bromo-1-chloro-3-indolyl phosphate ... 295
BSA. *See* Bovine serum albumin

C

Calmodulin kinase II ... 220
CAMKII. *See* Calmodulin kinase II
Cancer patient
data analysis and practical considerations ... 277–279
disease signatures imprinted in ... 279–282
microarrays technology, in B cell response monitoring ... 273–275
phage-displayed B cell epitopes, verification methods of ... 279
phage-displayed peptide
immunnoselection of ... 276–277
microarrays, characteristics of ... 275–276
microspots ... 277
technology, for protein detection ... 275
CAPS. *See* 3-cyclohexylamino-1-propanesulfonic acid
Carcinoembryonic antigen ... 59
CD1 molecules, in glycolipd antigens ... 352–353. *See also* T-cell receptors
CDRs. *See* Complementarity-determining regions
CD8+ T-cell antigens, in *Mycobacterium tuberculosis* ... 370–371
CD8+ T-cell clones and PBMCs ... 376. *See also* *Mycobacterium tuberculosis,* T-cell epitope mapping in
CD4+ T-helper cells, role of ... 417
CD8+ T-lymphocytes, role of ... 369
CE. *See* Conformational epitopes
CEA. *See* Carcinoembryonic antigen
Cellular immune responses, role of ... 427
CEP server, role in epitope conformation ... 341
Chemical modification. *See also* Epitope mapping
of amino acid residues ... 123
of antigen, epitope mapping
materials for ... 124
methodologies for ... 125–127
application of ... 104–105
of proteins, reagents for ... 106–109
Citrate/phosphate buffer ... 63
Combinatorial peptide libraries. *See also* Antibody epitope mapping, *de novo* approaches in
classification of ... 205–208
usage of ... 204
Compact reaction columns ... 93
Complementarity-determining regions ... 24, 348
Composite pixel intensity ... 423
Conformational and discontinuous epitopes, characterization of ... 122
Conformational epitopes ... 341. *See also* Discontinuous epitope, of protein
Continuous epitope, of proteins ... 4
fuzzy boundaries of ... 10
Conventional Ab2 antibodies ... 28
CPB. *See* Citrate/phosphate buffer

CPI. *See* Composite pixel intensity
CRC. *See* Compact reaction columns
Cross-protective immunogenicity 14
Cross-reactivity determination, peptide
microarrays in 225–229
materials in 229–230
methods used in 230–232
Cryptotopes, definition of 7
CTLs. *See* Cytotoxic T-lymphocytes
3-cyclohexylamino-1-propanesulfonic acid 81
Cytotoxic T-lymphocytes 417

D

Deconvolution process, of active
peptide mixtures 206–207
Dendritic cells, preparation of .... 375–376.
See also Mycobacterium tuberculosis, T-cell
epitope mapping in
de novo approaches, in antibody
epitope mapping 203–205
combinatorial peptide libraries 204–208
random generated peptide libraries 208–210
Deoxyinosine 5´-triphosphate 306
Deuterium-labeled antigen and antibody,
application of 123
Differential chemical modification,
in epitope mapping 120–124
materials for 124–125
methodologies for 125–131
Dimethyl sulfoxide 263
Discontinuous epitope, of protein 5–6, 16
DiscoTope server, in epitopes prediction 341–342
Disease signatures, imprinted
in patient sera 279–282
Dithiothreitol 302
dITP. *See* Deoxyinosine 5´-triphosphate
DMSO. *See* Dimethyl sulfoxide
DNase I partial digestion, random
fragments in .... 321–322.
See also Phage-display random fragment
libraries, in epitope mapping
DNA sequence analysis .... 311.
See also PCR-mutagenesis,
in epitope mapping region-specified
DTT. *See* Dithiothreitol
Dynamic filtering technique, usages of .... 38–41.
See also Antibody-antigen complexes

E

ECL. *See* Enhanced chemiluminescence
EDTA. *See* Ethylene diamine tetraacetic acid
Electrospray ionization 89
Electrospray ionization mass spectrometry 91
ELISA. *See* Enzyme-Linked ImmunoSorbent Assay
ELISPOT. *See* Enzyme-linked immunospot
Enhanced chemiluminescence 320
Enzyme-Linked ImmunoSorbent Assay 68
in linear B-Cell epitope mapping 137–138
materials for 139–140
methodologies for 140–142
Enzyme-linked immunospot 440, 441
assay, in T-cell epitopes identification
and validation 439–441
cells, preparation of 442–443
coat EliSPOT plates, block EliSPOT
plates and cells stimulation 443
count settings 444–445
materials used in 441–442
secondary antibody, addition of 443–444
specimen and sample acceptability 442
in *Mycobacterium tuberculosis* mapping 377–378
(*see also* T-cell epitope mapping,
in *Mycobacterium tuberculosis*)
Epitope-displaying phage clones,
characterization of .... 326–328.
See also Phage-display random fragment
libraries, in epitope mapping
Epitope mapping
of antibody–antigen complexes,
NMR spectroscopy 37–38
dynamic filtering approach 38–41
^{1}H-^{15}N HSQC peaks intensities,
epitope mapping 42–43
^{1}H-^{15}N NOE measurements, epitope
mapping by 44–45
materials in 47–48
methodology in 48–54
transverse relaxation time measurement,
epitope mapping by 43–44
$T_{1\rho}$ measurements, in epitope mapping 45–46
by differential chemical modification 120–124
materials for 124–125
methodologies for 125–131
homolog-scanning mutagenesis in 289–292
antibody-binding assays 298–300
hybrid genes, generation of 295–297
materials in 293–295
recombinant protein molecules,
production of 297–298
peptide libraries in 237–242
data processing and statistical
analyses 243–244
materials in 242
microarray incubation and signal
read-out 242–243
phage display peptide libraries in 181–183
materials for 183–186
methodologies for 186–198
phage-display random fragment libraries in 315–318

Epitope mapping (*Continued*)
epitope-displaying phage clones, characterization of ... 326–328
library construction ... 321–324
library screening ... 324–326
materials used in ... 318–321
proteolysis of antibody–antigen complexes in ... 88–92
materials for ... 93
methodologies for ... 93–98
proteolytic fragmentation for ... 77–79
materials for ... 79–81
methodologies for ... 81–84
region-specified PCR-mutagenesis in ... 305–308
DNA sequence analysis ... 311
immunoblotting experiments ... 310–311
lambda-gt11 vector, mutagenized DNA in ... 310
materials in ... 308–309
PCR ... 309–310
primers for PCR, design of ... 309
by surface plasmon resonance
analysis of ... 69–72
antibodies screening and characterization of ... 68–69
biacore systems ... 67–68
materials in ... 72–73
methodologies for ... 73–74
Epitope–paratope recognition phenomena ... 12
Epitope signatures, in autoimmune disease ... 247–248
data management and visualization ... 254–255
image aquisition ... 252
materials used in ... 248–250
peptide microarray, handling of ... 250–251
peptide microarray, staining of ... 251–252
perspectives of ... 255–256
primary data analysis ... 252–253
Epitope signatures of IgG antibodies, in human serum ... 247–248
data management and visualization ... 254–255
image aquisition ... 252
materials used in ... 248–250
peptide microarray, handling of ... 250–251
peptide microarray, staining of ... 251–252
perspectives of ... 255–256
primary data analysis ... 252–253
Epitopes of proteins
prediction of ... 15–16
relational entities ... 9–10
types of ... 4–7
Escherichia coli $V3_{MN}$ fusion protein, expression of ... 50–51
ESI. *See* Electrospray ionization
ESI-MS. *See* Electrospray ionization mass spectrometry
Ethylene diamine tetraacetic acid ... 47, 267
in IgE detection ... 267

F

FAST slide, limitations of ... 268
FBS. *See* Foetal bovine serum
FK506-binding proteins ... 226
FKBPs. *See* FK506-binding proteins
Fluorescent labeled secondary antibody, preparation of ... 262–263. *See also* Microarrayed allergen molecules, in allergy diagnosis
Foetal bovine serum ... 431, 433
Food allergy, diagnosis of ... 259
Functional epitope, definition of ... 4
Fv–peptide complex and NMR sample preparation. *See also* Antibody-antigen complexes
materials used in ... 48
methodologies for ... 52

G

GAL. *See* GenePix Array List
GenePix Array List ... 248, 422
in primary data analysis ... 252
GenePix Pro
analysis of tiff files in ... 422–424
(*see also* Major histocompatibility complex)
role of 243
GenePix "Results," 424
GenePix-Results-File ... 253
Genespotter™ software, in epitope mapping ... 243. *See also* Peptide library
Glutathione-S-Transferase ... 215
Glycolipid antigens, TCRs recognition of ... 352–353. *See also* T-cell receptors
GPR. *See* GenePix "Results"; GenePix-Results-File
GST. *See* Glutathione-S-Transferase

H

HA. *See* Hemagglutinin
HEL. *See* Hen egg lysozyme
Heligmosomoides polygyrus ... 238
Hemagglutinin ... 349
Hen egg lysozyme ... 27
Heteroclitic binding. *See* Heterospecificity binding
Heterospecificity binding ... 10–11
Hidden Markov models ... 340
High-content phosphopeptide microarrays, usage of ... 170
High-density human protein microarray technology ... 215
High-performance liquid chromatography ... 78
H-2Kb/S8L detection, flow cytometry in ... 413. *See also* T-cell epitope processing

HLA molecules, peptides binding to .... 424.
See also Major histocompatibility complex
HMM. *See* Hidden Markov models
^{1}H-^{15} N HSQC peaks intensities,
epitope mapping .... 42–43.
See also Antibody-antigen complexes
^{1}H-^{15} N HSQC spectra of Fv–peptide
complex, study of 42
^{1}H-^{15}N NOE measurements, epitope
mapping by .... 44–45.
See also Antibody-antigen complexes
HOHAHA. *See* Homonuclear Hartmann Hahn
Homolog-scanning mutagenesis
in epitope mapping 289–292
antibody-binding assays 298–300
hybrid genes, generation of 295–297
materials in 293–295
recombinant protein molecules,
production of 297–298
Homonuclear Hartmann Hahn 38
Homonuclear spectra, in epitope mapping .... 45–46.
See also Antibody-antigen complexes
Horseradish peroxidase 63, 293, 389
HPLC. *See* High-performance liquid
chromatography
HRP. *See* Horseradish peroxidase
HSA. *See* Human serum albumin
HSM. *See* Homolog-scanning mutagenesis
Human class I binding peptides identification,
iTOPIA™-epitope in 361–364
affinity assay 366–367
binding assay 365
materials used in 364–365
off-rate 365–366
Human cyclophilins, PPIase domains of 228
Human cyclophilin33, western blot analysis of 229
Human MHC class I binding peptides,
identification of 361–364
affinity assay 366–367
binding assay 365
materials used in 364–365
off-rate 365–366
Human proenkephalin 195
Human serum 373
epitope signatures of IgG antibodies in 247–248
data management and visualization 254–255
image aquisition 252
materials used in 248–250
peptide microarray, handling of 250–251
peptide microarray, staining of 251–252
perspectives of 255–256
primary data analysis 252–253
Human serum albumin 262
huPENK. *See* Human proenkephalin
HuS. *See* Human serum
Hybrid genes, generation of .... 295–297.
See also Epitope mapping, homolog-scanning
mutagenesis in
Hydrogen/deuterium exchange (H/D exchange).
See also Epitope mapping
epitope mapping by
materials for 125
methodologies for 127–128
usage of 123
Hydropathic complementarity,
in epitope 8

I

ICS. *See* Intracellular cytokine staining;
Intracellular staining
Idiotope, definition of 28
IFN-γ. *See* Interferon gamma
IFN-γ ELISPOT assay
in antigen detection 428
in T-cell epitopes identification
and validation 439–441
cells, preparation of 442–443
coat EliSPOT plates, block EliSPOT
plates and cells stimulation 443
count settings 444–445
materials used in 441–442
secondary antibody, addition of 443–444
specimen and sample acceptability 442
IgE binding, detection of .... 267.
See also Allergy diagnosis, microarrayed
allergen molecules in
IgE epitope mapping, in allergy
diagnosis 260
IHC. *See* Immunohistochemical
Immobilized antibody column, preparation
methods for 93–94
Immunoglobulin, antibody
nature of 9
Immunohistochemical 214
Interferon gamma 440
Intracellular cytokine staining 427, 430
Intracellular staining 433
Intravenous immunoglobulin 249
In vitro stimulation 442
IPTG. *See* Isopropyl-β-D-thiogalactoside
Isopropyl-β-D-thiogalactoside 47, 294
iTOPIA™-epitope, in human class I
binding peptides identification 361–364
affinity assay 366–367
binding assay 365
materials used in 364–365
off-rate 365–366
IVIG. *See* Intravenous immunoglobulin
IVS. *See* In vitro stimulation

L

Lactacystin 411
Lambda-gt11 vector, mutagenized DNA in .... 310. *See also* Epitope mapping, region-specified PCR-mutagenesis in
LC. *See* Lactacystin
LC/ESI-MS. *See* Liquid chromatography/electrospray ionization mass spectrometry
LCL. *See* Lymphoblastoid cell lines
LDA. *See* Linear discriminant analysis
Limited proteolysis epitope mapping principle 89
Linear and discontinuous epitopes, difference of 148
Linear B-Cell epitope mapping, enzyme-linked immunosorbent assay in 137–138
  materials for 139–140
  methodologies for 140–142
Linear B-cell epitopes, prediction of 335–336
  materials used in 336–337
  sequence information, methods in 337–340
  structural information, methods in 340–342
Linear discriminant analysis 238
  in epitope mapping 243
Linear protein epitope mapping, by MS 90, 120
Liquid chromatography/electrospray ionization mass spectrometry 122
L-1-Tosylamido-2-phenylethyl chloromethyl ketone 79
Lymphoblastoid cell lines 374

M

MAbs. *See* Monoclonal antibodies
MAbs epitope specificity determination, solid-phase mutual competition assay for 60–62
  materials for 62–63
  methodologies for 63–64
MAD. *See* Median absolute deviation
Major histocompatibility complex 347
  class II binding peptides, identification of 417–419
    GenePix Pro, analysis of tiff files in 422–424
    materials used in 420
    slide preparation and incubation 420–421
    slide scanning 421–422
    soluble HLA molecules, peptides binding to 424
  class I ligands, usage of .... 386–387 (*see also* T-cell epitope discovery, MHC peptide exchange in)
  guided epitope mapping, in antigenic epitopes detection 428–430
  monomers, multimerization of exchanged 399–401
  peptide exchange, in T-cell epitope discovery .. 383–387
    conditional ligands, synthesis of 390–391
    materials in 388–390
    MHC monomers, multimerization of exchanged 399–401
    MHC rescue measurement 395–399
    p*HLA A2.1 complexes, purification of 391–394
    UV-mediated peptide exchange 394–395
  rescue, peptide measuring mediated 395–399
MALDI. *See* Matrix assisted laser desorption ionization
MALDI-TOF MS. *See* Matrix-assisted laser desorpion ionization time of flight mass spectrometry
Mapitope strategy 16
Mass spectrometers (MS), in chemical modifications study 121
Mass spectro-metric analysis, methodology for 128–130
MATLAB 7.0, role of 243
Matrix-assisted laser desorpion ionization time of flight mass spectrometry 78–79
Matrix assisted laser desorption ionization 89
MBP. *See* Myelin basic protein
Mean fluorescence intensity 413
Median absolute deviation 268
Membrane-proximal external region 14
MFI. *See* Mean fluorescence intensity
MHC. *See* Major histocompatibility complex
MHC ELISA, in MHC rescue measurement. 395–399. *See also* T-cell epitope discovery, MHC peptide exchange in
Microarrayed allergen molecules, in allergy diagnosis 259–261
  fluorescent labeled secondary antibody, preparation of 262–263
  immunolabelling with patient serum 265–267
  materials in 261–262
  microarray scanning and slide alignment 267–268
  peptide microarrays, printing of 263–265
Microarray incubation and signal read-out .... 242–243. *See also* Epitope mapping, peptide libraries in
Microarray printing instrument. *See also* Allergy diagnosis, microarrayed allergen molecules in
  preparation of 261
  role of 263–264
Microarray scanning and slide alignment .... 267–268. *See also* Allergy diagnosis, microarrayed allergen molecules in
Microarrays technology
  advantages of 225
  in B cell response monitoring 273–275
Mimotope, definition of 7–9, 248
Molecular Devices Axon Instruments GenePix 4000B scanner, role of 252

Monoclonal antibodies....6, 25, 59, 289
MOPS. *See* 3-[*N*-morpholino]propanesulfonic acid
MPER. *See* Membrane-proximal external region
Murine IgM
  peptides sequences of....240
  P-SVM classification of....241
Mutual competition assay....62–64. *See also* Solid-phase mutual competition assay, for MAbs epitope specificity determination
Mutual competition assays, in group D MAbs....61–62
Mycobacterium tuberculosis
  enzyme-linked immunospot in....377–378
  T-cell epitope mapping in....369–373
    cells in....375–377
    elispot....377–378
    materials in....373–375
    peptide library....377
Myelin basic protein....351

N

NanoPrint™ microarray manager software, role of....264
Natural killer T....352
NBT. *See* Nitro blue tetrazolium
Neotopes, definition of....7
Neutralization epitopes....15
NHSB. *See* *N*-hydroxysuccinimidobiotin
N-hydroxysuccinimidobiotin....62
Nitro blue tetrazolium....295
NKT. *See* Natural killer T
3-[*N*-morpholino]propanesulfonic acid....79
NMR spectroscopy. *See* Nuclear magnetic resonance spectroscopy
NOE. *See* Nuclear Overhauser effect
NOE spectrometry....38
NOESY. *See* NOE spectrometry
Novel serum antibody capture molecules, in protein detection....275
^{15}N relaxation, application of....43
Nuclear magnetic resonance spectroscopy
  antibody-antigen complexes epitope mapping by....37–38
    dynamic filtering approach....38–41
    ^{1}H-^{15} N HSQC peaks intensities, epitope mapping....42–43
    ^{1}H-^{15}N NOE measurements, epitope mapping by....44–45
    materials in....47–48
    methodology in....48–54
    transverse relaxation time measurement, epitope mapping by....43–44
    $T_{1\rho}$ measurements, in epitope mapping....45–46
Nuclear Overhauser effect....38, 44

O

OPD. *See* *o*-phenylenediamine
o-phenylenediamine....60

P

Pairwise epitope mapping....69–70
Paraformaldehyde....431
PBMCs. *See* Peripheral blood mononuclear cells
PBS. *See* Phosphate buffered saline
PBST. *See* Phosphate buffered saline with Tween....20
PCA. *See* Principal component analysis
PCR-mutagenesis, in epitope mapping region-specified....305–308
  DNA sequence analysis....311
  immunoblotting experiments....310–311
  lambda-gt11 vector, mutagenized DNA in....310
  materials in....308–309
  PCR....309–310
  primers for PCR, design of....309
PDB. *See* Protein data bank
PEG. *See* Polyethylene glycol
Peptide antigens, recognition of....351–352. *See also* T-cell receptors
Peptide expression and purification. *See also* Antibody-antigen complexes
  materials used in....47–48
  methodologies for....50–51
Peptide library
  in epitope mapping....237–242
    data processing and statistical analyses....243–244
    materials in....242
    microarray incubation and signal read-out....242–243
  Mycobacterium tuberculosis genes in....372 (*see also* T-cell epitope mapping, in *Mycobacterium tuberculosis*)
Peptide–major histocompatibility complexes....349
Peptide microarray
  in autoimmune disease....247–248
    data management and visualization....254–255
    image aquisition....252
    materials used in....248–250
    peptide microarray, handling of....250–251
    peptide microarray, staining of....251–252
    perspectives of....255–256
    primary data analysis....252–253
  in cross-reactivity determination....225–229
    materials in....229–230
    methods used in....230–232
  data, EpiMap dynamic web visualization of....254

handling of 250–251
(*see also* Human serum, epitope signatures of IgG antibodies in)
printing of 263–265
(*see also* Microarrayed allergen molecules, in allergy diagnosis)
staining of 251–252
in state-specific antibodies profiling 170–174
materials for 174–176
methodologies for 176–178
Peptide mimics, phage-display random peptide libraries in .... 315. *See also* Epitope mapping, phage-display random fragment libraries in
Peptides identification, *de novo* approaches in 203–205
Peptide signals, diversity and reproducibility of 257
Peptide, TCRs recognition of. *See also* T-cell receptors
altered-self peptide antigens, recognition of 351–352
foreign peptide antigens, recognition of 349–350
self peptide antigens, recognition of 351
Peptidyl-prolyl *cis/trans* isomerases 226
Peripheral blood mononuclear cells 373
pET vector system, importance of 290
PFA. *See* Paraformaldehyde
PHA. *See* Phytohemagglutinin
Phage-displayed B cell epitopes, verification methods of 279
Phage-displayed peptide, in cancer patient
immunnoselection of 276–277
microarrays, characteristics of 275–276
microspots 277
technology, for protein detection 275
Phage display library amplification
materials of 183–184
methodologies for 186–188
Phage display peptide libraries, in epitope mapping 181–183
materials for
antibody-binding phage, affinity selection of 184–185
phage display library, amplification of 183–184
phage ELISA 185–186
phage sequencing 186
methodologies for
antibody-binding phage, affinity selection of 188–192
phage display library, amplification of 186–188
phage ELISA 192–193
phage sequencing 193
Phage-display random fragment libraries, in epitope mapping 315–318
epitope-displaying phage clones, characterization of 326–328
library construction 321–324
library screening 324–326
materials used in 318–321
Phage ELISA
materials for 185–186
methodologies for 192–193
Phage sequencing
materials for 186
methodologies for 193
P'HLA A2.1 complexes, purification of .... 391–394. *See also* Major histocompatibility complex
Phorbol 12-myristate 13-acetate 433
Phosphate buffered saline 261, 374
Phosphate buffered saline with Tween 20, 261
Phytohemagglutinin 375, 379
Plasmodium falciparum 210
PMA. *See* Phorbol 12-myristate 13-acetate
pMHCs. *See* Peptide–major histocompatibility complexes
Polyclonal and monoclonal antibodies, difference of 148
Polyclonal antibody, epitope mapping of 225
Polyethylene glycol 320
Potential support vector machines 238
in epitope mapping 243–244
PPB. *See* Protein printing buffer
PPIases. *See* Peptidyl-prolyl *cis/trans* isomerases
Principal component analysis 243
Protein antigens
epitopes, classification of 88
immunodominant regions determination in 225
Protein data bank 340
Protein detection, phage-displayed peptides technology for 275
Protein microarrays
antibody specificity profiling assay 217–221
antibody specificity profiling of 213–215
materials for 215–217
methodologies for 217–221
specificity profiling, for anti-CAMKII antibody 220
usage of 274
Protein printing buffer 261
Proteolytic fragmentation, for epitope mapping 77–79
materials for 79–81
methodologies for 81–84
Proteolytic fragments, HPLC purification of 84
ProtoArray® Protein Microarrays, manufacture of 215
P-SVM. *See* Potential support vector machines

R

Random generated peptide libraries.............182, 208–210. *See also* Antibody epitope mapping, *de novo* approaches in
RecA protein..307
Receiver operating characteristic...........................279, 338
Recombinant INF proteins, antigenic and biological properties of..........................292
Recombinant protein molecules, production of.....................................297–298. *See also* Epitope mapping, homolog-scanning mutagenesis in
Recurrent neural networks...337
RepliTope microarrays, in epitope mapping.................243. *See also* Peptide library
Replitope™ microarrays, role of......................................248
Rhodococcus equi...138
RNN. *See* Recurrent neural networks
ROC. *See* Receiver operating characteristic
ROESY. *See* Rotating-frame Overhauser enhancement spectroscopy
Rotating-frame Overhauser enhancement spectroscopy..38

S

SAGs. *See* Superantigens
SDS-PAGE. *See* SDS-polyacrylamide gel electrophoresis
SDS-polyacrylamide gel electrophoresis.........................78
SEA. *See* Staphylococcal enterotoxin A
SEB. *See* Staphylococcal enterotoxin B
SEC. *See* Staphylococcal enterotoxin C
Shared cross-reactivity...10
Site-specific chemical modification, usage of........104–105
SLE. *See* Systemic lupus erythematosus
SMB. *See* Streptavidin magnetic beads
Solid-phase mutual competition assay, for MAbs epitope specificity determination...60–62
- materials for...62–63
- methodologies for...63–64

Somatic hypermutation process................................30–32
Somatic hypermutations, quantity and cooperativity of.......................................32
SpeA. *See* Streptococcal pyrogenic exotoxin A
SPOT™ peptide arrays, in antibody epitope mapping
- materials for...149–153
- methodologies for
  - antibody binding detection........................162–163
  - library design..154–157
  - N-terminus acetylation.............................159–160
  - peptide arrays membrane and array formatting...153–154
  - peptide arrays screening............................161–162
  - peptide synthesis.......................................157–159
  - reutilization of peptide array.....................163–165
  - side-chain deprotection and automated SPOT™ synthesis...............................160–161

SPOT™ synthesis technique, principle of..............146–148
Staphylococcal enterotoxin A..433
Staphylococcal enterotoxin B................................354, 433
Staphylococcal enterotoxin C..354
Staphylococcus aureus...318, 353
State-specific antibodies profiling, peptide microarrays in.....................................170–174
- materials for...174–176
- methodologies for...176–178

Streptavidin magnetic beads..................295, 300, 320, 324
Streptococcal pyrogenic exotoxin A..............................355
Streptococcus pyogenes..353
Superantigens..353
- TCRs recognition of...353 (*see also* T-cell receptors)
  - T-cell signaling complexes.........................356–357
  - TCR specificity and cross-reactivity..........354–356

SuperNitro slide, limitations of....................................268
Support vector machine..243–244
Surface plasmon resonance, epitope mapping
- antibodies screening and characterization of..................................68–69
- biacore systems...67–68
- epitope mapping analysis....................................69–72
- materials in...72–73
- methodologies for...73–74

SVM. *See* Support vector machine
Systemic lupus erythematosus......................................249

T

Taq. See Thermus aquaticus
TB. *See* Terrific broth
TBS. *See* Tris-buffered saline
T-cell epitope discovery, MHC peptide exchange in...383–387
- conditional ligands, synthesis of......................390–391
- materials in..388–390
- MHC monomers, multimerization of exchanged.......................................399–401
- MHC rescue measurement.............................395–399
- p*HLA A2.1 complexes, purification of..........391–394
- UV-mediated peptide exchange.......................394–395

T-cell epitope mapping...427–431
- intracellular cytokine staining.........................435–436
- materials in..431–433
- in *Mycobacterium tuberculosis*............................369–373
  - cells in...375–377
  - elispot...377–378
  - materials in...373–375
  - peptide library..377
- whole blood assay..434–435

T-cell epitope processing 407–410
flow cytometry values, analysis of 413–414
H-2Kb/S8L detection 413
materials used in 408–409
S8L-vector, transfection with fugene 410–411
transfectants, cell surface of 411–413
T-cell epitopes, identification and validation of 439–441
cells, preparation of 442–443
coat EliSPOT plates, block EliSPOT plates and cells stimulation 443
count settings 444–445
materials used in 441–442
secondary antibody, addition of 443–444
specimen and sample acceptability 442
T-cell receptors 347
anatomy of 348–349
ligands molecular recognition by 347–348
recognition of glycolipid antigens
TCR/glycolipid-CD1 complexes, characteristics of 352–353
recognition of peptide
altered-self peptide antigens, recognition of 351–352
foreign peptide antigens, recognition of 349–350
self peptide antigens, recognition of 351
recognition of superantigens 353
T-cell signaling complexes 356–357
TCR specificity and cross-reactivity 354–356
T-cell receptor Vβ domain, superantigen engagement of 354
TCR/glycolipid-CD1 complexes, characteristics of .... 352–353. *See also* T-cell receptors
TCRs. *See* T-cell receptors
Template-coupled PCR
amplification, hybrid genes generation by 291
definition of 290
Terrific broth 320
Tetramer-based epitope mapping, role of .... 428. *See also* T-cell epitope mapping
TGEV. *See* Transmissible gastroenteritis virus
Thermus aquaticus 312
Thermus thermophilus 306
TMV. *See* Tobacco mosaic virus
T_2 ^{15}N relaxation time, measurements of 55
Tobacco mosaic virus 11
Toxic shock syndrome 353
Toxic shock syndrome toxin-1 355
TPCK. *See* L-1-Tosylamido-2-phenylethyl chloromethyl ketone
TPI. *See* Triosephosphate isomerase
Transmissible gastroenteritis virus 210
Transverse relaxation time measurement, epitope mapping by .... 43–44. *See also* Antibody-antigen complexes
Triosephosphate isomerase 351
Tris-buffered saline 230
$T_{1\rho}$ measurements, in epitope mapping .... 45–46. *See also* Antibody-antigen complexes
True cross-reactivity, definition of 10
TSS. *See* Toxic shock syndrome
TSST-1. *See* Toxic shock syndrome toxin-1
TST. *See* Tuberculin skin test
Tuberculin skin test 373

U

Ultraamp amplification detection system, in IgE detection .... 267. *See also* Allergy diagnosis, microarrayed allergen molecules in

V

VapA protein of *R. equi,* linear B-cell epitopes detection of 138
$V3_{IIIB}$ expression vector, construction of 51
$V3_{IIIB}$ peptide, 1H-^{15}N NOE measurements of 44–45
$V3_{MN}$ fusion protein in *E. coli*, expression of 50–51
$V3_{MN}$ peptide, cleavage and purification of 51

W

WBA. *See* Whole blood assay
Whatman slide, limitations of 268
Whole blood assay .... 427. *See also* T-cell epitope mapping
in T-cell epitope mapping 434–435

X

X-ray diffraction analysis, of antigen–antibody complex 88

Color Plates

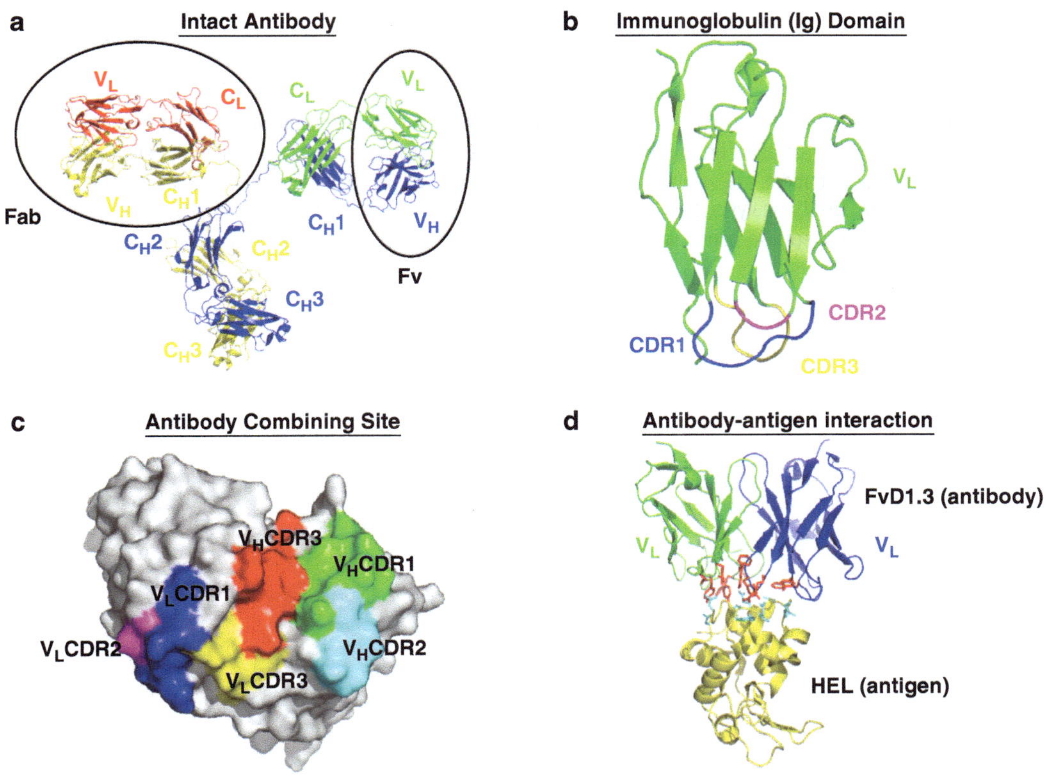

Chapter 2, Fig. 1. Structural overview of antibodies. (**a**) Structure of the intact murine IgG2a monoclonal antibody, Mab231 *(2)*, including two light chains, each composed of a variable (V_L) and a constant (C_L) immunoglobulin (Ig) domain (*red* and *green*), and two heavy chains, each composed of a variable (V_H) and three constant (C_H1, C_H2, and C_H3) domains (*blue* and *yellow*) *(3)*. The common fragments, Fab (*large oval*) and Fv (*small oval*), are indicated. (**b**) Ribbon diagram of a single Ig domain, V_L, of Mab231 highlighting its antiparallel β-sheet secondary structure. The complementarity-determining region loops are marked, CDR1 (*blue*), CDR2 (*magenta*), and CDR3 (*yellow*). (**c**) Molecular surface of the antibody-combining site of Mab231 formed by the intersection of the apical regions of V_L and V_H. The CDR loops provide a contiguous surface for antigen recognition. Colors are as follows: V_LCDR1 (*blue*), V_LCDR2 (*magenta*), V_LCDR3 (*yellow*), V_HCDR1 (*green*), V_HCDR2 (*cyan*), V_HCDR3 (*red*). (**d**) Ribbon diagram of the FvD1.3–hen egg lysozyme (HEL) antibody–antigen complex. Colors are as follows: HEL (*yellow*), D1.3 V_L domain (*green*), and D1.3 V_H domain (*blue*). Residues of HEL and D1.3 involved in interactions in the antigen–antibody interface are *cyan* and *red*, respectively.

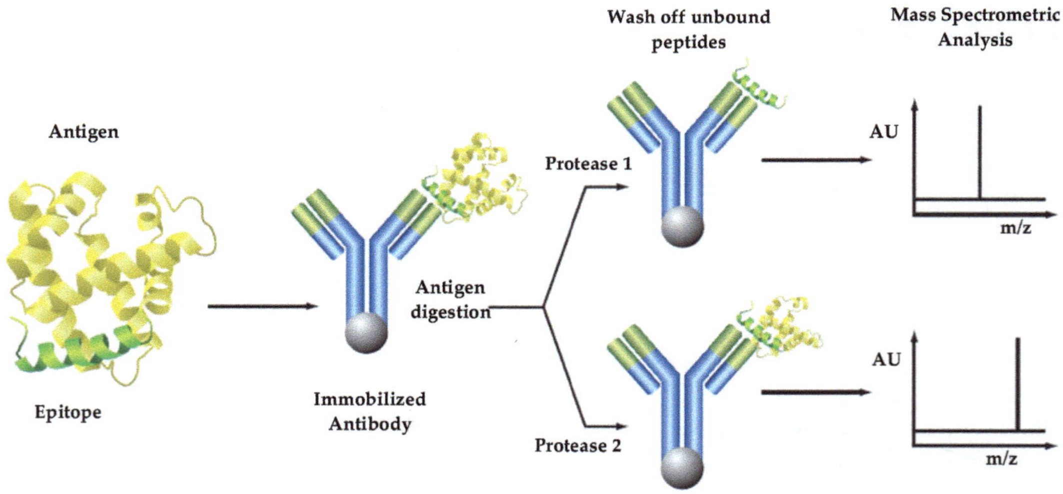

Chapter 7, Fig. 1. Summary of the epitope excision process. The antigen is bound to the immobilized antibody, followed by digestion with various proteases. Protease 1 and protease 2, which cleave the antigen at different sites, are used for the digestion. Unbound peptides are washed off the column and the affinity-bound peptides are analyzed using mass spectrometry. For MALDI-MS analysis, the immobilized antibody–epitope complex can be directly analyzed, while the ESI-MS requires elution of epitope-containing peptides from the immune complex prior to analysis.

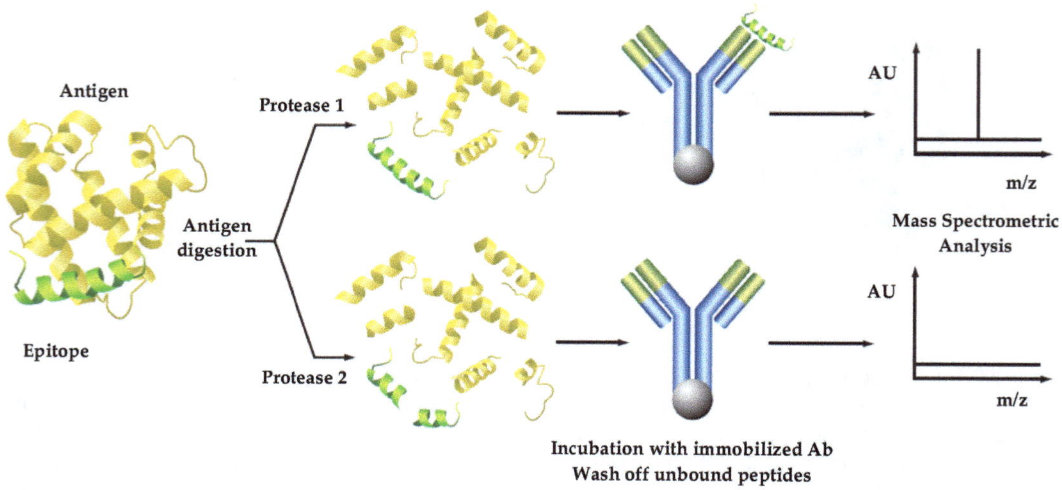

Chapter 7, Fig. 2. Summary of the epitope extraction process. Protease 1 and protease 2, which cleave the antigen at different sites, are used to digest the antigen. The antigen digest is passed through separate columns containing immobilized antibody. The antigenic determinants of the antigen remain bound to the antibody, while the unbound peptides are washed off from the column. The affinity-bound peptides are analyzed using mass spectrometry. For MALDI-MS analysis, the immobilized antibody–epitope complex can be directly analyzed, while the ESI-MS requires elution of epitope-containing peptides from the immune complex prior to analysis.

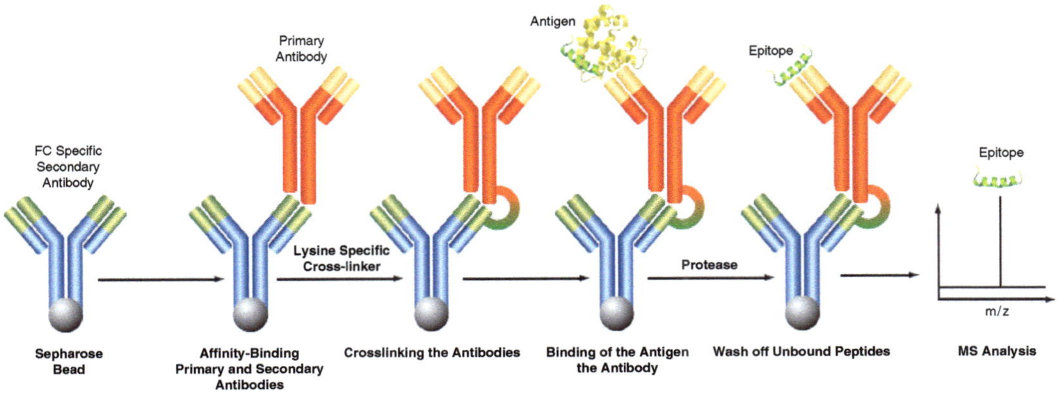

Chapter 7, Fig. 3. Summary of the indirect immunosorption procedure, followed by the epitope mapping process. Immobilization of a secondary Fc-specific antibody on Sepharose is followed by indirect immobilization of the primary antibody (i.e. antibody of interest). The assembly of the primary and the secondary antibodies is followed by cross-linking with a lysine-specific cross-linker. The antigen is bound to the immobilized antibody and digested with protease. For MALDI/MS analysis, the immobilized antibody–epitope complex can be directly analyzed, while the ESI/MS requires elution of epitope-containing peptides from the immune complex prior to analysis.

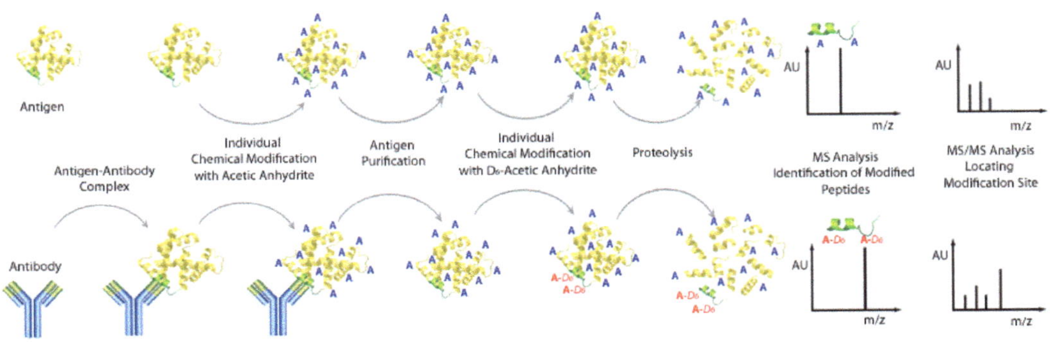

Chapter 9, Fig. 1. General approach for characterizing conformational and discontinuous epitopes by differential chemical modification of antigen and mass spectrometry. Free antigen and antigen–antibody complex are individually subjected to chemical modification. Modified antigens (free and antibody-bound) are purified and proteolytically digested. Peptides are analyzed by MS.

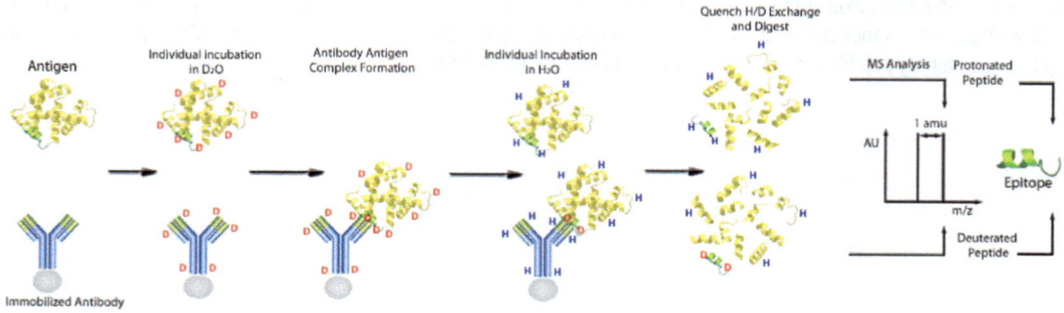

Chapter 9, Fig. 2. General approach for characterizing conformational and discontinuous epitopes by H/D exchange and mass spectrometry. Antigen and antibody are individually labeled with deuterium by incubating in D_2O. Immune complex is formed using deuterated antigen and antibody. Back exchange is performed on the free antigen and the immune complex. Following quenching of the exchange reaction, the antigen is proteolytically cleaved using immobilized pepsin at pH 2.25 and the peptides are analyzed by MS.

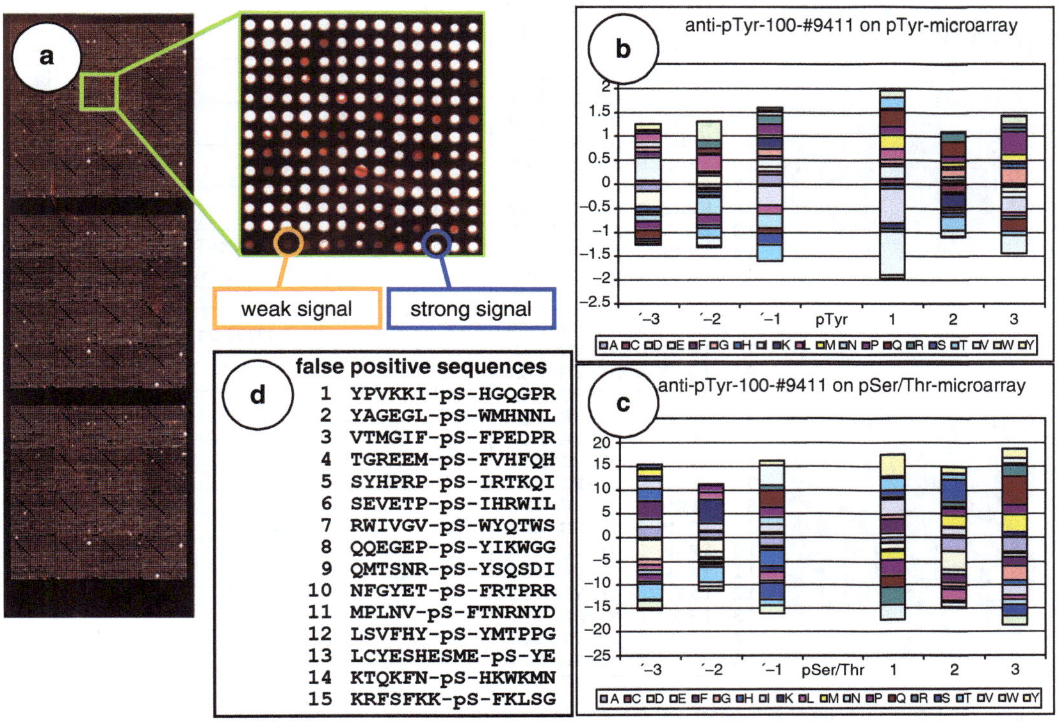

Chapter 12, Fig. 1. Profiling of anti-phosphotyrosine antibody. (**a**) Fluorescence image of peptide microarray displaying 6,262 phosphotyrosine-containing peptides derived from annotated human phosphorylation sites together with background controls and fluorescent landmarks. Three identical subarrays result in 3 × 6,262 = 18,786 data points. Microarray was incubated with anti-phosphotyrosine antibody followed by fluorescently labeled secondary antibody. *Enlarged image* shows strong signals (*white color* equal to saturated signal at used laser settings) and weak signals (*red color*). (**b**) Statistical analysis of detected strong binding phosphotyrosine-epitopes is shown. Probability of each amino acid residue relative to the phosphotyrosine moiety is compared with the probability of each residue relative to phosphotyrosine in all displayed peptides. If residue is preferred in the 5,665 strong binding phosphopeptide value is positive and proportional to preference. If residue is disfavored at given position values will be negative. Color code for amino acid residues is given in the legend of the graph. (**c**) Statistical analysis of detected strong binding phosphothreonine-epitopes is shown. Probability of each amino acid residue relative to the phosphothreonine moiety is compared with the probability of each residue relative to phosphothreonine in all displayed peptides. If residue is preferred in the 93 strong binders value is positive and proportional to preference. If residue is disfavored at given position values will be negative. Color code for amino acid resides is given in the legend of the graph. (**d**) Sequences of top 15 false-positive binders are given. Aromatic residues in +1 position relative to the phosphoserine are a hallmark of detected cross-reactivity.

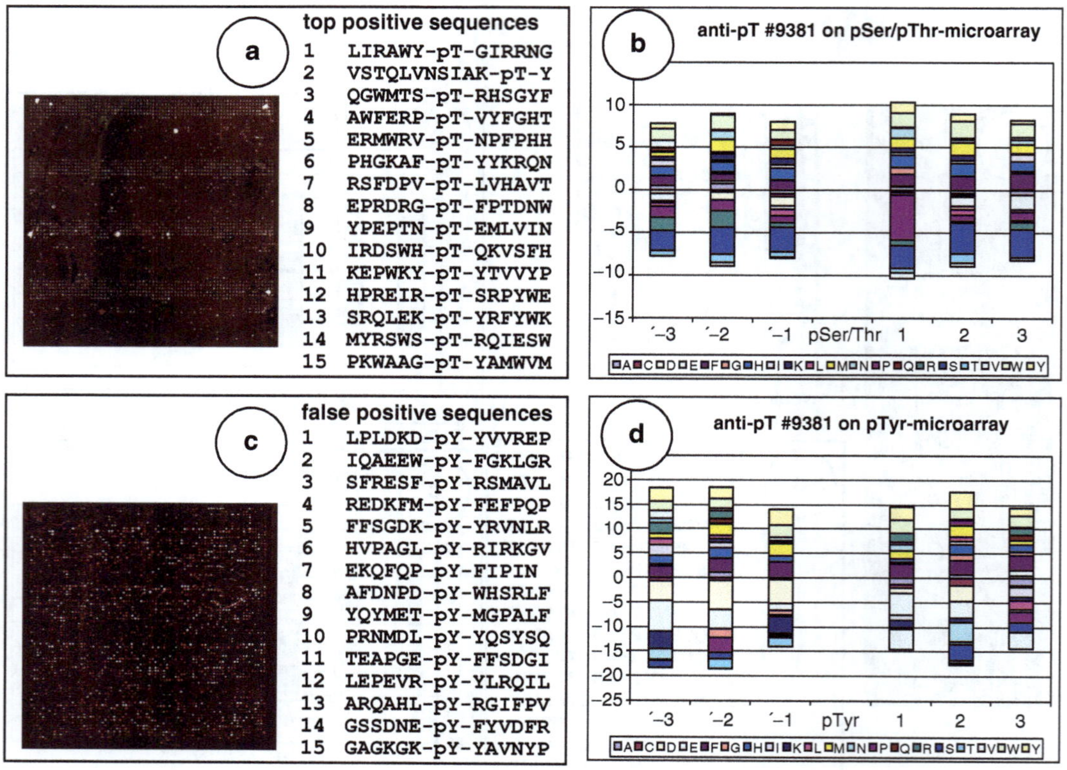

Chapter 12, Fig. 2. Profiling of anti-phosphothreonine antibody. (**a**) Fluorescence image of peptide microarray displaying 2,034 phosphoserine-containing peptides and 2,159 phosphothreonine-containing peptides both derived from annotated human phosphorylation sites together with background controls and fluorescent landmarks. One of the three identical subarrays is shown. Microarray was incubated with anti-phosphothreonine antibody followed by fluorescently labeled secondary antibody. *White color* indicates saturated signal at used laser settings and *red color* represents epitopes with weaker affinity. Sequences of top 15 identified phosphothreonine-epitopes are given. (**b**) Statistical analysis of detected strong binding phosphothreonine-epitopes is shown. Probability of each amino acid residue relative to the phosphothreonine/phosphoserine moiety in the 1,739 binders is compared with the probability of each residue relative to phosphothreonine/phosphoserine in all displayed peptides. If residue is preferred value is positive and proportional to preference. If residue is disfavored at given position values will be negative. Color code for amino acid residues is given in the legend of the graph. (**c**) Fluorescence image of peptide microarray displaying 6,262 phosphotyrosine-containing peptides derived from annotated human phosphorylation sites together with background controls and fluorescent landmarks. One of the three identical subarrays is shown. Microarray was incubated with anti-phosphothreonine antibody followed by fluorescently labeled secondary antibody. *White color* indicates saturated signal at used laser settings and *red color* represents epitopes with weaker affinity. Sequences of top 15 cross-reactive phosphotyrosine-epitopes are given. (**d**) Statistical analysis of detected strong binding cross-reactive phosphotyrosine-epitopes is shown. Probability of each amino acid residue relative to the phosphotyrosine moiety is compared with the probability of each residue relative to phosphotyrosine in all displayed peptides. If residue is preferred in the 1,844 cross-reactive binders value is positive and proportional to preference. If residue is disfavored at given position values will be negative. Color code for amino acid residues is given in the legend of the graph.

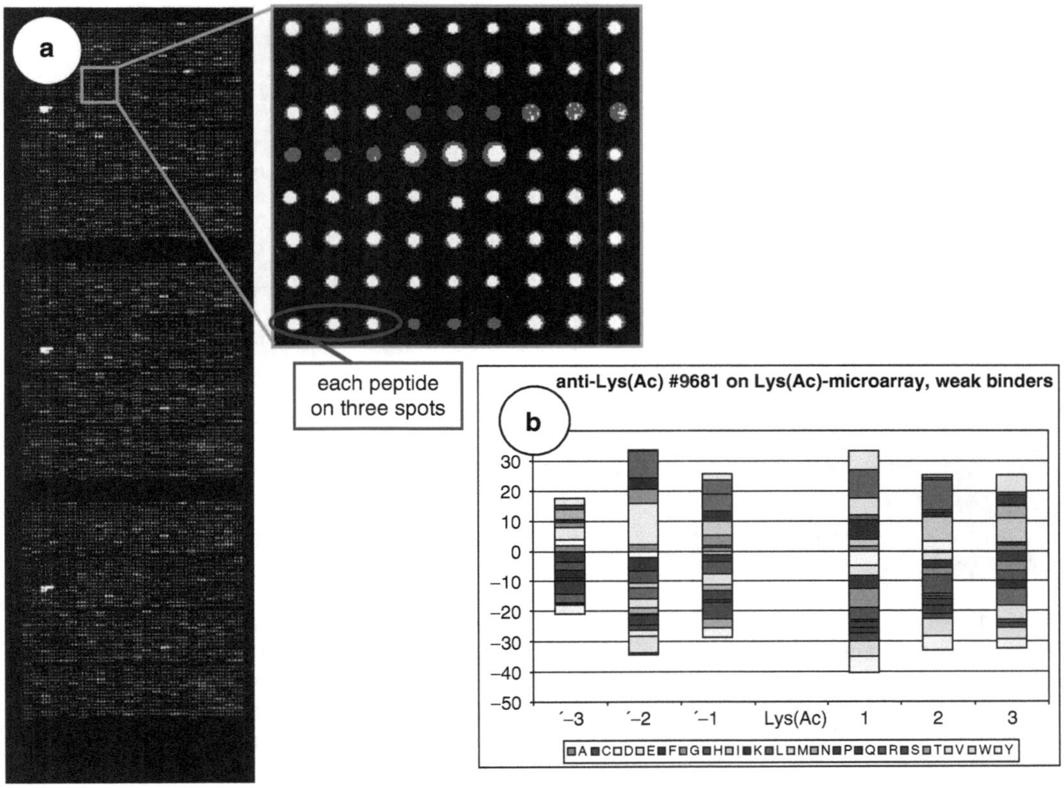

Chapter 12, Fig. 3. Profiling of anti-acetyllysine antibody. (**a**) Fluorescence image of peptide microarray displaying 1,536 acetyllysine-containing peptides of general structure hexapeptide-acetyllysine-hexapeptide. Peptide sequences were generated randomly using a software package allowing similar distribution of all amino acid residues in all positions. Each peptide represents a defined single peptide and is printed three times within each of the three identical subarrays resulting in 3 × 3 × 1,536 = 13,824 data points. Microarray was incubated with anti-acetyllysine antibody followed by fluorescently labeled secondary antibody. *Enlarged image* shows strong signals (*white color* equals to saturated signal at used laser settings) and weak signals (*red color*). (**b**) Statistical analysis of detected weak binding acetyllysine-containing peptides is shown. Probability of each amino acid residue relative to the acetyllysine moiety in the 68 weak binders is compared with the probability of each residue relative to acetyllysine in all displayed peptides. If residue is preferred in the weak binders value is positive and proportional to preference. If residue is disfavored at given position values will be negative. Color code for amino acid residues is given in the legend of the graph.

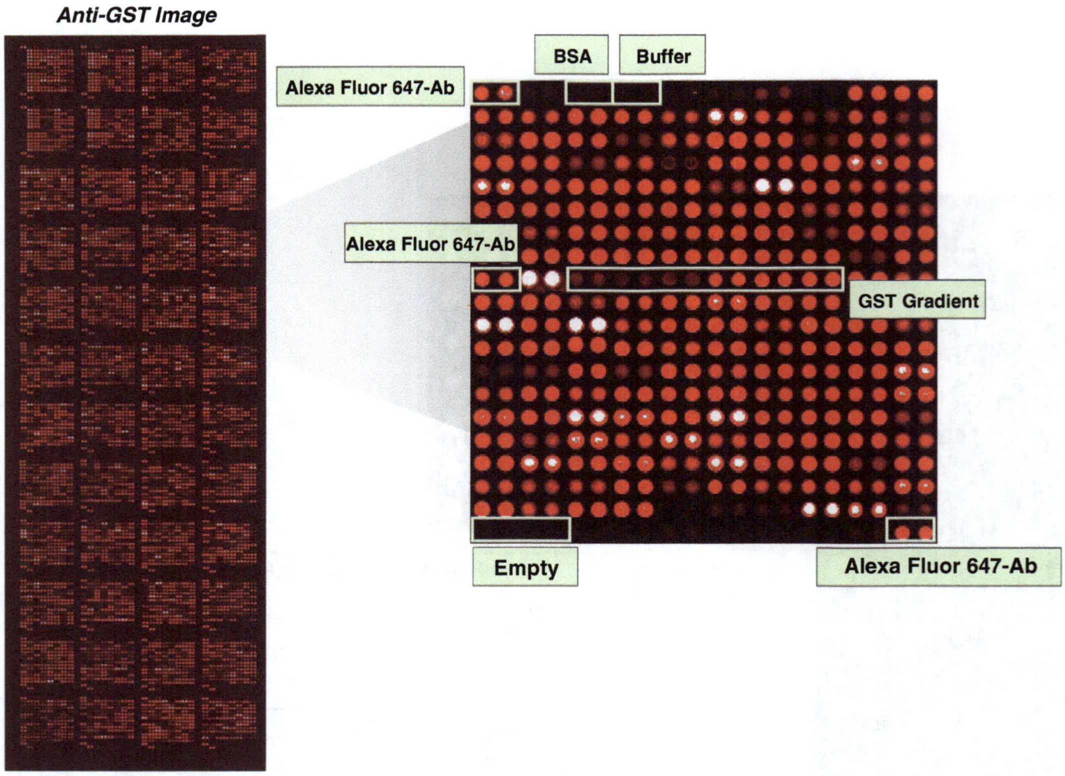

Chapter 15, Fig. 1. A ProtoArray® Human Protein Microarray was probed with an anti-GST antibody conjugated to Alexa Fluor 647. The microarray was dried and scanned at 635 nm on an Axon 4000B scanner. The ProtoArray® is divided up into 48 individual subarrays, each comprised of an identical set of negative and positive control elements, and variable human protein content. An enlarged image of a single subarray is shown on the right, with fluorescent positional mapping markers and a subset of the control elements highlighted. Note the fluorescence pattern associated with the spotting of features as adjacent duplicates.

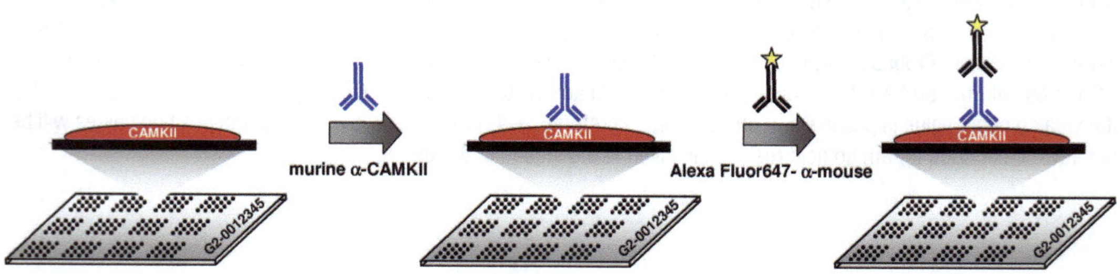

Chapter 15, Fig. 2. Protein microarray antibody specificity profiling workflow: Following blocking, the protein microarrays are probed with a dilute solution of the antibody of interest. The antibody binds to its cognate protein on the microarray, and potentially to additional cross-reactive proteins. Following incubation with primary antibody, the microarrays are washed, and incubated with a dye-labelled anti-species antibody. Following incubation with the secondary detection reagent, arrays are washed, dried, and imaged in a fluorescent microarray scanner.

a

	Signal - Background		Z-Score		Replicate Spot CV		
Database ID	α-CAMKII	neg	α-CAMKII	neg	α-CAMKII	neg	Protein Description
NM_012920.1	16818	296	37.6	0.6	10.7%	1.2%	calcium/calmodulin-dependent protein kinase (CaM kinase) II alpha, transcript variant 1
BC040457.1	5995	130	11.8	0.2	5.6%	5.7%	calcium/calmodulin-dependent protein kinase (CaM kinase) II alpha, transcript variant 2
NM_024316.1	3390	545	5.6	1.1	31.2%	15.7%	leukocyte receptor cluster (LRC) member 1 (LENG1)
NM_002738.5	1696	159	3.3	0.2	2.5%	19.2%	protein kinase C, beta 1 (PRKCB1), transcript variant 2

b c

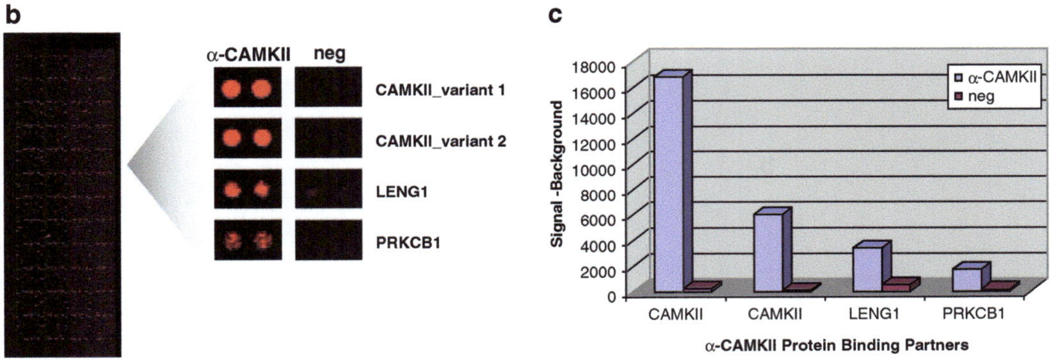

Chapter 15, Fig. 3. Protein microarray specificity profiling for anti-CAMKII antibody. (**a**) Murine monoclonal anti-calmodulin kinase II (α-CAMKII, Invitrogen) was probed on a ProtoArray at 1.0 μg/mL. Following completion of the antibody specificity profiling assay protocol, the arrays were scanned on a fluorescent microarray scanner, and pixel intensity data was used to calculate background subtracted signal values, *Z*-scores, and coefficients of variation (CV) for adjacent duplicate spots (**b**) The >8,000 human proteins were evaluated for features giving rise to a *Z*-score >3.0 in the α-CAMKII assay, with a corresponding *Z*-score <1.5 in the negative control (Alexa Fluor 647-anti-mouse antibody, detection reagent only) assay. Four proteins met these threshold criteria, with the two splice variants of CAMKII identified as the most robust binding partners. Duplicate spots from the α-CAMKII and negative control assays are shown. (**c**) background subtracted signal values are plotted.

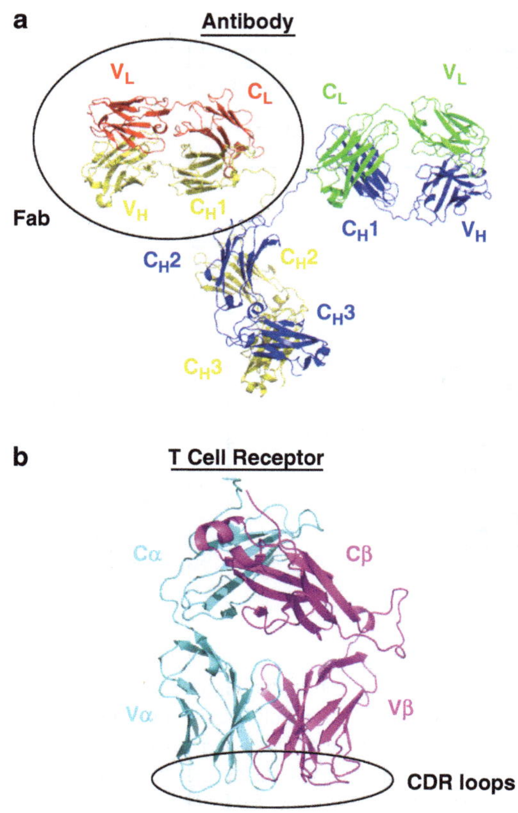

Chapter 25, Fig. 1. Structural similarities between antibodies and T-cell receptors. (**a**) An intact antibody molecule in which the Fab fragment is highlighted by the *oval*. (**b**) A T-cell receptor. The CDR loops are highlighted by the *oval*.

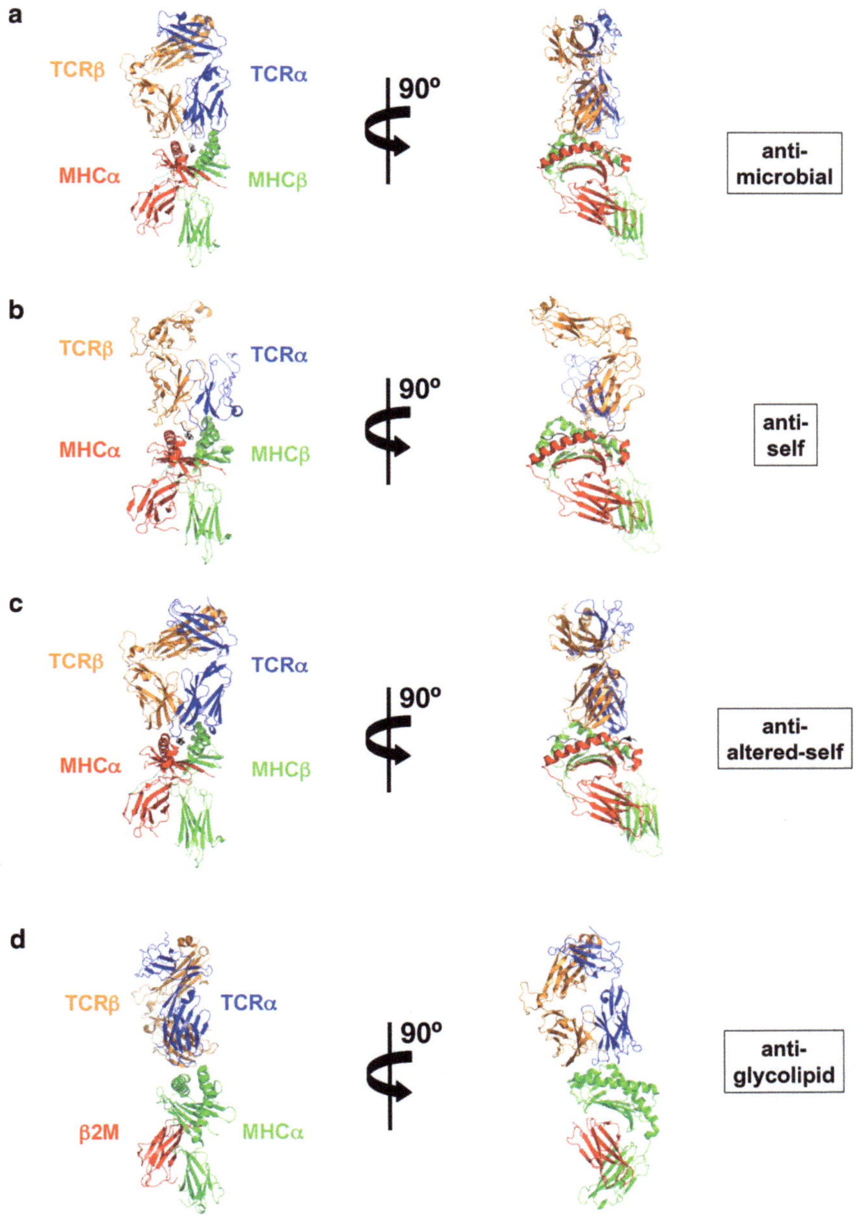

Chapter 25, Fig. 2. TCR recognition of peptidic and lipid-based antigens displayed by MHC and CD1 molecules. (**a**) A representative anti-microbial complex, the structure of TCR HA1.7 bound to an influenza hemagglutinin peptide displayed by the MHC class II molecule HLA-DR1. (**b**) A representative anti-self complex, the structure of TCR Ob.1A12 bound to the MBP_{85-99} peptide displayed by the MHC class II molecule HLA-DR2b. (**c**) A representative anti–altered-self complex, the structure of TCR E8 bound to the mutant TPI peptide displayed by the MHC class II molecule HLA-DR1. (**d**) A representative anti-glycolipid complex, the structure of NKT TCR bound to the α-GalCer glycolipid displayed by CD1d.

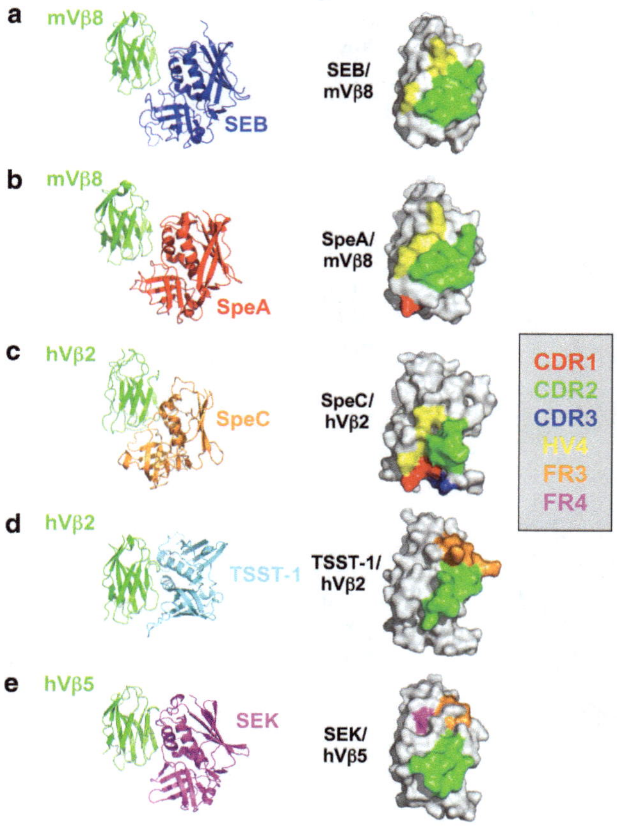

Chapter 25, Fig. 3. Superantigen engagement of the T-cell receptor Vβ domain. *Left*, structures of the (**a**) SEB/mVβ8, (**b**) SpeA/mVβ8, (**c**) SpeC/hVβ2, (**d**) TSST-1/hVβ2, and (**e**) SEK/hVβ5 complexes. The Vβ domains are aligned to one another to highlight the distinct orientations by which these SAGs engage their TCR ligands. *Right*, TCR Vβ domain molecular surface buried by various SAGs. Hypervariable and framework region surface residues buried in the interface formed by TSST-1, SEB, SpeC, and SEK are color-coded as follows: CDR1 (*red*); CDR2 (*green*); CDR3 (*blue*); HV4 (*yellow*); FR3 (*orange*); and FR4 (*magenta*). The Vβ domains in the *right* are rotated counter-clockwise approximately 90? about the vertical axis of the page relative to their positions in the *left*.

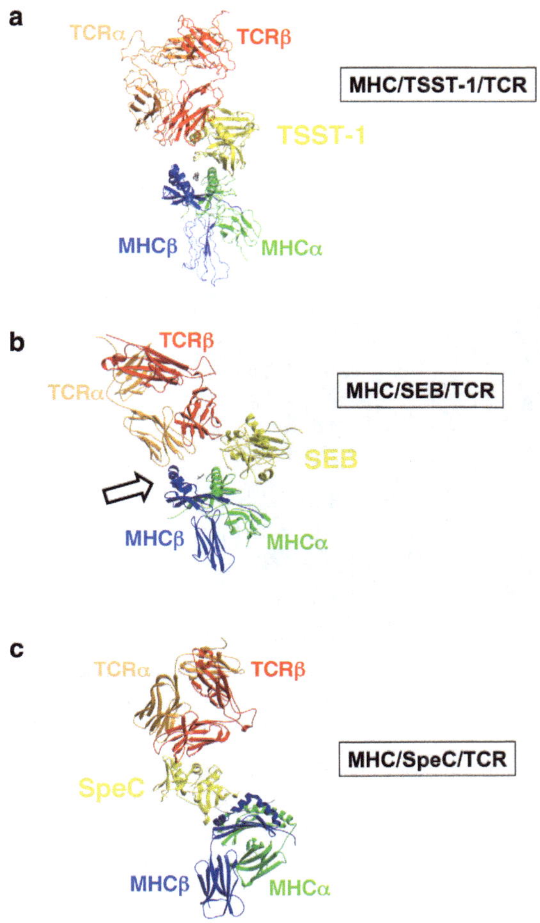

Chapter 25, Fig. 4. MHC/SAG/TCR ternary signaling complexes mediated by (**a**) TSST-1, (**b**) SEB, and (**c**) SpeC. Colors are as follows: MHC α subunit, *green*; MHC β subunit, *blue*; antigenic peptide, *gray*; TCR α chain, *orange*; TCR β chain, *red*; SAGs, *yellow*. For clarity, the MHC/SAG/TCR complexes mediated by SpeC (**c**) is rotated approximately 90? clockwise about the vertical axis of the page relative to those mediated by TSST-1 (**a**) and SEB (**b**).

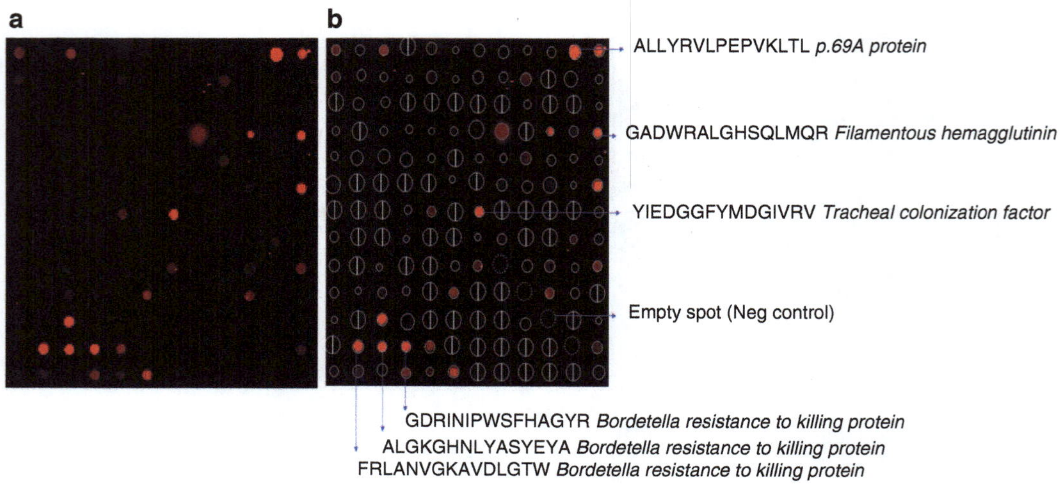

Chapter 30, Fig. 1. Example of a peptide library from *B. pertussis* proteins probed for binding to DRB*0101. (**a**) Reactive (*red*) and nonreactive (*dark*) spots signify individual peptide species binding to MHC class II molecules visualized by the anti-DR mAb. Nonreactive spots (*dark*) indicate the absence of a peptide/MHC class II molecule complex. (**b**) Overlay with the GAL file enables the identification of binding peptides to DR*0101 molecules.

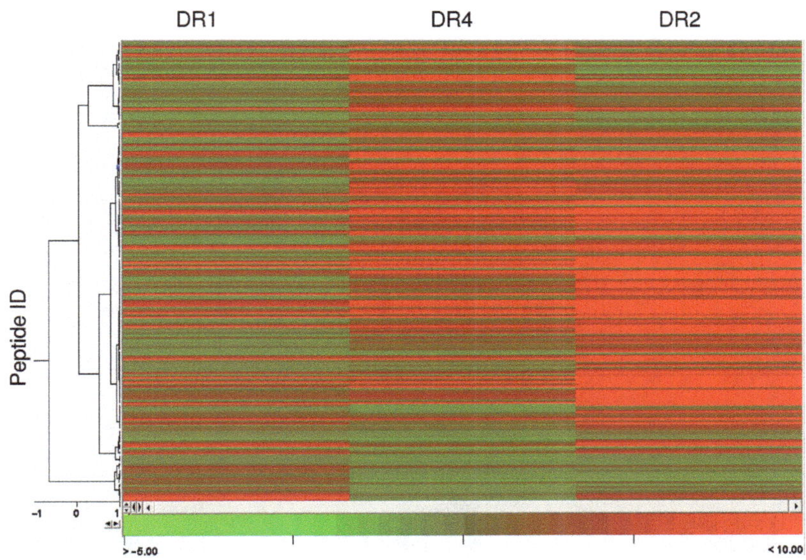

Chapter 30, Fig. 2. Example of hierarchical cluster analysis of HIV-1 peptides binding to different DR molecules. Note the different peptide binding profile to individual MHC class II molecules.

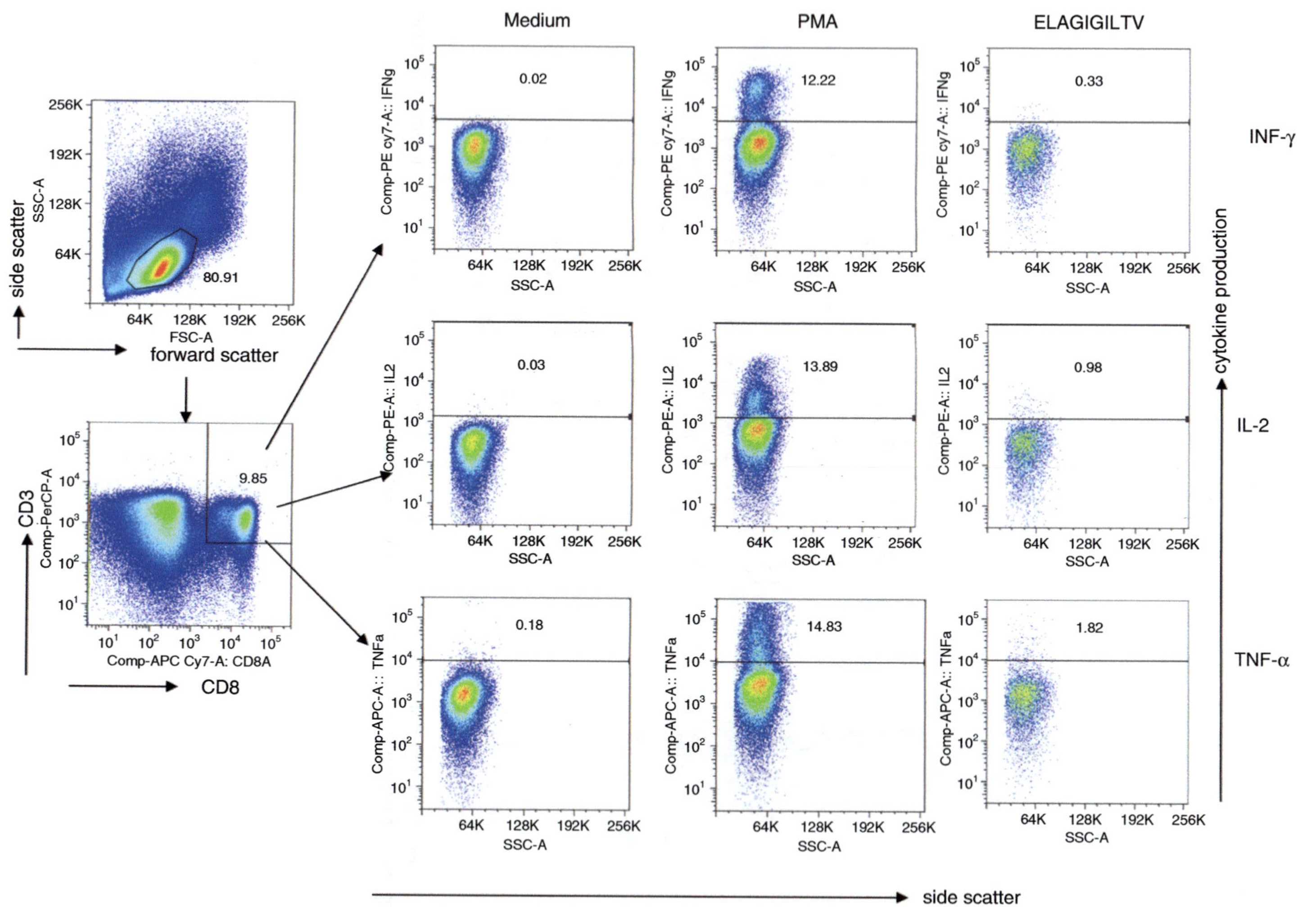

Chapter 31, Fig. 3. ICS: PBMCs from a patient with melanoma were stimulated either with medium (*left*), the positive control (PMA) or the test peptide (ELAGIGILTV from the tumor antigen Melan-A/MART-1) for 6 h. Cells were gated based on forward and side-scatter, followed by gating on CD3+ and CD8+ T-cells and single cytokine-producing cells were identified using intracellular staining for IL-2, TNF-α or IFN-γ using appropriate mAbs. Note that different gating strategies also allow (1) detection of CD4+ T-cells (in the presence of an appropriate target antigen) as well as (2) detection of T cells capable of producing simultaneously any of the tested cytokines (i.e., "multi- or polyfunctional T-cells").

FSC
www.fsc.org
MIX
Papier aus verantwortungsvollen Quellen
Paper from responsible sources
FSC® C105338

If you have any concerns about our products, you can contact us on
ProductSafety@springernature.com

In case Publisher is established outside the EU, the EU authorized representative is:
Springer Nature Customer Service Center GmbH
Europaplatz 3, 69115 Heidelberg, Germany

Printed by Libri Plureos GmbH
in Hamburg, Germany